Springer Proceedings in Physics

Volume 306

Indexed by Scopus

The series Springer Proceedings in Physics, founded in 1984, is devoted to timely reports of state-of-the-art developments in physics and related sciences. Typically based on material presented at conferences, workshops and similar scientific meetings, volumes published in this series will constitute a comprehensive up to date source of reference on a field or subfield of relevance in contemporary physics. Proposals must include the following:

- Name, place and date of the scientific meeting
- A link to the committees (local organization, international advisors etc.)
- Scientific description of the meeting
- List of invited/plenary speakers
- An estimate of the planned proceedings book parameters (number of pages/articles, requested number of bulk copies, submission deadline).

Please contact:

For Americas and Europe: Dr. Zachary Evenson; zachary.evenson@springer.com
For Asia, Australia and New Zealand: Dr. Loyola DSilva; loyola.dsilva@springer.com

Rajendra Singh · Madhusudan Singh · Ashok Kapoor
Editors

The Physics of Semiconductor Devices

Proceedings of IWPSD 2021

Springer

Editors
Rajendra Singh
Department of Physics
Indian Institute of Technology
New Delhi, India

Madhusudan Singh
Department of Physics
Indian Institute of Technology
New Delhi, India

Ashok Kapoor
DRDO
Solid State Physics Laboratory
New Delhi, India

ISSN 0930-8989 ISSN 1867-4941 (electronic)
Springer Proceedings in Physics
ISBN 978-981-97-1570-1 ISBN 978-981-97-1571-8 (eBook)
https://doi.org/10.1007/978-981-97-1571-8

This Springer imprint is published by the registered company Springer Nature Singapore Pte Ltd.
The registered company address is: 152 Beach Road, #21-01/04 Gateway East, Singapore 189721, Singapore

Paper in this product is recyclable.

Preface

This book disseminates the current knowledge of semiconductor physics and its applications across the scientific community. It is based on a biennial workshop that provides the participating research groups with a stimulating platform for interaction and collaboration with colleagues from the same scientific community.

The book discusses the developments in the field of III-nitrides; materials and devices, compound semiconductors, VLSI technology, optoelectronics, sensors, photovoltaics, crystal growth, epitaxy and characterization, graphene and other 2D materials and organic semiconductors.

The editors are grateful to all the colleagues who contributed in the preparation of this volume. Especially noteworthy is the contribution from Dr. Ms. Sheetal and Dr. Manikantha Babu.

New Delhi, India

Rajendra Singh
Ashok Kapoor

Contents

Contributors

Shah Masheerul Aalam Department of Physics, Jamia Millia Islamia, New Delhi, India

Vivek Adepu Department of Electrical and Electronics Engineering, Birla Institute of Technology and Science Pilani, Hyderabad Campus, Hyderabad, India

Tarun Kumar Agarwal Indian Institute of Technology Gandhinagar, Gandhinagar, Gujarat, India

Rajan Agrawal Hybrid Nanodevice Research Group, Department of Electrical Engineering, Indian Institute of Technology Indore, Indore, MP, India

Imran Ahmad Department of Electrical Engineering, Jamia Millia Islamia, New Delhi, India;
Department of Electrical Engineering, Indian Institute of Technology, Delhi, India

Rehan Ahmed Department of Physics, Indian Institute of Technology Bombay, Mumbai, Maharashtra, India

Javid Ali Department of Physics, Jamia Millia Islamia, New Delhi, India

Washim Reza Ali Silicon Austria Labs, Villach, Austria

Anupama Anand Department of Electronic Science, University of Delhi, New Delhi, India;
Solid State Physics Laboratory, DRDO, New Delhi, India

Sankaran Aniruddhan Department of Electrical Engineering, IIT Madras, Chennai, Tamil Nadu, India

Dhairya S. Arya Centre for Applied Research in Electronics (CARE), Indian Institute of Technology Delhi, New Delhi, India

Babita Beamline Development and Application Section, Bhabha Atomic Research Centre, Mumbai, India

Debopam Bhattacharya Department of Electronic Science, University of Calcutta, Kolkata, India

Bisma Bilal Department of Electronics and Communication Engineering, National Institute of Technology, Srinagar, J&K, India

Abhilasha Bora Department of Physics, Indian Institute of Technology Guwahati, Guwahati, India

Baishali Bora Department of Electronics and Communication Engineering, National Institute of Technology Agartala, Agartala, Tripura, India

Subhadip Chakraborty Department of Electronic Science, University of Calcutta, Kolkata, India;
Department of Chemical Biological and Macro-Molecular Sciences, S.N. Bose National Centre for Basic Sciences, Bidhannagar, Kolkata, India

Chanchal Department of Electronic Science, University of Delhi, New Delhi, India;
Solid State Physics Laboratory, New Delhi, India

Avik Chattopadhyay Department of Radio Physics and Electronics, University of Calcutta, Kolkata, India

Sanatan Chattopadhyay Department of Electronic Science, University of Calcutta, Kolkata, India

Nitanshu Chauhan Indian Institute of Technology Roorkee (IITR), Roorkee, India;
National Institute of Technology Uttarakhand, Srinagar, Garhwal, India

Surya Shankar Dan Department of Electrical and Electronics Engineering, Birla Institute of Technology and Science Pilani, Hyderabad Campus, Hyderabad, India

Abhijit Das Department of Electronics and Communication Engineering, National Institute of Technology Agartala, Agartala, Tripura, India

Om Prakash Das Department of Electronics and Instrumentation Engineering, National Institute of Technology Silchar, Silchar, Assam, India

Sanghamitra Das Department of ECE, Silicon Institute of Technology, Bhubaneswar, India

Shreyasi Das Institute of Radio Physics and Electronics, University of Calcutta, Kolkata, India

Suman Das Department of Radio Physics and Electronics, University of Calcutta, Kolkata, India;
Department of Electronics and Communication Engineering, Sikkim Manipal Institute of Technology, Sikkim Manipal University, Sikkim, India

Sudeb Dasgupta Indian Institute of Technology Roorkee (IITR), Roorkee, India

Taraprasanna Dash Department of ECE, R V College of Engineering, Bengaluru, India;
Department of ECE, Siksha 'O' Anusandhan (Deemed to be University), Bhubaneswar, India;
Department of Electronics and Communication Engineering, Odisha University of Technology and Research, Bhubaneswar, Odisha, India

Arnab Datta Indian Institute of Technology Roorkee (IITR), Roorkee, India

Emona Datta Electronic Science Department, University of Calcutta, Kolkata, India

Ambesh Dixit Advanced Materials and Device (A-MAD) Laboratory, Department of Physics, Indian Institute of Technology Jodhpur, Jodhpur, Rajasthan, India

V. K. Dixit Semiconductor Materials Laboratory, Materials Science Section, Raja Ramanna Centre for Advanced Technology, Indore, Madhya Pradesh, India;
Homi Bhabha National Institute, Mumbai, India

Soumya Dutta Department of Electrical Engineering, IIT Madras, Chennai, Tamil Nadu, India

Manu Garg Centre for Applied Research in Electronics (CARE), Indian Institute of Technology Delhi, New Delhi, India

P. K. Giri Centre for Nanotechnology, Indian Institute of Technology Guwahati, Guwahati, India;
Department of Physics, Indian Institute of Technology Guwahati, Guwahati, India

Anjali Goel Department of Electrical Engineering, Nanoelectronic Devices Laboratory (NANOLAB), Indian Institute of Technology, Rupnagar, Ropar, India

Khush Gohel Department of Electrical and Electronics Engineering, Birla Institute of Technology and Science Pilani, Hyderabad Campus, Hyderabad, India

Ankur Goswami Department of Materials Science and Engineering (DMSE), Indian Institute of Technology Mandi, Kamand, Himachal Pradesh, India

Vinay Goyal CSIR-Central Electronics Engineering Research Institute, Pilani, India;
Academy of Scientific and Innovative Research (AcSIR), Ghaziabad, India

Shrabani Guhathakurata Department of Electronics and Communication Engineering, NIST (Autonomous), Brahmapur, Odisha, India;
Department of Physics, Condensed Matter Physics Research Centre, Jadavpur University, Kolkata, India

Ankur Gupta Indian Institute of Technology, New Delhi, India

Mridula Gupta Semiconductor Device Research Laboratory, Department of Electronic Science, University of Delhi, New Delhi, India

Shradha Gupta Solid State Physics Laboratory, Timarpur, India;
Indian Institute of Technology, New Delhi, India

Fiheon Imroze Department of Electrical Engineering, IIT Madras, Chennai, Tamil Nadu, India

Tarikul Islam Department of Electrical Engineering, Jamia Millia Islamia, New Delhi, India

S. Sundar Kumar Iyer National Centre for Flexible Electronics, Indian Institute of Technology Kanpur, Kanpur, India;
Samtel Centre for Display Technologies, Indian Institute of Technology Kanpur, Kanpur, India;
Department of Electrical Engineering, Indian Institute of Technology Kanpur, Kanpur, India

Deepti Jain Solid State Physics Laboratory, Timarpur, New Delhi, Delhi, India

Nilesh Kumar Jaiswal Department of Electronics and Communication Engineering, Shri Vishnu Engineering College for Women, Bhimavaram, India

Devika Jena Department of ECE, R V College of Engineering, Bengaluru, India;
Department of ECE, Siksha 'O' Anusandhan (Deemed to be University), Bhubaneswar, India;
Department of Electronics and Communication Engineering, Odisha University of Technology and Research, Bhubaneswar, Odisha, India

S. N. Jha Beamline Development and Application Section, Bhabha Atomic Research Centre, Mumbai, India;
Homi Bhabha National Institute, Anushaktinagar, Mumbai, India

Khanjan Joshi Centre for Applied Research in Electronics (CARE), Indian Institute of Technology Delhi, New Delhi, India

Anoop Kamalasanan Indian Institute of Science Education and Research, Thiruvananthapuram, Kerala, India

Anupam Karmakar Department of Electronic Science, University of Calcutta, Kolkata, India

J. K. Kaushik Solid State Physics Laboratory, Timarpur, India

Mohd. Shahid Khan Department of Physics, Jamia Millia Islamia, New Delhi, India

Manisha Khurana Semiconductor Devices Research Laboratory, Department of Electronic Science, University of Delhi South Campus, New Delhi, India

Gargi Konwar Department of Electrical Engineering, Indian Institute of Technology Jodhpur, Jodhpur, Rajasthan, India

Ajay Kumar CSIR-Central Electronics Engineering Research Institute, Pilani, India;
Academy of Scientific and Innovative Research (AcSIR), Ghaziabad, India

Ajeet Kumar Advanced Photonics Simulation Research Laboratory, Delhi Technological University, New Delhi, India

Anil Kumar Department of Electronics and Communication Engineering, PDPM Indian Institute of Information Technology, Design and Manufacturing, Jabalpur, MP, India

Ashish Kumar National Institute of Technology, Raipur, India

Niraj Kumar Solid State Physics Laboratory, New Delhi, India

Pawan Kumar Hybrid Nanodevice Research Group, Department of Electrical Engineering, Indian Institute of Technology Indore, Indore, MP, India

Pramod Kumar Department of Physics, Indian Institute of Technology Bombay, Mumbai, Maharashtra, India

Ravi Kumar Semiconductor Materials Laboratory, Materials Science Section, Raja Ramanna Centre for Advanced Technology, Indore, Madhya Pradesh, India; Homi Bhabha National Institute, Mumbai, India

Sanjay Kumar Hybrid Nanodevice Research Group, Department of Electrical Engineering, Indian Institute of Technology Indore, Indore, MP, India

Sunil Kumar Solid State Physics Laboratory, New Delhi, India

Sushil Kumar Centre for Applied Research in Electronics (CARE), Indian Institute of Technology Delhi, New Delhi, India

Vivek Kumar Indian Institute of Technology Roorkee (IITR), Roorkee, India; National Institute of Technology Uttarakhand, Srinagar, Garhwal, India

Y. Kumar Beamline Development and Application Section, Bhabha Atomic Research Centre, Mumbai, India;
Homi Bhabha National Institute, Anushaktinagar, Mumbai, India

Vandana Kumari Department of Electronics, Maharaja Agrasen College, University of Delhi, New Delhi, India

Prasanta Kumbhakar Indian Institute of Science Education and Research Thiruvananthapuram, Kerala, India

Robert Laishram Solid State Physics Laboratory, New Delhi, India

Sanna Mairaj Department of Electronics and Communication Engineering, NIT Srinagar, Hazratbal, Srinagar, India

C. K. Maiti SouraNiloy, Kolkata, India

Abhijit Mallik Electronic Science Department, University of Calcutta, Kolkata, India

Sandipan Mallik Department of Electronics and Communication Engineering, NIST (Autonomous), Brahmapur, Odisha, India

Arun Mambra Indian Institute of Science Education and Research Thiruvananthapuram, Thiruvananthapuram, India

Nabin Baran Manik Department of Physics, Condensed Matter Physics Research Centre, Jadavpur University, Kolkata, India

Abdul kaium Mia Centre for Nanotechnology, Indian Institute of Technology Guwahati, Guwahati, India

Meena Mishra Solid State Physics Laboratory, DRDO, New Delhi, India

Yuvraj Misra Indian Institute of Technology Mandi, Mandi, Himachal Pradesh, India

Bhaskar Mitra Department of Electrical Engineering, Indian Institute of Technology, Delhi, India

J. Mitra Indian Institute of Science Education and Research Thiruvananthapuram, Thiruvananthapuram, India

Eleena Mohapatra Department of Electronics and Communication Engineering, R V College of Engineering, Bengaluru, Karnataka, India

Eleena Mohpatra Electronics and Communication Engineering, RV College of Engineering, Bengaluru, Karnataka, India

Anuraag Mukherjee Department of Electronic Science, University of Calcutta, Kolkata, India

Shaibal Mukherjee Hybrid Nanodevice Research Group, Department of Electrical Engineering, Indian Institute of Technology Indore, Indore, MP, India; Centre for Advanced Electronics (CAE), Indian Institute of Technology Indore, Indore, MP, India

Hakim Najeeb-ud-Din Department of Electronics and Communication Engineering, National Institute of Technology, Srinagar, J&K, India

Mangla Nand Beamline Development and Application Section, Bhabha Atomic Research Centre, Mumbai, India;
Homi Bhabha National Institute, Anushaktinagar, Mumbai, India

Kapil Narang Solid State Physics Laboratory, Timarpur, India

Rakhi Narang Department of Electronics, Sri Venkateswara College, University of Delhi, New Delhi, India

Sreevidya Narayanan Indian Institute of Science Education and Research, Thiruvananthapuram, Kerala, India

P. Pal MEMS and Micro/Nano Systems Laboratory, Department of Physics, Indian Institute of Technology Hyderabad, Hyderabad, India

A. K. Pandey SENACT Lab, Department of Mechanical and Aerospace Engineering, Indian Institute of Technology Hyderabad, Hyderabad, India

Saurabh Kumar Pandey Sensors and Optoelectronics Research Group, Department of Electrical Engineering, Indian Institute of Technology Patna, Patna, India

Shivendra Kumar Pandey Department of Electronics and Instrumentation Engineering, National Institute of Technology Silchar, Silchar, Assam, India

Ravi Pant Indian Institute of Science Education and Research Thiruvananthapuram, Thiruvananthapuram, India

Akash Patnaik Advance Semiconductor Materials and Nanoelectronics Research Group, PDPM Indian Institute of Information Technology, Design and Manufacturing, Jabalpur, MP, India

Meenakashi Patwal Advanced Photonics Simulation Research Laboratory, Delhi Technological University, New Delhi, India;
Solid State Physics Laboratory, Timarpur, New Delhi, Delhi, India

Shuvam Pawar Department of Electronics and Communication Engineering, National Institute of Technology, Warangal, Telangana, India

Rahul Prajesh CSIR-Central Electronics Engineering Research Institute, Pilani, India;
Academy of Scientific and Innovative Research (AcSIR), Ghaziabad, India

Chandra Prakash Advanced Materials and Device (A-MAD) Laboratory, Department of Physics, Indian Institute of Technology Jodhpur, Jodhpur, Rajasthan, India

Mahanth Prasad Central Electronics Engineering Research Institute, Pilani, Rajasthan, India

S. Purohit MEMS and Micro/Nano Systems Laboratory, Department of Physics, Indian Institute of Technology Hyderabad, Hyderabad, India;
SENACT Lab, Department of Mechanical and Aerospace Engineering, Indian Institute of Technology Hyderabad, Hyderabad, India

Saiganesh Puttur Department of Electrical Engineering, IIT Madras, Chennai, Tamil Nadu, India

Vivek Raghuwanshi Department of Electrical Engineering, Indian Institute of Technology Jodhpur, Jodhpur, Rajasthan, India

Sachin Rahi Department of Electrical Engineering, Indian Institute of Technology Jodhpur, Jodhpur, Rajasthan, India

S. K. Rai Synchrotrons Utilization Section, Raja Ramanna Centre for Advanced Technology, Indore, India

V. N. Ramakrishnan Department of Micro and Nanoelectronics, Vellore Institute of Technology, Vellore, India

Spandan Ranpariya Indian Institute of Information Technology Vadodara, Vadodara, Gujarat, India

Anshuman Raunak Defence Laboratory Jodhpur, Jodhpur, Rajasthan, India

D. S. Rawal Solid State Physics Laboratory, Timarpur, New Delhi, Delhi, India

Dipendra Singh Rawal Solid State Physics Laboratory, DRDO, New Delhi, India

Akhilesh Rawat Department of Electrical Engineering, Nanoelectronic Devices Laboratory (NANOLAB), Indian Institute of Technology, Rupnagar, Ropar, India

Brajesh Rawat Department of Electrical Engineering, Nanoelectronic Devices Laboratory (NANOLAB), Indian Institute of Technology, Rupnagar, Ropar, India

Madhu Rawat National Centre for Flexible Electronics, Indian Institute of Technology Kanpur, Kanpur, India;
Samtel Centre for Display Technologies, Indian Institute of Technology Kanpur, Kanpur, India

Mohammad Moeen Hasan Raza Department of Physics, Jamia Millia Islamia, New Delhi, India

R. Roychowdhury Semiconductor Materials Laboratory, Materials Science Section, Raja Ramanna Centre for Advanced Technology, Indore, Madhya Pradesh, India;
Homi Bhabha National Institute, Mumbai, India

Sachchidanand Advance Semiconductor Materials and Nanoelectronics Research Group, PDPM Indian Institute of Information Technology, Design and Manufacturing, Jabalpur, MP, India

Mohd Sadiq Department of Physics, Jamia Millia Islamia, New Delhi, India;
Department of Physics, ARSD College, University of Delhi, New Delhi, India

Soumi Saha Department of Electrical and Electronics Engineering, Birla Institute of Technology and Science Pilani, Hyderabad Campus, Hyderabad, India

Parikshit Sahatiya Department of Electrical and Electronics Engineering, Birla Institute of Technology and Science Pilani, Hyderabad Campus, Hyderabad, India

Rakesh Kumar Saini CSIR-Central Electronics Engineering Research Institute, Pilani, India;
Academy of Scientific and Innovative Research (AcSIR), Ghaziabad, India

Mitra Barun Sarkar Department of Electronics and Communication Engineering, National Institute of Technology Agartala, Agartala, Tripura, India

Mohd Sarvar Department of Physics, Jamia Millia Islamia, New Delhi, India

Rajeev Sawal Solid State Physics Laboratory, New Delhi, India

Manoj Saxena Department of Electronics, Deen Dayal Upadhyaya College, University of Delhi, New Delhi, India

Pulkit Saxena Department of Electrical Engineering, Indian Institute of Technology Jodhpur, Jodhpur, Rajasthan, India

Khushwant Sehra Department of Electronic Science, University of Delhi, New Delhi, India

Anusha Shanmugam Indian Institute of Science Education and Research Thiruvananthapuram, Kerala, India

Monika Sharma Semiconductor Device Research Laboratory, Department of Electronic Science, University of Delhi, New Delhi, India

Pankaj Sharma Advance Semiconductor Materials and Nanoelectronics Research Group, PDPM Indian Institute of Information Technology, Design and Manufacturing, Jabalpur, MP, India

R. K. Sharma Department of Chemistry, University of Delhi, Delhi, India

T. K. Sharma Semiconductor Materials Laboratory, Materials Science Section, Raja Ramanna Centre for Advanced Technology, Indore, Madhya Pradesh, India; Homi Bhabha National Institute, Mumbai, India

Amandeep Singh Department of Electronics and Communication Engineering, NIT Srinagar, Hazratbal, Srinagar, India

Pushpapraj Singh Centre for Applied Research in Electronics (CARE), Indian Institute of Technology Delhi, New Delhi, India

Dhirendra Kumar Sinha Indian Institute of Information Technology Vadodara, Vadodara, Gujarat, India

Srest Somay Department of Metallurgical and Materials Engineering, National Institute of Technology Warangal, Warangal, India

V. Swarnalatha MEMS and Micro/Nano Systems Laboratory, Department of Physics, Indian Institute of Technology Hyderabad, Hyderabad, India

Suchismita Tewari Department of Radio Physics and Electronics, University of Calcutta, Kolkata, India

Madhu Thalakulam Indian Institute of Science Education and Research, Thiruvananthapuram, Kerala, India

Shree Prakash Tiwari Department of Electrical Engineering, Indian Institute of Technology Jodhpur, Jodhpur, Rajasthan, India

S. Tripathi Beamline Development and Application Section, Bhabha Atomic Research Centre, Mumbai, India

Upasana Semiconductor Devices Research Laboratory, Department of Electronic Science, University of Delhi South Campus, New Delhi, India

Satish K. Verma Indian Institute of Technology Delhi, New Delhi, India

Mujeeb Yousuf Centre for Applied Research in Electronics (CARE), Indian Institute of Technology Delhi, New Delhi, India

Zaheeruddin Department of Electrical Engineering, Jamia Millia Islamia, New Delhi, India

Chapter 1
Impact of Various Device Parameters on the Series Resistance of Perovskite Solar Cell

Srest Somay and Saurabh Kumar Pandey

1.1 Introduction

Perovskite material has shown a tremendous potential in the field of photovoltaics [1]. Its exceptional optoelectronic properties and excellent performance have gained lots of attention worldwide. The rapid surge in the advancement of Perovskite solar cells (PSCs) and considering the pressing need for a reliable renewable source of energy, it is important to work towards commercialization of PSCs. Solar cells function like actual diodes and its I-V characteristics can be described by (1.1)

$$J = J_p - J_o^{\left(e^{\left(\frac{e(V-JR_S)}{nK_BT}\right)}-1\right)} + \frac{V - IR_s}{R_{sh}} \quad (1.1)$$

where J is the current in the circuit, Jp is the photo generated current, Jo is the reverse saturation current of diode, n is the ideality factor of diode, Rs is the internal series resistance, Rsh is the internal shunt resistance, Kb is the Boltzmann constant, T is the operating temperature and e is the charge of electron [2]. The dark current is often in charge of managing the cell's open circuit voltage [2], whereas n often denotes the type of recombination (first order or second order) that predominates in the cell [3]. Shunt and series resistance are in charge of controlling the device's fill factor and short circuit current density [4]. To get an insight on the performance of the device and

S. Somay (✉)
Department of Metallurgical and Materials Engineering, National Institute of Technology Warangal, Warangal, India
e-mail: srestsomay447@gmail.com

S. K. Pandey
Sensors and Optoelectronics Research Group, Department of Electrical Engineering, Indian Institute of Technology Patna, Patna, India
e-mail: saurabh@iitp.ac.in

R. Singh et al. (eds.), *The Physics of Semiconductor Devices*, Springer Proceedings in Physics 306, https://doi.org/10.1007/978-981-97-1571-8_1

to optimize their performance it is crucial to understand the effect of various material parameters on these diode parameters. Recently efforts have been made in direction to investigate diode parameters of the device to correlate it with their performance, Lu et al. have shown the effect of grain size on the diode parameters of the cell [5]. Further, You et al. showed dependence of the diode parameters on the annealing condition of their sample [6], Pietro et al. elucidated the dependence of ideality factor on the mechanism of recombination [3]. Apurba et al. showed the dependence of ideality factor on ageing time [7]. Despite these reports, a thorough examination of the relationship between diode parameters and device material attributes remains elusive. In this study we have focused on understanding the influence of active layer thickness, defect density and top contact work function on the internal series resistance (Rs) of the mixed cation perovskite solar cell. Series resistance is considered to be important in the engineering point of view of solar cells. The low fill factor and sometimes low short circuit current density is attributed to high series resistance of the solar cell [4] however the concrete evidence of the dependence of series resistance on the material attributes is missing in the case of perovskite solar cells. This study is expected to cover this aspect through comprehensive numerical simulation and modeling.

1.2 Device Architecture and Numerical Simulation

In this study we have used SCAPS (solar cell capacitance simulator-1-D) software. This software solves basic semiconductor equations and helps us to analyse the performance of the developed model. In this study we have simulated $FA_{0.8}MA_{0.2}Sn_{0.5}Pb_{0.5}I_3$ mixed cation perovskite solar cells [8]. We have used PCBM and C60 as an electron transport layer and PEDOT:PSS as a hole transport layer. The flat band condition was employed at the contacts. Mid-gap states neutral traps with Gaussian energetic distribution were considered in each layer. Further neutral mid-gap state defects with single energetic distribution were considered at the interfaces. Figure 1.1 shows the proposed device structure of the device. Table 1.1 shows the device parameters used in the simulation. All the parameters were taken from reported studies [8, 9]. Absorption curve of $FA_{0.8}MA_{0.2}Sn_{0.5}Pb_{0.5}I_3$ was taken from [8]. All the simulations were performed under the AM 1.5G 1 sun spectrum. The Series resistance was estimated by calculating the slope of the J-V curve at Voc. The slope was approximately calculated by interpolating the J-V curve up to 2000 points using Origin software. Although we appreciate the fact that there are various method to calculate the exact value of Series resistance [4–6, 10] but in this work we have focused on the relative change in the value of Series resistance on the various parameters and therefore we assume that this method is good enough to give us a qualitative picture of dependence of Series resistance on various device parameters.

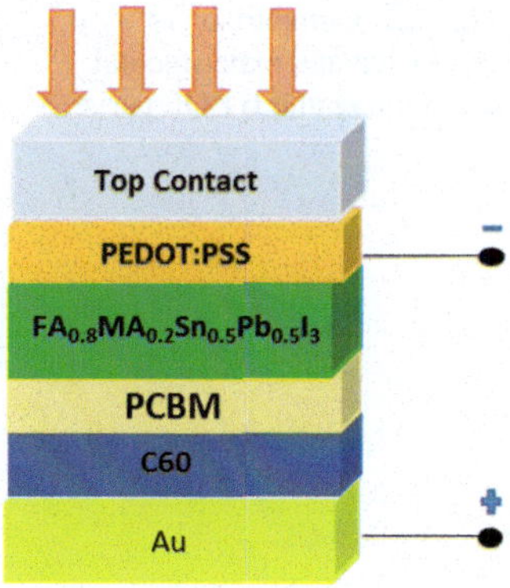

Fig. 1.1 Proposed device architecture

Table 1.1 Materials parameters used in the simulation

Parameters	$FA_{0.8}MA_{0.2}Sn_{0.5}Pb_{0.5}I_3$	PEDOT	PCBM	C60
Thickness (nm)	600	100	20	5
Band gap (eV)	1.270	1.80	2.00	2.0
Electron affinity (eV)	4.13	3.50	3.90	3.60
Dielectric	6.50	10.00	10.00	10.00
Conduction band density of states (DOS) (cm^{-3})	2.200×10^{18}	2.20×10^{18}	2.20×10^{18}	2.20×10^{18}
VB DOS ($1/cm^3$)	1.80×10^{19}	1.80×10^{19}	1.80×10^{19}	1.80×10^{19}
Electron mobility (cm^2/V s)	2.00	0.01	0.01	0.1
Hole mobility (cm^2/V s)	2.00	0.8	0.01	0.1
Defect density (cm^{-3})	2.500×10^{13}	1.00×10^{15}	1.00×10^{14}	1.00×10^{14}
Shallow acceptor density (cm^{-3})	5.00×10^{12}	1.00×10^{16}	–	0

1.3 Results and Discussions

1.3.1 Device Validation

It is important to validate the device parameters to assure a self-consistent model. In this section we have compared the calculated and experimental J-V and quantum efficiency curve (EQE) [8]. Figure 1.2a and b shows the comparative analysis between simulated and experimental J-V curve and EQE curve of the device. The simulated Voc of 0.78 V and Jsc of 29.9 mA/cm^2 and PCE of 18.6% matches well with the experimental Voc of 0.78 V and Jsc of 30 mA/cm^2 and PCE of 18.2%. Further the simulated EQE curve follows the same trend as the experimentally obtained curve. This assures that our constructed model aligns with the experimental model and

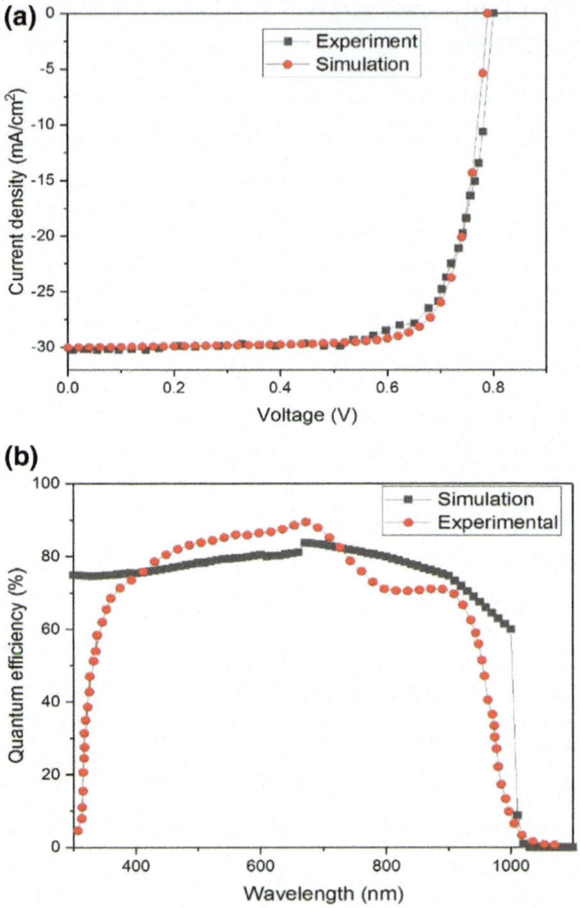

Fig. 1.2 Comparison of calculated and experimental **a** J-V curve and **b** EQE curve

therefore with this model we now will study the effect of various device parameters on the series resistance of the solar cells.

1.3.2 *Impact of Active Layer Thickness on Series Resistance*

An active layer's thickness is determined by its absorption edge and carrier diffusion length. A thick layer absorbs more photons, increasing the short circuit current, but due to the weak electric field, effective charge transportation is limited [2, 11]. In our study the absorption and carrier diffusion length both are 1000 nm respectively. In the section we have varied the thickness from 400 to 1200 nm and studied its effect on the series resistance in relation to its performance. Series resistance is the body resistance of the solar cell and is expected to increase with the increase

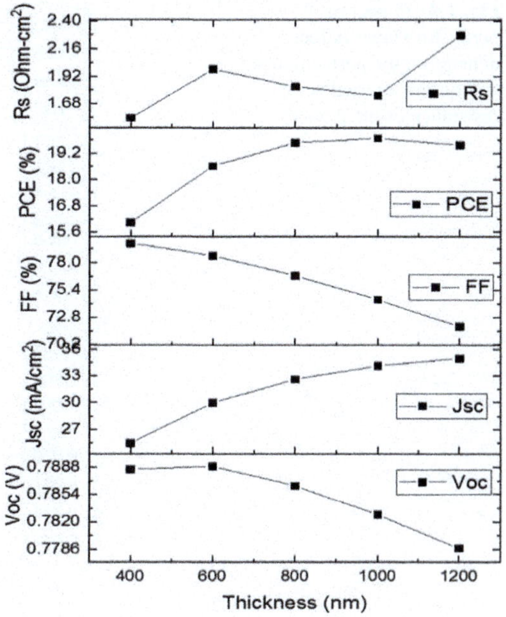

Fig. 1.3 Impact of perovskite layer thickness on the performance and on the internal series resistance of the device

in the thickness of the semiconductor layers [12] but it is evident from Fig. 1.3. that Series resistance doesn't increase uniformly with thickness of the active layer. We observed a low series resistance of 1.5527 Ω cm^2 at 400 nm which increased significantly to 1.9815 Ω cm^2 at 600 nm. While further increase in the thickness resulted in a subsequent decrease in the Series resistance and a low Series resistance of 1.7482 Ω cm^2 was observed for 1000 nm which is followed by a substantial increase in Series resistance at 1200 nm. The non-uniformity can be explained as follows; Series resistance depends on the thickness of the material as well as on the charge carriers generation and recombination [13], the increase in thickness directly leads to increase in the Series resistance while the more generation of charge carriers with the thickness decreases the Series resistance, thus we observe a cumulative effect with the increase in the thickness. It is interesting to note that while minimum Series resistance of 1.5527 Ω cm^2 was observed at 400 nm yet a 1000 nm thick layer with a Series resistance of 1.7482 Ω cm^2 performs the best.

1.3.3 Impact of Perovskite Layer Trap Density on the Series Resistance

It's crucial to look at the effect of active layer defect density on the device's diode parameters. Single crystal films have low trap density typically in range of 10^{10}–10^{12} due to its controlled deposition while polycrystalline film has large defect density

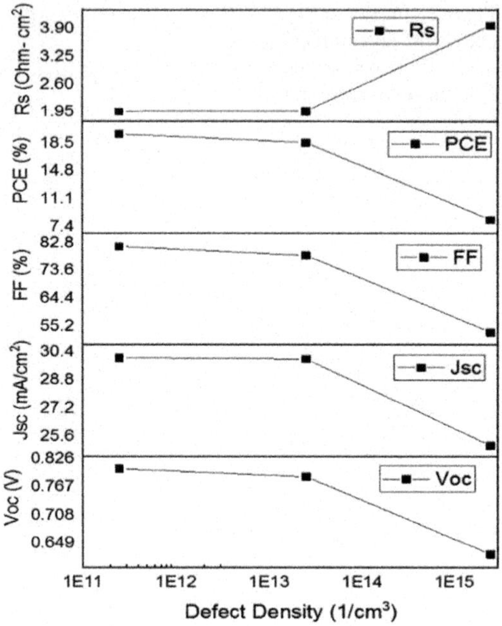

Fig. 1.4 Impact of perovskite layer defect density on the performance and on the internal series resistance of the device

typically up to 10^{15} due to significant grain boundaries trap which act as a site for non-radiative process [2]. In this section we have studied the effect of active layer trap density on the Series resistance. We varied the trap density from 2.500×10^{11} to 2.5×10^{15} keeping all the other parameters constant and observed a substantial increase in Series resistance from 1.9723 to 3.9808 Ω cm^2 with the increase in the defect density as shown in Fig. 1.4. The rise in Series resistance is linked to a rise in charge carrier recombination [14], resulting in a drop in efficiency from 19.77 to 8.37%.

1.3.4 Impact of Top Contact Work Function on Series Resistance

The impact of contact electrodes on the internal resistance of device is substantial. The Metal/semiconductor contacts are classified as Schottky and ohmic contact. Schottky contacts often leads to rectification characteristics while ohmic contacts is a low resistance contact with an insignificant potential drop. In this section, we looked at how the top contact work function affects the device's Series resistance. It's worth noting that, due to the simulation software's limitations, we treated the contacts as pure conductors and ignored their sheet resistance. We have varied the work function from 4.9 to 5.3 eV keeping all the parameters constant. We observed that Series resistance decreases from 5.35 to 1.29 Ω cm^2 and consequently PCE

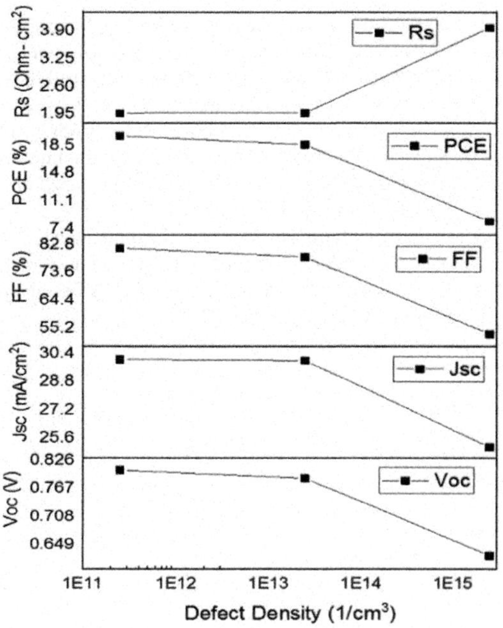

Fig. 1.5 Impact of top contact work function on the performance and on the internal series resistance of the device

increased for 15.75 to 19.21% as shown in Fig. 1.5. The decrease Series resistance is due to formation of ohmic contact at the top contact/PEDOT:PSS junction which leads to efficient extraction of charge carriers and thus results in high PCE [2].

References

1. NREL, Best Research-Cell Efficiencies [Online], Available at: https://www.nrel.gov/pv/cell-efficiency.html.20181221.pdf (2021)
2. S.K. Pandey, Numerical analysis of rGO/silver-nanowire-based single-crystal perovskite solar cell. IEEE Trans. Electron Devices **67**(10), 4321–4327 (2020)
3. P. Caprioglio, On the origin of the ideality factor in perovskite solar cells. Adv. Energy Mater. **10**(27), 2000502 (2020)
4. N. Mundhaas, Series resistance measurements of perovskite solar cells using Jsc-Voc measurements. Sol. RRL **3**(4), 1800378 (2019)
5. C.H. Lu, Effects of precursor composition on morphology and microstructure of hybrid organic–inorganic perovskite solar cells. J. Mater. Sci. **54**(19), 12758–12766 (2019)
6. J. You, Moisture assisted perovskite film growth for high performance solar cells. Appl. Phys. Lett. **105**(18), 183902 (2014)
7. A. Mahapatra, Changes in the electrical characteristics of perovskite solar cells with aging time. Molecules **25**(10), 2299 (2020)
8. A. Nejand, Vacuum-assisted growth of low-bandgap thin films ($FA_{0.8}MA_{0.2}Sn_{0.5}Pb_{0.5}I_3$) for all-perovskite tandem solar cells. Adv. Energy Mater. **10**(5), 1902583 (2020)
9. J. Madan, Device simulation of 17.3% efficient lead-free all-perovskite tandem solar cell. Sol. Energy **197**, 212–221 (2020)

10. D. Sera, Series resistance monitoring for photovoltaic modules in the vicinity of MPP, in *Proceedings of the 25th European Photovoltaic Solar Energy Conference and Exhibition: EU PVSEC 2010* (WIP Wirtschaft und Infrastruktur GmbH and Co Planungs KG, 2010)
11. S.K. Pandey, Device engineering approach toward stable, efficient, and eco-friendly perovskite solar cell. IEEE Trans. Electron Devices **68**(3), 1142–1148 (2021)
12. T. Ouslimane, Impact of absorber layer thickness, defect density, and operating temperature on the performance of MAPbI3 solar cells based on ZnO electron transporting material. Heliyon **7**(3), 06379 (2021)
13. L. Duan, Relationship between the diode ideality factor and the carrier recombination resistance in organic solar cells. IEEE J. Photovoltaics **8**(6), 1701–1709 (2018)
14. X. Zhou, Design and simulation of C2N based solar cell by SCAPS-1D software. Mater. Res. Express **7**(12), 126303 (2020)

Chapter 2
Performance Investigation of Alkali Metal Fluorides as Alternate Electron Selective Contacts for TOPCon Solar Cells

Bisma Bilal and Hakim Najeeb-ud-Din

Abbreviations

AMF	Alkali metal fluoride
CB	Conduction band
CSPC	Carrier selective passivating contact
ESL	Electron selective layer
HIT	Heterojunction with thin intrinsic layer
HSL	Hole selective layer
PERC	Passivated emitter and rear contact
TMO	Transition metal oxide
TOPCon	Tunnel oxide passivating contact
VB	Valance band

2.1 Introduction

A broad classification of solar cells divides them into two categories, one based on diffused p-n junctions called the homojunction cells and the other based on heterojunction designs. Silicon heterojunction solar cells are evolving as potential candidates to push the solar cell efficiencies beyond the Shockley Queisser limit. These designs are based on the concept of carrier selective passivating contacts (CSPC)

B. Bilal (✉) · H. Najeeb-ud-Din
Department of Electronics and Communication Engineering, National Institute of Technology, Srinagar, India
e-mail: bisma_03phd17@nitsri.net

R. Singh et al. (eds.), *The Physics of Semiconductor Devices*, Springer Proceedings in Physics 306, https://doi.org/10.1007/978-981-97-1571-8_2

which exploit the advantage of excellent interfaces c-Si forms with different materials [1, 2]. The silicon heterojunction designs include the cells using doped silicon and dopant free CSPC. The former include the amorphous and polysilicon based designs like HIT [3, 4] and TOPCon [5, 6] while the dopant free configurations encompass the different classes of materials like TMO, AMF etc. [7].

Carrier selectivity, passivation and differential conductivities have been identified as the main contributors to develop efficient heterojunction designs. More recently the polysilicon based TOPCon designs are gaining attention for industrial uptake since they are developed on PERC lines, making them choicest designs for large scale manufacturing [8]. They incur low additional costs and better efficiencies which in turn reduces the levelized cost of electricity. TOPCon cells use tunnel oxides separating the absorber and highly doped polysilicon layers to induce carrier selectivity and collection of charge carriers. The main transport mechanism in such architecture is the quantum mechanical tunneling [9, 10]. The oxide layer passivates the interface of defects to improve the open circuit voltage characteristics of the device and the highly doped polysilicon layers provide high conductivity to one type of carriers depending on its doping. An appropriate band bending occurs at the TOPCon/absorber interface to allow only one type of carriers to flow while blocking the opposite polarity. Although such a configuration seems the ultimate it is limited by high parasitic absorption of the polysilicon layers which restricts the use of TOPCon designs on the front of the solar cell [11].

Different materials forming efficient heterojunctions with c-Si are being actively explored to do away with the parasitic absorption losses at the sun exposed front of the solar cell. Figure 2.1 shows the energy band alignment of different materials with crystalline silicon. The materials with low conduction band offset form efficient electron selective layers while the materials with lower VB offset form the hole selective layers. Besides the CB/VB offset the work function of the material also defines the electron and hole collection capability at the heterojunction. The low work function materials form electron selective layers while high work function layers are best for hole collection. Many such materials have been extensively studied in the literature to obtain better performance parameters in silicon heterojunction solar cells [12, 13]. Alkali metal fluorides are materials of interest to form selective contacts in silicon solar cells. These layers have roots in organic electronics and show distinct electron and hole selective features. They include materials like LiF_x, KF_x and CsF_x where of notable interest is the electron selective contact, LiF_x forms with c-silicon. In this study, Lithium Fluoride AMF has been used as an electron selective layer.

2.2 Proposed Design

In order to capitalize on the different advantages of silicon based TOPCon and LiF_x AMF, we herein have investigated a design which uses TOPCon for the rear of the solar cell and profits from LiF_x front contact design. The modelled cell is shown in Fig. 2.2a. It consists of a 300 μm p-silicon absorber, thin LiF_x (2–10 nm) for

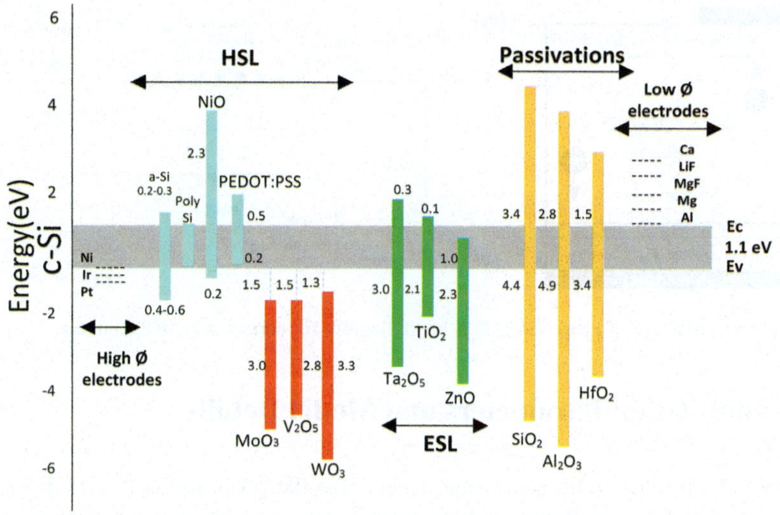

Fig. 2.1 Band offsets with respect to c-Si for different materials used to form CSPC [1, 12, 13]

electron selective layer, and a 50 nm p polysilicon-based TOPCon rear design with a 1.2 nm SiO_2 tunneling layer. Aluminum electrodes are modeled for the cell. Lithium Fluoride (LiF) is an AMF with a wide bandgap (> 10 eV) and low work function (< 4.5) which make it an apt candidate for front side solar cell as shown in Fig. 2.2b. LiF_x is a potential candidate for silicon cells due to its fabrication simplicity using methods like thermal evaporation and stability it offers in the design without incurring additional high costs for the production of the cells [14, 15]. The low work function of the LiF_x layers (< 4.5 eV) is particularly responsible for its ESL properties. The low work function enhances the downward band bending at the c-Si interface leading to the generation of an electric field which provides required barrier for the holes and allows electron transport. It also causes accumulation and thus high conductivity of electrons near the surface. The restriction of the holes therein reduces the possible recombination and loss of carriers thus allowing higher open circuit voltage for the designs using LiF_x ESL [16]. The presence of thin LiF_x layers between the c-Si and the electrode depins the fermi level, provides electrical isolation between them and also reduces the contact resistivity at the contact. Besides only thin films are needed to obtain the required improvement. These layers have also exhibited less degradation with time as shown in [17, 18]. The rear p-polysilicon layers provide high conductivity to holes which tunnel through the thin SiO_2 layer via quantum tunneling [5, 6].

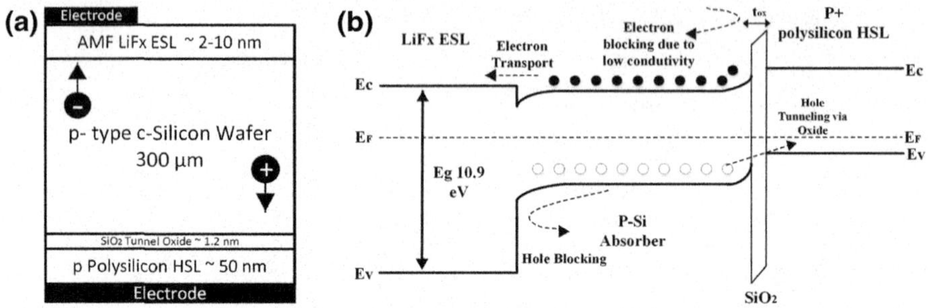

Fig. 2.2 **a** Schematic of proposed cell; **b** band diagram for carrier selective contacts

2.3 Simulation Parameters and Model Details

Numerical simulation are performed to analyze the proposed cell. AFORS-HET v 2.5 (Automat for simulation of heterostructures) tool is used for performing the simulations. It solves the Poisson, continuity and transport equations to obtain the characteristics of the design in one dimension and uses AM 1.5G for illumination [19, 20]. The carrier transport in the absorber silicon is modeled by drift diffusion and lambert beer model is used for optical calculations. The optical n,k data of LiF_x is extracted from the available experimental studies and incorporated into the simulator. Optical parameters like shading and grid losses are not taken into account. Thermionic emission and tunneling model is used for transport in TOPCon which uses SiO_2 tunnel layers. The contacts are chosen as Schottky contacts and flat bands are considered. The values for layer parameters and defects are taken from literature and appropriate references have been provided. Consequently the parameter values are summarized in Table 2.1.

2.4 Results and Discussion

2.4.1 JV Characteristics

The simulations of the proposed cell are performed under the simulation set up mentioned in Sect. 2.3 and the JV curve is extracted as shown in Fig. 2.3. For the preliminary simulations the doping of 1E12 and LiF_x thickness of 5 nm is considered. The figures of merit obtained for the design are shown in Table 2.2. The proposed cell is compared with the design with TOPCon configuration on both sides and amorphous silicon based cell available in literature [23, 24]. An explicit improvement for the proposed cell is obtained. It implies that using the AMF layer on the front side reduces the absorption losses at the front due to wide bandgap of the layer and allows better selectivity owing to the low work function.

Table 2.1 Details of material parameters used in simulations [15, 18, 21–23]

Parameter	LiF_xESL	c-Si (p)	p-poly HSL	a-Si (i)	SiO_2 interface
Thickness (cm)	Varied	Varied	Varied	Varied	1.2 nm
Dk	11.9	11.9	11.9	11.9	3.9
χ (eV)	3.9	4.05	4.05	3.9	1
Eg (eV)	10.9	1.124	1.12	1.72	8.9
DOS_{CB} (cm^{-3})	1E20	2.86E19	1E20	1E20	–
DOS_{VB} (cm^{-3})	1E20	2.86E19	1E20	1E20	–
μ_e ($cm^2\ v^{-1}\ s^{-1}$)	2.5	1041	20	20	–
μ_h ($cm^2\ v^{-1}\ s^{-1}$)	1	412.9	5	5	–
N_A (cm^{-3})	0	Varied	Varied	0	–
N_D (cm^{-3})	Varied	0	0	1000	–
$V_{th\text{-}electron}$ ($cm\ s^{-1}$)	1E07	1E07	1E07	1E06	–
$V_{th\text{-}hole}$ ($cm\ s^{-1}$)	1E07	1E07	1E07	1E06	–
R_{ae} ($cm^6\ s^{-1}$)	0	2.2E−31	0	0	–
R_{ah} ($cm^6\ s^{-1}$)	0	9.9E−32	0	0	–
Defects	–	Single defect at intrinsic energy	CB tail, VB tail, near constant region at bottom of CB	–	–

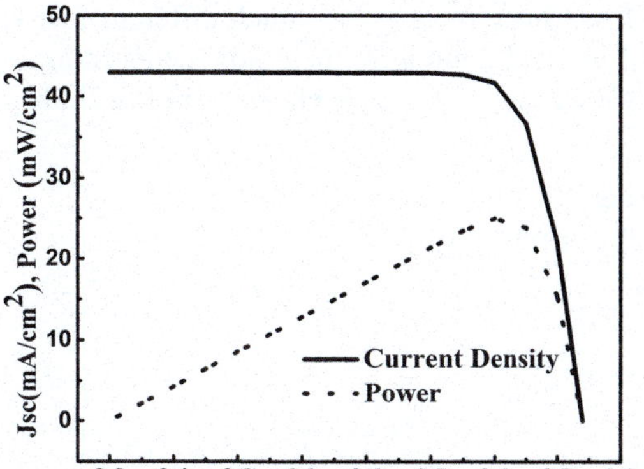

Fig. 2.3 JV characteristics of the proposed cell

Table 2.2 Comparison of proposed cell with double sided TOPCon

Cell	V_{oc} (mV)	J_{sc} (mA/cm^2)	FF (%)	η (%)
a-Si based cell [23]	741	38.73	82.12	23.74
Double sided TOPCon [24]	722.3	38.14	82.43	22.7
*Proposed cell (p wafer)	740.2	42.89	78.15	24.81

2.4.2 *Effect of Inclusion of Front Passivation*

The proposed cell simulations are performed for different parameter variations and their effect on the figures of merit are analyzed to obtain an optimal solar cell. First of all the performance of the cell is examined with the inclusion of the passivation layer between c-silicon and LiF_X. A direct contact between LiF_X and silicon results in the formation of dangling bonds and interface defects due to the lattice mismatch between the layers. To overcome this a thin layer of intrinsic amorphous silicon is added to the design to passivate the dangling bonds and thus reduce the possible carrier loss occurring thereby. The corresponding schematic and band diagram is shown in Fig. 2.4. Table 2.3 shows the comparative performance parameters for the cell with and without passivation for a LiF_x thickness and doping of 5 nm, 1E18 respectively and an a-Si thickness of 2 nm. It is explicit herein that the cell using the passivation improves the open circuit voltage and short circuit current density in the cell. This is credited to less number of defects at the LiF_X interface with silicon which reduces recombinations and thus improves the available extracted current.

The effect of passivation layer thickness on the efficiency of the cell is investigated in Fig. 2.5a. The results imply the efficiency of the cell increases till an optimal doping for all thicknesses of a-Si passivation. The a-si layer with less thickness however shows a slightly better performance than thicker passivation layers. This is attributed to the photon loss by parasitic absorption in amorphous silicon layers and is validated by OPAL simulations shown in Fig. 2.5b. It is evident that as the a-Si passivation layer thickness increases the parasitic absorption in a-Si leads to photon loss. Since there is not much difference between the figures of merit obtained for 1 nm and 2 nm a-Si, the 2 nm is chosen as a starting point for obtaining the trends over a range of a-Si thicknesses in Fig. 2.5b.

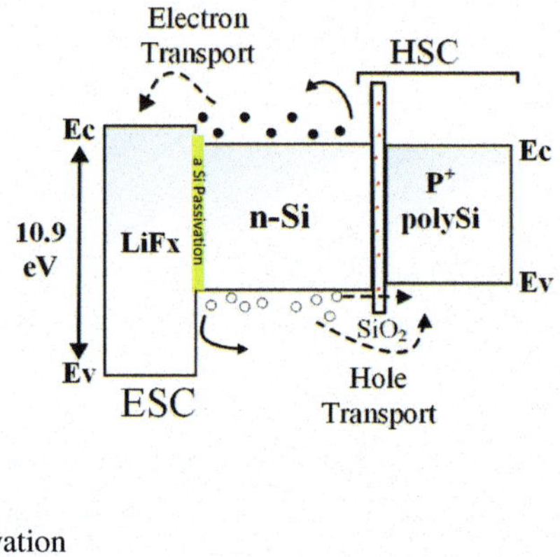

Fig. 2.4 Proposed cell with a-Si passivation layer

Table 2.3 Comparison of cell with and without passivation

Proposed cell	V_{OC} (mV)	J_{SC} (mA/cm^2)	FF (%)	η (%)
No passivation	739.5	41.53	82.6	25.3
With passivation	740.2	42.75	82.7	26.17

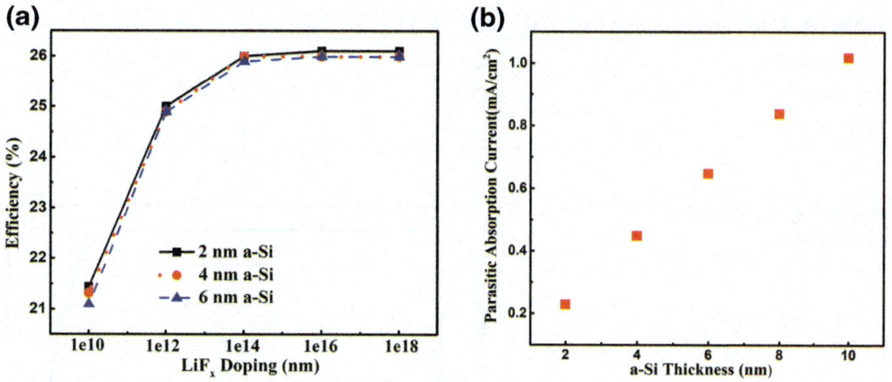

Fig. 2.5 **a** Efficiency as a function of LiFx doping for different a-Si thickness **b** OPAL validation

2.4.3 *Effect of LiF$_x$ ESL Doping*

The effect of increasing the LiF$_x$ doping for cell with and without passivation is shown in Fig. 2.6. It can be seen that as the doping of LiF$_x$ increases the efficiency of the cell also increases till it reaches an optimal level. This is credited to the better conductivity of electrons provided by doped layers which makes the contact highly selective. It is important to mention here that although LiF$_x$ comes under the category of dopant free contacts the doping herein is used to model the defect states in the LiF$_x$ layers which are responsible for transport through these layers. At low doping the number of such transport states is lower and as the number of such states is increased the transport of carriers improves till an optimal doping is reached. After an optimal doping the performance remains constant for the cell with passivation since a negligible barrier for electrons is reached but a slight degradation is observed for the cell without passivation owing to recombinations at the interface. This highlights the resilience of the device using passivation layers.

Fig. 2.6 Efficiency as a function of doping

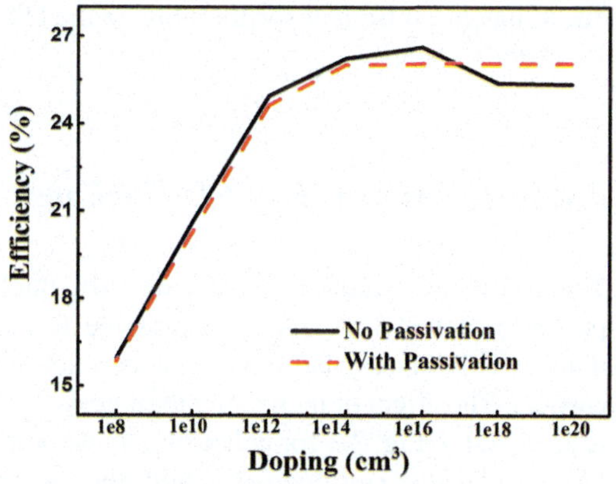

Table 2.4 Comparison of cell with and without TCO

Proposed cell	V_{oc} (mV)	J_{sc} (mA/cm^2)	FF (%)	η (%)
No TCO	740.2	42.89	78.15	24.81
With TCO	734	36.31	82.74	22.05

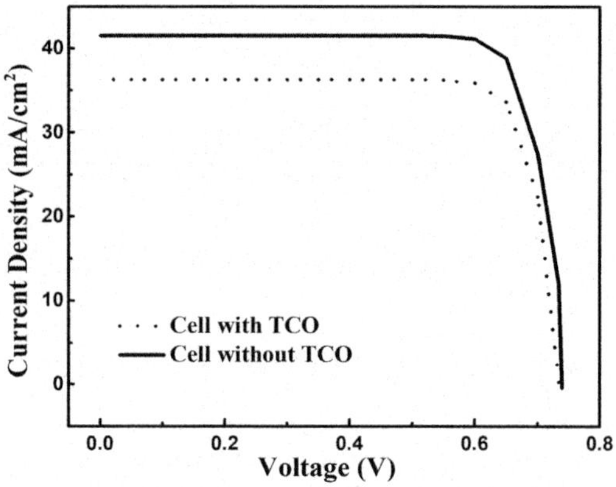

Fig. 2.7 Cell JV with and without TCO

2.4.4 *Effect of Front Transparent Conductive Oxide (TCO)*

The effect of using transparent conductive oxides like ITO (indium tin oxide) at the front side of the proposed cell is studied. For the existing designs based on amorphous silicon contacts, the need for using TCO arises due to low lateral conductivity of amorphous silicon. However, better conductivity and wide bandgap of the LiF_x layers averts the need of these layers. The cell was simulated using ITO and was compared with the cell without ITO layers. The ITO is included in the simulations as a front contact boundary with a work function of 4.2 eV. Further the n, k data for ITO has been obtained from SOPRA material database. The figures of merit noted for the cell are summarized in Table 2.4. It is clear that both Voc and Jsc of the cell deteriorate due to increased front absorption loss in the ITO layers. Figure 2.7 shows the relevant JV characteristics.

2.4.5 *Effect of LiF_x ESL Thickness*

The cell is also studied for varying LiF_x thickness. As discussed in Sect. 2.2, the presence of thin LiF_x layers depins the fermi level and improves the cell performance. Thin layers also reduce the contact resistivity allowing better collection of carriers. The effect of increasing thickness of LiF_x for different doping concentration is analyzed in Fig. 2.8 for cell without and with passivation. At low LiF_x doping, the efficiency of the cell decreases with the increase in the thickness of the LiF_x layer

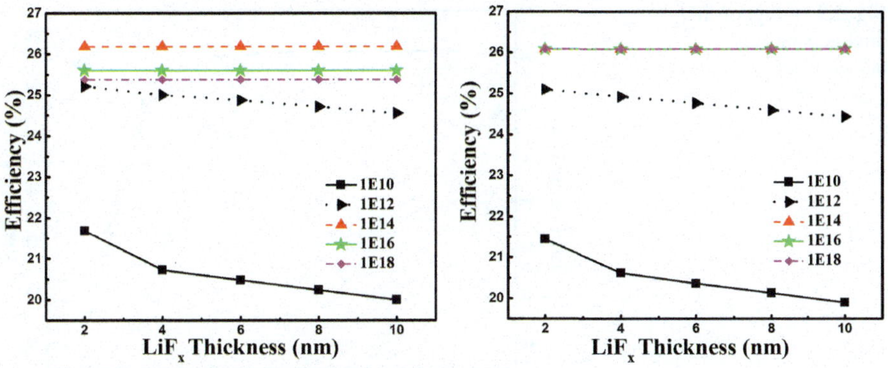

Fig. 2.8 Efficiency as a function of LiF_x thickness for different doping without and with passivation

for both with and without passivation designs. This is due to reduced possibilities of transport owing to less number of states for carrier tunneling through the layers. As the doping increases the large number of possible states allow good transportation thus improving the cell performance for both configurations. However beyond a certain optimal doping concentration the performance of the non-passivated design deteriorates while the passivated designs show resilience to higher doping concentrations. Also after the optimal doping the cell is independent of the LiF_x layer thickness.

2.4.6 Effect of Rear TOPCon Oxide Layer

The design is investigated for rear side TOPCon cell tunnel oxide parameters. We have used the TOPCon design in our cell as a hole selective contact by using highly doped p-polysilicon layers. Herein we have investigated the effect of tunnel oxide thickness on the overall performance of the cell. The inclusion of the SiO_2 layer improves the performance of the cell by allowing tunneling of carriers through the thin oxide layer [25]. The highly conductive polysilicon layers collects these holes. Increasing the thickness of the tunnel oxide initially does not affect the performance of the cell but beyond 1.4 nm thick tunnel layers the fill factor and efficiency of the cell start deteriorating owing to restriction in tunneling of the carriers by thicker layers as shown in Fig. 2.9. The use of tunnel layers of appropriate thickness is thus an important metric for overall performance of the proposed cell.

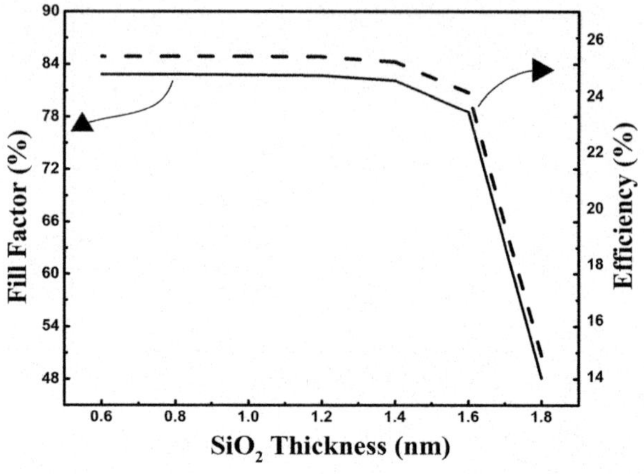

Fig. 2.9 Effect of SiO_2 thickness on cell performance

2.5 Conclusion

In this paper a class of materials called Alkali Metal Fluorides are investigated for use as electron selective contacts in silicon solar cells. These contacts are explored to overcome the limitation of parasitic absorption which arise by using TOPCon on the front side of the cell. For the simulation study, Lithium Fluoride is used due to its wide band gap and low work function. The proposed cell consist of a p wafer with a TOPCon rear design and a LiF_x front layer. The proposed cell is compared with a both sided TOPCon, amorphous silicon based designs and the simulation results imply that the use of LiF_x at the front reduces the absorption losses thus improving the overall performance parameters. Numerical simulations are also performed to analyze the effect of inclusion of passivation layers, LiF_x thickness and doping, transparent conductive oxides and tunnel oxide thickness. It is inferred that using alkali metal fluorides as alternate electron selective contacts in silicon heterojunction cells can increase the overall performance of the cell without incurring addition high costs.

Declaration of Interests

The authors declare that they have no known competing financial interests or personal relationships that could have appeared to influence the work reported in this paper.

References

1. T.G. Allen, J. Bullock, X. Yang, A. Javey, S.J.N.E. De Wolf, Passivating contacts for crystalline silicon solar cells. Nat. Energy **4**(11), 914–928 (2019)
2. J. Melskens, B.W. van de Loo, B. Macco, L.E. Black, S. Smit, W.M.M. Kessels, Passivating contacts for crystalline silicon solar cells: from concepts and materials to prospects. IEEE J. Photovoltaics **8**(2), 373–388 (2018)

3. M. Taguchi et al., 24.7% record efficiency HIT solar cell on thin silicon wafer. IEEE J. Photovoltaics **4**(1), 96–99 (2013)
4. K. Yoshikawa et al., Exceeding conversion efficiency of 26% by heterojunction interdigitated back contact solar cell with thin film Si technology. Sol. Energy Mater. Sol. Cells **173**, 37–42 (2017)
5. F. Feldmann, M. Bivour, C. Reichel, M. Hermle, S.W. Glunz, Passivated rear contacts for high-efficiency n-type Si solar cells providing high interface passivation quality and excellent transport characteristics. Sol. Energy Mater. Sol. Cells **120**, 270–274 (2014)
6. F. Feldmann, M. Bivour, C. Reichel, H. Steinkemper, M. Hermle, S.W.J. Glunz, Tunnel oxide passivated contacts as an alternative to partial rear contacts. Sol. Energy Mater. Sol. Cells **131**, 46–50 (2014)
7. Y. Liu et al., High-efficiency silicon heterojunction solar cells: materials, devices and applications. Mater. Sci. Eng. R. Rep. **142**, 100579 (2020)
8. B. Kafle, B.S. Goraya, S. Mack, F. Feldmann, S. Nold, J. Rentsch, TOPCon–technology options for cost efficient industrial manufacturing. Sol. Energy Mater. Sol. Cells **227**, 111100 (2021)
9. F. Feldmann, C. Reichel, R. Müller, M. Hermle, The application of poly-Si/SiO_x contacts as passivated top/rear contacts in Si solar cells. Sol. Energy Mater. Sol. Cells **159**, 265–271 (2017)
10. F. Feldmann, M. Simon, M. Bivour, C. Reichel, M. Hermle, S.W.J. Glunz, Efficient carrier-selective p- and n-contacts for Si solar cells. Sol. Energy Mater. Sol. Cells **131**, 100–104 (2014)
11. S. Reiter et al., Parasitic absorption in polycrystalline Si-layers for carrier-selective front junctions. Energy Procedia **92**, 199–204 (2016)
12. J. Schmidt, R. Peibst, R. Brendel, Surface passivation of crystalline silicon solar cells: present and future. Sol. Energy Mater. Sol. Cells **187**, 39–54 (2018)
13. B. Bilal, H. Najeeb-ud-Din, Fundamentals of and recent advances in carrier selective passivating contacts for silicon solar cells. J. Electron. Mater. **50**, 3761–3772 (2021)
14. M. Adeel, N.Z. Butt, Heterojunction solar cell with selective contacts, in *2019 International Conference on Electrical, Communication, and Computer Engineering (ICECCE)* (IEEE, 2019), pp. 1–6
15. M.Q. Khokhar et al., Role of electron carrier selective contact layer of lithium fluoride films with wide bandgap and low work function for silicon heterojunction solar cells. Mater. Sci. Semicond. Process. **134**, 105982 (2021)
16. K. Singh, D.K. Singh, V.K. Komarala, Numerical simulations of carrier-selective contact silicon solar cells: role of carrier-selective layers electronic properties. J. Comput. Electron. **20**(5), 1815–1821 (2021)
17. J. Bullock et al., Efficient silicon solar cells with dopant-free asymmetric heterocontacts. Nat. Energy **1**(3), 1–7 (2016)
18. J. Bullock et al., Lithium fluoride based electron contacts for high efficiency n-type crystalline silicon solar cells. Adv. Energy Mater. **6**(14), 1600241 (2016)
19. A. Froitzheim, R. Stangl, L. Elstner, M. Kriegel, W. Fuhs, AFORS-HET: a computer-program for the simulation of heterojunction solar cells to be distributed for public use, in *3rd World Conference on Photovoltaic Energy Conversion, 2003. Proceedings of*, vol. 1 (IEEE, 2003), pp. 279–282
20. R. Varache, C. Leendertz, M. Gueunier-Farret, J. Haschke, D. Muñoz, L. Korte, Investigation of selective junctions using a newly developed tunnel current model for solar cell applications. Sol. Energy Mater. Sol. Cells **141**, 14–23 (2015)
21. C. Quan et al., Computational analysis of a high-efficiency tunnel oxide passivated contact (TOPCon) solar cell with a low-work-function electron-selective-collection layer. Sol. Energy **170**, 780–787 (2018)
22. Z. Zhang et al., Carrier transport through the ultrathin silicon-oxide layer in tunnel oxide passivated contact (TOPCon) c-Si solar cells. Sol. Energy Mater. Sol. Cells **187**, 113–122 (2018)
23. M.Q. Khokhar et al., Simulation of silicon heterojunction solar cells for high efficiency with lithium fluoride electron carrier selective layer. Energies **13**(7), 1635 (2020)

24. S. Dasgupta, *Understanding Solar Cell Contacts Through Simulations* (Arizona State University, 2020)
25. B. Bilal, H. Najeeb-ud-Din, Towards low cost, industrially compatible silicon heterojunction solar cells using hybrid carrier selective passivating contacts. Opt. Mater. **124**, 111957 (2022)

Chapter 3
Computational Analysis of Antireflection Coatings for Semiconductor Solar Cells

Bisma Bilal and Hakim Najeeb-ud-Din

3.1 Introduction

With the availability of the conventional energy sources reaching its borderline, the need for increasing the efforts in making renewables the primary sources of energy is increasing more than ever [1]. One of the promising candidates in the renewables is solar energy. For the last few decades, consistent efforts are being made to develop materials that can assist the conversion of solar energy to its electrical counterpart with the maximum possible efficiency. Since the onset of research into photovoltaic devices, a large number of materials ranging from inorganic to organics are rigorously studied for use as absorbers in solar cell designs. The most important breakthrough was however achieved when semiconductors were investigated for photovoltaics [2]. This began with a simple p–n junction configuration and spread out to carrier selective, heterostructure and organic Photovoltaics. Currently, both organic and inorganic materials are analyzed to remove the efficiency and performance bottlenecks in solar cells.

Solar cell is an electronic device used for the conversion of sunlight to electrical energy via the process of complex interactions in the materials used for designing the cell. The basic concept of the conversion as shown in Fig. 3.1 is when photons (sunlight) fall on the absorber material, the interactions result in the generation of electron hole pairs after absorption of the photon. The number and range of absorption of the photons depends on the type of material used as absorber, its band gap and other electronic characteristics [3, 4]. The generated electrons and holes are further drifted to different electrodes via the process of charge separation which is an important consideration while designing the solar cell [5, 6]. According to the principle of detailed balance [7] the electrons and holes in the vicinity tend to combine and form

B. Bilal (✉) · H. Najeeb-ud-Din
Department of Electronics and Communication, National Institute of Technology, Srinagar, J&K 190006, India
e-mail: bisma_03phd17@nitsri.net

R. Singh et al. (eds.), *The Physics of Semiconductor Devices*, Springer Proceedings in Physics 306, https://doi.org/10.1007/978-981-97-1571-8_3

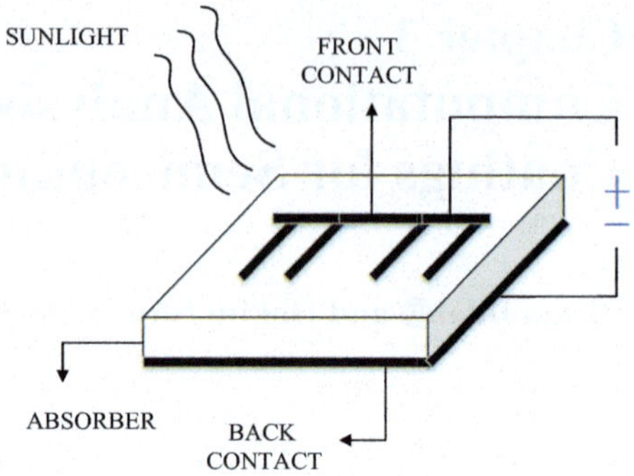

Fig. 3.1 The photovoltaic phenomena

an exciton which is a localized electron–hole pair with energy that dissipates via different mechanisms. The aim is to reduce this loss and allow maximum separation of carriers. For the details of such transport mechanisms the reader can refer to studies in [8–10]. Although numerous materials ranging from inorganic to organic have been investigated for designing solar cells, the silicon based cells continue to dominate the photovoltaic industry even after decades of its inception due to the enormous advantages of silicon as the material for photovoltaics. These include the cost factor, stability and the easy availability of silicon. The aim in solar cell design is to improve the various figures of merit which include the open-circuit voltage, Short circuit current, Fill Factor and the overall efficiency of the cell [11, 12]. The most important optimization technique identified for solar cells is the inclusion of layers of antireflection material to reduce the reflectance from the absorber materials used in the design. In the subsequent sections the need for antireflection coatings and their design considerations will be discussed.

3.2 Antireflection Coating for Solar Cells

The semiconductor materials used for the design of solar cells have their innate limitations due to which their capability of light absorption is limited. For example, the case of Silicon, in which almost 30% of the incident light is reflected because of the refractive index difference with the surrounding. This amount of loss is quite large when combined with other effects e.g. recombination which further reduces the number of carriers generated for the conversion and contribution to electrical energy. The need for avoiding these losses becomes of utmost importance to improve the overall device performance [13, 14]. This issue has been addressed by the addition of antireflection coating layers over the solar cells. Antireflection coatings are dielectric material layers which have the properties of better light trapping thus restricting the reflection of light from the surface. Figure 3.2 shows the simulation comparison of generation rates of a planar silicon solar cell with and without the antireflection

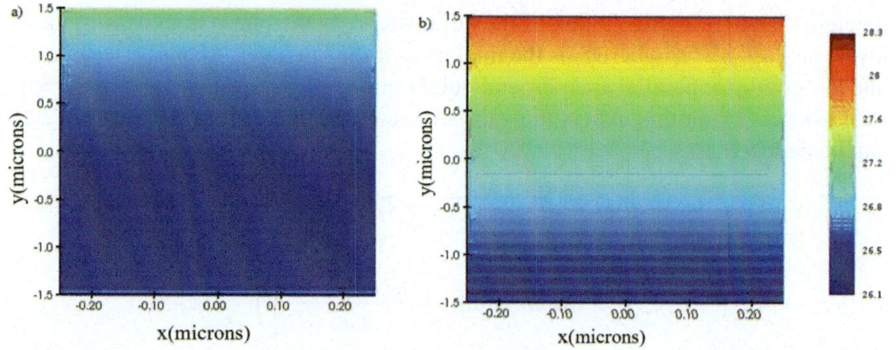

Fig. 3.2 Generation rate for solar cells without antireflection coating (**a**) and with antireflection coating (**b**)

coating. In the case of antireflection layer included the photons are trapped and travel further into the depth of the cell thus increasing the generation rate of the carriers.

The essential requirements for Anti-reflection in solar cells include:

I. Broadband anti reflectivity which focuses on the fact that the ARC should have consistent antireflection property over a broad range of the solar spectrum. Due to this property maximum light is trapped and used for the generation of carriers inside the absorber of the solar cell.
II. Omnidirectional characteristics target the impact of the angle of light incidence on the overall reduction of reflection from the surface of the material. This means that the ARC should have the capability of trapping the light coming from different grazing angles.
III. Polarization insensitivity is another important property since the light can be s or p polarized [14].

The use of ARC is also dependent on the absorber being used for the solar cell e.g. the ARC which gives the best light trapping for one semiconductor may not work that efficiently for some other material due to the refractive index gradients [15, 16]. The different classifications of ARC are based on layer composition which includes a single layer, double, triple-layer ARC and the surface topographies covering the porous structures and incorporation of various texturing techniques on the surface of the solar cell [17, 18].

Matrix methods are used for the reflectance calculations of ARC on photovoltaic devices [19]. The matrix relates the electromagnetic field components and each layer is represented by a 2×2 matrix as;

$$M_{eq} = \prod_{k=1}^{k=N} \begin{pmatrix} \cos \varphi_k & i/n_k^{eff} \sin \varphi_k \\ i n_k^{eff} \sin \varphi_k & \cos \varphi_k \end{pmatrix} \quad (3.1)$$

where n_k is the refractive index of the kth layer, φ_k is the phase thickness of the kth layer, $\varphi_k = \frac{2\pi}{\lambda} n_k^{eff} t_k$ where t_k is the thickness of the kth layer. Also $n_k^{eff} = n_k \cos\theta_k$ and $n_k / \cos\theta_k$ for parallel and perpendicular polarization respectively. θ_k is the angle of incidence. From this matrix the values of refractive index and thickness of ARC can be deduced as;

$$n = \sqrt{(n_0 n_s - (n_0 k_s)/(n_0 - n_s))} \tag{3.2}$$

$$t = (\lambda/2\pi)\text{arctg}\left(\left[\frac{n_0(n_0 - n_s)}{n_0 k_s}\right]\right) \tag{3.3}$$

where n_0 is the refractive index of the surrounding material (air), n_s is the absorber index and k_s is the extinction coefficient [19]. The aim in solar cells is to reduce the reflectance to a minimum value for a maximum photon flux and quantum efficiency. This is obtained from the calculation of weighted reflectance as;

$$R_w = \frac{\int_{\lambda_1}^{\lambda_2} Ø(\lambda) Q(\lambda) R(\lambda) d\lambda}{\int_{\lambda_1}^{\lambda_2} Ø(\lambda) Q(\lambda) d(\lambda)} \tag{3.4}$$

where Ø(λ) represents the photon flux, Q(λ) is the quantum efficiency and R(λ) is the reflectance. The optimum values for n and t need to be obtained for obtaining minimum reflectance which will be discussed in Sect. 3.3. Figures 3.3 and 3.4 show a comparison of reflectance and power absorption of the planar solar cell with and without the Antireflection coating. All the simulations were performed in Lumerical FDTD Solutions. The antireflection coating, MgF_2 is considered for this comparison. From the figures it is evident that for the cell with a coating of ARC material the overall reflectance decreases while the overall power absorbed by the cell increases. Also the ripples are observed at high values of lambda due to the device's Fabry Perot effect.

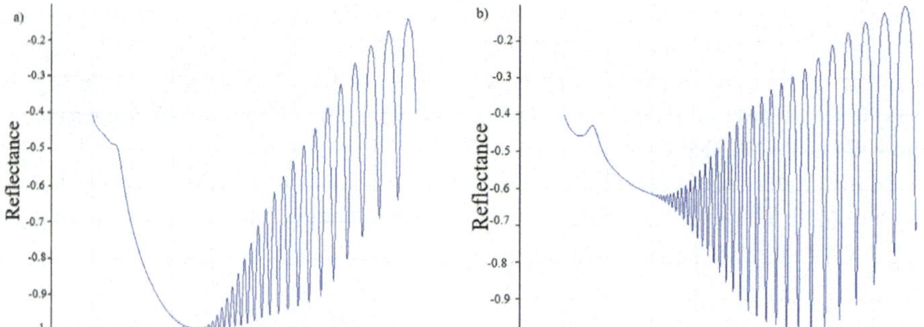

Fig. 3.3 Reflectance with respect to wavelength for a silicon solar cell with antireflection coating (**a**) and without antireflection coating (**b**)

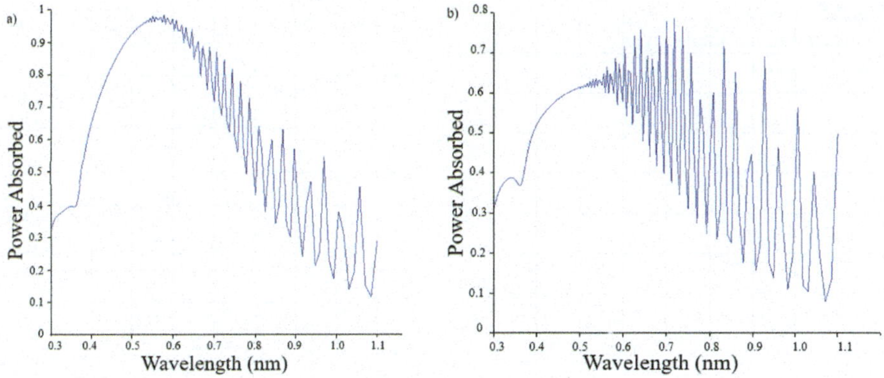

Fig. 3.4 Power absorbed with respect to wavelength for a silicon solar cell with antireflection coating (**a**) and without antireflection coating (**b**)

Increasing the number of layers in the ARC configuration reduces the overall reflectance from the device which apparently suggests inclusion of a larger number of layers. However multiple layers call for a higher design intricacy and time to market which does not assist the cost to benefit ratio of the devices [20, 21]. If the major design criteria in the solar cell is cost then the single layer coatings are most preferred and are by far the most employed ARC's for terrestrial solar cells. However when improving the efficiency and light trapping is the required optimization then multiple layer ARC's are the most viable option. Also the improvement in the performance after two layers of coatings is by far negligibly small. A number of techniques including chemical vapor deposition, plasma enhanced CVD, screen printing etc. have been used for these coating depositions [22]. For all our simulation studies single layer antireflection coatings have been analyzed for two major semiconductors i.e. silicon and gallium arsenide.

3.3 Antireflection Coating Design Considerations

The major design considerations for the ARC configuration consists of achieving minimum reflectivity with respect to wavelength in the given solar spectrum. This is based on two criteria of calculating the values of optimal ARC thickness and refractive index. The single layered ARC configuration is shown in Fig. 3.5, where n_0 is the refractive index of the surrounding material (air), n_{ar} is the refractive index for the antireflection coating under consideration and n_s is the absorber index. R_1, R_2 represent the reflected waves and T_1, T_2 are the transmitted waves at the two interfaces.

The basic mechanism of ARC design is based on the interference of light. The first consideration in the design of antireflection coating system is the refractive index of the materials used [21]. Since the refractive index undergoes changes at the two

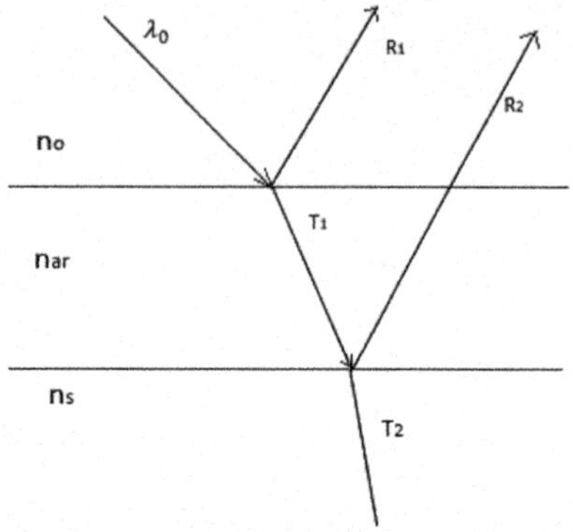

Fig. 3.5 Layered configuration for single layer ARC Solar cell

interfaces the consequent reflection coefficients can be expressed as;

$$r_1 = \frac{n_{ar} - n_0}{n_{ar} + n_0} = \frac{1 - \frac{n_0}{n_{ar}}}{1 + \frac{n_0}{n_{ar}}} \tag{3.5}$$

$$r_2 = \frac{n_s - n_{ar}}{n_s + n_{ar}} = \frac{1 - \frac{n_{ar}}{n_s}}{1 + \frac{n_{ar}}{n_s}} \tag{3.6}$$

For obtaining the maximum destructive interference

$$\frac{n_0}{n_{ar}} = \frac{n_{ar}}{n_s} \tag{3.7}$$

or

$$n_{ar} = \sqrt{n_0 n_s} \tag{3.8}$$

For a system of M layers, the refractive index of the mth layer is given as;

$$n_m = n_{sup}^{\left(\frac{M+1-m}{M+1}\right)} \cdot n_{sub}^{\left(\frac{m}{M+1}\right)} \tag{3.9}$$

where n_m is the refractive index of the mth layer, n_{sub} is the refractive index of the used substrate and n_{sup} is the refractive index of the superstrate [22].

The other factor is the thickness of the antireflection coating. The phase difference between the two reflected waves shown in Fig. 3.5 is $2n_{ar}t$, where t is the thickness of the antireflection coating. If this phase difference becomes equal to half the wavelength of the incoming wave then a destructive interference occurs and the resultant is the zero reflection from the device surface.

Mathematically this condition arises when

$$2n_{ar}t = \frac{\lambda_0}{2} \tag{3.10}$$

or

$$t = \frac{\lambda_0}{4n_{ar}} \tag{3.11}$$

This formulation suggests that the reflection via a layered configuration is a function of the thickness of the antireflection coating used. Choosing an optimal value of thickness is an essential requirement for obtaining the maximum possible performance. The value of thickness corresponds to a value where the interactions between the waves are out of phase, leading to a destructive interference and a zero reflected energy. This is called the quarter wavelength rule and is followed for choosing the thickness for minimum reflectance [20].

Further it can be shown that quarter wavelength value is mostly used as the optimum value of thickness but other minimum values correspond to the odd multiples i.e.

$$t = \frac{\lambda}{4n_{ar}}, \frac{3\lambda}{4n_{ar}}, \frac{5\lambda}{4n_{ar}}, \ldots \tag{3.12}$$

The other design consideration for antireflection coatings is to obtain the maximum photogenerated current density from any solar cell [21, 22]. The maximum photogenerated current density is given as;

$$J_{ph}^{ideal}(t, n) = \int_{\lambda_{min}}^{\lambda_{max}} q\phi(\lambda)(1 - R(\lambda, t, n))d\lambda \tag{3.13}$$

where λ_{min} and λ_{max} represent the minimum and maximum radiation wavelength, $\phi(\lambda)$ is the photon flux.

The photon flux is calculated from the solar spectrum of AM 1.5. The maximum value of photogenerated current is obtained when 100% internal quantum efficiency is considered. For practical cases we have;

$$J_{ph}^{real}(t, n) = \int_{\lambda_{min}}^{\lambda_{max}} q\phi(\lambda)n_q(1 - R(\lambda, t, n))d\lambda \tag{3.14}$$

where n_q represents the real internal quantum efficiency. For cells with high efficiencies the maximum and real values are close and the difference can be actuated to the uncertainty in the thickness of the deposited layers. The aim here is reducing the differentials $\frac{\partial J_{ph}^{ideal}}{\partial k} = 0$, $\frac{\partial J_{ph}^{real}}{\partial k} = 0$ where k is the adjusting parameter [22].

Considering the above discussed design considerations this study is directed to understand the effect of thickness of the antireflection coating layer on the reflection coefficient/reflectance for a wavelength of 632 nm. This value of wavelength is selected since it corresponds to the peak power of the solar spectrum i.e. there is a maximum photon flux around this range of the spectrum. The antireflection coatings MgF_2, Al_2O_3, Si_3N_4, ZnO, TiO_2 and SiC with refractive indices of 1.37, 1.77, 2, 2.05,

2.48 and 2.65 respectively are used for the Silicon and Gallium Arsenide semiconductor absorbers. The refractive indices of ARC have been chosen according to the selected value of wavelength (632 nm) at which the simulations are performed. This is because the refractive index of the material is not constant at all the wavelengths and varies slightly at different values of lambda. The greater the wavelength change from one medium to other, more is the refraction and thus refractive index.

It is important to mention here that although from the above mathematical formulations the optimal values of t and n can be calculated, the deposition processes often result in variations in the thicknesses of the deposited layers. For this the reflectance trends for varying thickness are obtained for different antireflection coating layers. These trends provide a pathway towards the understanding of the behavior of the solar cell subjected to processing techniques which are often not optimal and have good amount of diversion from the ideal figures of deposition required. The antireflection coating of choice must show little change with the random thickness variations and this low sensitivity allows the choice material to perform to the best ability for a range of deposited layer thicknesses.

3.4 Numerical Simulations

The FDTD-Solutions suite provided by Lumerical Inc. [23] is used to study the effect of the thickness of the ARC on the reflectance for the absorber materials. FDTD (Finite Difference Time Domain) method is used for obtaining the solution of 2D and 3D Maxwell equations in various structures. Since the technique is a time-domain technique, all solutions are obtained as a function of time. The FDTD optical simulation results in the spatial absorption in the substrate region from which the generation rate can be calculated. This rate can then be used further for different electrical measurements etc. The Drift–Diffusion model provides self-consistent solutions to the Poisson and the continuity equations [24]. In our study, a 2-dimensional optical simulation of a planar cell is considered. The semiconductor absorbers (Si and GaAs) are considered to be 3 μm thick with the surface area of 100×100 μm. Emitter and base are connected to the top and the bottom of the film respectively as shown in Fig. 3.6 (Table 3.1).

Now, in the case of the optical simulations the shadowing effects i.e. the effect of contacts is neglected and a smaller area of 0.5 μm is considered for keeping the simulation time to the minimum. The periodic boundary conditions are used to obtain the generation rate for the entire structure. However, PML (perfectly matched layer) boundary condition is considered for the top and bottom to avoid any unwanted reflection effects.

A plane wave light source with the wavelength from 300 to 1100 nm is used and monitors are used for the calculation of the amount of light reflected and transmitted through the solar cell. Further, different ARC are considered and the optimum values are investigated. Figure 3.7 shows the optical simulation structure designed in Lumerical FDTD-Solutions. For a particular ARC and absorber combination, the reflectance

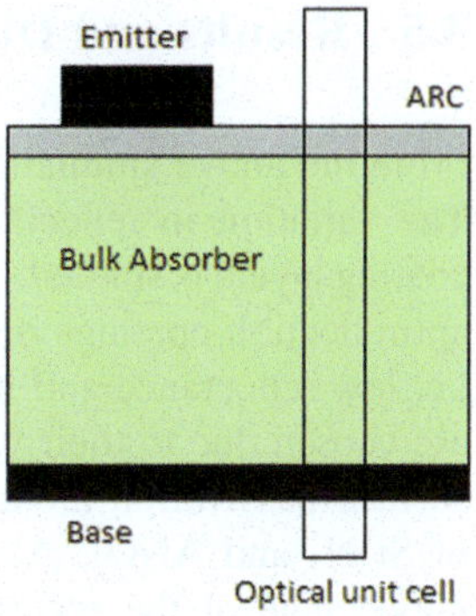

Fig. 3.6 Design of the silicon solar cell

Table 3.1 Design considerations

Parameter	Value
Semiconductor thickness	3 μm
Optically simulated region	0.5 μm
Light source	Plane wave source
Spectrum	AM 1.5 G
Boundary conditions	Periodic

at 632 nm wavelength is obtained from the simulation. The antireflection coating thickness is incremented and corresponding reflectance with changing thickness is observed. The simulations are performed for all the absorber/ARC combinations and the trends are analyzed.

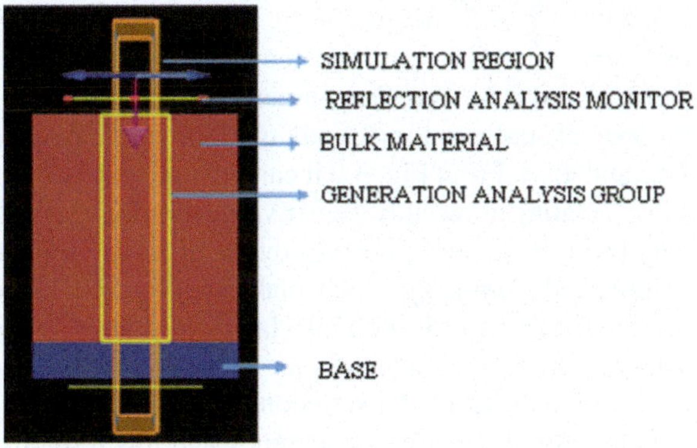

Fig. 3.7 Simulation schematic of solar cell

3.5 Results and Discussion

From the above simulations, the results were compiled as shown in figures below. The variation in reflectance as a function of thickness of different antireflection coating layers is studied. The simulations emphasize the importance of designing the antireflection coatings of appropriate thickness for obtaining the best performance i.e. low reflectance and maximum absorption. The different antireflection coatings are chosen due to their wide use in solar cells or multiple functions like providing surface passivation along with antireflection properties. This is true for coatings of Si_3N_4 and Al_2O_3. Aluminum trioxide also has an excellent transparency for a good range of the spectrum and good resistance to mechanical strain. The TiO_2 films form the stable choice for antireflection coatings and ZnO layers grown on the surface of semiconductors act as the active layer along with antireflection agent. Further ZnO also provides radiation damage resistance ensuring longer lifetime for the device. MgF_2 antireflection coating has a better transparency and low refractive index [25–30].

Although the optimal values of t and n can be mathematically obtained from equations discussed in section III, our simulations, form a proof of concept for such theoretical results. From Figs. 3.8, 3.9, 3.10, 3.11, 3.12 and 3.13, it is evident that for a particular ARC, the minimum value of reflectance is obtained at the same value of thickness for all semiconductor absorbers. This is because the values are observed at the wavelength of 632 nm which is related to the thickness by a factor of 4π as discussed earlier. Also the points of optimal thickness which correspond to lowest reflectance values are in accordance with those mathematically obtained from the above equations. Considering e.g. the case of Si_3N_4 the mathematically calculated optimal thickness is 0.079 μm which is reflected in Fig. 3.11. The other low reflectance point values as calculated from Eq. 3.12 are 0.237 μm and 0.395 μm for 3λ and 5λ respectively. Similarly for the case of TiO_2 the optimal value of thickness is 0.06 μm and the other values form the mathematical calculations are 0.19 and 0.31 μm. Our numerical simulations are in correspondence with these theoretical analysis. Similar calculations can be done for other antireflection coatings for validation. From Fig. 3.9 it can be seen that the performance of the solar cell with MgF_2 coating is weakly sensitive to the thickness of the MgF_2 layer between 50 and 150 nm. There is no improvement in the collection or current for the thicker or thinner layers [22] which makes MgF_2 one of the preferred material for ARC design. In other words the MgF_2 layers are almost independent of the process induced thickness variations which is an important consideration to be taken care of given the non-idealities of the semiconductor materials and the equipment used therein. Additionally the better transparency and low refractive index of MgF_2 are the other reasons for its good performance.

The graphical results also emphasize that the ARC materials with lower refractive index show minimum reflectance at higher thickness values as for MgF_2 and Al_2O_3 coatings. However the interest here is to analyze the trends with respect to random thickness variation which is clearly the lowest for MgF_2 coating in the solar cell.

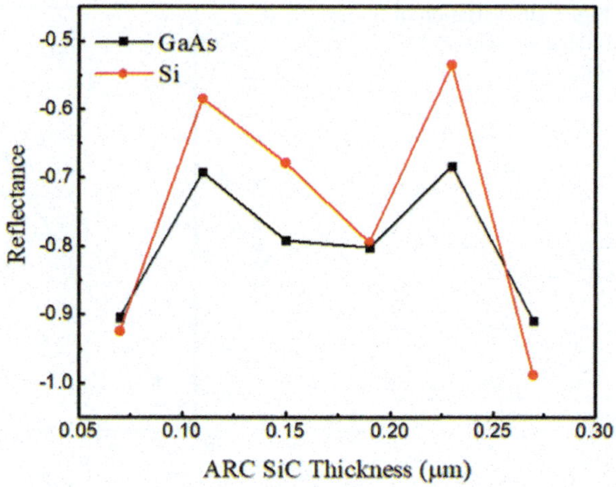

Fig. 3.8 Variation of reflectance with SiC thickness

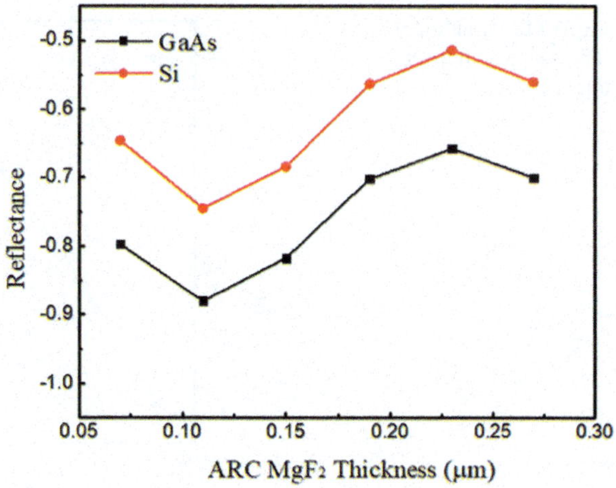

Fig. 3.9 Variation of reflectance with MgF_2 thickness

As the refractive index of the antireflection coating increases the high/low peaks and the sensitivity of the cells towards the process induced thickness variation is seen to be increasing. This high sensitivity of these layers to the random thickness variation of the antireflection coatings limits the use of these materials owing to the non-idealities of the deposition processes. Also, it should be noted that we have not plotted the power absorption graphs since power and reflectance are inversely related. Lower reflectance means more power absorbed. Further it was observed that the best results i.e. lowest Reflectance and maximum power absorption values were obtained for GaAs based solar cells. This is because better performance is presented by III–V direct band gap semiconductors solar cells as compared to solar cells made from other materials. However the restrictions like availability and cost factors limit the use of GaAs based devices.

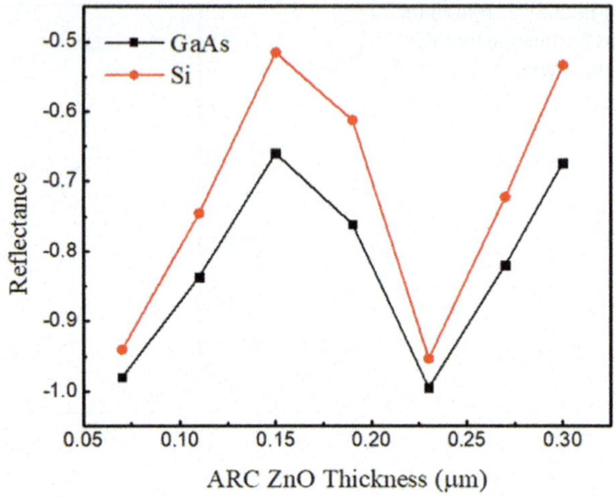

Fig. 3.10 Variation of reflectance with ZnO thickness

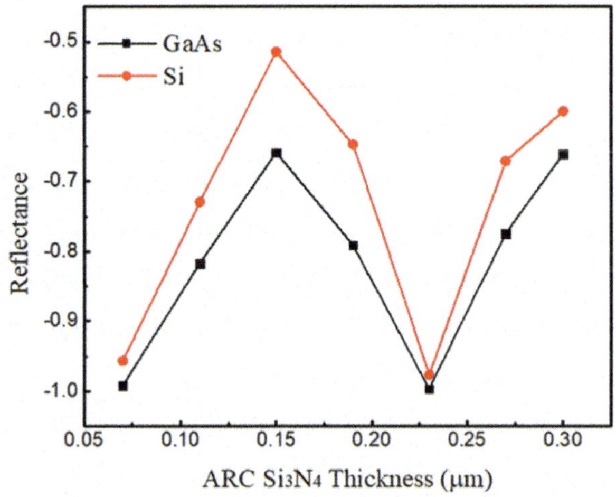

Fig. 3.11 Variation of reflectance with Si_3N_4 thickness

Figure 3.14 plots the differentiated values of reflectance for a silicon absorber with respect to the ARC thickness. This again proves the stability of MgF_2 layers for the random thickness process variations. As we can see, the trends of all the other antireflection coatings are highly sensitive to the thickness variations showing peaks of highs and lows but the MgF_2 coating shows a close to zero slope with the changing thickness. This emphasis the utility of MgF_2 layers as the best option for single layer ARC design for terrestrial solar cells.

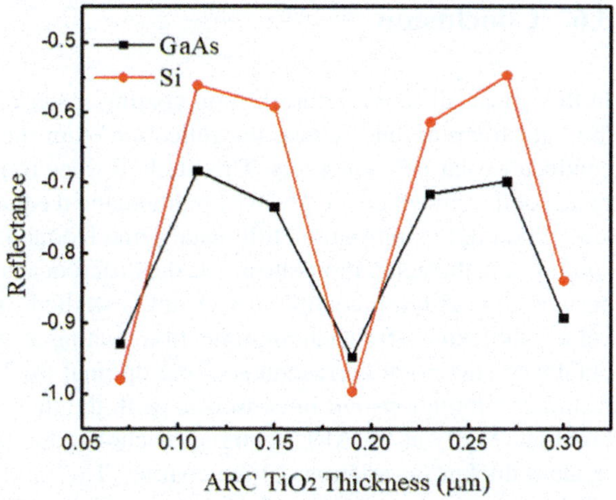

Fig. 3.12 Variation of reflectance with TiO_2 thickness

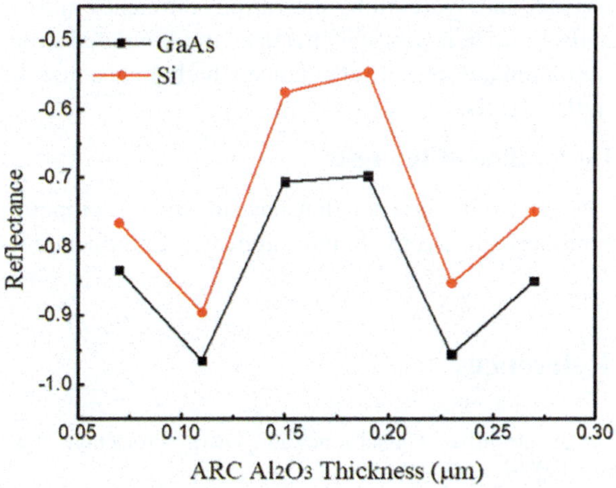

Fig. 3.13 Variation of reflectance with Al_2O_3 thickness

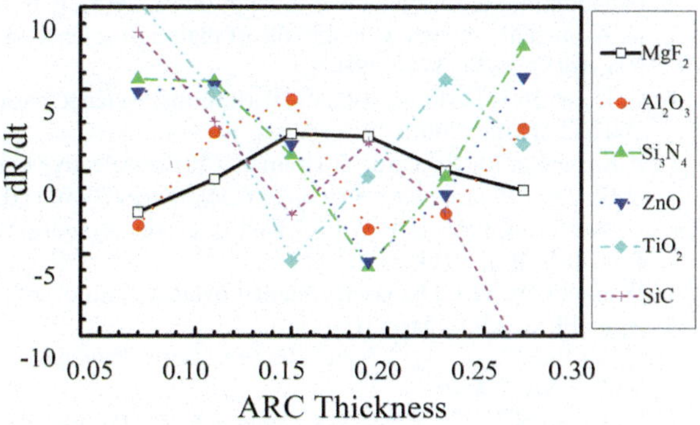

Fig. 3.14 Variation of reflectance with ARC thickness for a silicon absorber

3.6 Conclusion

In this paper, different antireflection coating materials are investigated to improve the light trapping and reduce the reflection from the surfaces of prominent semiconductor solar cell absorbers. This study is done to numerically prove the optimal points of design which are given by mathematical equations. The optical simulations are carried out using Finite Difference Time Domain analysis for obtaining optical solution and the generation rate inside the different absorbers. The effect of the thickness of antireflection coating at 632 nm is studied. It was observed that there is a value of thickness for each antireflection coating at which minimum reflectance is obtained. This point corresponds to the optimal thickness and the results obtained from our simulations are in consensus with the theoretical results. The trends also highlight MgF_2 as an ARC material of choice due to its lower variation with the random thickness variations of the coating. This study thus emphasizes on the role of choosing the best ARC and its optimal thickness for obtaining ultimate characteristics and overcoming the various efficiency and performance bottlenecks in solar cells. Further, it was seen that in all the simulated structures the GaAs solar cell shows minimum amount of reflectance which means that the maximum power is absorbed in these cells.

Declaration of Interests

The authors declare that they have no known competing financial interests or personal relationships that could have appeared to influence the work reported in this paper.

References

1. A. Demirbas, Global renewable energy projections. Energy Sources Part B **4**(2), 212–224 (2009)
2. D.M. Chapin, C. Fuller, G. Pearson, A new silicon p-n junction photocell for converting solar radiation into electrical power. J. Appl. Phys. **25**(5), 676–677 (1954)
3. N. Asim et al., A review on the role of materials science in solar cells. Renew. Sustain. Energy Rev. **16**(8), 5834–5847 (2012)
4. G. Knier, How Do Photovoltaics Work? https://science.nasa.gov/science-news/science-at-nasa/2002/solarcells (2002)
5. P. Koswatta, M. Boccard, Z. Holman, Carrier-selective contacts in silicon solar cells, in *2015 IEEE 42nd Photovoltaic Specialist Conference (PVSC)* (IEEE, 2015), pp. 1–4
6. U. Würfel, A. Cuevas, P. Würfel, Charge carrier separation in solar cells. IEEE J. Photovoltaics **5**(1), 461–469 (2014)
7. W. Shockley, H.J. Queisser, Detailed balance limit of efficiency of p-n junction solar cells. J. Appl. Phys. **32**(3), 510–519 (1961)
8. P. Wurfel, *Physics of Solar Cells: from Basic Principles to Advanced Concepts* (Wiley, New York, USA, 2009)
9. S.M. Sze, K.K. Ng, *Physics of Semiconductor Devices*, 3rd ed. (Wiley, 2007)
10. D.D. Smith, P. Cousins, S. Westerberg, R. De Jesus-Tabajonda, G. Aniero, Y.-C. Shen, Toward the practical limits of silicon solar cells. IEEE J. Photovoltaics **4**(6), 1465–1469 (2014)

11. M.A. Green, K. Emery, Y. Hishikawa, W. Warta, E.D. Dunlop, Solar cell efficiency tables (Version 45). Prog. Photovoltaics Res. Appl. **23**(1), 1–9 (2015)
12. V. Tyagi, N.A. Rahim, N.A. Rahim, A. Jeyraj, L. Selvaraj, Progress in solar PV technology: research and achievement. Renew. Sustain. Energy Rev. **20**, 443–461 (2013)
13. K.W. Chee, Z. Tang, H. Lü, F. Huang, Anti-reflective structures for photovoltaics: numerical and experimental design. Energy Rep. **4**, 266–273 (2018)
14. B. Kumaragurubaran, S. Anandhi, Reduction of reflection losses in solar cell using anti reflective coating, in *International Conference on Computation of Power, Energy, Information and Communication (ICCPEIC), Chennai, India*, 2014
15. S. Chhajed, M.F. Schubert, J.K. Kim, E.F. Schubert, Nanostructured multilayer graded-index antireflection coating for Si solar cells with broadband and omnidirectional characteristics. Appl. Phys. Lett. **93**(25), 251108 (2008)
16. M. Moayedfar, M.K. Assadi, Various types of anti-reflective coatings (ARCS) based on the layer composition and surface topography: a review. Rev. Adv. Mater. Sci. **53**(2), 187–205 (2018)
17. M.S. Mozumder, A.-H.I. Mourad, H. Pervez, R. Surkatti, Recent developments in multifunctional coatings for solar panel applications: a review. Sol. Energy Mater. Sol. Cells **189**, 75–102 (2019)
18. H.K. Raut, V.A. Ganesh, A.S. Nair, S. Ramakrishna, Anti-reflective coatings: a critical, in-depth review. Energy Environ. Sci. **4**(10), 3779–3804 (2011)
19. D. Bouhafs, A. Moussi, A. Chikouche, J.M. Ruiz, Design and simulation of antireflection coating systems for optoelectronic devices: application to silicon solar cells. Sol. Energy Mater. Sol. Cells **52**(1–2), 79–93 (1998)
20. T. Markvart, L. Castañer, Principles of solar cell operation, in *McEvoy's Handbook of Photovoltaics* (Elsevier, 2018), pp. 3–28
21. S. Al-Turk, *Analytic Optimization Modeling of Anti-reflection Coatings for Solar Cells* (2011)
22. P. Nubile, Analytical design of antireflection coatings for silicon photovoltaic devices. Thin Solid Films **342**(1–2), 257–261 (1999)
23. Lumerical, FDTD Solutions 6.5 [Online]. Available at: http://www.lumerical.com/tcad-products/fdtd
24. Y. Fang et al., Drift-diffusion InGaN/GaN solar cell simulator with optical management, in *2017 IEEE 44th Photovoltaic Specialist Conference (PVSC)* (IEEE, 2017), pp. 1603–1605
25. L.A. Dobrzański, M. Szindler, A. Drygała, M. Szindler, Silicon solar cells with Al_2O_3 antireflection coating. Open Phys. **12**(9), 666–670 (2014)
26. B. Hussain, A. Ebong, I. Ferguson, Zinc oxide as an active n-layer and antireflection coating for silicon based heterojunction solar cell. Sol. Energy Mater. Sol. Cells **139**, 95–100 (2015)
27. C. Yu, S. Xu, J. Yao, S. Han, Recent advances in and new perspectives on crystalline silicon solar cells with carrier-selective passivation contacts. Crystals **8**(11), 430 (2018)
28. T.G. Allen, J. Bullock, X. Yang, A. Javey, S. De Wolf, Passivating contacts for crystalline silicon solar cells. Nat. Energy **1** (2019)
29. B. Bilal, H. Najeeb-ud-Din, Towards low cost, industrially compatible silicon heterojunction solar cells using hybrid carrier selective passivating contacts. Opt. Mater. **124**, 111957 (2022)
30. B. Bilal, H. Najeeb-ud-Din, Fundamentals of and recent advances in carrier selective passivating contacts for silicon solar cells. J. Electron. Mater. **50**, 3761–3772 (2021)

Chapter 4
Effect of O_2 Partial Pressure on Valence Band Maxima of HfO_2 Thin Film

Mangla Nand, Y. Kumar, S. Tripathi, Babita, S. K. Rai, and S. N. Jha

4.1 Introduction

The size of metal oxide semiconductor devices decreases with the increase in density of integrated circuit. The nano-scaling of such devices is highly influenced by gate dielectric in complementary metal oxide semiconductor (CMOS) devices. HfO_2 is found to be one of the best contenders for high k gate dielectric materials to replace SiO_2 in CMOS transistors and this reduces the leakage current and increase the gate capacitance in miniaturizing the devices [1]. Apart from the relatively high dielectric constant of HfO_2 (k $\sim$ 25) in comparison to SiO_2 (k $\sim$ 3.9), it possesses the large bandgap (Eg $\sim$ 5.68 eV), and thermodynamically stable when grown on Si [1, 2]. The electronic properties of microelectronic devices were determined by the stoichiometry of transition metal oxides. Also, there are reports that band gap changes due change in valence band maxima (VBM) [3], and also the band gap of HfO_2 thin film changes with O_2 partial pressure [4]. Shifting of the VBM may be the one reason for this band gap change in HfO_2 thin films. The oxygen defects in HfO_2 thin films influence the functional properties in semiconductor devices [5]. Although VBM of oxide thin films plays an important role in charge transport of semiconductor devices and strongly dependent on oxygen vacancies, there is lack of understanding on the effect of oxygen vacancies on VBM. Therefore, it is desired to study the VBM of stoichiometric and sub-stoichiometric HfO_2 thin films. Aiming at

M. Nand (✉) · Y. Kumar · S. Tripathi · Babita · S. N. Jha
Beamline Development and Application Section, Bhabha Atomic Research Centre, Mumbai 400085, India
e-mail: kappu001@gmail.com

M. Nand · Y. Kumar · S. N. Jha
Homi Bhabha National Institute, Anushaktinagar, Mumbai 400094, India

S. K. Rai
Synchrotrons Utilization Section, Raja Ramanna Centre for Advanced Technology, Indore 452013, India

R. Singh et al. (eds.), *The Physics of Semiconductor Devices*, Springer Proceedings in Physics 306, https://doi.org/10.1007/978-981-97-1571-8_4

this, in the present study we have investigated the effect of the O_2 partial pressure on the VBM of the HfO_2 thin films deposited using pulsed laser deposition (PLD) technique.

4.2 Experimental Details

Two HfO_2 thin films were deposited using PLD on Si(100) substrate at 6.78×10^{-5} mbar without O_2 ambience (sample A) and at 5.53×10^{-4} mbar with O_2 (sample B). 3rd harmonic of Nd:YAG laser ($\lambda = 355$ nm) with fluence of ~ 1 J/cm^2 was used in deposition. The substrate temperature was kept at 700 °C during the deposition. The crystalline phase of the thin films was determined by GIXRD measurements using a Bruker D8 Discover Diffractometer with Cu K_α X-ray source. Stoichiometry, oxidation states, oxygen vacancy, and VBM were determined using PES measurements, which were performed at ARPES beamline (BL-3) [6] of Indus-1 synchrotron radiation source using PHOIBOS 150 hemispherical electron analyzer. Core levels were recorded using Al Kα X-ray source with 30 eV pass energy, while the valence band spectra were recorded using 100 eV photon of Indus-1 SR source with 20 eV pass energy. All the spectra were calibrated by taking C 1s (284.6 eV) core level as reference peak.

4.3 Results and Discussion

Figure 4.1 shows GIXRD spectra of both films, which confirms that the deposited films are in monoclinic polycrystalline phase and peaks are in accordance with the PDF card #04-004-3850. The XRD pattern showed a high degree of crystallinity with increased intensity of (− 111) diffraction peaks for samples B with increased oxygen content.

Figure 4.2a depicts the O 1*s* core level spectra of sample-A, in which no O_2 flow was used during the deposition. The spectrum was deconvoluted using two peaks. The main peak was observed at 530.23 eV binding energy, which corresponds to O-Hf bond in HfO_2 lattice [2]. Second peak V_o was observed at 531.75 eV and assigned to O_2 vacancy peak as reported in literature [7]. As shown in Fig. 4.2b, similar features were observed in sample-B (deposited in O_2 ambience) also. However, the intensity of vacancy peak V_o in sample-B decreases to 17% from 35% in compare to sample A. This infers that the presence of O_2 gas during deposition reduces the oxygen vacancy in deposited thin films.

Hf 4*f* core level spectra are shown in Fig. 4.2c. In both films, the observed spin orbit splitting (~ 1.63 eV) and peak positions ($4f_{7/2}$—17.32 eV, $4f_{5/2}$—18.95 eV), correspond to HfO_2 peaks and matches well with reported values [2]. The inset in Fig. 4.2c is showing the deconvoluted Hf 4*f* spectrum for sample-A. The deconvoluted symmetric peaks ($4f_{7/2}$ & $4f_{5/2}$) also confirm the presence of only Hf^{4+}

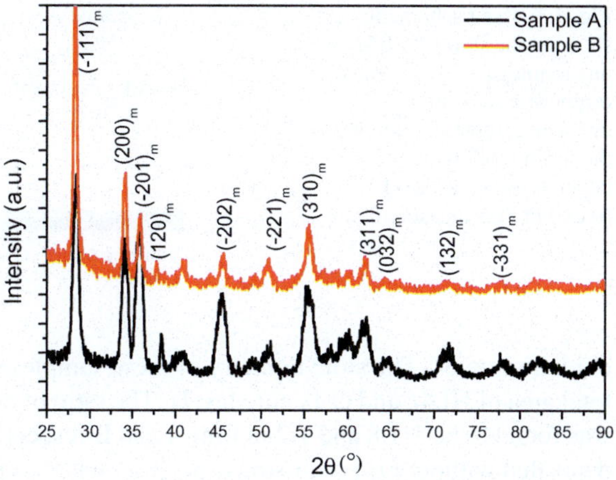

Fig. 4.1 GIXRD patterns of both the deposited HfO_2 thin films recorded with Cu k-alpha lab source

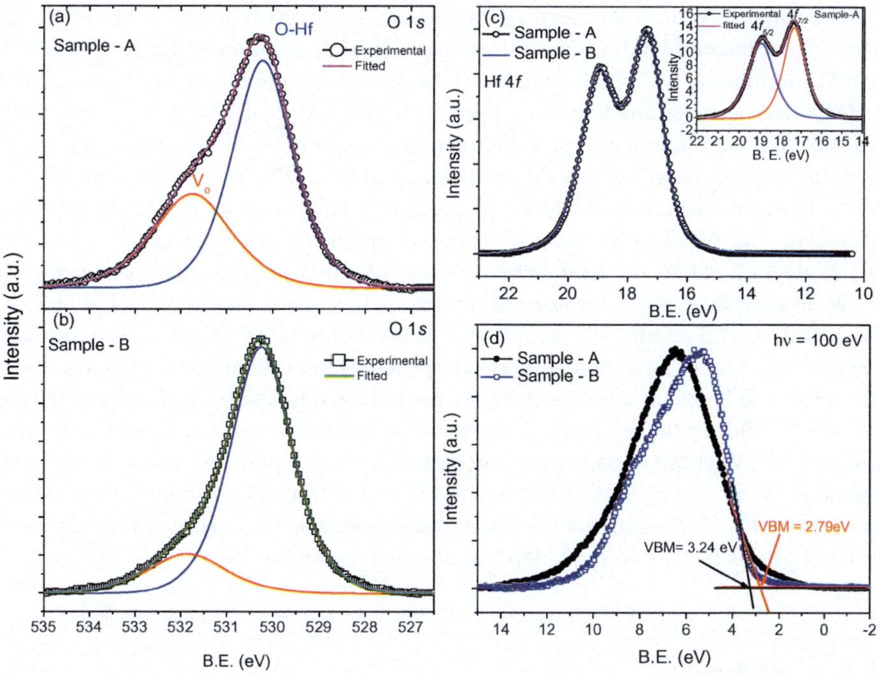

Fig. 4.2 XPS spectra of O 1*s*, Hf 4*f* core level recorded with Al k-alpha X-ray source, and valence band spectra recoded with 100 eV photon energy form Indus-1 synchrotron source. **a** O 1*s* core level spectra showing O-Hf main and V_o vacancy peak recorded on sample-A, **b** O 1*s* core level spectrum recorded on sample-B, **c** Hf 4*f* core level spectra. Inset of **c** is showing $4f_{7/2}$ and $4f_{5/2}$ deconvoluted spectrum for sample-A, and **d** valence band spectrum and VBM of film A and B recorded by 100 eV photon energy

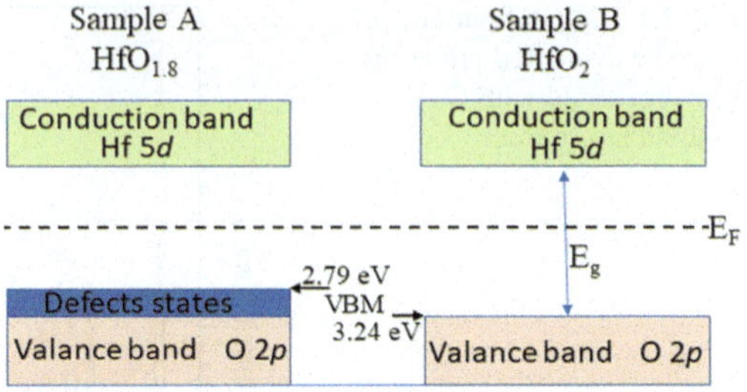

Fig. 4.3 Band schematic: showing the shift of VBM due to midgap defects states originated in oxygen deficient sample. E_F denotes the fermi level, which is taken as reference level (0 eV) in PES measurements

oxidation states. The stoichiometry of these samples was calculated considering the total area of Hf 4*f* and O 1*s* core levels. The ratio of atomic concentration of Hf to O was found to be 1:1.8 and 1:2 in film A and B respectively. This shows that the films deposited without O_2 is sub-stoichiometric, whereas the films deposited in presence of O_2 is stoichiometric.

The valence band (VB) spectra of both samples A and B are shown in Fig. 4.2d. The valance band of HfO_2 is mainly dominated by O 2*p* orbitals, while the conduction band is dominated by Hf 5*d* orbitals. VBM determines the top of the valance band. VBM was determined by linear fitting of the leading edge of VB and baseline. VBM was taken as the binding energy values at the intercept of two lines, extrapolating the baseline and the main intensity drop (leading edge). VBM of sample A and B was found to be at 2.79 eV and 3.24 eV respectively. This suggests that in the presence of oxygen, the VBM shifts toward higher binding energy as observed by others also [8]. Figure 4.3 shows the band schematic of both the samples. The lower values of VBM in sample A arises because of midgap defect states as shown in Fig. 4.3. As discussed in XPS results, the sample A is sub stoichiometric ($HfO_{1.8}$) in nature due oxygen deficiency. This oxygen deficiency creates the midgap defects states above the valance O 2*p* band, whereas there are no defects states for stoichiometric (HfO_2) sample B. Midgap defect states were proposed in other oxygen deficient oxide thin film also [9]. We have already mentioned that there are reports that the increase in O_2 partial pressure increases the band gap of HfO_2 thin films [4]. In connection, we can predict that the shifting of valence band maxima may be responsible for the increase of band gap of HfO_2 thin films deposited in oxygen ambience.

4.4 Conclusions

HfO_2/Si(100) thin films were prepared using PLD with and without O_2 partial pressure. The XRD pattern of both samples showed that the films exist in polycrystalline monoclinic phase. The O 1*s* and Hf 4*f* core levels were measured to determine the variation in stoichiometry and oxygen vacancy with O_2 partial pressure. The sample deposited with oxygen ambience is non stoichiometric in nature. It has been observed that the VBM shifted towards higher binding energy from 2.79 to 3.24 eV whereas O

1*s* core level spectra showed that the oxygen vacancy reduces in O_2 ambience. This increase in VBM with higher O_2 pressure may be the reason for increase in band gap of HfO_2 film.

Acknowledgements We thank Dr. Parasmani Rajput, BDAS, BARC, Mumbai, and Dr. Archna Sagdeo, SUS, RRCAT, Indore for providing high temperature furnace and powder XRD facilities, utilized during the PLD target preparation.

References

1. J.H. Choi, Y. Mao, J.P. Chang, Development of hafnium based high-k materials—a review. Mater. Sci. Eng. R **72**, 97–136 (2011)
2. Z.J. Yan, R. Xu, Y.Y. Wang, S. Chen, Y.L. Fan, Z.M. Jiang, Thin HfO_2 films grown on Si(100) by atomic oxygen assisted molecular beam epitaxy. Appl. Phys. Lett. **85**, 85 (2004)
3. T. Yang, J. Cao, X. Wang, Structural, electronic, magnetic, mechanic and thermodynamic properties of the inverse Heusler alloy Ti_2NiIn under pressure. Crystals **8**, 429 (2018)
4. M. Haque, P.R. Sagdeo, S. Balaji, K. Sridhar, S. Kumar, D. Bhattacharyya, D. Bhattacharyya, N.K. Sahoo, Effect of substrate bias and oxygen partial pressure on properties of RF magnetron sputtered HfO_2 thin films. J. Vac. Sci. Technol. B **32**(3), 03D104 (2014)
5. S. Papernov, M.D. Brunsman, J.B. Oliver, B.N. Hoffman, A.A. Kozlov, S.G. Demos, A. Shvydky, F.H.M. Cavalcante, L. Yang, C.S. Menoni, B. Roshanzadeh, S.T.P. Boyd, L.A. Emmert, W. Rudolph, Optical properties of oxygen vacancies in HfO_2 thin films studied by absorption and luminescence spectroscopy. Opt. Express **26**, 17609 (2018)
6. M. Nand, Y. Kumar, A. Dwivedi, S. Tripathi, R.J. Choudhary, S.N. Jha, Development of UHV pulsed laser deposition set-up for in-situ photoelectron spectroscopic study at ARPES beamline, Indus-1 synchrotron radiation source, India. Nucl. Instr. Methods Phys. Res. Sect. A **986**, 164805 (2021)
7. S. Kumar, C. Rath, Oxygen vacancy mediated stabilization of cubic phase at room temperature and resistive switching effect in Sm- and Dy-doped HfO_2 thin film. Phys. Status Solidi (a) **217**(1), 1900756 (2020)
8. A.K. Rumaiz, B. Ali, A. Ceylan, M. Boggs, T. Beebe, S. Ismat Shah, Experimental studies on vacancy induced ferromagnetism in undoped TiO_2. Solid State Commun. **144**, 334–338 (2007)
9. J. Cui, Y. Zhang, J. Wang, Z. Zhao, H. Huang, W. Zou, M. Yang, R. Peng, W. Yan, Q. Huang, Z. Fu, Y. Lu, Oxygen deficiency induced strong electron localization in lanthanum doped transparent perovskite oxide $BaSnO_3$. Phys. Rev. B **100**, 165312 (2019)

Chapter 5
Dependence of Analog Performance and Linearity with Channel Doping Concentration for an InGaAs MOSFET

Emona Datta, Avik Chattopadhyay, and Abhijit Mallik

5.1 Introduction

In recent years, the Si-based metal–oxide–semiconductor (MOS) systems have been scaled down as stated by Moore's law. Due to this rigorous scaling to nanoscale regime in device dimensions, Si-CMOS faces various difficulties as it reaches its fundamental physical limits. Although some of them have already been addressed, e.g., by the use of high-*k* gate dielectrics, employing 3-D FinFET architecture, etc., quite a few have remained unresolved until now. One of them is to find ways to increase the drive current for improved performance. In the recent past, Ge for p-channel and III-V compound semiconductors for n-channel transistors have been aggressively investigated as alternative channel materials to Si. It is revealed that electron mobility and saturation velocity of InGaAs are ten times and two times greater than that of Si, respectively [1–4]. Due to better transport properties of electrons, especially for their high mobility, InGaAs MOSFET gives better performance in terms of transconductance, cut-off frequency, gate delay, etc. [5–9].

5.2 Aim(s) of the Study

The aim of this study is to observe the effect of different channel doping concentration values on the analog performance and the linearity of an inversion-mode enhancement InGaAs MOSFET.

E. Datta (✉) · A. Mallik
Electronic Science Department, University of Calcutta, Kolkata, India
e-mail: emonadatta123@gmail.com

A. Chattopadhyay
Radio Physics and Electronics Department, University of Calcutta, Kolkata, India

R. Singh et al. (eds.), *The Physics of Semiconductor Devices*, Springer Proceedings in Physics 306, https://doi.org/10.1007/978-981-97-1571-8_5

5.3 Methods

A 2D schematic of the inversion-mode enhancement type n-channel InGaAs MOSFET, used in this study, is shown in Fig. 5.1. The device specifications include 2-μm gate length, 300-nm channel layer, and 500-nm buffer layer. The buffer layer and the channel region doping concentrations are 4×10^{17} cm^{-3} and 1×10^{17} cm^{-3}, respectively. Al_2O_3 is used as gate dielectric. To analyze the performance, SILVACO Atlas device simulator, version 5.18.3.R [10] is used. The Fermi–Dirac statistics is used in our simulations. Because of the alloy channel material, its dielectric constant and bandgap are calculated using the values of its components, such as GaAs and InAs [11–13]. Dependences of mobility on concentration, temperature, parallel field, and transverse field are accounted for by the use of Lombardi CVT model. The Shockley–Read–Hall and the Auger models are also activated to calculate carrier lifetime related to radiative and non-radiative recombination processes, respectively. Our simulation models are benchmarked against the experimental data in [9]. A good agreement of the simulated data with the experimental characteristic is evident from Fig. 5.2.

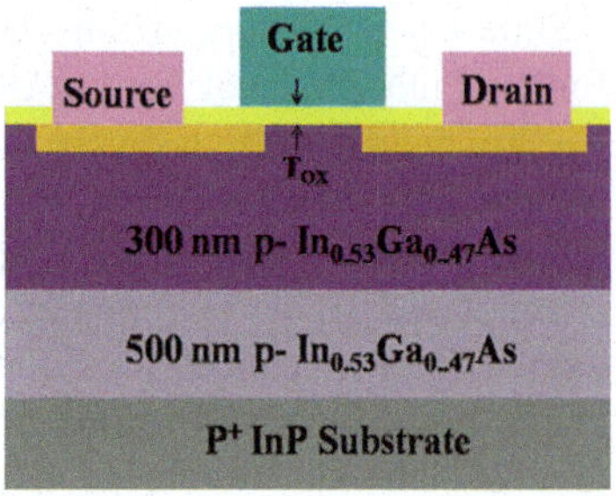

Fig. 5.1 2D schematic of the device structure

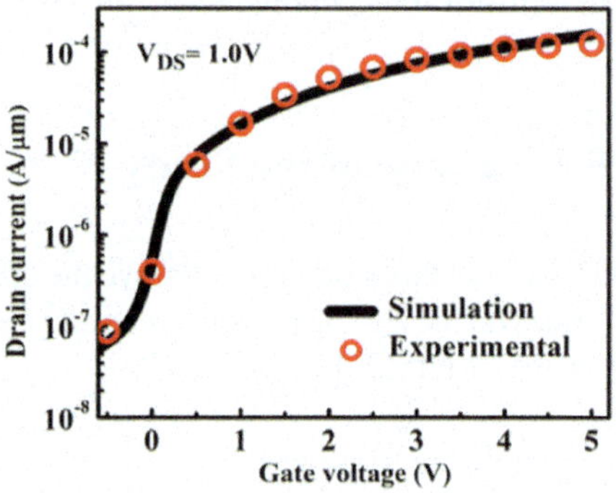

Fig. 5.2 Calibration of the simulated I_{DS}–V_{GS} with the experimental data [9]

5.4 Results and Discussions

5.4.1 Analog Performance Analysis

The transfer characteristics and the g_m-V_{GS} curves for different channel doping concentration values are shown in Fig. 5.3a and b, respectively. It can be observed that lower channel doping concentration gives high I_{ON}, but it simultaneously produces high value of I_{OFF}. The nature of g_m depends upon the I_{DS}–V_{GS} characteristics following the relation,

$$g_m = \frac{\partial I_{DS}}{\partial V_{GS}} \tag{5.1}$$

It can be seen that lower doping concentration gives larger values of drain current and g_m, which is due to higher effective mobility of the carriers [9].

To get an insight, the values of electron mobility in the horizontal direction of the channel region at a distance of 1 nm from the top oxide–semiconductor interface is plotted in Fig. 5.4 for different channel concentrations at $V_{GS} = 0.1$ V and $V_{DS} = 1.0$ V.

It can also be observed from Fig. 5.5a that output resistance R_O is higher for higher channel doping concentrations though the difference is small. Due to such results obtained from both g_m and R_O, the intrinsic gain, as shown in Fig. 5.5b, is found to decrease with increase in channel doping.

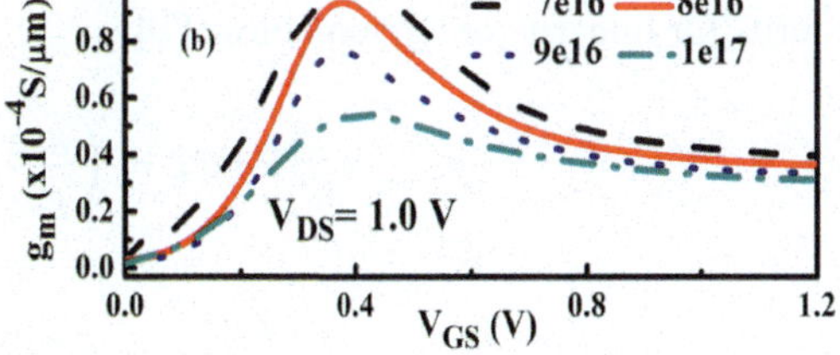

Fig. 5.3 **a** I_{DS}-V_{GS} curve and **b** g_m-V_{GS} curve for different InGaAs-channel doping concentrations at $V_{DS} = 1$ V

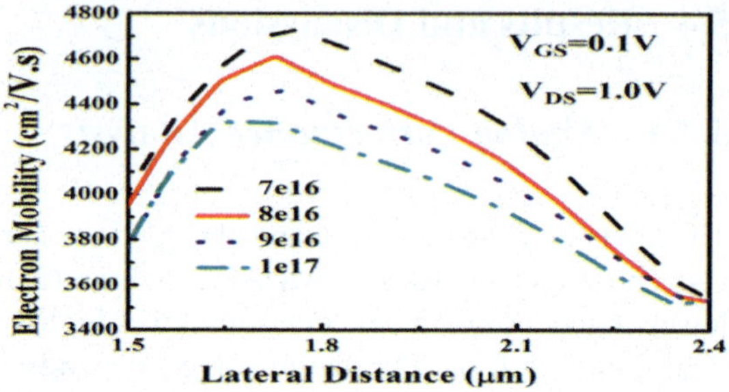

Fig. 5.4 Variations of electron mobility in the channel region due to different InGaAs-channel doping concentrations

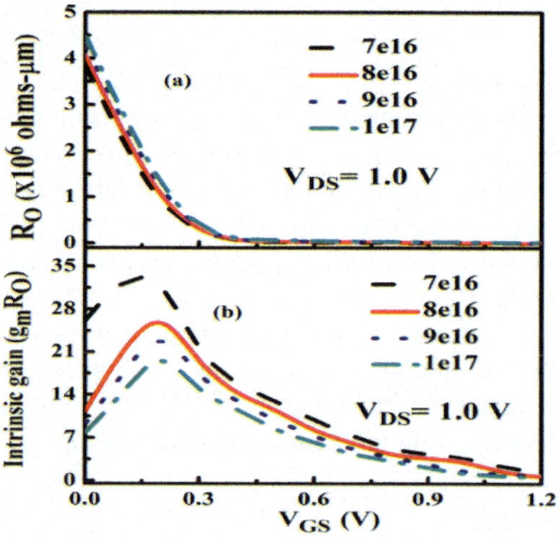

Fig. 5.5 Variations of **a** R_O and **b** intrinsic gain for different InGaAs-channel doping concentrations at $V_{DS} = 1$ V

5.4.2 *Linearity Analysis*

To achieve minimum higher-order harmonics at the output of a circuit, one of the most reasonable criteria for a device is to have better linearity. Various figure of merits for linearity are given below [14],

$$VIP2 = \left(\frac{g_{m1}}{g_{m2}}\right) \times 4 \tag{5.2}$$

$$VIP3 = \sqrt{24 \times \left(\frac{g_{m1}}{g_{m3}}\right)} \tag{5.3}$$

$$IIP3 = \frac{2}{3} \times \left(\frac{g_{m1}}{R_s \times g_{m3}}\right) \tag{5.4}$$

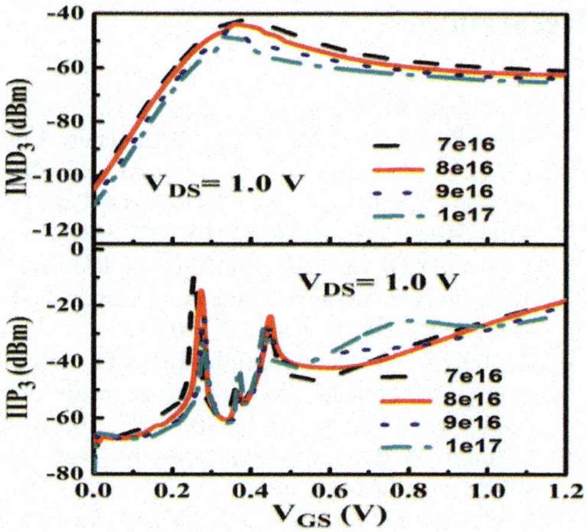

Fig. 5.6 Variations of **a** IIP3 and **b** IMD3 for different InGaAs-channel doping concentration values

$$IMD3 = \left[\frac{9}{2} \times (g_{m3}) \times (VIP3)^3\right]^2 \times R_s \tag{5.5}$$

where, g_{m1}, g_{m2}, etc. can be expressed using the equation,

$$g_{mn} = \frac{1}{n!}\frac{\partial^n I_{DS}}{\partial V_{GS}^n} \tag{5.6}$$

Here, $R_s = 50\ \Omega$ as usually used for RF applications [15]. VIP2 denotes the extrapolated input voltage where the first- and second-order harmonic voltages are equal; VIP3 signifies the extrapolated input voltage at which the third- and first-order harmonic voltages are the same; IIP3 indicates the extrapolated input power at which third- and first-order harmonic powers are equal; IMD3 represents the third-order intermodulation distortion at which the power of the third- and first-order intermodulation components are equal. It can be seen from Fig. 5.6 that the values of IIP3 are higher than that of IMD3 for all doping concentrations. This indicates better linearity and less distortion for all of them. From Fig. 5.6a it can also be observed that for the lowest doping concentration the peak of IIP3 is obtained at the lowest voltage and for which it gives slightly better linearity performance.

5.5 Conclusion

Analog performance and linearity for an experimentally demonstrated inversion-type enhancement-mode InGaAs MOSFET over a certain range of channel doping concentration values have been studied. Our observation, however, reveals that a lower channel doping yields better analog performance. It has also been observed that the lowest doping concentration leads to slightly better value for linearity.

References

1. W. Wang, J.C.M. Hwang, Y. Xuan, P.D. Ye, Analysis of electron mobility in inversion-mode Al_2O_3/In_xGa_{1-x} As MOSFETs. IEEE Trans. Electron Devices **58**(7), 1972–1978 (2011)
2. T.D. Lin et al., Self-aligned inversion-channel $In_{0.75}Ga_{0.25}As$ metal–oxide–semiconductor field-effect-transistors using UHV$Al_2O_3/Ga_2O_3(Gd_2O_3)$ and ALD-Al_2O_3 as gate dielectrics. Solid-State Electron. **54**(9), 919–924 (2010)
3. Y. Xuan, Y.Q. Wu, H.C. Lin, T. Shen, P.D. Ye, Submicrometer inversion-type enhancement-mode InGaAs MOSFET with atomic-layer deposited Al2O3 as gate dielectric. IEEE Electron Device Lett. **28**(11), 935–938 (2007)
4. A. Lubow, S. Ismail-Beigi, T.P. Ma, Comparison of drive currents in metal-oxide-semiconductor field-effect transistors made of Si, Ge, GaAs, InGaAs, and InAs channels. Appl. Phys. Lett. **96**(12), 122105-1–122105-3 (2010)
5. S. Tewari, A. Biswas, A. Mallik, Study of InGaAs-channel MOSFETs for analog/mixed-signal system-on-chip applications. IEEE Electron Device Lett. **33**(3), 372–374 (2012)
6. L. Morassi, G. Verzellesi, H. Zhao, J.C. Lee, D. Veksler, G. Bersuker, Error limiting split-*CV* mobility extraction accuracy in buried-channel InGaAs MOSFETs. IEEE Electron Device Lett. **35**(4), 440–442 (2014)
7. J. Lin, X. Zhao, D.A. Antoniadis, J.A.D. Alamo, A novel digital etch technique for deeply scaled II–V MOSFETs. IEEE Trans. Electron Devices **59**(4), 1068–1075 (2012)
8. M. Egard et al., High-frequency performance of self-aligned gate-last surface channel $In_{0.53}Ga_{0.47}As$ MOSFET. IEEE Electron Device Lett. **33**(3), 369–371 (2012)
9. Y. Xuan, Y.Q. Wu, T. Shen, T. Yang, P.D. Ye, High performance submicron inversion-type enhancement-mode InGaAs MOSFETs with ALD Al_2O_3, HfO_2 and HfAlO as gate dielectrics. IEDM Tech. Dig. 637–640 (2007)
10. *ATLAS User's Manual, A Device Simulation Software Package* (SILVACO Int., Santa Clara, CA, USA, 2015)
11. I. Strzalkowski, S. Joshi, C.R. Crowell, Dielectric constant and its temperature dependence for GaAs, CdTe, and ZnSe. Appl. Phys. Lett. **28**(6), 350–352 (1976)
12. O.G. Lorimor, W.G. Spitzer, Infrared refractive index and absorption of InAs and CdTe. J. Appl. Phys. **36**(6), 1841–1844 (1965)
13. W.Q. Chen, S.K. Hark, Strain-induced effects in (111)-oriented InAsP/InP, InGaAs/InP, and InGaAs/InAlAs quantum wells on InP substrates. J. Appl. Phys. **77**(11), 5747–5750 (1995)
14. B. Razavi, *RF Microelectronics* (Prentice-Hall, New York, NY, USA, 1998)
15. P.H. Woerlee et al., RF-CMOS performance trends. IEEE Trans. Electron Devices **48**(8), 1776–1782 (2001)

Chapter 6
Effect of Channel Epilayer Thickness on Low Power Analog Operation of Asymmetric U-Shaped pTFET

Suman Das, Avik Chattopadhyay, and Suchismita Tewari

6.1 Introduction

The leading trend of device miniaturization, in search of system performance benefits, in terms of multi-functionality, package density, easy portability, etc. demands the individual device to perform under extreme low power (LP). During supply voltage scaling, together with device dimensions, conventional Metal Oxide Semiconductor Field-Effect Transistor (MOSFET) faces challenges due to its limitation in achieving sub-60-mV/dec subthreshold swing (SS) under room temperature and hence, suffers from an enhanced OFF-state leakage current (I_{OFF}), the peril for low power operation. However, the band-to-band tunneling (BTBT) mechanism enables Tunnel FET (TFET) to meet those challenges [1–5] and, thus, to extend Moore's law further. But, TFET has got its drawbacks of producing low ON-current (I_{ON}) and showing an undesirable ambipolar nature in its transfer characteristic. Several studies have been carried out to improve the I_{ON} of TFET. In the field of architectural engineering, L-shaped gate [6], elevated drain/source (ED/S) [7, 8] are worth mentioning. The ED structure reduces SS and I_{OFF}, while the ES structure enhances tunneling current by increasing tunnel area. For this work, an amalgamation of ED and ES has been adopted in order to get the benefits of both from a single device. However, the novelty of this work lies in the introduction of an asymmetricity in terms of thickness of gate oxide as well as vertical orientation of channel region in that adopted elevated S-D architecture and, thus, projecting an asymmetric U-shaped pTEFT (AU-pTFET). The former helps reduce the undesirable ambipolar nature of current, while the latter enhances the ON-current by effectively increasing the channel

S. Das (✉) · A. Chattopadhyay · S. Tewari
Department of Radio Physics and Electronics, University of Calcutta, Kolkata 700009, India
e-mail: das.suman194@gmail.com

S. Das
Department of Electronics and Communication Engineering, Sikkim Manipal Institute of Technology, Sikkim Manipal University, Sikkim 737136, India

R. Singh et al. (eds.), *The Physics of Semiconductor Devices*, Springer Proceedings in Physics 306, https://doi.org/10.1007/978-981-97-1571-8_6

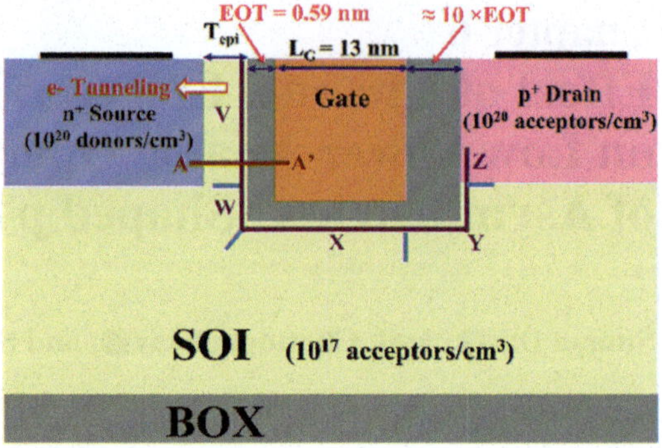

Fig. 6.1 A 2D schematic of AU-pTFET

length vertically (hence, the tunnel area) without affecting the device footprint much, ensuring a high package density. Predominantly, the unique feature of recessed or trench or grooved or L-/U-/V-shaped gate MOS device structure lies in exploiting the advantage of increased effective channel length (L_{eff}) without increasing gate length, leading to an unaffected device footprint plus improved channel electrostatics (as geometric screening length (Λ) $\ll$ L_{eff}) [4, 5, 9, 10]. Here, for the first time, the effect of epilayer—the vertical portion of the asymmetric channel getting overlapped with the source region—thickness (T_{epi}) has been thoroughly explored for the proposed AU-pTFET of gate length 13 nm in order to improve its analog performance under LP operation.

6.2 Device Structure and Simulation Set-Up

A 2D schematic of the TFET structure in question is shown in Fig. 6.1. A 70-nm-thick body, a 13-nm-long gate with an over-60-nm-long effective channel, a 0.59-nm-EOT gate oxide, along with an n^+-source (10^{20} donors/cm^3), a p^+-drain (10^{20} acceptors/cm^3), and a p-body (10^{17} acceptors/cm^3) regions, are the basic constituents of the device. For the purpose of simulating the AU-pTFET on Silvaco TCAD platform [11], the models invoked are: non-local BTBT, CVT Lombardi, position dependent Bohm Quantum Potential (BQP), band gap narrowing (BGN), F-D statistics, Shockley–Read–Hall (SRH), and Auger [11, 12].

6.3 Results and Discussion

Figure 6.2a and b shows comparisons between the experimental [7, 8] and the simulated characteristics. It has been observed that our simulated results tallied well with the experimental results [7, 8]. This claims the validity of our simulation scheme.

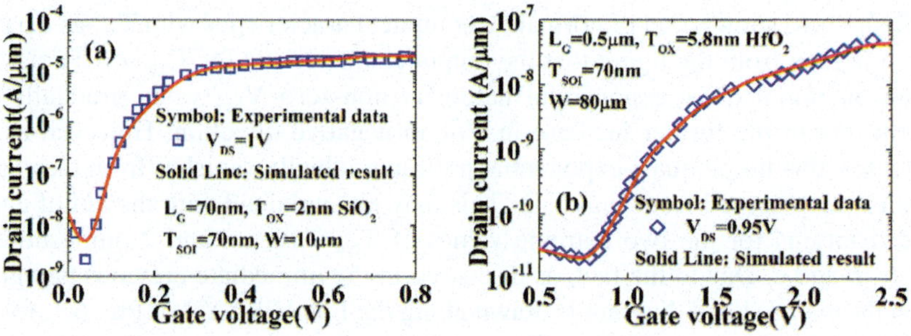

Fig. 6.2 Calibration of simulated results with experimental data [7, 8]

Figure 6.3a shows a set of transfer characteristics of the projected AU-pTFET device at $V_{DS} = -0.8$ V with T_{epi} as a parameter. It is found that with the decrease in T_{epi} value, the *knee point* ($V_{GS(knee)}$)—at which a sudden jump of current due to BTBT is observable—of the corresponding characteristic curve shifts towards left, indicating the commencement of BTBT at a higher $|V_{GS}|$. This may be explained with the aid of a set of suitably chosen band diagrams shown in Fig. 6.3b. For a fixed value of V_{GS}, the band bending in the vertical channel epilayer region (or, simply 'epi') is more in case of $T_{epi} = 10$ nm than that in case of $T_{epi} = 2$ nm. This results in overlapping of some portions of the source conduction-band (CB) and the epi valance-band (VB) at a lesser magnitude of V_{GS} for the former case, causing relatively early occurrence of BTBT, compared to the latter case. Moreover, it is to be noted that this early occurrence of tunneling in case of higher T_{epi} does not give a sharp rise in drive current with the increase of $|V_{GS}|$. In fact, it is the other way round, i.e., lower value of T_{epi} shows relatively sharper super-knee swing (SKS) and better drive current, as seen from Fig. 6.3a. This is due to the fact that low T_{epi}, in fact, decreases the total tunnel width, which gives rise to a rapid carrier tunneling beyond the onset of BTBT (i.e., the *knee point*) and hence, the sharp characteristic and the high drive current can be witnessed.

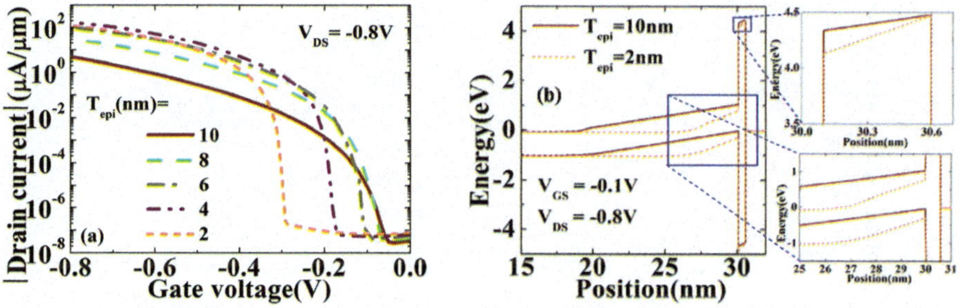

Fig. 6.3 **a** Transfer characteristics (for five different values of T_{epi}) and **b** energy band diagrams (for $T_{epi} = 10$ nm and 2 nm) at $V_{DS} = -0.8$ V

Figure 6.4a shows a set of normalized output characteristics with T_{epi} as a parameter. It is seen from the figure that the output characteristic for T_{epi} = 10 nm which shows an initial quasi-exponential nature at non-zero V_{DS} value gradually gets saturated with the further increase in V_{DS} in negative direction. However, as T_{epi} decreases, this initial quasi-exponential portion gradually vanishes from the characteristics, as observed from Fig. 6.4a. This may be explained with the aid of energy band diagrams for the two extreme values of T_{epi}—10 nm and 2 nm—shown in Fig. 6.4b and c. Due to the fact, as discussed in the immediate previous paragraph, when the energy band diagram is drawn along the line VWXYZ of Fig. 6.1, for T_{epi} = 10 nm, the bands are almost aligned between the channel epilayer (V) and the rest of the channel (WXYZ) portion, without having any prominent energy step in between. But, for T_{epi} = 2 nm, a pronounced energy step exists in between the said two portions—effectively electrically isolates (through the formation of depletion layer) the V-region from the rest of the channel. Now, For T_{epi} = 10 nm, at V_{DS} = 0 V the entire channel VWXYZ is almost equipotential and is naturally filled up with a large number of holes supplied by the drain. Thus, electrons in the V-region get enough number of holes to recombine with, effectively resisting (and hence, affecting) the epi-to-source electron tunneling. This gives rise to the initial quasi-exponential portion in the corresponding output characteristic curve. As $|V_{DS}|$ increases (beyond ~ 0.2 V), these holes get dragged out from the V-region with the increased drain field, generating sufficient number of electrons in the V-region to tunnel freely from the epi to the source. But, in case of T_{epi} = 2 nm, due to the presence of the aforesaid energy step, the holes accumulated in the rest of the channel are not able to surmount that and hence cannot interfere in the said BTBT process. Consequently, no initial quasi-exponential nature exists in the characteristic concerned, as clearly seen in Fig. 6.4a.

Figure 6.5a shows variations of transconductance (g_m) as a function of gate overdrive voltage ($V_{OV} = V_{GS} - V_{GS(knee)}$) with T_{epi} as a parameter at V_{DS} = − 0.8 V.

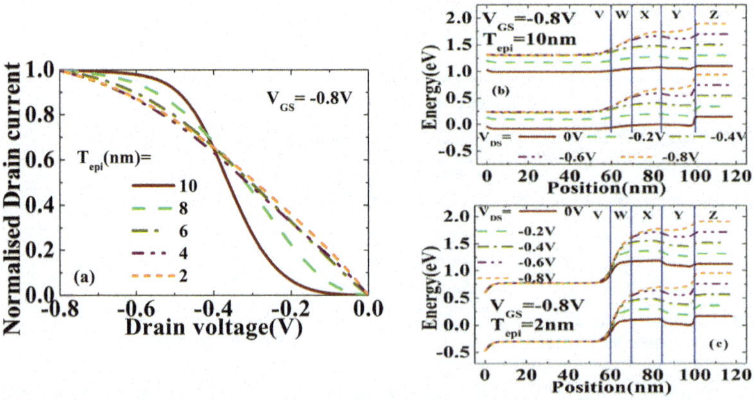

Fig. 6.4 **a** Normalized output characteristics (for five different values of T_{epi}) and **b** and **c** energy band diagrams (for T_{epi} = 10 nm and 2 nm, respectively) at V_{GS} = − 0.8 V

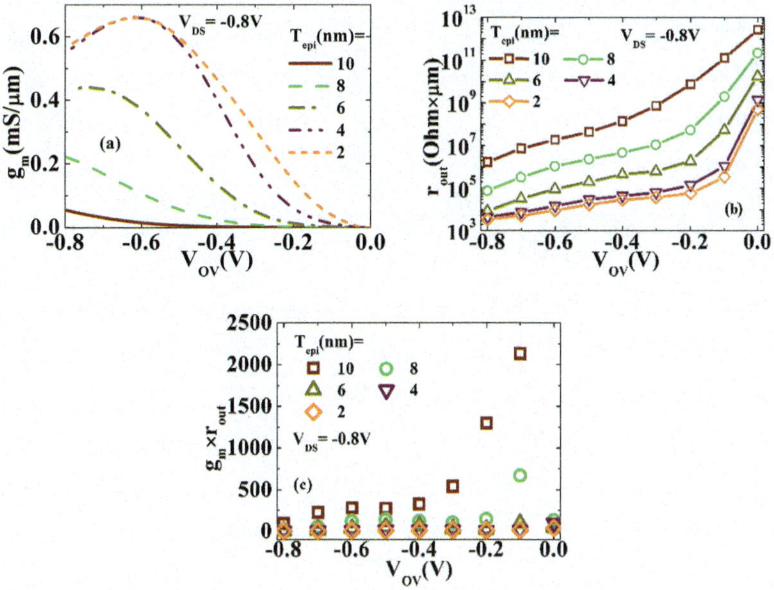

Fig. 6.5 **a** g_m-V_{OV}, **b** r_{out}-V_{OV}, and **c** gain-V_{OV} plots for five different values of T_{epi} at $V_{DS} = -0.8$ V

It is observed that lower the value of T_{epi}, higher the value of g_m almost throughout the entire range of V_{OV}. But, on the other hand, output resistance (r_{out}), when plotted against V_{OV} with T_{epi} as a parameter in Fig. 6.5b, projects an exact opposite trend and is sporting higher values for higher values of T_{epi} throughout the entire range of V_{OV}. The amalgamated effect of these two parameters determines the nature of device gain ($g_m \times r_{out}$)—plotted against V_{OV} with T_{epi} as a parameter in Fig. 6.5c. It is observed from Fig. 6.5c that with the increase in T_{epi}, device gain increases throughout the entire operating range. More percentage increment of r_{out} compared to the percentage decrement of g_m, as a function of T_{epi} for the same V_{OV} value, ultimately gives the increasing trend of device gain with the increase in T_{epi} value. Nearly similar trends could be followed at $V_{DS} = -50$ mV, as shown in Fig. 6.6a–c.

Some key performance metrics are tabulated in Table 6.1 for the high and the low V_{DS} values. It is observed that the proposed device with $T_{epi} = 10$ nm outperforms its other counterparts having lower T_{epi} values in terms of device gain, OFF-current, and leakage power. But, from the device active area point-of-view, the projected AU-pTFET with $T_{epi} = 2$ nm excels above all the rest of the structures, boosting enhanced package density.

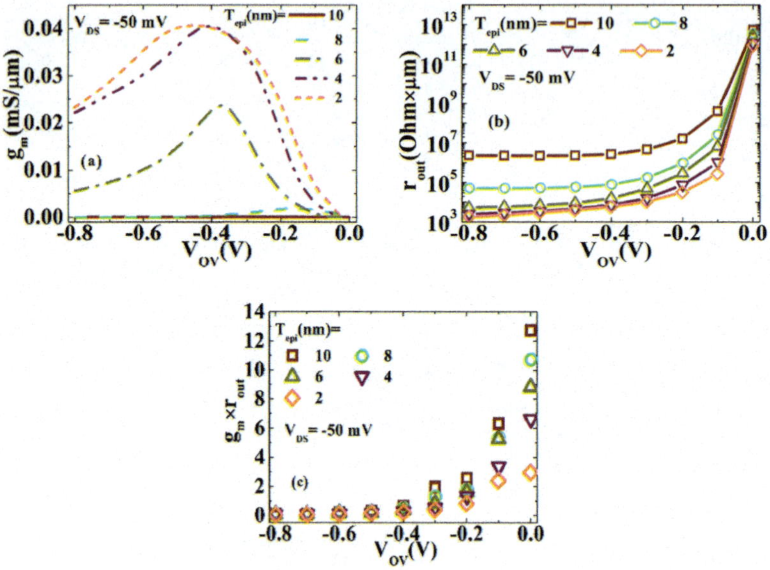

Fig. 6.6 **a** g_m-V_{OV}, **b** r_{out}-V_{OV}, and **c** gain-V_{OV} plots for five different values of T_{epi} at $V_{DS} = -$ 50 mV

Table 6.1 Comparison of a few key performance metrics at $V_{DS} = -$ 0.8 V and $-$ 50 mV

	$T_{epi} = 2$ nm		$T_{epi} = 6$ nm		$T_{epi} = 10$ nm	
	$V_{DS} = -$ 0.8 V	$V_{DS} = -$ 50 mV	$V_{DS} = -$ 0.8 V	$V_{DS} = -$ 50 mV	$V_{DS} = -$ 0.8 V	$V_{DS} = -$ 50 mV
Gain improvement over $T_{epi} = 2$ nm	–	–	7 times	3 times	156 times	4 times
Device active area footprint reduction from $T_{epi} = 10$ nm	27%	27%	13.5%	13.5%	–	–
Device I_{OFF} reduction from $T_{epi} = 2$ nm	–	–	79.6%	78.7%	84.7%	87.2%
Device $P_{leakage}$ reduction from $T_{epi} = 2$ nm	–	–	79.6%	78.7%	84.7%	87.2%

6.4 Conclusion

In this article, for the first time, the effect of channel epilayer thickness has been thoroughly explored for an asymmetric U-shaped pTFET of gate length 13 nm in terms of its analog performance while maintaining its suitability for LP operation. The choice of the ultimate architecture/structure remains application specific—depending on whether high drive current coupled with high package density is needed or high gain coupled with low leakage is needed. However, the device with a T_{epi} value falling in the midway of the two extremes (i.e., 6 nm) could be chosen as the optimized device structure.

References

1. K. Boucart, A.M. Ionescu, Double-gate tunnel FET with high-κ gate dielectric. IEEE Trans. Electron Devices **54**(7), 1725–1733 (2007). https://doi.org/10.1109/TED.2007.899389
2. Q. Zhang, W. Zhao, A. Seabaugh, Low-subthreshold-swing tunnel transistors. IEEE Electron Device Lett. **27**(4), 297–300 (2006). https://doi.org/10.1109/LED.2006.871855
3. C. Wu, Q. Huang, Y. Zhao, J. Wang, Y. Wang, R. Huang, A novel tunnel FET design with stacked source configuration for average subthreshold swing reduction. IEEE Trans. Electron Devices **63**(12), 5072–5076 (2016). https://doi.org/10.1109/TED.2016.2619694
4. S. Das, A. Chattopadhyay, S. Tewari, Minimization of drain-end leakage of a U-shaped gated tunnel FET for low standby power (LSTP) application, in *Nanoelectronics, Circuits and Communication Systems 2021*, ed. by V. Nath, J. Mandal, LNEE, vol. 692 (Springer, Singapore, 2021), pp. 393–402. https://doi.org/10.1007/978-981-15-7486-3_36
5. S. Das, A. Chattopadhyay, S. Tewari, Asymmetric-elevated-source-drain TFET: a fairly scalable and reliable device architecture for sub-400-mV low-stand-by-power digital applications. IETE Tech. Rev. https://doi.org/10.1080/02564602.2020.1852120
6. Z. Yang, Tunnel field-effect transistor with an L-shaped gate. IEEE Electron Device Lett. **37**(7), 839–842 (2016). https://doi.org/10.1109/LED.2016.2574821
7. W.Y. Choi, B. Park, J.D. Lee, T.K. Liu, Tunneling field-effect transistors (TFETs) with subthreshold swing (SS) less than 60 mV/dec. IEEE Electron Device Lett. **28**(8), 743–745 (2007). https://doi.org/10.1109/LED.2007.901273
8. S.W. Kim, J.H. Kim, T.J.K. Liu, W.Y. Choi, B.G. Park, Demonstration of L-shaped tunnel field-effect transistors. IEEE Trans. Electron Devices **63**(4), 1774–1778 (2016). https://doi.org/10.1109/TED.2015.2472496
9. K. Tian, A. Hallén, J. Qi, S. Ma, X. Fei, A. Zhang, W. Liu, An improved 4H-SiC trench-gate MOSFET with low ON-resistance and switching loss. IEEE Trans. Electron Devices **66**(5), 2307–2313 (2019). https://doi.org/10.1109/TED.2019.2905636
10. A. Sudarsanan, K. Nayak, TCAD-based investigation of statistical variability immunity in U-channel FDSOI n-MOSFET for sub-7-nm technology. IEEE Trans. Electron Devices **68**(6), 2611–2617 (2021). https://doi.org/10.1109/TED.2021.3074116
11. *ATLAS User's Manual, A Device Simulation Software Package* (SILVACO Int., Santa Clara, CA, USA, 2016).
12. D. Abdi, M.J. Kumar, Controlling ambipolar current in tunneling FETs using overlapping gate-on-drain. IEEE J. Electron Devices Soc. **2**(6), 187–190 (2014). https://doi.org/10.1109/JEDS.2014.2327626

Chapter 7
Effect of Helium Gas Addition to SF_6/O_2 Chemistry for SiC Dry Etching in AlGaN/GaN/SiC HEMTs

Chanchal, Sunil Kumar, Rajeev Sawal, Niraj Kumar, Robert Laishram, D. S. Rawal, and Manoj Saxena

7.1 Introduction

In power electronics applications, Wide bandgap (WBG) Semiconductors GaN, SiC, Ga_2O_3, AlN, etc. play a very important role as these material devices are smaller, faster, and more reliable with higher efficiency as compared to conventional materials like silicon and gallium arsenide [1, 2]. WBG semiconductor allows devices to operate at high voltage, and high frequency with high-temperature applications [3, 4]. For GaN/AlGaN HEMT MMICs, sapphire or SiC is used as the substrate. SiC is preferred as a substrate due to its minimal lattice deviation (3.5%) compared to sapphire (16%) and, also has very high thermal conductivity. GaN HEMTs are widely used in space and satellite communications, and radar applications where the performance and reliability of equipment with minimal dimensions are key factors. Over the past decade, backside via has received great attention and is in high demand for MMICs design. Via contacts that provide low source inductance and contribute to the uniform distribution of heat, which increases device gain and efficiency. Via connection also increases the packing density and provides flexibility in circuit designing [5, 6].

SiC via hole etching is a crucial backside process because SiC is the second hardest material which is very difficult to etch due to its strong binding energy between the

Chanchal (✉)
Department of Electronic Science, University of Delhi, New Delhi 110021, India
e-mail: chanchal0210saraswat@gmail.com

Chanchal · S. Kumar · R. Sawal · N. Kumar · R. Laishram · D. S. Rawal
Solid State Physics Laboratory, New Delhi 110054, India

M. Saxena
Department of Electronics, Deen Dayal Upadhyaya College, University of Delhi, New Delhi 110078, India

R. Singh et al. (eds.), *The Physics of Semiconductor Devices*, Springer Proceedings in Physics 306, https://doi.org/10.1007/978-981-97-1571-8_7

silicon and carbon (4.25 eV). SiC wet chemical etching is not controllable therefore plasma etching techniques are required for the desired pattern [7–12]. Silicon carbide has been successfully etched using fluorinated plasmas by Charbet [13] and Myeong et al. [14]. Both fluorinated and chlorinated plasmas have been investigated by Leerungnawarat et al. [15] and Khan et al. [16]. The etch rates achieved using fluorinated plasmas have been several times faster than those obtained in chlorine-based plasmas. Researchers report SiC dry etch rates as high as 2 μm/min [17]. For the fabrication of vias with good connectivity, micro masking, trenching and pillar formation are the main problems.

In the present study, we have used fluorine-based etching chemistries SF_6/O_2 and SF_6/O_2/He for via-hole fabrication using the Inductively Coupled Plasma Reactive Ion Etching (ICP-RIE) technique. Helium (He) is an inert gas that acts as a plasma diluent and stabilizer [18] and improves the anisotropic etching profile and also contributes to the heat transfer mechanism in the chamber. Also, when helium is added to other gas mixtures, it increases the total pressure in the chamber while keeping the partial pressures of the other gases constant and enhancing the dissociation of other gases [18–20]. In the present study, we have discussed the impact of SF_6/O_2/He gas chemistry on the Via profile, etch rate, and selectivity. Also, issues related to the formation of via-hole grounds in GaN HEMTs for source grounding are discussed.

7.2 Experimental Details

After front-side device fabrication, GaN/SiC wafer was temporarily bonded face down to a 4 inch carrier by using a liquid wax that can withstand temperatures up to 120 °C. Bubble-free mounting has been carried out to prevent the wafer from any kind of damage. Si and sapphire are frequently used as carrier wafers for the etching process. The grinding process is used in a controlled manner to achieve a thickness of 100 μm with ± 10% variation. This was followed by photolithographic patterning of (90 μm × 60 μm) oval holes, using a positive photoresist or thick electro-plated metal hard mask. SiC etching was done using the ICP-RIE technique in Oxford Plasma Pro 100 system with SF_6/O_2/He gas chemistries in ratio 5:1:1. SF_6, O_2, and Helium are given in the chamber through Mass Flow Control (MFCs). After completion of the etching process, the wafer was wet cleaned to strip the metal mask and clean the etching byproducts, Fig. 7.1a. After that, the seed metal layer of (Ti/Au) is deposited by RF sputtering to provide a conductive path for subsequent electroplating steps. A thick gold layer is then electrodeposited on the seed metal layer to connect the front side metal pads of the device to the backside ground plane. This gold layer provides mechanical strength to the thinned wafer and also acts as a backside ground plane. After that carrier wafer was demounted using heated solvent and then the wafer is cleaned to free from any trace of residues, the process flow chart is shown in Fig. 7.1b.

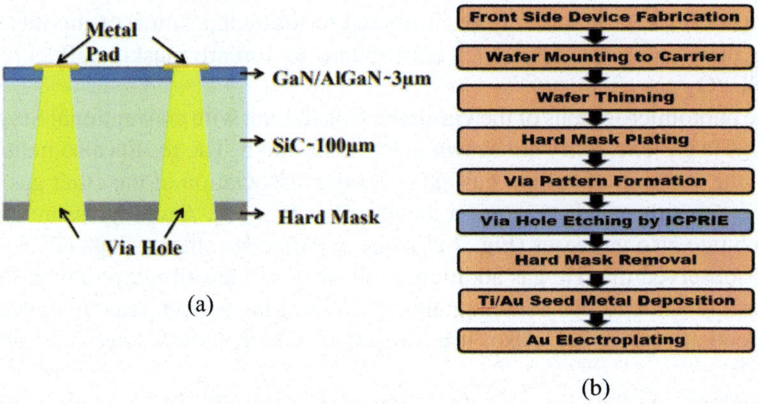

Fig. 7.1 **a** Schematic diagram and **b** Process flow chart of fabrication of GaN/SiC backside via hole process in AlGaN/GaN HEMT device

7.3 Results and Discussion

The experiments have been carried out for SiC via hole etching using different fluorine gas chemistries (SF_6/O_2 and SF_6/O_2/He) for achieving a high etch rate with smooth sidewalls.

In starting of the process, DC bias is high due to an increase in the concentration of reactive Fluorine (F) species in the plasma, and second due to a less volatile component formation of F species CF, CF_2, CF_3, CO, CO_2, SiO_2, SiF, etc. [14]. As the concentration of reactive volatile component species increases, DC bias starts decreasing which results in a decrease in etch rate as shown in Fig. 7.2a [15, 16].

Figure 7.2b shows etch rate variation with both the chemistries, SF_6/O_2 and SF_6/O_2/He. We have observed with the addition of helium gas, etch rate is high as compared to conventional SF_6/O_2. The presence of defects near the edges of via

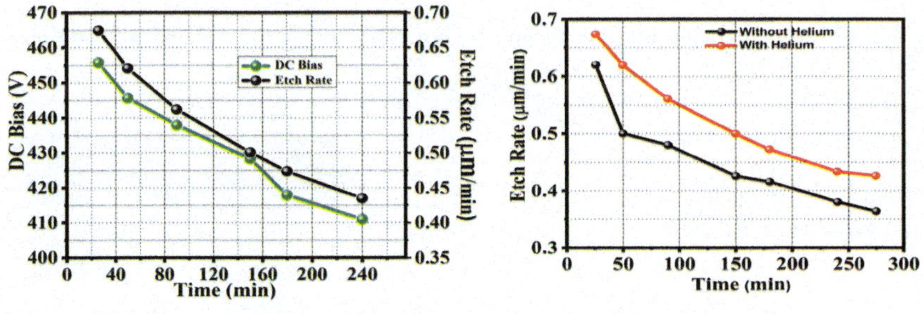

Fig. 7.2 **a** Etch rate and DC bias variation with etching time and **b** Effect of addition of Helium gas on Etch rate with SF_6/O_2 gas chemistry

is high in SF_6/O_2, which may be attributed to the redeposition of the metal mask during the etching process, which corresponds to a micro masking effect observed with SF_6/O_2 gas chemistry.

The photomicrographs of the via after 4 h of etching with conventional SF_6/O_2 and SF_6/O_2/He gas chemistry are shown in Fig. 7.3a and b. The results also indicate that due to the addition of helium gas increases the dissociation of the other gas species and increases the total pressure in the chamber [18–20]. As the pressure increases the etch rate also increases (Fig. 7.2b), also reported by other groups [15–18].

It is observed that He gas addition improves the sidewall morphology of the via etch profile and has less trenching and micro masking as compared to conventional gas chemistries Fig. 7.4. Also, it has resulted in better mask selectivity, shown in Table 7.1.

The addition of He gas has resulted in 100 μm deep via hole etching with a high rate but with an oversized footprint which is undesirable. Also observed, dimensions of the via hole pattern have been increased due to higher mask erosion with an increased flow of Helium in the gas mixture, shown in Fig. 7.5.

The Optimization of the flow rate of He in a controlled manner is critical for achieving desired via hole foot pattern without any damage during the process. The

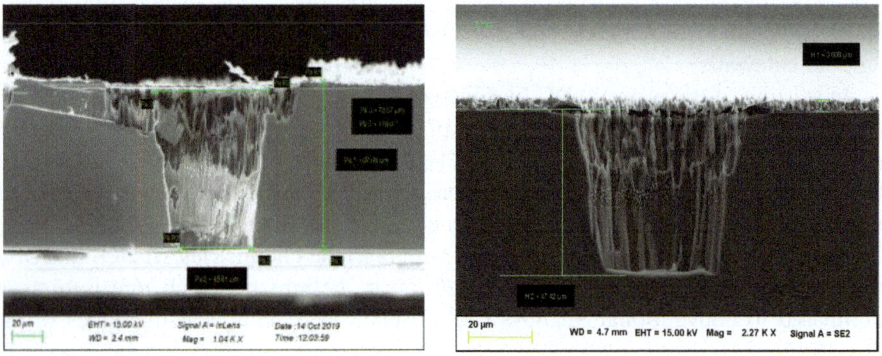

(a) (b)

Fig. 7.3 SEM micrographs after SiC etching with **a** conventional SF_6/O_2 and **b** SF_6/O_2/He gas chemistries

Fig. 7.4 SEM micrographs of via profile of 100 μm depth after 4 h SiC etching using gas chemistry SF_6/O_2/He

Table 7.1 List of parameters with gas chemistry SF_6/O_2 and SF_6/O_2/Helium

Parameter	SF_6/O_2	SF_6/O_2/He
Micromasking and Trenching	High	Low
Selectivity (to hard mask)	Good (16:1)	Very good (22:1)
Etch rate	0.35 μm/min	0.45 μm/min
Surface morphology	Rough	Smooth

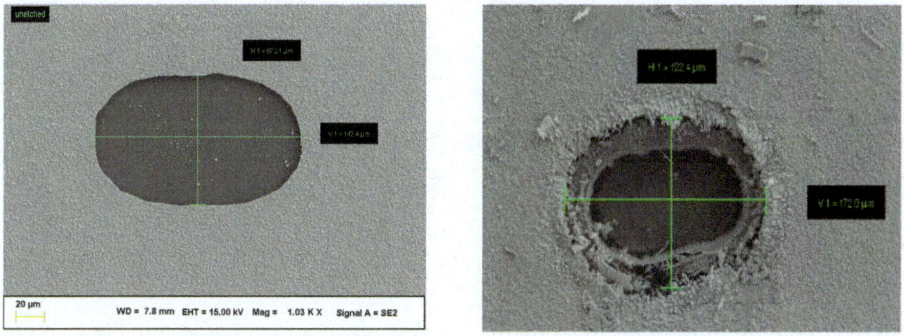

(a) (b)

Fig. 7.5 SEM images of via dimensions **a** before etching and **b** after etching, oversized footprint observed with high He flow in the gas mixture

effect of He on SiC via hole etching parameters such as micro masking and trenching, selectivity, etch rate, and % increase via dimension is shown in Table 7.1.

The measured I-V characteristics of AlGaN/GaN/SiC/HEMT devices before and after backside via hole etching using SF_6/O_2/He gas chemistry is presented in Fig. 7.6a and b. There are no significant changes in DC I-V characteristics before and after via etching which indicates good process compatibility. The actual effect of via-hole source grounding can be assessed by RF characterization.

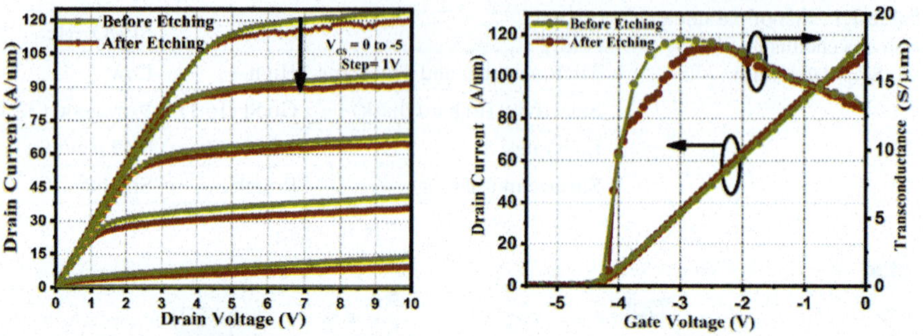

(a) (b)

Fig. 7.6 DC I-V measurement of AlGaN/GaN/SiC device, a width of 150 μm before and after the etching with SF_6/O_2/He gas, **a** drain current characteristics **b** transfer characteristics

7.4 Conclusion

SiC via hole etching in an inductively coupled plasma reactive ion etching system with the addition of Helium gas to SF_6/O_2 chemistry is investigated. We have observed a significant increase in the etch rate, smooth sidewalls with less micro masking, and trenching of Via hole using SF_6/O_2/He gas chemistry as compared to the conventional SF_6/O_2. The addition of Helium gas also improves the selectivity of the etching process. At the same time, it increases via dimensions as a result of an increase in the undercut. Electrical parameters like source ground connectivity, drain current, transconductance, and have been characterized before and after via hole etching process and have not found any significant changes in the parameters.

Acknowledgements The authors are thankful to the Director of Solid State Physics Laboratory (SSPL), DRDO for providing the opportunity to carry out this work and permission to publish this work. The authors would also like to thank GaN Team for their valuable support in the Fabrication and characterization of the devices.

References

1. U.K. Mishra, L. Shen, T.E. Kazior, Y.F. Wu, GaN-based RF power devices and amplifiers. Proc. IEEE **96**(2), 287–305 (2008). https://doi.org/10.1109/JPROC.2007.911060
2. R.S. Pengelly, M. Simon, J. Wood, W. Milligan, T. Scott, L. William, A review of GaN on SiC high electron-mobility power transistor and MMICs. IEEE Trans. Microw. Theory Tech. **60**(6), 1764–1783 (2012). https://doi.org/10.1109/TMTT.2012.2187535
3. N. Keshmiri, D. Wang, B. Agarwal, R. Hou, A. Emadi, Current status and future trends of GaN HEMTs in electrified transportation. IEEE Access **8**, 70553–70571 (2020). https://doi.org/10.1109/ACCESS.2020.2986972
4. T. Ueda, GaN power devices: current status and future challenges. Jpn. J. Appl. Phys. **58**, SC0804 (2019). https://doi.org/10.7567/1347-4065/ab12c9

5. Y.H. Hwang, T.S. Kang, F. Ren, S.J. Pearton, Novel approach to improve heat dissipation of AlGaN/GaN high electron mobility transistors with backside Cu via. ECS Trans. Electrochem. Soc. **66**(1), 223–230 (2015)
6. S.J. Pearton, E.A. Douglas, Z.J. Shul, F. Ren, Plasma etching of wide bandgap and ultra-wide bandgap semiconductors. J. Vac. Sci. Technol. A **38**, 020802 (2020)
7. K.R. Szmidt, B. Stonio, J. Zelazko, M. Filipiak, M. Sochacki, A review: inductively coupled plasma reactive ion etching of silicon carbide. Materials **15**, 123 (2022). https://doi.org/10.3390/ma15010123
8. S. Tanaka, K. Rajanna, T. Abe, M. Esashi, Deep reactive ion etching of silicon carbide. J. Vac. Sci. Technol. B **19**, 2173–2176 (2001)
9. D.S. Rawal, H. Arora, B.K. Sehgal, R. Murlidharan, J. Vac. Sci. Technol. A **32**, 031301 (2014)
10. D. Zhuang, J.H. Edgar, Wet etching of GaN, AlN, and SiC: a review. Mater. Sci. Eng. Rep. **48**(1), 1–46 (2005)
11. N.O.V. Plank, M.A. Blauw, J.M. Van Der Drift, R. Cheung, The etching of silicon carbide in inductively coupled SF_6/O_2 plasma. J. Phys. D **36**, 482 (2003)
12. S.H. Kuah, P.C. Wood, Inductively coupled plasma etching of poly-SiC in SF_6 chemistries. J. Vac. Sci. Technol. A **23**, 947–952 (2005)
13. P. Chabert, Deep etching of silicon carbide for micromachining applications: etch rates and etch mechanisms. J. Vac. Sci. Technol. B **19**, 1339 (2001)
14. S. Myeong, S. Lim, T.N. Jackson, Fast, smooth, and anisotropic etching of SiC using SF_6/Ar. J. Vac. Sci. Technol. B **17**, 2055 (1999)
15. P. Leerungnawarat, H. Cho, S.J. Pearton, C.-M. Zetterling, M. Ostling, J. Electron. Mater. **29**, 342 (2000)
16. F.A. Khan, B. Roof, L. Zhou, I. Adesida, J. Electron. Mater. **30**, 212 (2001)
17. N. Okamoto, T. Ohki, S. Masuda, M. Kanamura, Y. Inoue, K. Makiyama, K. Imanishi, H. Shigematsu, T. Kikkawa, K. Joshin, N. Hara, SiC Backside Via-Hole Process for GaN HEMT MMICs Using High Etch Rate ICP Etching, CS MANTECH Conference, Tampa, Florida, USA, May 18th–21st, 2009
18. M.J. Lercel, D. Dang, N. Marmillion, W. Mlynko, Role of gas feed delivery and dilutent on oxide etching in an inductively coupled plasma etch system. J. Vac. Sci. Technol. A **16**, 1459 (1998). https://doi.org/10.1116/1.581169
19. Stockman P: Gasworld Helium in Electronics, Special Feature (2016)
20. L.F. Voss, K. Ip, S.J. Pearton, R.J. Shul, M.E. Overberg, A.G. Baca, C. Sanchez, J. Stevens, M. Martinez, M.G. Armendariz, G.A. Wouters, SiC via fabrication for wide-band gap high electron mobility transistor/microwave monolithic integrated circuit devices. J. Vac. Sci. Technol. B **26**(2), 487–494 (2008)

Chapter 8
Modeling of High and Low Resistant States in Single Defect Atomristors

Yuvraj Misra and Tarun Kumar Agarwal

8.1 Introduction

The successful fabrication of a memristor in 2008 at Hewlett Packard Research Labs has opened a window of opportunities toward more energy-efficient computing systems. The term "memristor" is derived from "memory-resistor" and rightly so, as it can work as a non-volatile electronic memory device with multiple memory states [1]. This new class of memory storage unit is a potential candidate for a universal memory that can address the requirements of a future memory device i.e, non-volatile capabilities with fast-switching, long retention and switching cycles [2]. Metal-insulator-metal (MIM) structures are at the heart of these developing memristive devices, and modeling the electrical characteristics of these nanoscale devices using quantum transport can provide insights into their switching mechanism [3].

The MIM structure consists of electrodes, similar to capacitors, that sandwich an insulating layer, and are usually made of metals like Au, Pt, Al and Ag. The insulating layer in the structure governs many crucial aspects of the device, viz. switching mechanism, conduction mechanism, working voltage polarity, working speed, and resistance ratio [4]. Recently, 2D sheets of transitional metal dichalcogenides (TMD) such as MoS_2, $MoSe_2$, WS_2, WSe_2 etc. are being explored to replace transition metal oxides like SiO_2, HfO_2, NiO etc., in the switching layer of the future memristive devices [5]. It is due to the promise of atomically-thin 2D materials or "atomristors" that can mitigate the variability associated with these devices by providing more controlled filament shapes than the bulk oxides [6]. Interestingly, dense memory fabrics, formed by replacing the electrodes in atomristors with graphene, are also crucial to enhancing brain-inspired or neuromorphic computing [2].

Y. Misra (✉)
Indian Institute of Technology Mandi, Mandi, Himachal Pradesh, India
e-mail: B19225@students.iitmandi.ac.in

T. K. Agarwal
Indian Institute of Technology Gandhinagar, Gandhinagar, Gujarat, India

R. Singh et al. (eds.), *The Physics of Semiconductor Devices*, Springer Proceedings in Physics 306, https://doi.org/10.1007/978-981-97-1571-8_8

In atomristors, the switching mechanism has been attributed to the presence of defects in the 2D sheets and metal diffusion through the switching layer, similar to conductive-bridge resistance-change random access memory (CBRAM) cells [3]. It thus becomes imperative to gain an insight into the origin of high resistance states (HRS) and low resistance states (LRS) in single defect memristors as such an analysis can aid in designing optimum memristive devices through precision defect engineering techniques [5]. In this work, we present a model that can capture the transport in the two resistance states with low computational complexity. Firstly, we present the methodology to calculate the electrical characteristics of MIM structures using an in-house 1D quantum transport framework, which can solve the 1D Poisson and Schrodinger equation self-consistently within the Non-equilibrium Green's Function (NEGF) framework [7, 8]. Next, the model is validated for the experimentally demonstrated metal—2D material—metal structures and subsequently used to shed light on the non-volatile resistance switching (NVRS) mechanism [9]. Finally, the model is applied to illustrate the origin of NVRS in atomristors arising due to the presence of a single defect [5]. In doing so, we confirm the efficacy of the model by validating obtained results with the experimental data [5].

8.2 Methodology

We use the effective-mass tight-binding Hamiltonian approach to model the MIM device, where the eigenstate representation and the real-space representation were used for the transverse and longitudinal dimensions respectively [7]. Hence, the Hamiltonian matrix was separated into a transverse and a longitudinal part. In working with the NEGF, the matrix of concern was the longitudinal matrix. The first step was, therefore, to write this matrix for our system. This was done using the finite-difference method. The resulting matrix, consisting of the potential energy and hopping term, was then written such that it suitably represented the properties of the device [7]. This involves accounting for the change in the effective masses of the electrons and hence, the hopping factor in the matrix, according to whether the lattice point lies in the metal or the insulator region. For the points that lay on the metal-insulator boundary, a new hopping term t_{par} was described to model the metal-insulator interaction (Fig. 8.1). The potential energy for each point was substituted from an estimated potential profile.

The longitudinal Hamiltonian H can be expressed as:

1. If i=j

$$H_{ij} = \begin{cases} 2t_m & 1 \le i \le x-2 \\ 2t_i & x+2 \le i \le x+y-2 \\ 2t_m & x+y+2 \le i \le 2x+y \\ 2t_{par} & \text{otherwise} \end{cases}$$

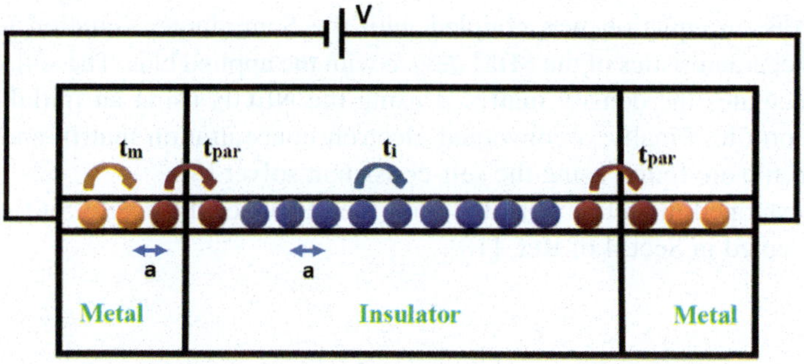

Fig. 8.1 Schematic of a MIM structure with a finite grid. The arrows indicate the hopping terms in the metal (t_m), junction (t_{par}) and insulator (t_i) regions. The orange, red and blue points represent the metal, coupled, and insulator grid points respectively. The grid point spacing a equals 0.05 nm

2. If $i = j \pm 1$

$$H_{ij} = \begin{cases} -t_m & 1 \le \frac{i+j-1}{2} \le x-1 \\ -t_i & x+1 \le \frac{i+j-1}{2} \le x+y-1 \\ -t_m & x+y+1 \le \frac{i+j-1}{2} \le 2x+y-1 \\ -t_{par} & \text{otherwise} \end{cases}$$

3. Else $H_{ij} = 0$

Here, x and y represent the number of metal and insulator grid points; i and j represent the row and column indices of the Hamiltonian matrix; and t_m, t_i and t_{par} represent the metal, insulator and metal-insulator junction hopping terms respectively.

The MIM device was taken to be connected to two contacts. Since the system being studied is an open system, the external effects on the device due to the contacts were considered by adding self-energy matrices, $\Sigma_1(E)$ and $\Sigma_2(E)$, to the Hamiltonian for both the contacts. This enables us to solely focus on the device's working regime [7]. Next, the local density of states (LDOS) for each lattice point was calculated by computing the spectral function $A(E)$, which was then used to derive the transmission $T(E)$. The quantities were derived using the Green's function method as shown below [7].

$$G(E) = [EI - H - \Sigma_1 - \Sigma_2]^{-1} \tag{8.1}$$

$$\Gamma_{1,2} = i[\Sigma_{1,2} - \Sigma_{1,2}^T] \tag{8.2}$$

$$A(E) = A^1 + A_2 = G\Gamma_1 G^T + G\Gamma_2 G^T \tag{8.3}$$

$$T(E) = trace(\Gamma_1 A_2) = trace(\Gamma_2 A_1) \tag{8.4}$$

The Poisson equation was coupled with the Schrodinger equation to find the electrical characteristics of the MIM device with the applied bias. The self-consistent solver calculates the density matrix ρ using the NEGF, using an initial estimated potential profile. Finally, a converged electron concentration matrix and potential energy profile are found using the self-consistent solver [8].

The developed model is validated with the results of the single-SiO_2-layer gate stack presented in Sect. 4 of Ref. [10].

8.3 Application and Analysis

8.3.1 Non-volatile Resistance Switching via Quantum Tunneling Effects

This section focuses on the study of the stable NVRS phenomenon in 2-D transitional metal dichalcogenide (TMD) atomic sheets that arises due to the presence of multiple vacancy defects [9]. The metal atoms from the electrodes get substituted in these defects and switch the device's resistance from the HRS to the LRS. This switch in resistance can be explained by the increase in tunneling probability in the LRS that occurs due to potential energy distortion in the energy band diagram of the structure. We model this distortion around the defect region by averaging the potential energy profile along the transverse direction. By resorting to an averaged potential profile in 1D, the transport at the metal-2D material surface can be reasonably captured using the proposed model.

In this section, we consider a simplified MoS_2 based memristor (Fig. 8.2) containing multiple defects to illustrate how the simulator can be used to describe its switch from the HRS to LRS using a given potential energy profile. However, the model can be generalized for any material that follows the same HRS and LRS relationships with respect to the applied biases and temperatures. For instance, h-BN, a semi-insulator, has roughly the same HRS and LRS temperature dependant I–V characteristics as MoS_2 [11].

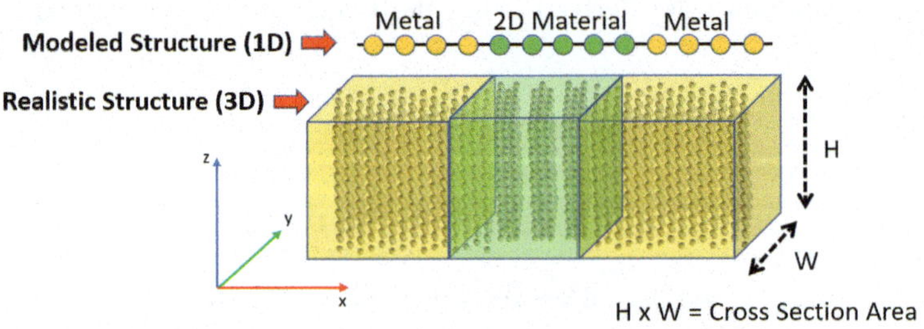

Fig. 8.2 The 1D and 3D visual of a generic 2D materials memristor

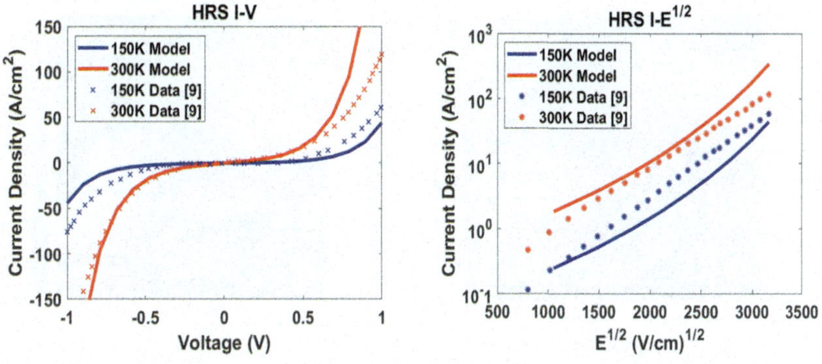

Fig. 8.3 The high resistance state I–V characteristics for 150 K and 300 K are shown in (**a**). The corresponding characteristics as a function of electric field (E) are shown in (**b**)

The High Resistance State The HRS for MoS_2 can be modeled using a potential profile similar to that of the metal-SiO_2-metal structure [10]. Given the experimental data for any general MoS_2 memristor with an unknown potential profile, the simulator can be used to determine a rough estimate of the metal-2D material coupling t_{par} and the grid-point spacing a, which allows it to successfully capture the temperature-dependent HRS I–V characteristics. Furthermore, it can be used to study the NVRS for a specific 2D materials memristor whose characteristics need to be determined from a known initial potential energy profile. Here, we present a phenomenological analysis of a memristor with multiple defects along two transverse planes to show how the model can capture the device's characteristics.

In Fig. 8.3, the current characteristics of the HRS are illustrated for different temperature values. The simulator is calibrated using the parameters t_{par} and a. The coupling parameter (t_{par}) is chosen between a range determined by the hopping parameter for the metal and the hopping parameter for the 2D material. The grid point spacing is calibrated between a range of 10 and 100 pm (1–5% of the length of the device). The MoS_2 layer is sandwiched between two Au metal electrodes with an electron effective mass of 1.1 m_o. Here, m_o is the rest electron mass. The plot fits in Fig. 8.3 were obtained for a transverse dielectric constant of 4 [12], an onset potential barrier of 1 eV, a tunneling electron effective mass of m_o at the metal-MoS_2 interface, and switching material (MoS_2) thickness of 1.5 nm [9].

For the HRS we observe that the current increases with the temperature. This can be explained by recalling the dependency of the Fermi function on temperature. It is known that as the temperature increases, the probability of the available density of states to be occupied increases for a given energy. As a result, the current increases. This semiconductor physics also holds true here and hence, the simulator is able to provide reasonable results for the HRS.

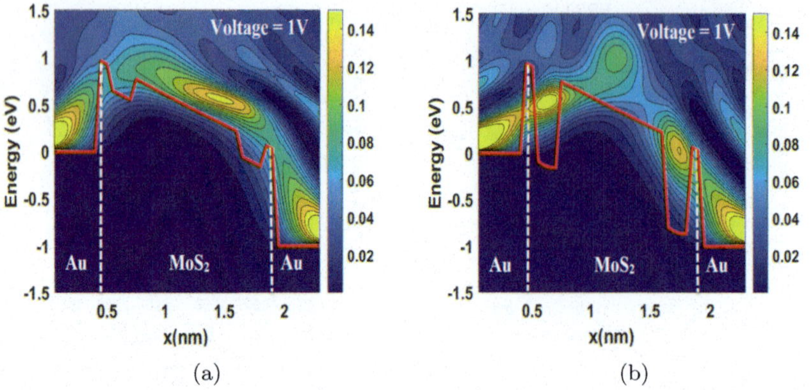

Fig. 8.4 The local density of states for Au-MoS_2-Au structure **a** in HRS **b** and LRS (after metal substitution) at 1 V. The averaged effect of defects in MoS_2 is highlighted by the potential energy profile shown in red

The Switching Mechanism The plots in Fig. 8.4 show the switching mechanism from the HRS to the LRS in the simplified MoS_2 memristor due to the substitution of sulfur vacancies by gold atoms at 300 K. The presence of defects within the MoS_2 bandgap lowers the potential energy as shown in Fig. 8.4a. When the electric field is sufficiently high, the wells get substituted by metal atoms switching to the LRS. Due to the presence of metal atoms within MoS_2, the band diagram gets distorted, resulting in deeper wells as shown in Fig. 8.4b. Beyond a certain extent of depth distortion, energy eigenstates close to the Fermi level are introduced, which allow high transmission through the barrier between the Au-MoS_2 junctions and the wells. This can be seen by the increase in the magnitude of the density of states in the LRS as compared to the HRS in Fig. 8.4. When the wells become deeper they also increase the tunneling probability by reducing barrier width at the Fermi level, which increases the current dramatically.

The Low Resistance State The LRS can be modeled from the simulator by replacing the vacancy-defects in MoS_2 with Au atoms, and subsequently, making the onset at the metal substituted points lower. As discussed in the last subsection, this leads to the creation of a potential well that allows the electrons at the Fermi level to pass through the allowed energy levels of the potential well, thus improving the tunneling probability. For the case under consideration, we replace MoS_2 atoms with Au and set the onset potential to 0 V. The LDOS is then plotted (Fig. 8.4b) to observe high-density state waves form in the wells. As we increase the doping of the metal in the dielectric at different grid points, we observe more states that the electrons can transmit through. Further, it is observed that the closer the metal substituted defect is to the metal-2D material junction, the higher the density of states is in the potential well. This is because tunneling probability is directly affected by the barrier width and hence, a defect closer to the metal-2D material interface will witness a smaller barrier to transmit through. As a result, the current would increase.

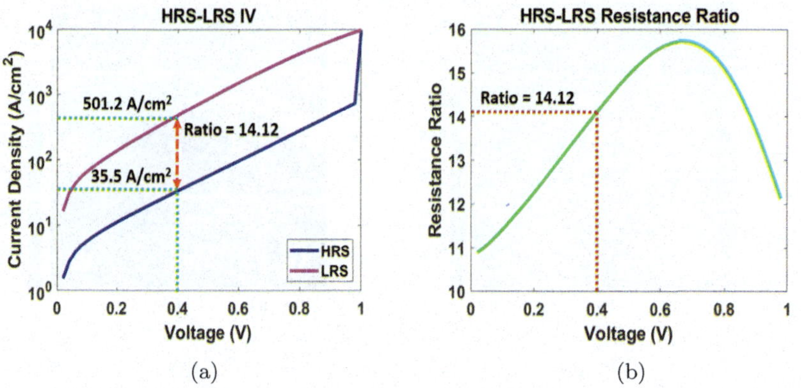

Fig. 8.5 **a** The electrical characteristics (current density vs. voltage) of HRS and LRS. **b** HRS (R_{off}) to LRS (R_{on}) resistance ratio for different applied biases

It is, however, important to mention that the LRS model cannot capture the temperature dependence of the LRS I–V characteristics as the model only considers ballistic transport.

The HRS and LRS IV Characteristics After getting an insight into the switching mechanism in the previous section, it is now imperative to take a more quantitative approach to explain the resistance states. In Fig. 8.5a, the current-voltage characteristics are plotted for the HRS and the LRS. Firstly, the device is assumed to be operating in the HRS. This is represented by the blue line. As the applied bias is increased, the current increases gradually until a voltage of 1 V, where the current is assumed to increase dramatically due to a change in the resistance state which is modeled as mentioned in the previous section. If the applied bias is decreased from this stage, it follows the magenta curve in the figure, which represents the device's operation in the LRS. Furthermore, Fig. 8.5b shows the resistance ratio for different applied biases in the device. The expected increase in the ratio with applied bias is observed. The ratio is contained in the range 11–15.75.

8.4 Transport in Single Defect Memristors

After gaining an insight into the NVRS of multi-defect 2D materials memristors in the previous section, we now analyze a more specific case—the NVRS of a single defect memristor. Here, the current is passed directly through a defect resulting from a sulfur (S) monovacancy in MoS_2. The model captures the single defect MoS_2 layer by assuming a 1 nm thick insulating layer—0.35 nm van der Waals distance on either side of a 0.32 nm thick MoS_2 layer. All other parameters are the same as calibrated in the previous section from the multi-defect memristor data. In this section, we first study the effect of changes in well parameters such as depth and location on the

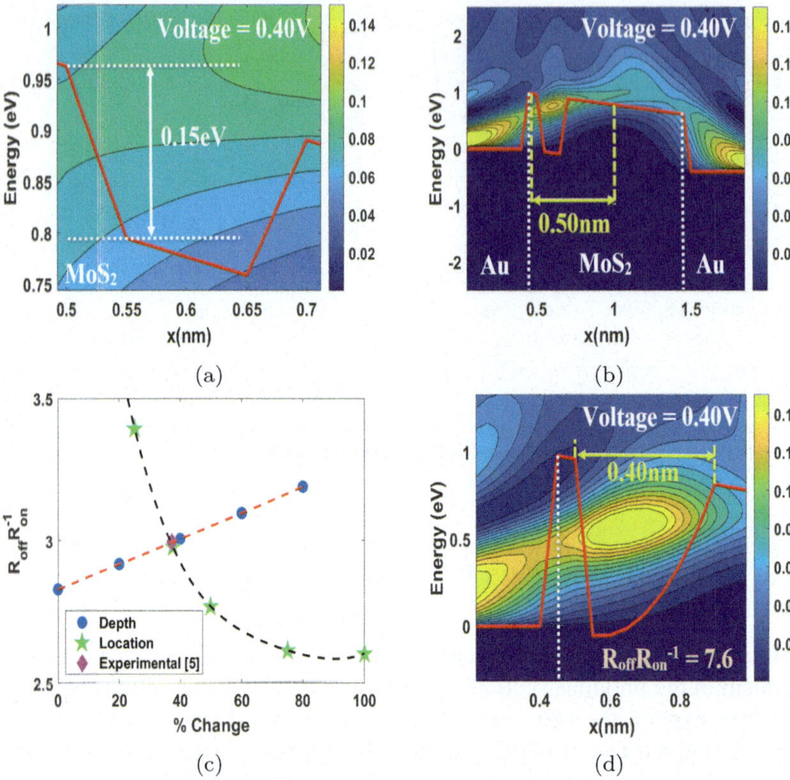

Fig. 8.6 **a** The zoomed-in HRS quantum well. Here, the depth is varied in steps of 0.05 eV. **b** The variation in well location within 0.50 nm owing to Mo-S bond length uncertainty. **c** Model calibration by varying well depth and location. The experimental ratio is obtained from Fig. 2f of Ref. [5]. The experimental HRS depth is determined to be 0.10 eV. The well location is found to be at 0.18 nm from the left electrode. **d** A possible LRS potential distortion resulting from Au atom substitution

NVRS ratio. The well depth is calibrated in the HRS case as it cannot be ascertained directly from the data, and the well location is varied to capture the rearrangement of atoms that occurs due to metal atom substitution in the defects. Second, we show the single defect NVRS ratio obtained from our model and compare it with the experimental one.

The well depth parameter corresponds to the depth of the well arising from the monovacancy in the HRS. An initial value of 0.15 eV is chosen for the depth, as shown in Fig. 8.6a, which is then incremented by steps of 0.05 eV. Each step gives an NVRS ratio as plotted in Fig. 8.6c. The initial value and the step size were chosen such that the trend between the ratio and depth on either side of the point of interest i.e., the experimental ratio, could be appropriately captured. Here, we observe that it is almost perfectly linear for the chosen values and that the ratio ranges between 2.8 and 3.2. Next, we discuss the calibration of the well location. This parameter corresponds to the site in MoS_2 where there is a monovacancy. We observe that the NVRS ratio is sensitive to changes in the monovacancy location (Fig. 8.6c). In our model, such a variation is considered to be within 0.50 nm i.e. the distance between the left electrode and the Mo atom (Fig. 8.6b). The change in the ratio with location is roughly exponential and ranges between 2.6 and 3.4.

After a relationship between the resistance ratio and the parameters is identified, we calibrate our model with the help of experimental observations. We begin by calculating the average switching ratio arising due to Au atom substitution in an S monovacancy from the experimental data given in Fig. 2f of Ref. [5]. The extracted experimental resistance ratio comes out to be approximately 3 at 0.40 V. The final calibrated model is then obtained through parameter variation.

Further, we consider the variation in the LRS potential distortion profile. So far, we had assumed that the sole effect of a gold substitution is the deepening of the wells. However, varying the potential profile after distortion is a parameter that can be experimented with. In our model, we see that distorting the LRS profile as shown in Fig. 8.6d gives an NVRS ratio of 7.6 indicating that the model is highly sensitive to change in the electrostatic potential profile after the metal substitution. The widening of the well lowers the energy eigenstates. Consequently, the electrons being injected at the Fermi level transmit through the device using these states instead of going over the barrier. Thus, the LRS current increases which results in a higher switching ratio. Therefore, the proposed model can provide physical insight into the HRS and LRS of 2D memristors with low defect densities.

8.5 Conclusion

A one-dimensional simulator is described to study quantum transport in two-dimensional material memristors by solving the 1D Poisson and Schrodinger equation self-consistently within the Non-equilibrium Green's Function Formalism. It is observed that the model provides a decent fit to the experimental HRS data of multi-defect memristors [9] and can also capture the temperature dependency of this state. Next, we can use it to study the non-volatile resistance switching ratio (NVRS) of a single defect memristor. For this purpose, a set of two parameters are varied to calibrate the model and gain insight into the ratio. The efficacy of the calibrated model is proved by capturing the experimental NVRS ratio [5] of the single defect memristor. The final model can be used to find the location at which the defect introduces energy eigenstates in the device's potential profile while also providing insight into the nature of its distortion. It follows from this study that the developed model can be an essential tool in studying the monolayer 2D material memristors with low defect densities especially when alternate simulation techniques are computationally extensive.

References

1. J. Joshua Yang, M.D. Pickett, X. Li, D.A.A. Ohlberg, D.R. Stewart, R.S. Williams, Memristive switching mechanism for metal/oxide/metal nanodevices. Nat. Nanotechnol. **3**(7), 429–433 (2008)
2. D. Akinwande, Memory, memristors, and atomristors. IEEE Micro **38**(5), 50–52 (2018)
3. D. Ielmini, V. Milo, Physics-based modeling approaches of resistive switching devices for memory and in-memory computing applications. J. Comput. Electron. **16**(4), 1121–1143 (2017)
4. S.G. Hu, S.Y. Wu, W.W. Jia, Q. Yu, L.J. Deng, Y.Q. Fu, Y. Liu, T.P. Chen, Review of nanostructured resistive switching memristor and its applications. Nanosci. Nanotechnol. Lett. **6**(9), 729–757 (2014)
5. S.M. Hus, R. Ge, P.-A. Chen, L. Liang, G.E. Donnelly, W. Ko, F. Huang, M.-H. Chiang, A.-P. Li, D. Akinwande, Observation of single-defect memristor in an MoS_2 atomic sheet. Nat. Nanotechnol. **16**(1), 58–62 (2021)
6. A. Chen, M.-R. Lin, Variability of resistive switching memories and its impact on crossbar array performance, in *2011 International Reliability Physics Symposium* (IEEE, 2011), pp. MY-7
7. S. Datta, Nanoscale device modeling: the Green's function method. Superlattices Microstruct. **28**(4), 253–278 (2000)
8. R. Lake, G. Klimeck, R.C. Bowen, D. Jovanovic, Single and multiband modeling of quantum electron transport through layered semiconductor devices. J. Appl. Phys. **81**(12), 7845–7869 (1997)
9. R. Ge, X. Wu, M. Kim, J. Shi, S. Sonde, L. Tao, Y. Zhang, J.C. Lee, D. Akinwande, Atomristor: nonvolatile resistance switching in atomic sheets of transition metal dichalcogenides. Nano Lett. **18**(1), 434–441 (2018)
10. M. Moreau, D. Munteanu, J.-L. Autran, Simulation of gate tunneling current in metal-insulator-metal capacitor with multi layer high-κ dielectric stack using the non-equilibrium green's function formalism. Jpn. J. Appl. Phys. **48**, 11 (2009)
11. X. Wu, R. Ge, P.-A. Chen, H. Chou, Z. Zhang, Y. Zhang, S. Banerjee, M.-H. Chiang, J.C. Lee, D. Akinwande, Thinnest nonvolatile memory based on monolayer h-BN. Adv. Mater. **31**(15), 1806790 (2019)
12. Y. Kang, D. Jeon, T. Kim, Local mapping of the thickness-dependent dielectric constant of MoS_2. J. Phys. Chem. C **125**(6), 3611–3615 (2021)

Chapter 9
Design of Vertical Superjunction AlGaN/GaN HEMT: A TCAD-Based Approach

Devika Jena, Taraprasanna Dash, Eleena Mohpatra, and Sanghamitra Das

9.1 Introduction

III-Nitrides are the foremost choice for next-generation high-power electronic devices [1, 2]. GaN-based power devices are started penetrating into the power conversion market in terms of battery chargers, smartphones, computers, servers, automotive, lighting systems and photovoltaics, etc. [1]. Therefore, extensive research and innovative design prospective are open in GaN-based device areas in the last decades. GaN has been in focus due to its specific properties like wide bandgap, high electron saturation velocity, high critical breakdown field, and ability to form lattice-matched heterojunctions with other group III-Nitride materials [3–6].

There are extensive research works going on GaN HEMT to scale down the device to achieve miniaturization and cost-effectiveness with high power applications [7–9]. In this regard, two possible solutions can be adopted, (1) to switch to a nonplanar device like tri-gate and gate all around structures and (2) to innovate design with planar structures. While considering the first possible option, there is a technological shift that needs further cost involvement, however, the later solution can be adapted with existing technology. Most of the reported planar GaN devices are lateral ones with horizontal 2DEG in nature [8]. However, for power switching applications,

D. Jena · T. Dash (✉)
Department of ECE, Siksha 'O' Anusandhan (Deemed to be University), Bhubaneswar 751030, India
e-mail: tpdiitkgp@gmail.com

Department of Electronics and Communication Engineering, Odisha University of Technology and Research, Bhubaneswar, Odisha 751003, India

S. Das
Department of ECE, Silicon Institute of Technology, Bhubaneswar 751024, India

E. Mohpatra
Electronics and Communication Engineering, RV College of Engineering, Bengaluru, Karnataka 560059, India

R. Singh et al. (eds.), *The Physics of Semiconductor Devices*, Springer Proceedings in Physics 306, https://doi.org/10.1007/978-981-97-1571-8_9

vertical planar devices are more desirable due to their advantages in reduced chip size and package convenience. Vertical GaN transistors, such as super junction HEMT (SJ HEMT) and CAVETs (current aperture vertical electron transistors), have intriguing properties for high-power switching, making them a viable choice for comparison with Si and SiC power devices [10, 11].

Superjunction (SJ) concept was first introduced by Fujihira in 1997 [12]. To create the optimal power devices with good $R_{on;sp}$ and BV, SJ structures and wide-band-gap (WBG) materials, as well as their combinations, have been introduced [10, 12–17]. Li in 2013 proposed SJ HEMT of half-pillar width 3 μm with 4.2 mΩ cm^2 [10]. H. Huang, et al. proposed a unified model for the improvement of $R_{on:sp}$ and BV of vertical doped SJ (d-SJ) and polarized SJ (p-Sj). An ununiformly doped SJ vertical heterojunction field-effect transistor (UDSJ-VHFET) with BV 6863 V and Ron 4.25 mΩ cm^2 had been presented by Zhu et al. [17]. Although the theoretical benefits of GaN devices are enormous, significant technological advancement is still necessary. There are also possible scopes for the suitable design of SJ HEMT. In this work, we have proposed a structure with doping pillar SJ HEMT (DP-SJ HEMT) to improve the $R_{on:sp}$ and BV. The structure has been adapted and modified after [10]. The electrostatics and characteristics of DP-SJ HEMT have been predicted using TCAD tools.

There are four sections in this manuscript. The brief introduction is presented in Sect. 9.1. In Sect. 9.2, the device parameters and simulation environment are discussed. In the Sect. 9.3, the simulated results of DP-SJ HEMT are presented and analyzed. The concluding remarks are presented in Sect. 9.4.

9.2 Device Simulation

The proposed device schematic is shown in Fig. 9.1. The length of n and p-pillar is 3 μm. The gate and p + GaN cap layer length is 4 μm. The cap layer is highly doped with a concentration of 1×10^{18} cm^{-3}. Below the cap layer, there is a 25 nm $Al_{0.1}Ga_{0.9}N$ layer. At the interface of $Al_{0.1}Ga_{0.9}N$ and GaN channel layer 2-D electron gas (2DEG) is created which flows the current from source to drain. The epi layer width and length are 60 μm and 6 μm respectively. The length of the epi layer is divided into 3 μm Mg-doped p-pillar and 3 μm Si-doped n-pillar. In between the p-pillar and source, there is a p + GaN current barrier layer (CBL) with doping 1×10^{18} cm^{-3} for the prevention of punch-through of the p-pillar. The width of the n-pillar is divided into 5 equal parts (N1, N2, N3, N4, N5) with a consistent doping gradient. The doping of n-pillar starts from N1 $= 1 \times 10^{16}$ cm^{-3} to N5 $= 5 \times 10^{16}$ cm^{-3}. Where the doping of the p-pillar is equal to the doping of the middle pillar N3 $= 3 \times 10^{16}$ cm^{-3}. The device is simulated using Silvaco TCAD tool [18]. Different models like Polarization (for 2DEG), Auger recombination (for the direct transition of carriers), Selberherr's Impact Ionization (for BV), and Nitride field-dependent models are used for simulation. In this work, we have used Farahmand

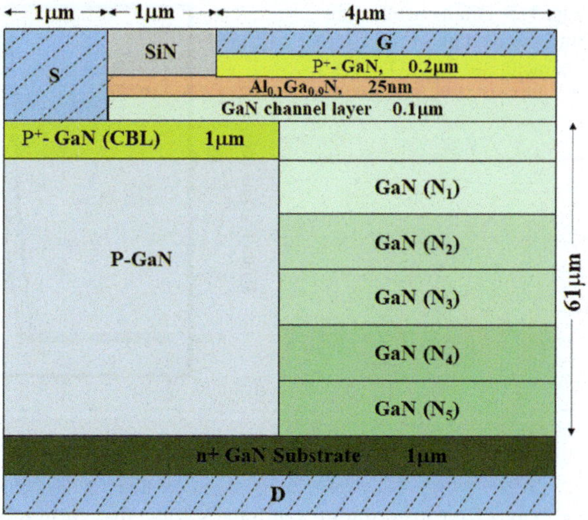

Fig. 9.1 Schematic of the doping pillar SJ HEMT (DP-SJ HEMT) structure

Modified Caughey Thomas (FMCT) model along with the High Field Mobility and Impact Ionization model to minimize the error in the software tool.

The following parameter are used in FMCT model in the different regions [10]:

Parameters	Electrons	Holes
Max. mobility $\mu_{max,k}$	1000	55
Min. mobility $\mu_{min,k}$	170	3
Ng,i	2×10^{17}	3×10^{17}
γ_i	1.0	2.0

For numerical simulation of the Impact Ionization model the impact ionization coefficients for electron and hole (α_n and α_p respectively) are calculated using [10]:

$$a_n = 2.81 \times 10^{18}\,\text{cm}^{-1},\ b_n = 3.43 \times 10^7\,\text{V/cm},$$
$$a_p = 5.41 \times 10^6\,\text{cm}^{-1}\ \text{and}\ b_p = 1.96 \times 10^7\,\text{V/cm}.$$

9.3 Results and Discussions

In Fig. 9.2 the transfer characteristics of the proposed DP-SJ HEMT are shown at a drain voltage (V_D) of 0.1 V. In the absence of gate bias, p + GaN cap with highly Mg-doped on top of the AlGaN layer depletes the 2-DEG, achieving normally-off and enhancement mode. The DP-SJ HEMT threshold voltage (V_{th}) and the specific on-resistance ($R_{on,sp}$) are 1.4 V and 4.052 mΩ cm^{-3} respectively. The transconductance

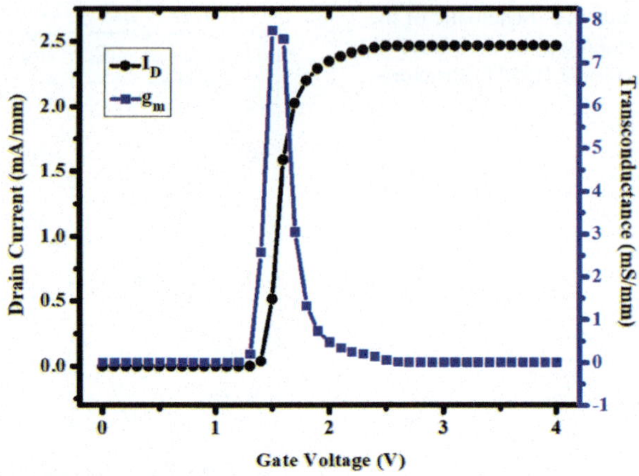

Fig. 9.2 Drain current and transconductance characteristics at VD = 0.1 V

(g_m) as a function of gate voltage (V_G) is also shown in Fig. 9.1. At $V_G = 1.5$ V there is a sharp increase in g_m and then a decrease. The maximum g_m is 7.75 mS/mm in the proposed device.

The gate voltage regulates the current which passes through the 2-DEG channel in on-state and then the current spreads into the n-pillar. In Fig. 9.3 the output characteristic (I_D-V_D) of the device is shown for different $V_G = 0, 2, 3, 4$ V. At $V_G = 4$ V maximum drain current flows in the device. In Fig. 9.4 we have presented the comparison of I_D-V_D of reported similar SJ HEMT [10] with DP-SJ HEMT. There is a very slight increase in V_{th} of DP-SJ HEMT from reported similar SJ HEMT [10] i.e., 1.4 V in DP-SJ HEMT and 1.37 V in reported similar SJ HEMT. Whereas the $R_{on;sp}$ of DP-SJ HEMT (4.0 mΩ cm^2) is less than that of reported similar SJ HEMT (4.2 mΩ cm^2). The I_{MAX} of DP-SJ HEMT is 2.47 mA/mm and reported similar SJ HEMT is 1.38 mA/mm.

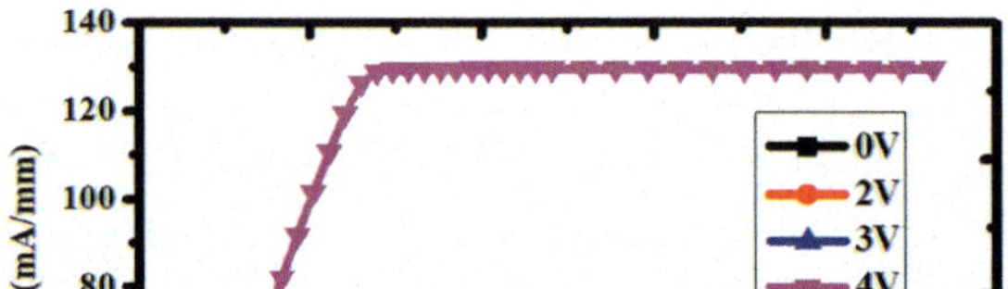

Fig. 9.3 Transfer characteristics I_D-V_D of the device at different gate voltage $V_G = 0, 2, 3, 4$ V

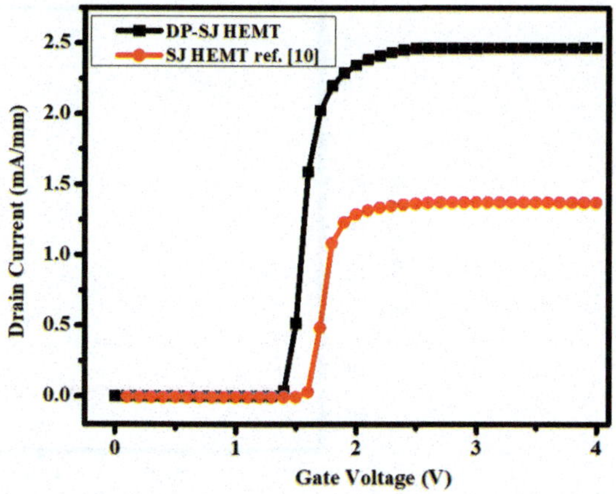

Fig. 9.4 Comparison of I_D-V_G of the proposed DP-SJ HEMT with reported similar SJ HEMT

In Fig. 9.5 we have presented the distribution of 2-D electric field of DP-SJ HEMT across the junction of n-pillar and p-pillar of the device. The distribution of the electric field is helpful for BV analysis. $R_{on:sp}$ is directly proportional to the BV and inversely proportional to the electric field so there is always a trade-off between $R_{on:sp}$ and BV. The $R_{on:sp}$ of a vertical device can be calculated with the help of an electric field and BV [19].

$$R_{on;sp} = \frac{2w_B V_{BV}}{\mu \varepsilon E_c^2} \tag{9.1}$$

where V_{BV}, w_B, μ, ε and Ec represent the BV, width of n-type or p-type pillar, carrier mobility, permittivity, and critical electric field respectively of SJ HEMT.

Figure 9.6 presents the BV of the DP-SJ HEMT device at $V_G = 1$ V. The BV of the device is 8.05 kV. The equivalent potential distribution of the device is shown in

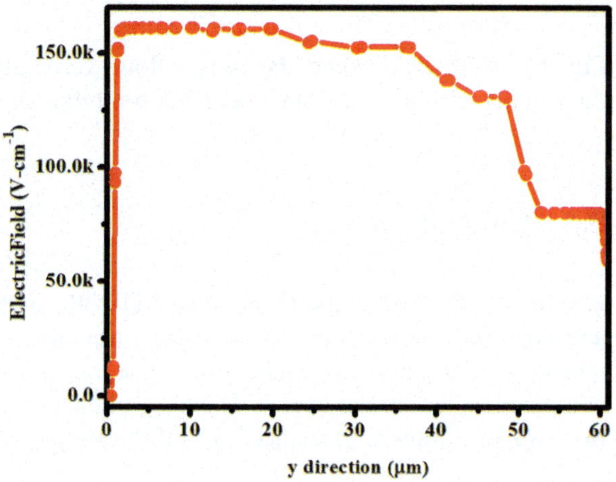

Fig. 9.5 Distribution of electric field along the junction of p-n pillar of the DP-SJ HEMT

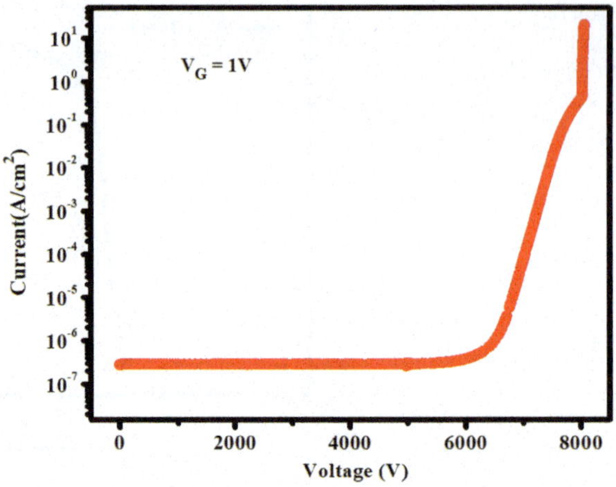

Fig. 9.6 Simulated breakdown voltage at $V_G = 1$ V of the DP-SJ HEMT

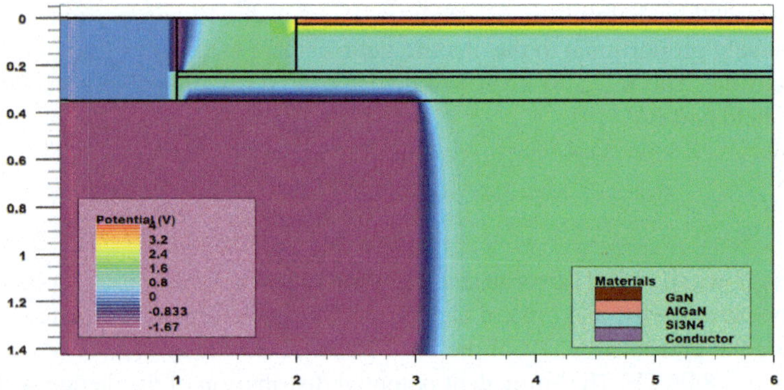

Fig. 9.7 On-state potential profile of DP-SJ HEMT

Fig. 9.7. As in the on-state, the current flows vertically from source to drain through the n-pillar and in the off-state depleted p-n pillar supports the voltage.

9.4 Conclusion

A vertical GaN-based superjunction HEMT with composite doping pillar is proposed and analyzed with Silvaco TCAD. The structure is proposed to improve the BV and $R_{on;sp}$ for high-power applications. The transfer characteristics are analyzed and compared with a similar device. The I_{MAX} of the proposed DP-SJ HEMT is 2.47 mA/mm with max transconductance 7.75 mS/mm at $V_D = 0.1$ V. The I_D-V_D of the device is also analyzed at different V_G. The on-state DC characteristics are analyzed. The $R_{on;sp}$ of the device reduced from the reported similar SJ HEMT.

References

1. H. Amano, Y. Baines, E. Beam, M. Borga, T. Bouchet, P.R. Chalker, M. Charles, K.J. Chen, N. Chowdhury, R. Chu, C. de Santi, M.M. de Souza, S. Decoutere, L. di Cioccio, B. Eckardt, T. Egawa, P. Fay, J.J. Freedsman, L. Guido, O. Häberlen, G. Haynes, T. Heckel, D. Hemakumara, P. Houston, J. Hu, M. Hua, Q. Huang, A. Huang, S. Jiang, H. Kawai, D. Kinzer, M. Kuball, A. Kumar, K.B. Lee, X. Li, D. Marcon, M. März, R. McCarthy, G. Meneghesso, M. Meneghini, E. Morvan, A. Nakajima, E.M.S. Narayanan, S. Oliver, T. Palacios, D. Piedra, M. Plissonnier, R. Reddy, M. Sun, I. Thayne, A. Torres, N. Trivellin, V. Unni, M.J. Uren, M. van Hove, D.J. Wallis, J. Wang, J. Xie, S. Yagi, S. Yang, C. Youtsey, R. Yu, E. Zanoni, S. Zeltner, Y. Zhang, The 2018 GaN power electronics roadmap. J. Phys. D Appl. Phys. **51**, 163001 (2018). https://doi.org/10.1088/1361-6463/AAAF9D
2. S. Sheikh, A.A. Shuvo, A.G. Bhuiyan, in *2021 5th International Conference on Electrical Engineering and Information and Communication Technology, ICEEICT 2021*. Effects of Structural Variation for Improved Performance of a Vertical AlGaN/GaN Superjunction HEMT (2021). https://doi.org/10.1109/ICEEICT53905.2021.9667937
3. S.L. Selvaraj, A. Watanabe, A. Wakejima, 1.4-kV breakdown voltage for AlGaN/GaN high-electron-mobility transistors on silicon substrate. IEEE Electron Dev. Lett. **33**, 1375–1377 (2012). https://doi.org/10.1109/LED.2012.2207367
4. Y. Wu, W. Zhang, J. Zhang, Au-Free Al0.4Ga0.6N/Al0.1Ga0.9N HEMT on Silicon Substrate With High Reverse Blocking Voltage of 2 kV. IEEE Trans. Electron Dev. **68**(9), 4543–4549 (2021). https://doi.org/10.1109/TED.2021.3093839
5. K.H. Hamza, D. Nirmal, A.S.A. Fletcher, L. Arivazhagan, J. Ajayan, R. Natarajan, Highly scaled graded channel {GaN} {HEMT} with peak drain current of 2.48 A/mm. AEU Int. J. Electron. Commun. **136**, 153774 (2021). https://doi.org/10.1016/j.aeue.2021.153774
6. R. Tomita, S. Ueda, T. Kawada, H. Mitsuzono, K. Horio, Analysis of dependence of breakdown voltage on gate-drain distance in AlGaN/GaN HEMTs with high-k passivation layer. IEEE Trans. Electron Devices **68**, 1550–1556 (2021). https://doi.org/10.1109/TED.2021.3060353
7. A. Khusro, M.S. Hashmi, A.Q. Ansari, in *Proceedings of the 2nd International Conference on Computing and Network Communications, CoCoNet 2018*. Empirical Device Scaling and RF Performance Perspective: A Small Signal Model for GaN High Electron Mobility Transistor (2018), pp. 45–49. https://doi.org/10.1109/COCONET.2018.8476895
8. R. Hou, J. Xu, D. Chen, in *2017 IEEE 5th Workshop on Wide Bandgap Power Devices and Applications, WiPDA 2017*. A Multivariable Turn-On/Turn-Off Switching Loss Scaling Approach for High-Voltage GaN HEMTs in a Hard-Switching Half-Bridge Configuration (2017), pp. 171–176. https://doi.org/10.1109/WIPDA.2017.8170542
9. R. Soligo, D. Guerra, D.K. Ferry, S.M. Goodnick, M. Saraniti, in *2012 15th International Workshop on Computational Electronics, IWCE 2012*. Cellular Monte Carlo Study Lateral Scaling Impact of on the DC-RF Performance of High-Power GaN HEMTs (2012). https://doi.org/10.1109/IWCE.2012.6242863
10. Z. Li, T.P. Chow, Design and simulation of 5–20-kV GaN enhancement-mode vertical superjunction HEMT. IEEE Trans. Electron Devices **60**, 3230–3237 (2013). https://doi.org/10.1109/TED.2013.2266544
11. D. Ji, Y. Yue, J. Gao, S. Chowdhury, Dynamic modeling and power loss analysis of high-frequency power switches based on GaN CAVET. IEEE Trans. Electron Devices **63**, 4011–4017 (2016). https://doi.org/10.1109/TED.2016.2601559

12. T. Fujihira, Theory of semiconductor superjunction devices. Jpn. J. Appl. Phys. Part 1 Regular Papers Short Notes Rev. Papers **36**, 6254–6262 (1997). https://doi.org/10.1143/JJAP.36.6254/XML
13. Z. Li, T.P. Chow, in *Proceedings of the International Symposium on Power Semiconductor Devices and ICs*. Robustness of GaN Vertical Superjunction HEMT (2013), pp. 217–220. https://doi.org/10.1109/ISPSD.2013.6694484
14. H. Huang, J. Cheng, B. Yi, W. Zhang, W.T. Ng, A unified model for vertical doped and polarized superjunction GaN devices. Appl. Phys. Lett. **116** (2020). https://doi.org/10.1063/1.5142855
15. M. Zhang, Z. Guo, Y. Huang, Y. Li, J. Ma, X. Xia, X. Tan, F. Xia, H. Sun, Study of AlGaN/GaN vertical superjunction HEMT for improvement of breakdown voltage and specific on-resistance. IEEE Access **9**, 9895–9902 (2021). https://doi.org/10.1109/ACCESS.2021.3049374
16. Y. Ma, M. Xiao, R. Zhang, H. Wang, Y. Zhang, Superjunction power transistors with interface charges: a case study for GaN. IEEE J. Electron Devices Soc. **8**, 42–48 (2020). https://doi.org/10.1109/JEDS.2019.2959713
17. C. Zhu, X. Zhou, Z. Feng, Z. Zhao, Z. Wei, Z. Zhao, Theoretical investigation of high-voltage superjunction GaN-based vertical hetero-junction field effect transistor with ununiformly doped buffer to suppress charge imbalance effect. Semicond. Sci. Technol. **34** (2019). https://doi.org/10.1088/1361-6641/ab2013
18. D.S. Software, *ATLAS User's Manual*. II, 567–1000 (1998)
19. M. Xiao, R. Zhang, D. Dong, H. Wang, Y. Zhang, Design and simulation of gan superjunction transistors with 2-deg channels and fin channels. IEEE J. Emerg. Sel. Topics Power Electron **7**, 1475–1484 (2019). https://doi.org/10.1109/JESTPE.2019.2912978

Chapter 10
Effect on Photo-Absorbance and Optical Energy-Gap of Al_2O_3 Thin Film After Deposition of Ag Thin Film

Abhijit Das, Baishali Bora, Shuvam Pawar, and Mitra Barun Sarkar

10.1 Introduction

Thin-film science has advanced many new fields of study in solid-state physics and chemistry [1–5]. When material properties are examined as thin films, they vary significantly which is why it is important to study the electrical, morphological and optical characteristics of materials when used as thin films [6–9]. Moreover, the possibility to integrate functionality in nanostructures, within a single nano-device or arrays, enables large scale integration for use in optical interconnects, transceivers etc. [10–12]. Al_2O_3 is a refractory ceramic oxide that is widely used in many industries closed packing structure leads to good mechanical and thermal properties making it preferable for many applications [13, 14]. Aluminium has a dielectric constant of 11 and a refractive index (RI) of 1.766 at wavelength 630 nm which makes it a desirable candidate in replacing SiO_2 as a gate oxide [15]. Single crystal Al_2O_3 known as sapphire is one of the hardest natural minerals after diamond, which is also used as substrate material for blue LEDs [16]. Al_2O_3 exists in several metastable structures like γ, η, θ and δ phases (gamma series) and k, and α phases (alpha series) [17]. Normally, the amorphous phase is formed at low temperatures [18]. Alumina is one of the most important oxide ceramic materials, finds large scale applications in the fields of optoelectronics, dielectrics, optics, sensors etc. [19, 20]. Silver (Ag) shows low absorption and high reflectance in the visible region [21, 22], Ag is widely used in high-performance optoelectronic devices such as solar cells, optical waveguides, organic light-emitting diodes (OLED) etc. [23, 24]. Physical

A. Das (✉) · B. Bora · M. B. Sarkar
Department of Electronics and Communication Engineering, National Institute of Technology Agartala, Agartala, Tripura 799046, India
e-mail: dasece01@gmail.com

S. Pawar
Department of Electronics and Communication Engineering, National Institute of Technology, Warangal, Telangana 506004, India

R. Singh et al. (eds.), *The Physics of Semiconductor Devices*, Springer Proceedings in Physics 306, https://doi.org/10.1007/978-981-97-1571-8_10

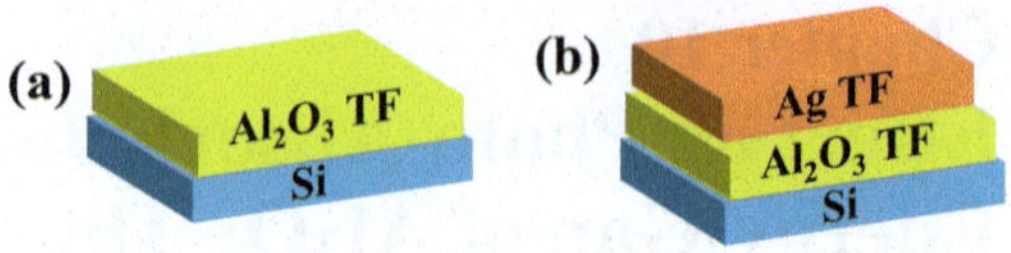

Fig. 10.1 Schematic structure of samples **a** Al_2O_3 TF and **b** Ag TF/Al_2O_3 TF

vapour deposition, chemical vapour deposition, pulsed laser deposition, magnetron sputtering deposition are a few deposition techniques for the deposition of Ag Thin Films [25, 26]. In the present work, we have used the physical vapour deposition technique (using an e-beam evaporator) to make an Al_2O_3 TF on an n-Si substrate and an Ag TF/Al_2O_3 TF sample and have examined the changes in optical absorption characteristics and change in optical energy-gap or bandgap, post-deposition of Ag TF on bare Al_2O_3 TF.

10.2 Experiment

10.2.1 Fabrication Process

At First, an n-type Si wafer cutting is done with the help of a Diamond Edge Cutter, we kept the dimension of each Si substrate as $1 \times 1\ cm^2$ then RCA cleaning of each sample is done for removing organic, metallic residue and films from silicon wafers. Then the substrates were placed at the vacuum coating unit of the e-beam evaporator. After that, Al_2O_3 TF (100 nm) was deposited over the n-type Si substrate as shown in Fig. 10.1a. The ambient temperature inside the coating unit was 30 °C and pressure inside the chamber was maintained at 1×10^{-5} mbar. Then, Ag TF (30 nm) was deposited over the Al_2O_3 TF (100 nm) as shown in Fig. 10.1b, by the same process with the ambient temperature inside the coating unit 30 °C and chamber pressure maintained at 1×10^{-5} mbar. The schematic structures of the samples, which we have fabricated are shown in Fig. 10.1.

10.2.2 Characterization

FESEM (field emission scanning electron microscopy) (ZEISS, Sigma 300) and EDAX (energy dispersive X-ray analysis) (ZEISS, Sigma 300) were done on the fabricated samples. Then the samples were characterized by XRD (X-ray diffraction) (Bruker, D8-Advance) method using Cu Kα radiation. The optical absorption measurement was done for both samples by UV–VIS spectroscopy (Shimadzu).

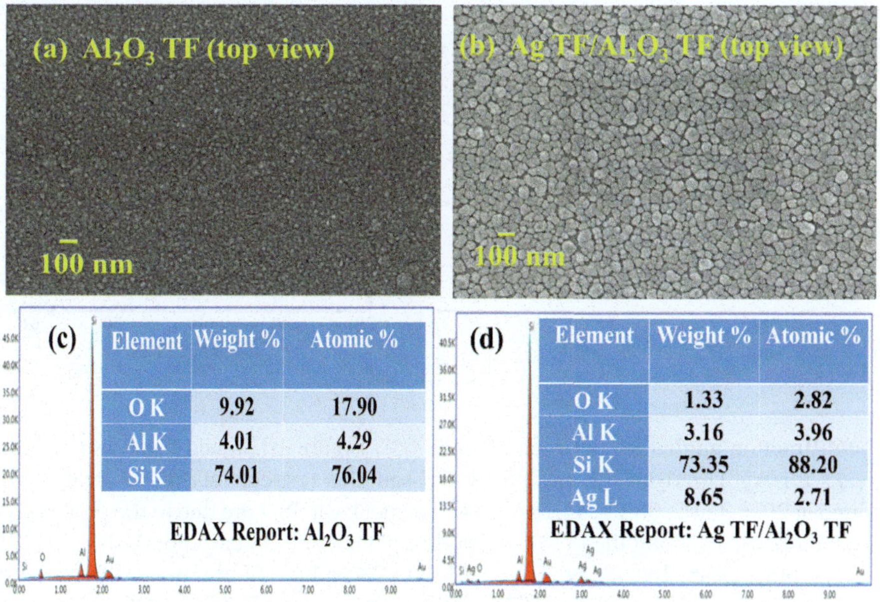

Fig. 10.2 **a** FESEM image of top view of Al_2O_3 TF, **b** FESEM image of top view of Ag TF/Al_2O_3 TF, **c** EDAX analysis report of Al_2O_3 TF and **d** EDAX analysis report of Ag TF/Al_2O_3 TF

10.3 Results and Discussion

10.3.1 FESEM and EDAX Analysis

The FESEM images were shown to study the morphological characteristics of the fabricated samples. Figure 10.2a shows the top view of FESEM images of Al_2O_3 TF (100 nm) deposited on the n-type silicon substrate. Figure 10.2b shows the top view of FESEM images of Ag TF (30 nm) deposited over Al_2O_3 TF (100 nm) on the n-type silicon substrate.

To further investigate the diffusion of Ag and Al_2O_3, EDAX analysis was performed. Figure 10.2c and d show the results of the EDAX analysis. Figure 10.2c and d inset image shows the percentage of the composition of the constituent elements for the respective structures. The peaks indicate the presence of Ag, Al, O, and Si.

10.3.2 XRD Analysis

To get further information about the phases, crystal orientations, and morphology of the bare Al_2O_3 TF and Ag TF/Al_2O_3 TF samples can be analyzed from XRD (Bruker D8 Advanced using Cu-Kα target source) measurements, which has been shown for

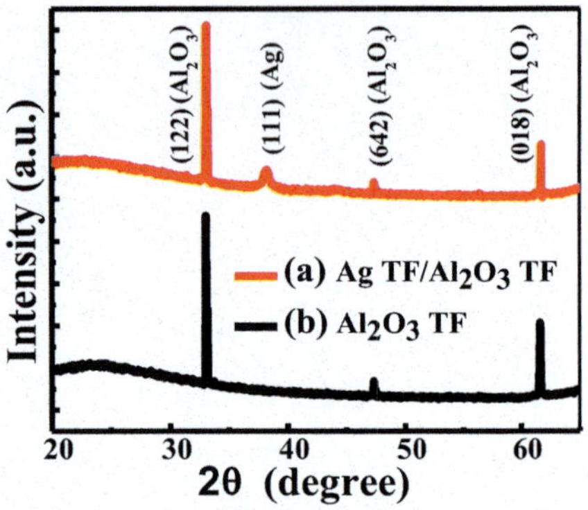

Fig. 10.3 XRD patterns: **a** Ag TF/Al_2O_3 TF and **b** bare Al_2O_3 TF

the diffraction angle (2θ) between 20° and 65°. Figure 10.3a and b shows the XRD peaks for Ag TF/Al_2O_3 TF and Al_2O_3 TF samples respectively. XRD Peaks were found at 2θ = 32.02, 38.27, 47.28, 61.76. The peaks at 38.3 are due to the presence of Ag nanoparticles along the (111) diffraction plane (JCPDS card no: 04-0783) [27].

The other peaks located at 2θ = 32.27, 47.28 and 61.77 are due to the presence of Al_2O_3 nanoparticles along (122), (642), (018) plane (JCPDS card no: 81-1667 and 02-1421) [28, 29]. Therefore, the XRD patterns confirm the presence of Ag and Al_2O_3 in the fabricated samples. The crystallite size was calculated using the Debye-Scherer equation (Eq. 10.1).

$$D = \frac{0.9\lambda}{\beta \times \mathrm{COS}\theta} \tag{10.1}$$

where

D crystallite size.
λ 1.54178 Å (wavelength of radiation used).
θ Bragg's angle in radians.
β Full-width half maximum (FWHM) in radians.

The average crystallite size calculated from the XRD peaks of the bare Al_2O_3 TF before deposition of Ag was ~ 144.45 nm. The crystallite size of the bare Ag TF was ~ 14.45 nm. So the deposited Ag nanoparticles has less crystallite size compared to bare Al_2O_3 nanoparticles in the fabricated Ag TF/Al_2O_3 TF samples.

10.3.3 Optical Absorption Analysis

UV–VIS optical absorption measurement of the as-deposited Al_2O_3 TF and Ag TF/Al_2O_3 TF samples were done by UV–VIS spectrophotometer (Shimadzu). The samples were examined at room temperature ambience under the wavelength range

200–700 nm. Optical absorption measurement of the as- deposited Al_2O_3 TF and Ag TF/Al_2O_3.

TF samples were done by UV–VIS spectrophotometer (shimadzu). The samples were examined at room temperature ambiance under the wavelength range 200–700 nm. Figure 10.4a shows the absorption spectrum for both samples. Improvement in photo absorption has been observed for Ag TF/Al_2O_3 TF sample compare to bare Al_2O_3 TF. Averagely ~ 1.2 fold enhanced photo-absorption was observed in the 430–500 nm region, and averagely ~ 1.4 fold enhanced photo-absorption was exhibited in the 260–330 nm region of the light spectra by the Ag TF/Al_2O_3 TF sample compared to bare Al_2O_3 TF sample. At the interface of Ag/Al_2O_3 an inorganic Ag–O compound forms. Formation of such type of inorganic Ag–O compound were also reported by Dwivedi et al. [30], Xie et al. [31], and Laskri et al. [32]. This inorganic Ag–O compound forms due to oxygen molecules attached with the Ag molecules. These oxygen molecules acts as trap state. Under light irradiation these trap states absorbs more photons [33] and hence the light absorption gets increased for Ag TF/Al_2O_3 TF sample compare to bare Al_2O_3 TF.

The optical bandgap of the Ag TF/Al_2O_3 TF sample and bare Al_2O_3 TF sample were then estimated from Tauc Plot [34]. Figure 10.4b represents $(\alpha h\upsilon)^2$ versus $(h\upsilon)$ plot (where α is the absorption coefficient, $h\upsilon$ is the photon energy). The optical bandgap at ~3.5 eV is due to the main band transition for Ag TF/Al_2O_3 TF sample. The main optical bandgap transition was observed at ~ 4.15 eV for the bare Al_2O_3 TF sample. We have found a shift of main optical bandgap transition from ~ 4.15 to ~ 3.5 eV post-deposition of Ag TF on Al_2O_3 TF. This shift in optical bandgap is due to the trapped electrons at oxygen related trap states (available at the Ag/Al_2O_3 interface) and electron transmission from Ag TF to the conduction band of the Al_2O_3 TF. Kumar et al. [35, 36], Bindushree et al. [37], Mahour et al. [38] reported such kind of change in optical bandgap related to trap state.

The change in the optical bandgap for Ag TF/Al_2O_3 TF sample compare to bare Al_2O_3 TF sample can also be explained by Urbach's empirical rule [39, 40].The

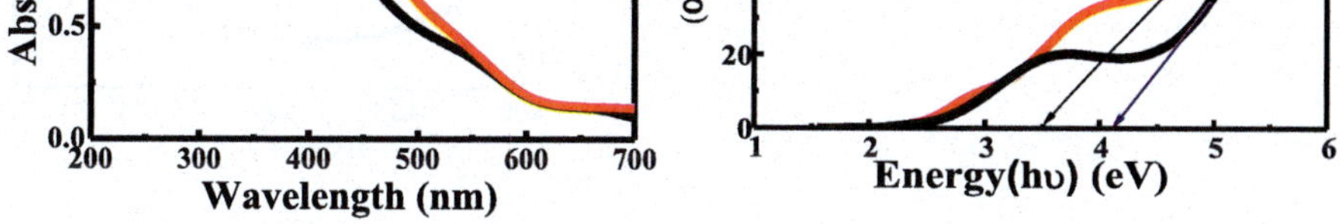

Fig. 10.4 **a** Optical absorption characteristics (at room temperature) of bare Al_2O_3 TF and Ag TF/Al_2O_3 TF and **b** $(\alpha h\upsilon)^2$ versus photon energy $(h\upsilon)$ characteristics (Tauc Plot)

presence of oxygen related trap states leads to add the additional energy levels in the optical bandgap, known as the Urbach tail or Urbach energy [41]. When a sample is irradiated with light of particular wavelength, the available trap states traps the photo-excited carriers, which was the reason for the appearance of the Urbach energy [42]. Urbach energy and optical bandgap are inversely related, i.e. sample having a less optical band gap is expected to have a more Urbach energy [43]. According to Urbach's empirical rule the Urbach energy for the samples were obtained using the following equations (Eqs. 10.2 and 10.3) [39, 40]:

$$\alpha = \alpha_0 \exp\left(\frac{h\upsilon}{EU}\right) \tag{10.2}$$

$$\Rightarrow \ln(\alpha) = \ln(\alpha_0) + \left(\frac{h\upsilon}{EU}\right) \tag{10.3}$$

where

α	Absorption coefficient.
α_0	Constant.
$h\upsilon$	Incident photon energy
EU	Urbach energy.

The value of EU for both samples were obtained from the reciprocal of the slope value of the linear part in ln (α) versus ($h\upsilon$) Characteristics (Fig. 10.5) [39]. The value of EU for bare Al_2O_3 TF and Ag TF/Al_2O_3 TF were found to be 3.3 eV and 7.1 eV respectively. The enhancement in the value of EU indicates the presence of large no of trap states [41] in Ag TF/Al_2O_3 TF compare to bare Al_2O_3 TF. These trap states introduce localized states near the conduction band [41], leading to an increase in Urbach energy and eventually a decrease in optical bandgap.

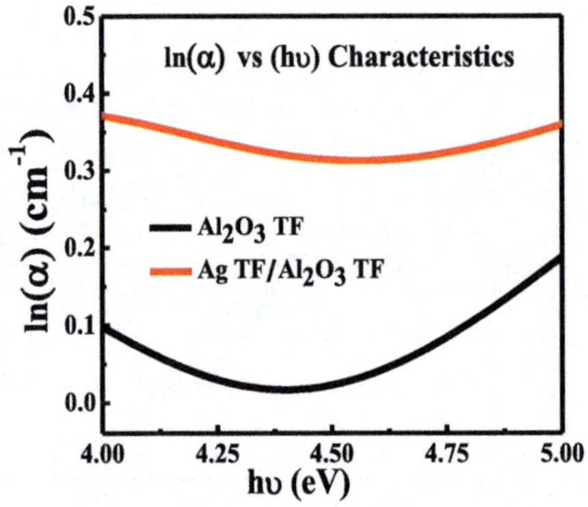

Fig. 10.5 ln (α) versus energy ($h\upsilon$) characteristics

10.4 Conclusion

In conclusion, Ag TF has been successfully deposited by using E-beam physical vapour deposition on Al_2O_3 TF. The post-deposition effect in optical absorption of Ag TF on Al_2O_3 TF with the bare Al_2O_3 TF structure has been compared. The optical absorption measurement was carried out at room temperature within the wavelength range between 200 and 700 nm. From the Tauc Plot, i.e., $(\alpha h\upsilon)^2$ versus Energy ($h\upsilon$) graph, the optical energy-gap or bandgap for both the samples has been calculated. The main direct band transition of Ag TF/Al_2O_3 TF was obtained close to the optical bandgap of 3.5 eV. Similarly, the main direct band transition of Al_2O_3 TF was obtained at close to 4.15 eV. The Ag TF/Al_2O_3 TF compared to as-deposited Al_2O_3 TF shows ~ 1.4 fold photon absorption enhancement in 260 nm-330 nm UV region and ~ 1.2 fold photon absorption enhancement in 430–500 nm visible region, which is mainly due to the formation of large numbers of trap states after the deposition of Ag TF on Al_2O_3 TF. The Urbach energy calculated from ln (α) versus ($h\upsilon$) characteristics is 3.3 eV for the bare Al_2O_3 TF sample and 7.1 eV for the Ag TF/Al_2O_3 TF sample. The higher value of Urbach energy also validates the formation of large numbers of trap states for the Ag TF/Al_2O_3 TF sample, which is the main reason for the reduction in optical energy-gap or bandgap of Ag TF/Al_2O_3 TF compared to bare Al_2O_3 TF. Thus, it can be concluded that in this work, the Ag TF in the Ag TF/Al_2O_3 TF structure played an important role in enhancing the optical absorption and reducing the optical energy-gap of bare Al_2O_3 TF and the Ag TF/Al_2O_3 TF structure can be used for efficient UV–VIS photo detection purposes.

Acknowledgements The authors are thankful to Central Instrumentation Centre (CIC), Tripura University, INDIA for providing the FESEM and EDAX facility.

References

1. Frank, Device scaling limits of Si MOSFETs and their application dependencies. Proc. IEEE **89**, 259–288 (2001)
2. A.R. West, *Solid State Chemistry*, 2nd edn. (Willey, Singapore, 2003)
3. M.C. Rao, A brief survey on basic properties of thin films for device application. Int. J. Mod. **22**, 576–582 (2013)
4. A.N. Cleland, Very low noise photodetector based on the single electron transistor. Appl. Phys. Lett. **61**, 2820 (1992)
5. S.S. Kalarickal, Ferromagnetic resonance linewidth in metallic thin films: comparison of measurement methods. J. Appl. Phys. **99**, 093909 (2006)
6. A. Beck, Reproducible switching effect in thin oxide film for memory applications. Appl. Phys. Lett. **26**, 351 (2000)
7. S. Su, Nanomaterials-based sensors for applications in environmental monitoring. J. Mater. Chem. **22**(35), 18101 (2012)
8. H. Yoon, Periplasmic binding proteins as optical modulators of single-walled carbon nanotube fluorescence: amplifying a nanoscale actuator. Angew. Chem. **123**(8), 1868–1871 (2011)

9. R.D. McMichael, Ferromagnetic resonance studies of NiO-coupled thin films of $Ni_{80}Fe_{20}$. J. Phys. Rev. B **58**, 8605 (1998)
10. M.D. Stiles, Model for exchange bias in polycrystalline ferromagnet-antiferromagnet bilayers. Phys. Rev. B **59**, 3722 (1999)
11. N. Lewis, Toward cost-effective solar energy use. Science **315**(5813), 798–801 (2007)
12. S.S. Mao, Selected nanotechnologies for renewable energy applications. Int. J. Energy Res. **31**, 619–636 (2007)
13. V.K. Sethi, Use of nanotechnology in solar PV cell. Int. J. Chem. Eng. Appl. **2**(2), 77–80 (2011)
14. G.A. Parks, Free energies of formation and aqueous solubilities of aluminum hydroxides and oxide hydroxides at 25°C. Am. Min. **57**(7–8), 1163–1189 (1972)
15. P.V. Patil, D.M. Bendale, R.K. Puri, V. Puri, Refractive index and adhesion of Al_2O_3 thin films obtained from different processes—a comparative study. Thin Solid Films **288**(1–2), 120–124 (1996)
16. I.H. Malitson, Refraction and dispersion of synthetic sapphire. J. Opt. Soc. Am. **52**(12), 1377–1379 (1962)
17. A. Khanna, Nanocrystalline gamma alumina coatings by inverted cylindrical magnetron sputtering. Surf. Coat. Technol. **201**(1–2), 168–173 (2006)
18. V.A.C. Haanappel, Properties of alumina films prepared by low-pressure metal-organic chemical vapour deposition. Surf. Coat. Technol. **72**(1–2), 13–22 (1995)
19. Y.S. Chang, Reactions at the aluminum oxide-ferrite interface. Thin Solid Films **193**(194), 959–964 (1990)
20. M. Kawamura, Thermally stable Ag thin films modified with very thin Al oxide layers. Jpn. J. Appl. Phys. **47**, 8917–8920 (2008)
21. Z. Fekkai, Optical, morphological and electrical properties of silver and aluminium metallization contacts for solar cells. Am. J. Mod. Phys. **3**(2), 45–50 (2014)
22. S.-K. Kim, High-efficiency vertical GaN slab light-emitting diodes using self-coherent directional emitters. Opt. Exp. **18**(11), 11025–11032 (2010)
23. M.P. Nezhad, Room-temperature subwavelength metallo-dielectric lasers. Nat. Photonics **4**, 395–399 (2010)
24. Y. Chi, Deposition of silver thin films using the pyrazolate complex $[Ag(3,5\text{-}(CF_3)_2C_3HN_2)]_3$. Chem. Vap. Deposition **11**(4), 206–212 (2005)
25. P. Kumar, Dynamic force microscopy study of the microstructural evolution of pulsed laser deposited ultrathin Ni and Ag films. J. Mater. Res. **23**(7), 1826–1839 (2008)
26. L. Wang, Investigation of the properties of silver thin films deposited by dc magnetron sputtering. Surf. Rev. Lett. **24**(04), 1750053 (2017)
27. K. Jemal, Synthesis, characterization, and evaluation of the antibacterial activity of allophylus serratus leaf and leaf derived callus extracts mediated silver nanoparticles. J. Nanomater. **2017**(4213275), 1–11 (2017)
28. M. Becerra, Green light emission in aluminium oxide powders doped with different terbium concentrations. Rev. Mex. Fis. **62**(4), 47–53 (2016)
29. M. C. Naik, CeO_2–Al_2O_3 nanocomposite by chemical combustion method for NO_2 gas–sensing application. J. Mater. Sci. Mater. Electron. **32**, 19925–19937 (2021)
30. S.M.M.D. Dwivedi, Detection technique for vitamin D3 using Er-doped TiO_2 nanowire-based UV photodetector. J. Nanophoton. **14**(4), 046001 (2020)
31. W. Xie, Surface modification of ZnO with Ag improves its photocatalytic efficiency and photostability. J. Photochem. Photobiol. A **216**(2–3), 149–155 (2010)
32. A. Laskri, Investigation of microstructural and optical properties of Ag_3O_4 thin films sprayed onto glass substrate. J. Nano Res. **58**, 90–101 (2019)
33. B.K. Mahajan, Enhanced photodetection with crystalline Si nanoclusters. J. Nanosci. Nanotechnol. **20**(4), 2344–2350 (2020)
34. A. Nath, Bandgap modulation of glancing angle deposition aided Ag nanoparticles covered TiO_2 thin film by high temperature annealing. J. Nanosci. Nanotechnol. **20**(12), 7636–7643 (2020)

35. R. Kumar, Infrared photodetectors based on multiwalled carbon nanotubes: insights into the effect of nitrogen doping. Appl. Surf. Sci. **538**, 148187 (2021)
36. A. Kumar, Effect of the band gap and the defect states present within band gap on the non-linear optical absorption behaviour of yttrium aluminium iron garnets. Opt. Mater. **108**, 110163 (2020)
37. N. Bindushree, Role of composition in enhancing heat transfer behavior of carbon nanotube-ethylene glycol based nanofluids. Electron. Mater. Lett. **16**, 595–603 (2020)
38. L.N. Mahour, Structural, optical and Mössbauer spectroscopic investigations on the environment of Fe in Fe-doped ZnO ($Zn_{1-x}Fe_xO$) ceramics synthesized by solution combustion method. Ceram. Int. **45**, 24625–24634 (2019)
39. A.S. Hassanien, Influence of composition on optical and dispersion parameters of thermally evaporated non-crystalline $Cd_{50}S_{50-x}Se_x$ thin films. J. Alloys Compd. **648**, 280–290 (2015)
40. S. Mondal, An experimental and theoretical understanding of a UV photodetector based on Ag nanoparticles decorated Er-doped TiO_2 thin film. Ceram. Int. **47**(10), 14879–14891 (2021)
41. A. Nath, Silver nanoparticles textured oxide thin films for surface plasmon enhanced photovoltaic properties. Plasmonics **17**, 193–201 (2021)
42. V.R. Akshay, Visible range optical absorption, Urbach energy estimation and paramagnetic response in Cr-doped TiO_2 nanocrystals derived by a sol–gel method. Phys. Chem. Chem. Phys. **21**, 12991–13004 (2019)
43. A. Dhanaraj, A study of the optical band gap energy and Urbach energy of fullerene (C_{60}) doped PMMA nanocomposites. AIP Conf. Proc. **2270**, 110040 (2020)

Chapter 11
Thermometry Across Switching Oxide Layer in ReRAM Device

Om Prakash Das and Shivendra Kumar Pandey

11.1 Introduction

The resistive switching technology is a commanding candidate in the emerging non-volatile memory for its potential utility in data storage applications, neuromorphic circuits, and cryptographic circuits [1, 2]. The resistive switching phenomena in the ReRAM device are based on filament formation and dissolution owing to the defects (oxygen vacancies, oxygen ions) migration within the oxide layer. These defects migration is controlled by the electric field (due to applied voltage) and triggered by local temperature by virtue of induced Joule heat. Though different models are reported for thermal analysis [2–7] and the current conduction mechanism of the ReRAM device [8], there is a lack of clarity in the rise of local temperature and heat conduction across the oxide layer. In this study, an electro-thermal model is proposed to trace the local temperature rise, heat conduction, and local temperature variation over a complete voltage sweep for high and low resistive regimes.

11.2 Experimental Details

Fabrication of W/HfO_2/Cu stack device was started with the deposition of BE (W thin layer) on the glass substrate by DC sputtering with working power of 15 W. A ~ 100 nm thick HfO_2 layer was RF sputtered on the W layer with sputtering power of 45 W using an HfO_2 target (99.99% pure, *ACI Alloys*). Then, the TE (Cu) was

O. P. Das (✉) · S. K. Pandey
Department of Electronics and Instrumentation Engineering, National Institute of Technology Silchar, Silchar, Assam 788010, India
e-mail: om_rs@ei.nits.ac.in

S. K. Pandey
e-mail: skpandey@ei.nits.ac.in

R. Singh et al. (eds.), *The Physics of Semiconductor Devices*, Springer Proceedings in Physics 306, https://doi.org/10.1007/978-981-97-1571-8_11

DC sputtered using a circular shadow mask with a working power of 19.5 W. The depositions were accomplished at room temperature with a working pressure of ~ 0.006 mbar with an Ar flow of 30 SCCM. The I-V characteristics of the fabricated W/HfO_2/Cu device were procured by electrical measurements using *Keithley 4200-SCS* semiconductor parameter analyzer. Applying a voltage sweep (0 V → 0.8 V → 0 V → − 0.8 V → 0 V) at TE, bipolar switching is observed at $V_{SET} = 0.5$ V and $V_{RESET} = -0.6$ V, respectively, explained later.

11.3 Modeling and Analysis

This study considers a filament with a uniform cylindrical shape with radius r_f for modeling simplicity. During the low resistive state (LRS), a complete filament with a height of oxide thickness ($H = t_{ox}$) is assumed connecting TE and BE, while V_a of 0.5 V ($\geq V_{SET}$) [4], shown in Fig. 11.1a. Likewise, in the high resistive state (HRS), a 5 nm gap is assumed near to BE with a V_a of 0.1 V ($< V_{SET}$) [1, 4], displayed in Fig. 11.1b. When TE is connected with V_a, current conduction from the TE to BE through the nano-filament (also, trap-assist conduction through a gap in HRS in small amount) induces Joule heat owing to electric field ($E = V_a/t_{ox}$) across the thin oxide. This induced Joule heat generates the local temperature rise inside the filament, surrounding oxide [4, 5]. The local temperature rise is influenced by factors, namely the magnitude of V_a, vacancies density, heating time. Here, the current heating time is estimated on a nanosecond scale (~1 ns) using a transient heat transfer equation [2], as presented in Fig. 11.1c.

The local temperature in the filament is evaluated by solving a 1D steady-state heat equation, $k\frac{d^2T(z)}{dz^2} + E^2\sigma = 0$, with boundary conditions $T(0) = T(t_{ox}) = T_0 = 300$ K. The solution for maximum local temperature (T_{max}) at the filament center ($z = t_{ox}/2$) is evaluated from Eq. (11.1):

$$T_{max} = T_0 + \frac{V_a^2}{8\rho k} \tag{11.1}$$

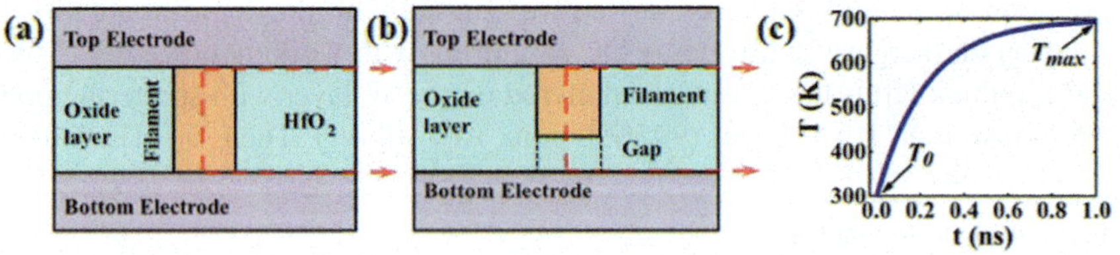

Fig. 11.1 ReRAM device geometry in **a** SET state, **b** RESET state and **c** current heating time of filament to achieve T_{max}. The segments under red-dotted lines are the area understudy for the 2D heat conduction model

where, ρ and k are the filament resistivity (400 $\mu\Omega$ cm) and thermal conductivity (20 W m^{-1} K^{-1}). This local temperature rise stimulates heat conduction along the vertical (from filament center to electrodes) and radial (filament center to surrounding oxide) directions. The heat conduction inside the ReRAM device is calculated by solving steady-state heat equations, such as $Q_z = -kA\frac{dT}{dz}$ (vertical heat conduction) and $Q_r = -kA\frac{dT}{dr}$ (radial heat conduction). For the thermodynamic simulation, the oxide layer (also filament) is divided into $n \times m$ unit blocks, with i (1, 2, 3 … N_x) and j (1, 2, 3 … N_y) as the radial and vertical indices, respectively. At the corner of each unit block, the elemental temperature (T (i, j)) is calculated in MATLAB by solving a 2D steady-state heat equation $\left(\frac{d^2T}{dz^2} + \frac{d^2T}{dr^2} = 0\right)$, as Eq. (11.2).

$$T(i, j) = \{T(i+1, j) + T(i-1, j) + T(i, j+1) + T(i, j-1)\}/4 \tag{11.2}$$

In the LRS, the heat conduction is simulated using Eq. (11.2), where 2D oxide matrix consists of the T_{max} from Eq. (11.1) in the filament, T_0 (300 K) at the sidewalls, and the thermal profile is shown in Fig. 11.2a. The T_{max} of 690 K is observed at the filament center with a parabolic thermal profile. In the HRS, the gap height of 5 nm and local temperature of 400 K is evaluated where the T_{max} appears at the filament tip nearer to BE [6], validated with [7]. The local temperature and thermal profile obtained from this simulation are validated with reported work [5, 7] simulated in COMSOL multiphysics for LRS and HRS, shown in Fig. 11.2b and e, respectively.

In addition, the temperature distribution from filament to surrounding oxide is traced by solving radial heat conduction equation, T $(i) = \{T$ $(i+1) + T$ $(i - 1)\}/2$ with T $(0) = T_{max}$ (at filament) and T $(L/2) = T_0$ (at sidewall) as boundary

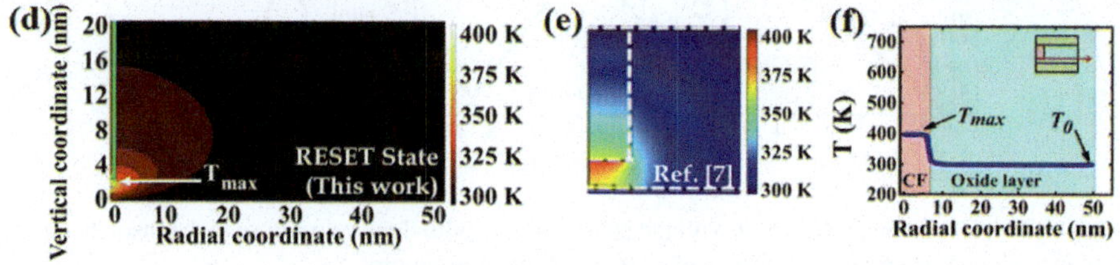

Fig. 11.2 **a** Simulated LRS thermodynamics (MATLAB) is validated with (**b**) reported FEM thermal map [5], **c** LRS radial temperature distribution profile, **d** simulated HRS thermodynamics (MATLAB) is validated **e** reported FEM thermal map [5], and **f** HRS radial temperature distribution profile

conditions. A declination of 52.93% (from $T_{filament} = 690$ K to $T_{junction} = 324.78$ K) and 23.44% (from $T_{filament} = 400$ K to $T_{junction} = 308$ K) in local temperature is observed at the filament-oxide junction in LRS and HRS, respectively, displayed in Fig. 11.2c and f. This declination in local temperature in oxide is due to its lower thermal conductivity (0.5 W Km^{-1}) than filament (22 W Km^{-1}) [5].

Following the above study that traced the local temperature rise and heat conduction for a particular V_a (individually in LRS and HRS), the model is upgraded to trace the local temperature for every point of V_a (total voltage sweep) in time scale. Figure 11.3a demonstrates the measured I-V characteristics from the experimental electrical measurement of W/HfO_2/Cu ReRAM device (Sect. 11.2), where the applied voltages sweep (both positive and negative bias) is shown in Fig. 11.3b in a time scale. Here, the analytical model is based on the formation and dissolution of filament due to defect (vacancies) migration in the influence magnitude and polarity of applied voltage [1–3]. In the LRS (SET transition), the radial growth of filament is calculated using Arrhenius law [8], $d\emptyset/dt = Ae^{-(E_A/k_B T_{max})}$, where, Ø is the filament diameter ($2 \times r_f$). The gap height (from the growth velocity) and the local temperature are calculated [1] using Eq. 11.3–11.6.

$$\frac{dg}{dt} = -\upsilon_0\left[\exp\left(\frac{-qE_{ag}}{kT}\right)\exp\left(\frac{\gamma a_0 qV}{LkT}\right) - \exp\left(\frac{-qE_{ar}}{kT}\right)\exp\left(\frac{-\gamma a_0 qV}{LkT}\right)\right] \tag{11.3}$$

$$g(t + dt) = g(t) + dg \tag{11.4}$$

$$\frac{dT}{dt} + \frac{T - T_0}{\tau_{th}} = \frac{|V \times I|}{C_{th}} \tag{11.5}$$

$$T(t + dt) = T(t) + dT \tag{11.6}$$

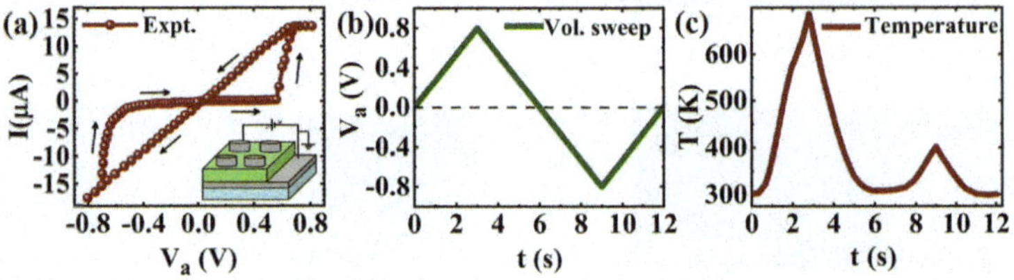

Fig. 11.3 **a** Experimental current–voltage characteristic with device structure in the inset, **b** applied voltage sweep with respect to time and **c** current heating time plot

where, v_0, γ, are the fitting parameters, the material parameters used in this model are taken from a reported HfO_2 based ReRAM [1]. The device parameters and operation variables (like I and V) are taken from experimental electrical measurement (Sect. 11.2).

The estimated temperature versus time curve is represented in Fig. 11.3c, corresponding to respective V_a in Fig. 11.3b. In the positive bias domain, an increment in temperature from 300 to 690 K (in LRS) is observed, and while the polarity of V_a changes, the temperature declines down to ~ 300 K with a decrease in V_a magnitude. In the negative bias condition, the local temperature rises up to 400 K (in HRS) is obtained (in HRS). The obtained temperature versus time trend in Fig. 11.3c is in line with reported works [2, 8] for a complete voltage sweep.

11.4 Conclusion

In this work, a numerical 2D heat conduction model is developed to trace the local temperature rise and heat conduction inside a ReRAM device. The T_{max} of 690 K and 400 K are observed at the filament center and filament tip near BE for LRS and HRS conditions, respectively. The radial heat conduction informed the declination of local temperature from the filament to the device sidewall (T_0), which reduces the possibilities of thermal cross-talk with neighboring cells. Further, a ReRAM device with W/HfO_2/Cu stack structure is fabricated, and a bipolar switching is observed with V_{SET} of 0.5 V and V_{RESET} of − 0.6 V. In addition, an analytical model is adopted to estimate the local temperature rise for the complete voltage sweep, where a maximum temperature of 700 K and 400 K are calculated for positive and negative bias domain, respectively. The observations and findings from this study offer a clear insight into the local temperature rise due to Joule heat and heat conduction across a ReRAM device, which will be helpful to understand the device functionality for better device performance in data storage applications.

Acknowledgements SKP acknowledges funding from Science & Engineering Research Board (File No. ECR/2018/002928) for the magnetron sputtering system.

References

1. P.Y. Chen, S. Yu, Compact modeling of RRAM devices and its applications in 1T1R and 1S1R array design. IEEE Trans. Electron Devices **62**(12), 4022–4028 (2015)
2. J.B. Roldán, G. González-Cordero, R. Picos, E. Miranda, F. Palumbo, M.E. Jiménez-Molinos, D. Maldonado, S.B. Baldoma, M.M.A. Chawa, C. Benito, S.G. Stavrinides, J. Sune, L.O. Chua, On the thermal models for resistive random access memory circuit simulation. Nanomaterials **11**(5), 1261 (2021)
3. D. Panda, P.P. Sahu, T.Y. Tseng, A collective study on modeling and simulation of resistive random access memory. Nanoscale Res. Lett. **13**(1), 1–48 (2018)

4. O.P. Das, S.K. Pandey, in *2020 IEEE International Symposium on Smart Electronic Systems (iSES) (Formerly iNiS)*. Influence of Conducting Filament Dimension on the Performance of ReRAM Device in the SET state (IEEE, 2020), pp. 13–16
5. O.P. Das, S.K. Pandey, Effect of conducting filament radius on local temperature and activation power of ON-state ReRAM device. Semicond. Sci. Technol. **36**(9), 095039 (2021)
6. E. Yalon, I. Riess, D. Ritter, Heat dissipation in resistive switching devices: comparison of thermal simulations and experimental results. IEEE Trans. Electron Devices **61**(4), 1137–1144 (2014)
7. E. Yalon, A.A. Sharma, M. Skowronski, J.A. Bain, D. Ritter, I.V. Karpov, Thermometry of filamentary RRAM devices. IEEE Trans. Electron Devices **62**(9), 2972–2977 (2015)
8. S. Ambrogio, S. Balatti, D.C. Gilmer, D. Ielmini, Analytical modeling of oxide-based bipolar resistive memories and complementary resistive switches. IEEE Trans. Electron Devices **61**(7), 2378–2386 (2014)

Chapter 12
Effective Mobility Extraction of GaN-HEMT Using S-Parameter

Anupama Anand, Rakhi Narang, Dipendra Singh Rawal, Meena Mishra, Manoj Saxena, and Mridula Gupta

12.1 Introduction

The conventional GaN based high electron mobility transistors (HEMT) enables high power, high temperature, high voltage and high frequency operation. The advantages of AlGaN/GaN heterostructures have been achieved because of high critical electric field, a larger conduction band discontinuity between AlGaN and GaN, and the presence of polarization fields which allow a two-dimensional electron gas (2DEG) to be confined at the interface. The Hall measurement technique is used to characterize the 2DEG carrier concentration and mobility. But, the electron relaxation time for two-dimensional carriers in these devices are limited by various scattering mechanisms (related to charged impurities, dislocations, interface roughness, phonon scattering etc.) which affects the carrier mobility. Hence, the effective mobility (μ_{eff}) plays a key role to characterise the carrier transport in high electron mobility transistor. One of the well-known method for effective mobility (μ_{eff}) extraction is the split capacitance-voltage (C-V) technique. Reference [1] proposed the accurate extraction of effective gate length and mobility using gate-to-channel capacitance (C_{gc}) and I_d-V_g. The effective channel length calculation for accurate μ_{eff} extraction, radio-frequency split C-V measurements of short-channel Fin-FET (Field Effect Transistor) was used in [2]. In [3] the problem of C-V characterization for ultra-thin gate oxides is addressed, especially for the extraction of effective mobility by split

A. Anand · M. Gupta (✉)
Department of Electronic Science, University of Delhi, South Campus, New Delhi, India
e-mail: mridula@south.du.ac.in

A. Anand · D. S. Rawal · M. Mishra
Solid State Physics Laboratory, DRDO, New Delhi, India

M. Saxena
Department of Electronics, Deen Dayal Upadhyaya College, University of Delhi, New Delhi, India

R. Narang
Department of Electronics, Sri Venkateswara College, University of Delhi, New Delhi, India

R. Singh et al. (eds.), *The Physics of Semiconductor Devices*, Springer Proceedings in Physics 306, https://doi.org/10.1007/978-981-97-1571-8_12

capacitance technique. For effective mobility (μ_{eff}) extraction in MOSFETs (Metal Oxide Field Effect Transistors), a novel technique is suggested in [4] which uses difference in drain bias correction during I-V and C-V measurements. In [5] a unified extraction procedure is proposed which accurately captures the parasitic series resistance and parasitic gate capacitance components and then, investigated the impact of parasitic components on the extracted effective mobility in $In_{0.7}Ga_{0.3}As$ quantum-well (QW) MOSFET. In [6], a subthreshold swing of 61 mV/decade and effective mobility of 11,900 cm^2/Vs for long-channel InAlAs/InGaAs/InAlAs single-quantum-well MISFETs (Metal Insulator Semiconductor Field Effect Transistor) is reported. For submicron MOSFETs, a novel method is proposed to directly extract the effective channel mobility from measured S-parameters in [7]. Based on the measured high-frequency S-parameters, the extraction of effective mobility (μ_{eff}) of $In_{0.52}Al_{0.48}As/In_{0.7}Ga_{0.3}As/In_{0.52}Al_{0.48}As$ single-quantum-well MISFETs is proposed in [8].

In this work, the extraction of effective mobility (μ_{eff}) of AlGaN/GaN HEMT using the measured S-parameter data is proposed. The external parasitic gate capacitance (C_{gs_ext} and (C_{gd_ext}) cannot be ignored, as it can affect the accuracy of the effective mobility value. Hence, these parasitics are calculated using the measured S-parameter and de-embedded from the intrinsic S-parameters from which the effective mobility is calculated.

12.2 Device Fabrication

The MOCVD AlGaN/GaN HEMT structure grown on SI SiC substrate is used for device fabrication. The device consists of a nucleation layer, GaN buffer layer, AlN spacer layer, AlGaN barrier layer, a thin GaN cap layer and a 100 nm thick passivation layer. Electron mobility of $\sim$1800 cm^2/Vs and two-dimensional electron gas (2DEG) density of $\sim 1.0 \times 10^{13}\,cm^{-2}$ were obtained using Hall-effect Measurement. This is followed by mesa formation for device isolation using BCl_3/Cl_2 based inductively coupled plasma reactive ion etching (ICP-RIE) for dry etching. Then 0.4 μm Ni/Au gate Schottky contacts are defined using e-beam evaporation followed by lift-off. GaN HEMT Devices are then passivated by a thin SiN layer using PECVD process. Surface passivation is used to reduce the dispersion/current collapse/current slump and microwave power degradation of AlGaN/GaN HEMTs. The traps are present at the surface and in deep levels. When a large negative bias is applied, electrons may leak from the gate region and fill the trap states in the un-gated region forming a virtual gate which modulates the depletion region and partially depletes the channel of electrons leading to reduction of overall output current. In order to minimize this effect surface passivation is used. For SiN passivation, Si atoms acts as shallow donors in trap sites which prevents virtual gate formation. Figure 12.1 shows the device structure and Fig. 12.2 shows the SEM image of 2×75 μm AlGaN/GaN HEMT.

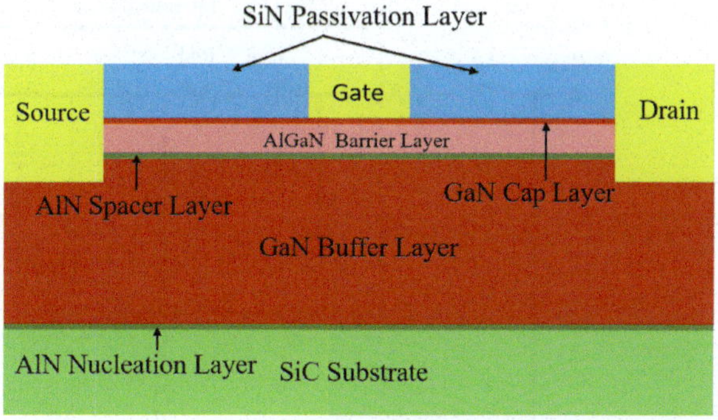

Fig. 12.1 AlGaN/GaN HEMT device structure with $L_g = 0.4\ \mu m$, $L_{sd} = 4.0\ \mu m$ and $W_g = 2 \times 75\ \mu m$

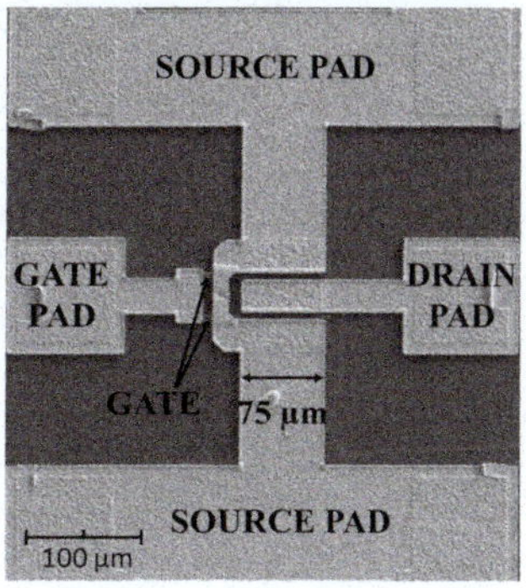

Fig. 12.2 SEM image of AlGaN/GaN HEMT with $L_g = 0.4\ \mu m$, $L_{sd} = 4.0\ \mu m$ and $W_g = 2 \times 75\ \mu m$

12.3 Effective Mobility (μ_{eff}) Calculation

Figure 12.3 shows the 16-element equivalent circuit of the high electron mobility transistor. The gate-source capacitance (C_{gs}) and gate-drain capacitance (C_{gd}) in the equivalent circuit, consist of intrinsic (C_{gs_i} and C_{gd_i}) and extrinsic (C_{gs_ext} and C_{gd_ext}) components, respectively, which is represented in Fig. 12.4. Equations 12.1 and 12.2 represent these capacitance.

$$C_{gs} = C_{gs\_i} + C_{gs\_ext} \tag{12.1}$$

$$C_{gd} = C_{gd\_i} + C_{gd\_ext} \tag{12.2}$$

All the extrinsic parasitic components should be de-embedded to get the accurate intrinsic matrix of a device. Hence the extrinsic gate-source and gate-drain capacitance (C_{gs_ext} and C_{gd_ext}) should be extracted and de-embedded including the other extrinsic components. The extrinsic gate-source and gate-drain capacitance (C_{gs_ext} and C_{gd_ext}) are extracted using S-parameters at pinch off ($V_{gs} < V_T$) and $V_{ds} = 0\,V$ [9]. Equations 12.3–12.5 show the Y-parameter equations for extrinsic capacitance.

$$Im(Y_{11}) = j\omega(C_{pg} + C_{gs\_ext} + C_{gd\_ext}) \tag{12.3}$$

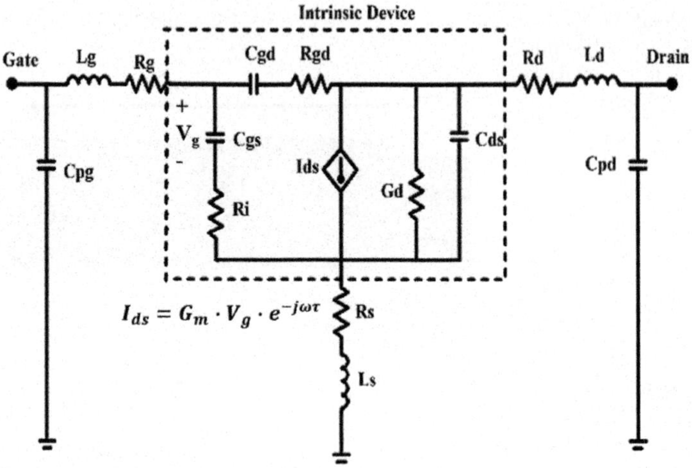

Fig. 12.3 Small-signal equivalent circuit of AlGaN/GaN high electron mobility transistor (HEMT)

Fig. 12.4 Small-signal equivalent circuit with extrinsic gate-to-source and gate-to-drain capacitance (C_{gs_ext} and C_{gd_ext})

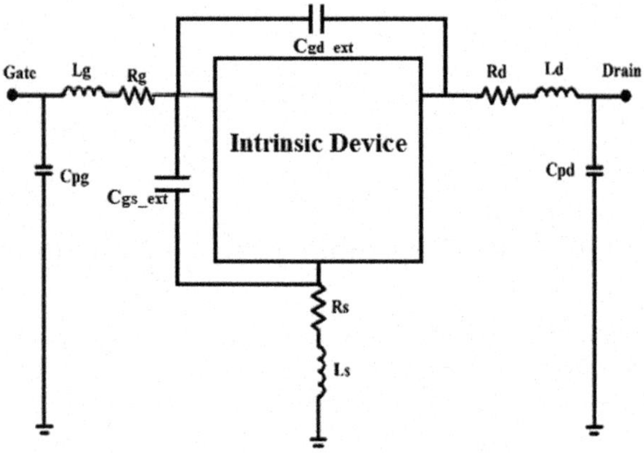

$$Im(Y_{12}) = Im(Y_{21}) = -j\omega(C_{gd\_ext}) \tag{12.4}$$

$$Im(Y_{22}) = j\omega(C_{pd} + C_{gd\_ext}) \tag{12.5}$$

The other extrinsic components are obtained using [10, 11]. The extrinsic components are then de-embedded from the S-parameter data above pinch-off ($V_{gs} > V_T$). Following the de-embedding procedure in [9], the parasitic pad capacitance (C_{pg} and C_{pd}), extrinsic inductance (L_s, L_d and L_g) and resistances (R_s, R_d and R_g) are de-embedded. The intrinsic part is left with the extrinsic gate-source and gate-drain capacitance (C_{gs_ext} and C_{gd_ext}). To de-embed these components, the procedure is further extended as shown in Fig. 12.5. And finally the intrinsic Y-parameter matrix of the device is obtained. In terms of Y-parameters of the device intrinsic part, effective mobility (μ_{eff}) can be expressed as:

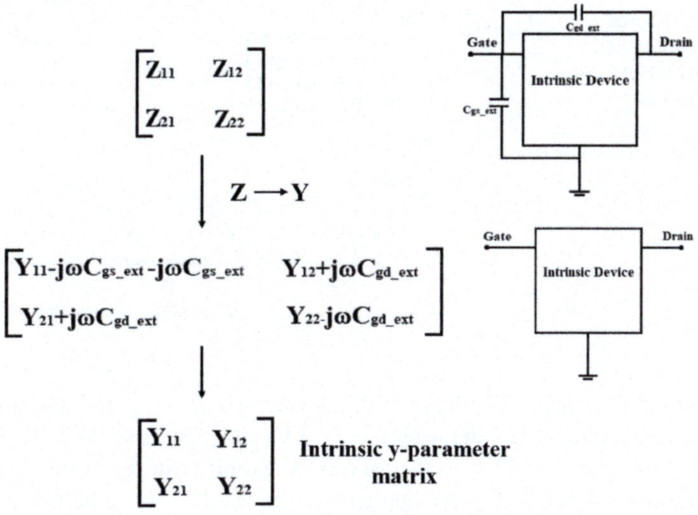

Fig. 12.5 De-embedding of extrinsic gate-to-source and gate-to-drain capacitance (C_{gs_ext} and C_{gd_ext})

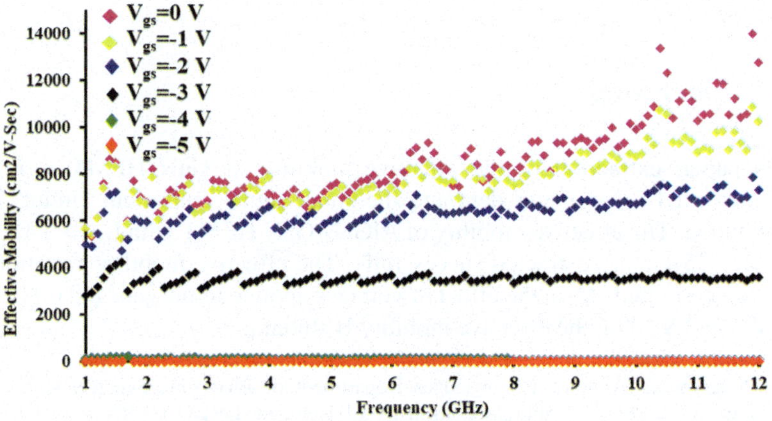

Fig. 12.6 Extracted effective mobility of AlGaN/GaN HEMT with L_g = 0.4 μm and W_g = 2 × 75 μm as a function of measured frequency, for V_{gs} = 0 to − 5 V

$$\mu_{eff} = \frac{L_g}{W_g} \cdot \frac{(Re(Y_{22}\_int)}{\int_{\infty}^{V_{gs}} \frac{Im(Y_{11}\_int)}{\omega} \cdot dV} \tag{12.6}$$

Figure 12.6 shows the extracted effective mobility (μ_{eff}) as a function of frequency for different V_{gs} of AlGaN/GaN HEMT. For calculating the effective mobility shown in Fig. 12.6, the drain-source voltage is kept at 0 V (V_{ds} = 0 V) to avoid the electric field dependency. It can be observed that at V_{gs} = − 5 and − 4 V, the μ_{eff} is very

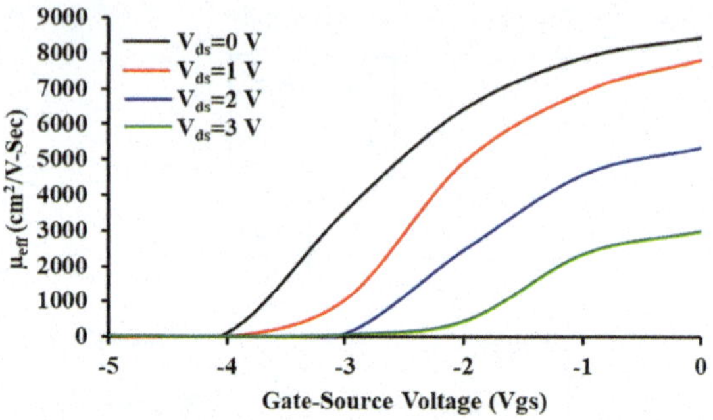

Fig. 12.7 Extracted effective mobility of AlGaN/GaN HEMT with $L_g = 0.4\ \mu m$ and $W_g = 2 \times 75\ \mu m$ as a function of V_{gs}, for $V_{ds} = 0$ to 3 V

low as it is below pinch-off voltage of the device. The μ_{eff} of the device increases drastically above pinch-off voltage ($V_{gs} = -3$ V) and there is slight increase (almost saturated) for higher V_{gs} ($V_{gs} = -1$ and 0 V). Mobility depends on electric field, which is strongly dependent on the applied voltages at the gate and drain terminals of the device. Figure 12.7 shows the extracted effective mobility (μ_{eff}) as a function of V_{gs} for different V_{ds} of AlGaN/GaN HEMT. As drain-source voltage (V_{ds}) increases, the electric field also increases. As a result, mobility gets reduced.

12.4 Conclusion

In this paper, extraction of the effective mobility of GaN HEMT is reported using measured S-parameter data and the small-signal equivalent circuit modeling technique. The effective mobility of AlGaN/GaN HEMT with $L_g = 0.4\ \mu m$ and $W_g = 2 \times 75\ \mu m$ are extracted successfully. The effective mobility variation with frequency is studied. Also, the effect of voltages applied to the gate and drain terminals (V_{gs} and V_{ds}) on the effective mobility is studied.

Acknowledgements Authors acknowledges Department of Electronic Science, University of Delhi; Solid State Physics Laboratory, Ministry of Defence, DRDO CARS project Ref. No.: 1115/TS/SPL/CARS-95/2022; DBT STAR College Laboratory at Deen Dayal Upadhyaya College, University of Delhi; University of Delhi IoE Grant Ref. No.: IoE-DU/MRP/2022/056 and Department of Science and technology project Ref. No.: SPG/2021/003067 for providing necessary tools and financial assistance during the course of this work.

References

1. K. Romanjek, F. Andrieu, T. Ernst, G. Ghibaudo, Improved split C-V method for effective mobility extraction in sub-0.1-μm Si MOSFETs. IEEE Electron Device Lett. **25**(8), 583–585 (2004)

2. J. Ramos, S. Severi, E. Augendre, C. Kerner, T. Chiarella, A. Nackaerts, T. Hoffmann, N. Collaert, M. Jurczak, S. Biesemans, Effective mobility extraction based on a split RF CV method for short-channel FinFETs, in *Solid-State Device Research Conference* (2006), pp. 363–366
3. F. Lime, C. Guiducci, R. Clerc, G. Ghibaudo, C. Leroux, T. Ernst, Characterization of effective mobility by split C(V) technique in N-MOSFETs with ultra-thin gate oxides. Solid-State Electron. **47**(7), 1147–1153 (2003)
4. S.M. Thomas, T.E. Whall, E.H.C. Parker, D.R. Leadley, R.J.P. Lander, G. Vellianitis, J.R. Watling, Improved effective mobility extraction in MOSFETs. Solid-state Electron. **53**(12), 1252–1256 (2009)
5. J.H. Park, D.K. Kim, S.W. Son, S.H. Shin, T.W. Kim, J.H. Lee, D.H. Kim, A new unified mobility extraction technique of In0.7Ga0.3As QW MOSFETs. IEEE Electron Device Lett. **37**(9), 1096–1099 (2016)
6. I.G. Lee, J.B. Jo, D.Y. Yun, C.S. Shin, J.H. Lee, T.W. Kim, D.H. Ko, D.H. Kim, Long-channel InAlAs/InGaAs/InAlAs single-quantum-well MISFETs with subthreshold swing of 61 mV/decade and effective mobility of 11,900 $cm^2\ V^{-1}\ s^{-1}$. J. Appl. Phys. Express **12**(6), 064003 (2019)
7. S. Lee, H.K. Kim, A new technique to extract channel mobility in submicron MOSFETs using inversion charge slope obtained from measured S-parameters. IEEE Trans. Electron Devices **48**(4), 784–788 (2001)
8. T.B. Rho, H.B. Jo, T.W. Kim, D.H. Kim, A unified method to extract the effective mobility in InGaAs metal-insulator-semiconductor field-effect-transistors using scattering-parameters. Solid-State Electron. **162**, 107644 (2019)
9. G. Dambrine, A. Cappy, F. Heliodore, E. Playez, A new method for determining the FET small-signal equivalent circuit. IEEE Trans. Microwave Theory Tech. **36**(7), 1151–1159 (1988)
10. R. Anholt, S. Swirhun, Equivalent-circuit parameter extraction for cold GaAs MESFET's. IEEE Trans. Microwave Theory Tech. **39**(7), 1243–1247 (1991)
11. A. Miras, E. Legros, Very high-frequency small-signal equivalent circuit for short gate-length InP HEMTs. IEEE Trans. Microwave Theory Tech. **45**(7), 1018–1026 (1997)

Chapter 13
Stress Dependent Electrical Characteristics of Flexible a-IGZO TFTs

Taraprasanna Dash and C. K. Maiti

13.1 Introduction

Amorphous oxide semiconductors (AOS) have gained a lot of interest for a variety of electronic device applications, including transparent electrodes, flat panel displays, solar cells, touch screen panels, optoelectronic devices, and thin-film transistors (TFTs) [1]. Since its inception, several structures have been reported for the planar TFTs. Four main structures currently in use are: "top-gate staggered", "bottom-gate staggered" (or "inverted staggered"), "top-gate coplanar" and "bottom-gate coplanar" (or "inverted coplanar"). Staggered structures have the source and drain electrodes on one side of the semiconductor layer, and the gate on the other side, whereas in the "coplanar" structures the three electrodes are on the same side.

There are several works reported the properties of flexible TFT under the influence of mechanical stress [2–4] due to the growing interest in flexible devices. The manufacturability at low temperatures [5] and stability under applied mechanical stress [6] are the two important properties of a-IGZO over other active layer materials. Many research groups have looked into how the electrical properties of a-IGZO TFTs alter under mechanical stress in a variety of situations, including bending type [7, 8], radius [9], and direction [10, 11]. The TFTs in the real circuit can bend in two ways, one parallel to the current direction and the other perpendicular to the current direction, depending on the design and application. However, so far simulation research has been carried out primarily to compare the degradation caused by various bending radii and cycles [12].

T. Dash (✉)
Department of ECE, Siksha 'O' Anusandhan (Deemed to be University), Bhubaneswar 751030, India
e-mail: tpdiitkgp@gmail.com

C. K. Maiti
SouraNiloy, Flat I/2W, 1 Kailash Ghosh Road, Kolkata 700008, India

R. Singh et al. (eds.), *The Physics of Semiconductor Devices*, Springer Proceedings in Physics 306, https://doi.org/10.1007/978-981-97-1571-8_13

The TFTs in the real circuit might be bent in both directions depending on the design. The change in electrical properties of TFTs for applied tensile strain along the parallel axis (along the channel length) and perpendicular axis (bending axis along the width) has been studied using simulation. For the tensile bending axis along the channel length shows a higher threshold voltage change than the device under stress with a bending axis along the width [13]. Strain applied perpendicular to the channel, on the other hand, causes a small change, and TFT properties restore when measured again in the flat state [13]. However, the dependency of the degradation of TFT characteristics in rectangular TFTs on bending direction is not clearly understood [8, 14, 15], and it needs further investigation.

13.2 Process Simulation

To stay connected with an industrial approach, in this work, we have chosen to focus on TFTs with a "bottom-gate staggered" structure. We start with a process simulation to obtain the 3D device. Some of the main fabrication steps of TFTs with the bottom-gate staggered structure are listed below. The manufacturing steps followed in process simulation are:

- initialize glass substrate with a mask
- gate metal deposition of aluminum 0.05 μm
- gate pattern etch aluminum 0.05 μm
- double gate insulator depo with uniform deposition
- deposit Si_3N_4 thick = 0.4 μm conformal
- deposit SiO_2 thick = 0.05 μm conformal
- active a-IGZO = 50 nm
- oxide Etch stopper 50 nm deposit oxide thick = 0.05 μm conformal
- S/D metal deposition
- deposit aluminum thick = 0.2 μm conformal
- etch aluminum thick = 0.2 μm
- passivation
- deposit oxide thick = 0.1 μm conformal
- deposit nitride thick = 0.1 μm conformal
- S/D contacts.

The final device structure is shown in Fig. 13.1 along with its 2D cutline. Figure 13.2 shows the typical transfer curves of the bottom-gate staggered TFTs. The simulation is made under Vds = 0.1 V and Vds = 10.1 V. The threshold voltage and subthreshold swing values were retrieved from the Id-Vg characteristics obtained under low drain bias (0.1 V) and high bias settings (10.1 V). We used the slope of the transconductance curve to calculate the sub-threshold swing value, which was then retrieved using 1/Gm. The device shows a threshold voltage V_{TH} of 5.78439971 V and a subthreshold slope SS of 0.369119824.

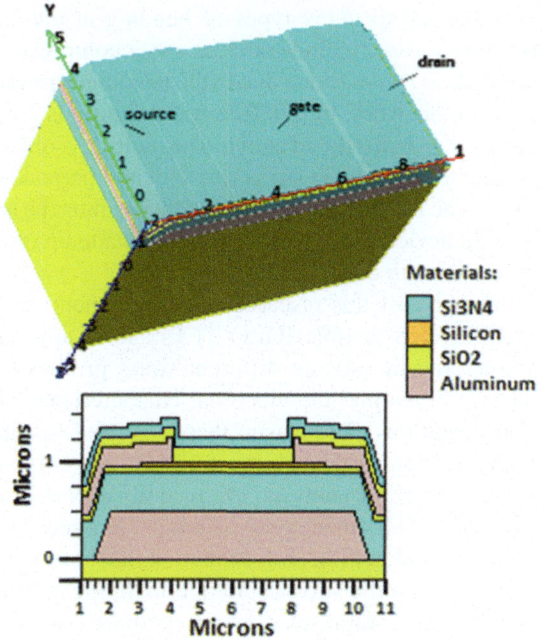

Fig. 13.1 Final 3D TFT device structure was obtained from process simulation along with its 2D cutline

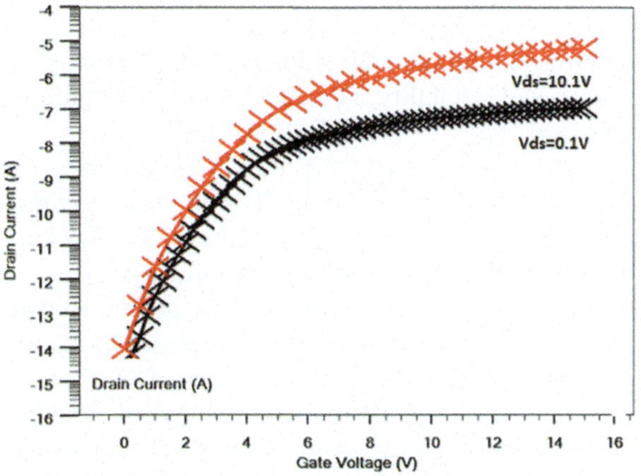

Fig. 13.2 Id-Vg transfer curves of the bottom-gate staggered TFTs. The simulation is made under Vds = 0.1 V and Vds = 10.1 V

13.3 Impact of Bending on TFT Electrical Performance

Thin film transistors are employed in a variety of flexible and stretchable circuit applications. Flexible devices are mainly multi-layered. In stacked multi-layered thin films of different sizes, thicknesses, and materials, intrinsic stress gets generated. For the use in flexible/wearable electronic devices, the impact of bending (mechanical deformation) has recently piqued interest. In a typical application, the device may be subjected to many types of bending at the same time. Commonly considered bending stresses include devices under compression or tension. A large deformation takes place in the device due to the bending, stretching, or twisting of flexible devices [16]. In this work, the TFT device structure is created directly from VictoryProcess simulation as discussed above. The multi-layered top part of the 3D device (Fig. 13.1) is used for the tensile stress (parallel and perpendicular) bending simulation experiment. The nonlinear deformation theory must be used to obtain a stress/strain profile when a device (like a-IGZO TFT) is made up of several stacked layers of different materials with arbitrary geometry.

The goal of this research is to implement TCAD-based approach to (1) design multi-layered flexible IGZO TFTs and to apply large deformations, (2) simulate, extract and investigate different stress profiles across different axes of geometry and (3) determine the electrical characteristics of the device under two deformation conditions. The elastic theory is used to study mechanical stress analyses in microelectronics. The elastic equations are computed using the finite element technique (FEM) for analysis. We recently reported modeling and simulation of stress in a-IGZO TFTs using process parameters such as temperature mismatch among the materials and nonlinear deformation theory to work out the stress distribution in the device. When the stress is not too high, it can be estimated using Young's modulus and Poisson's ratio, as stress is proportional to strain in linear elasticity theory. In this paper, we use the framework that was previously reported [16]. For stress/strain mapping in a-IGZO TFTs, a geometric nonlinear finite element technique was applied. The electrical characteristics of strained a-IGZO thin-film transistors are investigated concerning bending deformations. To test structural integrity, performance, and reliability, as well as predict structural failures, a 3D TCAD framework was used.

In this work, a predictive TCAD framework is used to simulate flexible a-IGZO TFTs with large deformations. The finite element method (FEM) is used to solve the elastic equations for analysis. The linear theory of elasticity can be used to calculate the stress when stress/strain is not too large. This simulation study was carried out to compare the electrical performance under two bending directions viz., bending in parallel and perpendicular to the direction of current flow (see Fig. 13.3). The 2D cut plane was used to analyze the von Mises stress profile inside the devices (see Fig. 13.4). TFTs under applied tensile strain with bending axis perpendicular to the channel length show a higher drain current due to higher stress than the device under stress bending axis parallel to the current direction (see Fig. 13.5).

13.4 Conclusion

We have proposed a novel 3D TCAD simulation technique that may provide design guidelines to investigate the bending axis dependency for a-IGZO TFTs. Using 3D process simulation, a-IGZO device was generated and electrical characterization was performed. Bending experiments were performed for two bending directions viz.,

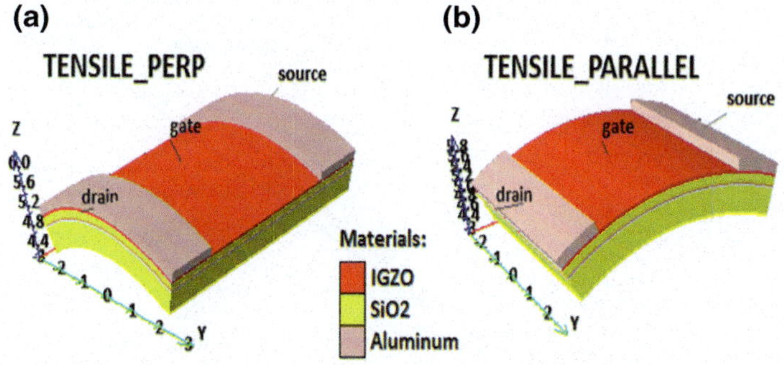

Fig. 13.3 Device structures under tensile strain **a** perpendicular and **b** parallel to the current direction

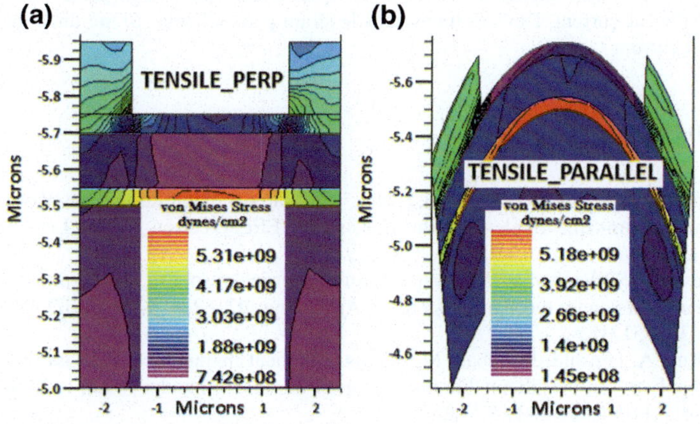

Fig. 13.4 Von Mises stress profile in devices under tensile strain **a** perpendicular and **b** parallel to the current direction

bending in parallel and perpendicular to the direction of current flow. TCAD-based mechanical stress tuning may be applied for the design of flexible a-IGZO TFTs.

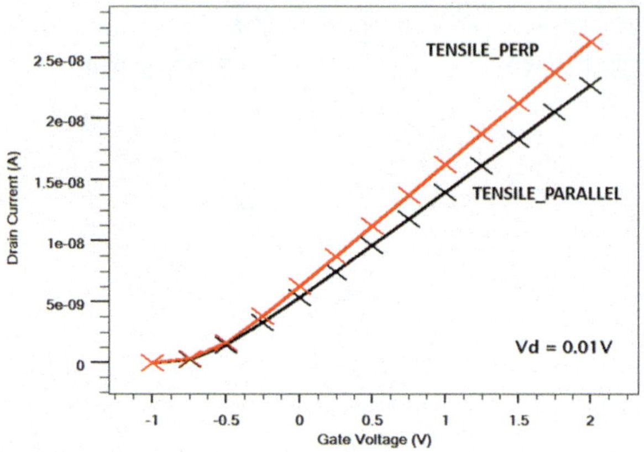

Fig. 13.5 Id-Vg transfer characteristics of TFTs under tensile strain in perpendicular and parallel to the direction of current. Devices under tensile strain perpendicular to the direction of current show higher current

References

1. M. Mativenga, M.H. Choi, J.W. Choi, J. Jang, Transparent flexible circuits based on amorphous-indium–gallium–zinc–oxide thin-film transistors. IEEE Electron Device Lett. **32**, 170–172 (2011). https://doi.org/10.1109/led.2010.2093504
2. H. Gleskova, S. Wagner, W. Soboyejo, Z. Suo, Electrical response of amorphous silicon thin-film transistors under mechanical strain. J. Appl. Phys. **92**, 6224–6229 (2002). https://doi.org/10.1063/1.1513187
3. P.-C. Kuo, A. Jamshidi-Roudbari, M. Hatalis, Electrical characteristics and mechanical limitation of polycrystalline silicon thin film transistor on steel foil under strain, 1–2 (2009). https://doi.org/10.1109/isdrs.2009.5378058
4. N. Munzenrieder, K.H. Cherenack, G. Troster, The effects of mechanical bending and illumination on the performance of flexible IGZO TFTs. IEEE Trans. Electron Devices **58**, 2041–2048 (2011). https://doi.org/10.1109/ted.2011.2143416
5. K. Nomura, H. Ohta, A. Takagi, T. Kamiya, M. Hirano, H. Hosono, Room-temperature fabrication of transparent flexible thin-film transistors using amorphous oxide semiconductors. Nature **432**, 488–492 (2004). https://doi.org/10.1038/nature03090
6. P. Heremans, A.K. Tripathi, A. de Jamblinne de Meux, E.C.P. Smits, B. Hou, G. Pourtois, G.H. Gelinck, Mechanical and electronic properties of thin-film transistors on plastic, and their integration in flexible electronic applications. Adv. Mater. **28**, 4266–4282 (2015). https://doi.org/10.1002/adma.201504360
7. P.-Y. Liao, T.-C. Chang, W.-C. Su, Y.-J. Chen, B.-W. Chen, T.-Y. Hsieh, C.-Y. Yang, Y.-Y. Huang, H.-M. Chang, S.-C. Chiang, Effect of mechanical-strain-induced defect generation on the performance of flexible amorphous In–Ga–Zn–O thin-film transistors. Appl. Phys. Express **9**, 124101 (2016). https://doi.org/10.7567/apex.9.124101
8. M.M. Billah, M.M. Hasan, J. Jang, Effect of tensile and compressive bending stress on electrical performance of flexible a-IGZO TFTs. IEEE Electron Device Lett. **38**, 890–893 (2017). https://doi.org/10.1109/led.2017.2707279
9. A.K. Tripathi, K. Myny, B. Hou, K. Wezenberg, G.H. Gelinck, Electrical characterization of flexible InGaZnO transistors and 8-b transponder chip down to a bending radius of 2 mm. IEEE Trans. Electron Devices **62**, 4063–4068 (2015). https://doi.org/10.1109/ted.2015.2494694

10. M. Mativenga, H. Jun, Y. Choe, J.G. Um, J. Jang, Circular structure for high mechanical bending stability of a-IGZO TFTs. IEEE J. Electron Devices Soc. **5**, 453–457 (2017). https://doi.org/10.1109/jeds.2017.2751651
11. Y.-Y. Choi, J.-S. Park, H.-S. Shin, Analysis of hump effect in tensile-stressed a-IGZO TFT using TCAD simulation (2021).https://doi.org/10.1109/iceic51217.2021.9369787
12. M. Billah, M. Chowdhury, M. Mativenga, J. Um, R. Mruthyunjaya, G. Heiler, T. Tredwell, J. Jang, Analysis of improved performance under negative bias illumination stress of dual gate driving a-IGZO TFT by TCAD simulation. IEEE Electron Device Lett. 1 (2016). https://doi.org/10.1109/led.2016.2557358
13. J.S. Lee, S. Chang, S.-M. Koo, S.Y. Lee, High-performance a-IGZO TFT with ZrO_2 gate dielectric fabricated at room temperature. IEEE Electron Device Lett. **31**, 225–227 (2010). https://doi.org/10.1109/led.2009.2038806
14. S. Lee, D. Jeong, M. Mativenga, J. Jang, Highly robust bendable oxide thin-film transistors on polyimide substrates via mesh and strip patterning of device layers. Adv. Funct. Mater. **27**, 1700437 (2017). https://doi.org/10.1002/adfm.201700437
15. M. Mativenga, D. Geng, B. Kim, J. Jang, Fully transparent and rollable electronics. ACS Appl. Mater. Interfaces **7**, 1578–1585 (2015). https://doi.org/10.1021/am506937s
16. T.P. Dash, E. Mohapatra, C.K. Maiti, Deformation-induced stress/strain mapping and performance evaluation of a-IGZO thin-film transistors for flexible electronic applications. J. Soc. Inf. Disp. 130–142 (2020). https://doi.org/10.1002/jsid.963

Chapter 14
Modeling of Negative Capacitance Field Effect Transistors Based on Different Ferroelectric Materials

Sanna Mairaj and Amandeep Singh

14.1 Introduction

Modern digital circuits consume power, P, as a function of capacitance at the output node, C, frequency of operation, f, and supply voltage, V_{DD}, and is related to the three by the relationship $P = fCV_{DD}^2$. However, the power consumption of these devices has reduced due to the lowering of the capacitances of these devices which is caused by the reduction in the dimensions of the devices according to Dennard scaling rules. According to these scaling rules however, the supply voltage cannot be scaled which is a major drawback in achieving high speed low power devices. The main reason behind this is that the Boltzmann statistics controls the carriers in the drain and source region of the MOSFET and thus restricts the increase in the rate of drain current in regards to voltage at 60 mV/dec [1]. In order to ensure higher gate drive, the threshold voltage V_{th} also needs to go down as there is reduction in supply voltage V_{DD}. But the reduction in of threshold voltage causes an exponential increase in subthreshold leakage. However, if the FET operation principle is considered then issue of the standby power in traditional MOSFET structure cannot be solved. The reason for this phenomenon is that at room temperature, the Subthreshold swing (S) of conventional MOSFET is not scalable below 60 mV/decade. The idea of negative capacitance is to lower Boltzmann limit ($S_B = 60\,\text{mV/dec}$) so as to realize high frequency low power circuits. Salahuddin and Datta [2] proposed that if the traditional insulating material is replaced with ferroelectric materials, the switching energy of a device can be reduced. This led to the phenomenon called "Negative Capacitance".

S. Mairaj (✉) · A. Singh
Department of Electronics and Communication Engineering, NIT Srinagar, Hazratbal, Srinagar, India
e-mail: sanna.mairaj@gmail.com

R. Singh et al. (eds.), *The Physics of Semiconductor Devices*, Springer Proceedings in Physics 306, https://doi.org/10.1007/978-981-97-1571-8_14

Being one of the most propitious looming technology, the negative capacitance field-effect transistor may soon replace the existing CMOS technology. NCFET subsumes a ferroelectric layer that provides the voltage amplification. The layer is subsumed inside the gate stack of the transistor and it functions as negative capacitance. The voltage amplification provided allows to deal with the fundamental constraint of sub-threshold swing of 60 mV at normal room temperature. The negative capacitance phenomenon arises because of dipole interactions that creates positive feedback. The reason for the formation of this positive feedback is the applied electric field being smaller than the one caused by the dipoles [3]. Various modeling approaches for negative capacitance field effect transistors that are based on different device structures have been proposed [4–8]. To optimize the electrical performance of NCFETs, ferroelectric material engineering could be further employed which however is yet to be explored. This work shows a model for negative capacitance-based transistors that considers the impact of different ferroelectric materials on the NCFET characteristics which has been ignored in the earlier works [4–8].

14.2 Device Structure

NCFET device structure is shown in Fig. 14.1a that has two main components the NC dielectric and the Conventional MOSFET. Therefore, it can be modeled as an NC dielectric together with a Conventional MOSFET. NCFET introduces a ferroelectric layer over the existing MOSFET structure. The main advantage is that it provides same I_{DS} at lower V_{DD} values than conventional Mosfet because of the internal voltage amplification provided by negative capacitance of the ferroelectric layer. Thus, a steep slope device can be achieved using negative capacitance that provides a subthreshold slope < 60 mV/decade.

The device structure is based on MFMIS (Metal Ferroelectric Metal Insulator Semiconductor) configuration. MFMIS consists of an internal metal layer between ferroelectric and insulator. The metal layer provides an equipotential surface at the intrinsic gate thereby simplifying the modeling of NCFET. Thus, the structure can be considered as 2 separate entities ferroelectric and the underlying MOSFET due to the spatially constant internal voltage provided by the metal layer. Figure 14.1b shows the equivalent capacitive divider model of NCFET. The NC dielectric is considered to be a capacitor in series with the gate of a Conventional MOSFET.

Equation (14.1) gives the reverse of the slope (S) of I–V characteristics in the subthreshold region:

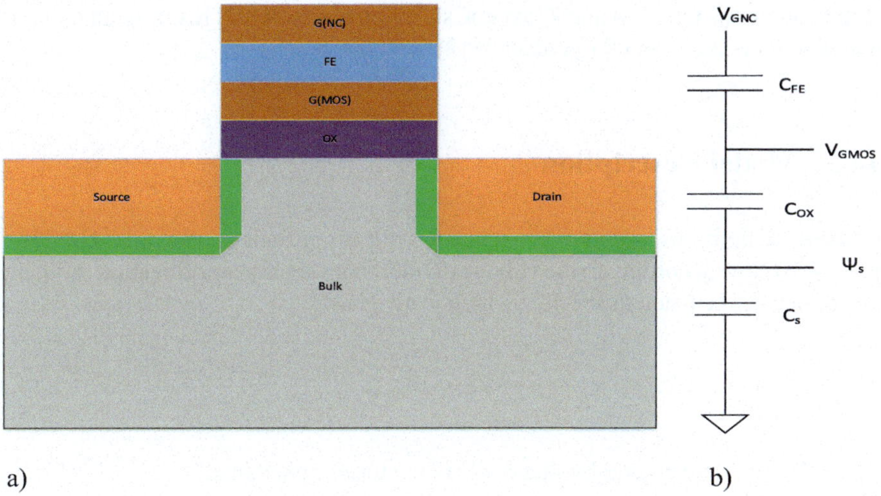

Fig. 14.1 **a** Schematic of NCFET. **b** Equivalent capacitive divider model

$$S = \frac{\partial V_g}{\partial (log_{10} I_D)} = \frac{\partial V_g}{\partial \psi_s} \frac{\partial \psi_s}{\partial log_{10} I_D}$$

$$S \cong \frac{\partial V_g}{\partial \psi_s} \frac{K_B T}{q} \ln 10 \tag{14.1}$$

where V_g, I_D, ψ_s, K_B, T, and q respectively denote gate voltage, drain current surface potential, Boltzmann constant, the temperature and the electron unit charge. The second term in Eq. (14.1) $\frac{K_B T}{q} \ln 10$ has constant value 60 mV/decade at 300 K. Equation 14.1 can also be written as

$$\mathrm{S} = m \times \frac{60\,\mathrm{mV}}{decade} \tag{14.2}$$

$$m = \frac{\partial V_g}{\partial \psi_s} = 1 + \frac{C_s}{C_{ox}} \tag{14.3}$$

where C_S is the semiconductor capacitance and C_{ox} is the oxide capacitance. The term m represents the body factor of the transistor, Ψ_s denotes the conduction process in the channel and n characterizes the change of drain current with surface potential. A sharper subthreshold slope (lower value of S) is required for the swift operation of the transistor. Reducing the body factor (m) below 1 can also help in decreasing the S value. The newly suggested negative-capacitance FET (NCFET) can be used to achieve this [2]. The transistor body factor m could also be made to less than unity,

If the conditions Cox < 0 and |Cox| < |Cs|, are met, which will result in an overall S that would be less than 60 mV/decade [3].

14.3 Model Description

Consider a fictitious negative capacitance with a constant capacitance C_{FE} as in Fig. 14.1b Now according to the voltage divider rule we have amplification in V_{GMOS} due to negative capacitance effect which is given as

$$V_{GMOS} = \frac{C_{GMOS}^{-1}}{C_T^{-1}} V_{GNC} = \frac{V_{GNC}}{1 + \frac{C_{GMOS}}{C_{NC}}} = \frac{V_{GNC}}{m_{MN}} \tag{14.4}$$

where V_{GNC} is the external voltage, C_T is the total capacitance

$$C_T^{-1} = C_S^{-1} + C_{OX}^{-1} + C_{FE}^{-1} \tag{14.5}$$

C_{OX} and C_s are the oxide and semiconductor capacitances respectively and the internal gate capacitance

$$C_{GMOS}^{-1} = C_{ox}^{-1} + C_s^{-1} \tag{14.6}$$

The amplification factor due to negative capacitance effect is given as $C_{GMOS}^{-1}/C_T^{-1} > 1\, or\, m_{MN} < 1$ that will lead to amplification in V_{GMOS}. The S and V_{th} are improved by this m_{MN} factor. The I–V, C–V, P–V_{FE} characteristics are obtained as follows.

14.3.1 *Total Capacitances*

The total capacitance of NCFET will be considered as series combination of MOSFET Capacitance C_{GMOS} and the ferroelectric capacitance C_{FE}.

The gate capacitance C_{GMOS} of a mosfet is dependent on the oxide capacitance C_{OX}, semiconductor capacitance C_s such as the accumulation capacitance C_{acc}, depletion capacitance C_D and the inversion capacitance C_{inv}. C_D in turn depends on the depletion width W_{dep}. W_{dep} can be computed from KVL relation of gate bias V_{GMOS} as in [10]

$$\begin{aligned} V_{GMOS} &= V_{FB} + \psi_s + V_{OX} \\ &= V_{FB} + \frac{qN_A W_{dep}^2}{2\varepsilon_s} + \frac{qN_A W_{dep}}{C_{OX}} \end{aligned} \quad (14.7)$$

where V_{FB} is the flat band potential, ψ_s is the surface potential, V_{OX} is the gate oxide potential across C_{OX}, q is the electron charge, N_A is the doping density, ε_s is the permittivity of semiconductor. From 7 C_D can be calculated as

$$C_D = \frac{dQ_{dep}}{d\psi_s} = \frac{\varepsilon_s}{W_{dep}} \quad (14.8)$$

where Q_{dep} is the semiconductor depletion charge.

The inversion capacitance is calculated by using the equation for inversion charges Q_{inv} [11]

$$Q_{inv} = C_{OX}(mk_BT)\exp\big[\big(V_{GMOSdep} - V_{th}\big)/(mk_BT)\big] \quad (14.9)$$

where $V_{GMOSdep}$ represents gate applied potential in the depletion and inversion region and m $= \left(1 + \frac{C_D}{C_{OX}}\right)$. Therefore, inversion capacitance $C_{inv} = \frac{dQ_{inv}}{dV_{GMOSdep}}$.

Also, the gate capacitance in accumulation is $C_{GMOS} = C_{OX}$. From C_{GMOS} obtained we can compute the gate charge as $Q_{GMOS} = \int C_{GMOS} dV_{GMOS}$ with Q_{GMOS} at V_{FB} equal to zero.

Effect of Parasitic Capacitances

The overlap and fringing capacitances between gate source (C_{OVS}) and gate drain (C_{OVD}) also affect the performance of the conventional MOSFET. Capacitance matching becomes particularly important in NCFET and for that we need to have C_{GMOS} and C_{FE} to be constant and equal. For better matching of capacitances these parasitic capacitances are taken into account in this work. The gate charge and gate capacitance due to these components gets modified as [11].

$$Q_{GMOSov} = Q_{GMOS} + C_{OVS}(V_{GMOS} - V_S) + C_{OVD}(V_{GMOS} - V_D) \quad (14.10)$$

And the modified gate capacitance is given as

$$C_{GMOSov} = C_{GMOS} + C_{OVS} + C_{OVD} \quad (14.11)$$

Current–Voltage (I–V) Characteristics

The I–V characteristics can now be obtained by considering the field dependent mobility model. The drain current defined as [9]

$$I_{DS} = Q_{inv} v_x F_s \tag{14.12}$$

where v_x is the saturation velocity of carriers and Q_{inv} is the inversion charge represented as

$$Q_{inv} = C_{ox}(mk_BT)\ln\left[1 + \exp[(V_{GMOS} - V_{th})/(mk_BT)]\right] \tag{14.13}$$

where m $= \left(1 + \frac{C_{GMOS}}{C_{OX}}\right)$. The field dependent factor F_s can be represented as

$$F_s = \frac{V_{DS}/V_{DSAT}}{\left[1 + \left(\frac{V_{DS}}{V_{DSAT}}\right)^{\beta}\right]^{1/\beta}} \tag{14.14}$$

Here β is a saturation-transition-region fitting parameter. V_{DS} represents drain bias and V_{DSAT} is the drain bias for current saturation due to velocity saturation. $V_{DSAT} = v_x L_{CH}/\mu$ where L_{CH} is gate length and μ is the effective peak mobility.

Negative Capacitance and P–V$_{\text{FE}}$ Characteristics

Ferroelectric materials exhibit negative capacitance [2]. This is evident from the double well energy curve. The dynamics of ferroelectric capacitor can be explained by LK Theory [12, 13].

$$\rho_{FE}\frac{dQ_{GMOS}}{dt} + \nabla Q_{GMOS} U_{FE} = 0 \tag{14.15}$$

The Gibbs free energy of ferroelectric material given as [12]

$$U_{FE} = \alpha_{FE} Q_{GMOS}^2 + \beta_{FE} Q_{GMOS}^4 + \gamma_{FE} Q_{GMOS}^6 - \frac{V_{FE}}{t_{FE}} Q_{GMOS} \tag{14.16}$$

where t_{FE} is the thickness of ferroelectric. The potential $(V_{FE}) - (Q_{GMOS})$ polarization relation of ferroelectric can be expressed from Eqs. 14.15 and 14.16 as

$$\begin{aligned} V_{FE} &= 2\alpha_{FE} t_{FE} Q_{GMOS} + 4\beta_{FE} t_{FE} Q_{GMOS}^3 \\ &\quad + 6\gamma_{FE} t_{FE} Q_{GMOS}^5 + \rho_{FE} t_{FE} \frac{dQ_{GMOS}}{dt} \end{aligned} \tag{14.17}$$

From Eq. 14.17 we can write $C_{FE} - Q_{GMOS}$ relation of ferroelectric material in steady state as:

$$C_{FE} = \frac{1}{2\alpha_{FE}t_{FE} + 12\beta_{FE}t_{FE}Q_{GMOS}^2 + 30\gamma_{FE}t_{FE}Q_{GMOS}^4} \quad (14.18)$$

Ferroelectric Materials

In this work different ferroelectric materials like Hafnium Silicon Oxide (HfSiO), Hafnium Zirconate ($HfZrO_2$) and SBT (Strontium Bismuth Tantalate) are used. The parameters used in this work are listed in Table 14.1. The parameters were chosen based on different studies present in the literature [10, 11, 21]. The coefficients of different materials used are listed in Table 14.2. α, β, γ are the Landau coefficients as seen in Eq. 14.16. These are anisotropy constants, β, γ are temperature independent and alpha is temperature dependent quantity as it further depends on a_0a_0. $\alpha = a_0(T - T_C)$ where a_0a_0 is temperature independent positive quantity, T and T_C are temperature and Curie Temperature respectively. As a result, $\alpha < 0$ below Curie temperature results in a negative curvature of energy landscape of ferroelectric materials around $P = 0$ and hence a double well energy landscape. The temperature dependence of alpha results in temperature dependent behavior of ferroelectrics. These coefficients also specify the first and second order transition. β is positive and negative respectively for second order and first order transition. γ is always positive for stability reasons.

Materials like lead Zirconate Titanate (PZT), Barium Titanate ($BaTiO_3$) have very high dielectric constant that puts a restriction on their scaling limit. The fabrication also will be more difficult because of its compatibility issues with silicon. SBT has certain disadvantages such as having very high dielectric constant and needs scaling down in order to meet the demands of the current CMOS industry standards.

Table 14.1 Parameters used in this work

Symbol	Description	Parameter value	Unit
L_G	Gate length	50	nm
W	Channel Width	1	μm
ε_{OX}	Oxide dielectric constant	3.9	
ε_0	Free space permittivity	8.854×10^{-14}	F/cm
N_{sub}	Substrate doping density	10^{19}	cm^{-3}
V_{th}	Threshold voltage	0.34	V
μ	Carrier mobility	295	cm^2/VS
v_x	Saturation velocity	1.6×10^7	cm/S
β	Saturation transition region fitting parameter	1.8	
EOT	Effective oxide thickness	0.7	nm
t_{FE}	Ferroelectric thickness	11	nm

Table 14.2 Coefficients of different ferroelectric materials used in this work

S. No	Ferroelectric material	α(cm/F)	$\beta(\text{cm}^5/\text{F}/Coul^2)$)	$\gamma(\text{cm}^9/\text{F}/Coul^4)$)	References
1	Hafnium silicon oxide (HfSiO)	-8.7×10^{10}	1.92×10^{20}	0	[15, 16]
2	Hafnium zirconate ($HfZrO_2$)	-4.8×10^{10}	5×10^{19}	0	[19]
3	Strontium bismuth tantalate (SBT)	-3.25×10^9	9.375×10^{18}	0	[17, 18]

Therefore, hafnium-based oxides are preferred as scaling down of Silicon doped hafnium oxide, to current technology nodes can be carried out because of small value of dielectric constant. Also, it has good interface properties with silicon [14].

14.4 Results and Discussion

The Current versus Voltage (I–V), Capacitance versus Voltage (C–V), Polarization versus Voltage (P–V_{FE}) have been computed for three different ferroelectric materials Hafnium Silicon Oxide (HfSiO), Hafnium Zirconate ($HfZrO_2$) and Strontium Bismuth Tantalate (SBT). These are plotted for ferroelectric thickness (t_{FE}) value of 11nm. Figure 14.2 shows the obtained I_D–V_G characteristics. It can be observed that Hafnium Silicon Oxide (HfSiO) shows better performance in terms of ON current in comparison to the Hafnium Zirconate ($HfZrO_2$) and Strontium bismuth tantalate (SBT). Figure 14.3 shows the obtained CV characteristics of NCFET. The total gate capacitance C_{total} is plotted as function of external gate voltage V_{GNC} for three different materials. The peak obtained in the graph corresponds to the Negative Capacitance region. HfSiO shows the highest peak value of 6.2 μF/cm^2 because of the better capacitance matching in comparison to other materials as evident from the graph. The total capacitance decreases with the increase in V_{GNC} and reaches its minimum value and then starts increasing until end of NC region. HfSiO shows a sharper curve in comparison to other materials.

The graphs obtained are in accordance with previously simulated data [20]. Figure 14.4a–c shows the Polarization (P) versus applied bias (V_{FE}) characteristics of NCFET. The S shaped curve confirms the ferroelectricity in the materials.

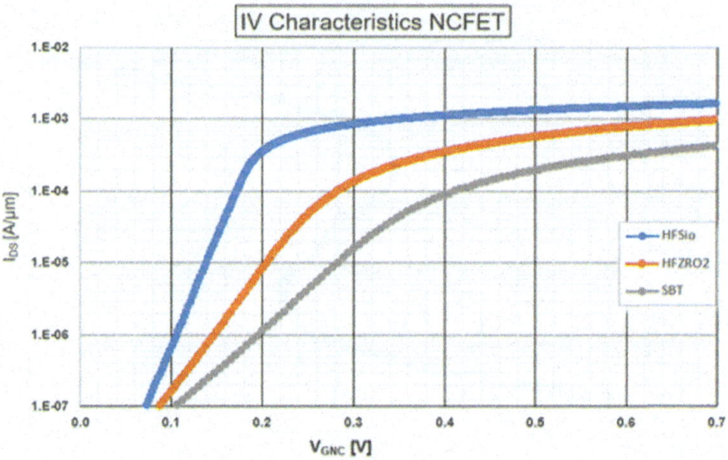

Fig. 14.2 Current Voltage (I–V) characteristics of NCFET obtained for different ferroelectric materials at $t_{FE} = 11$ nm

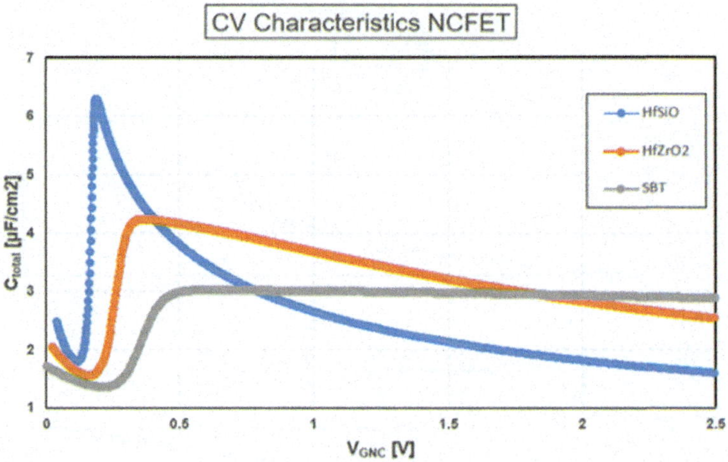

Fig. 14.3 Capacitance–Voltage (CV) characteristics of NCFET for different ferroelectric materials

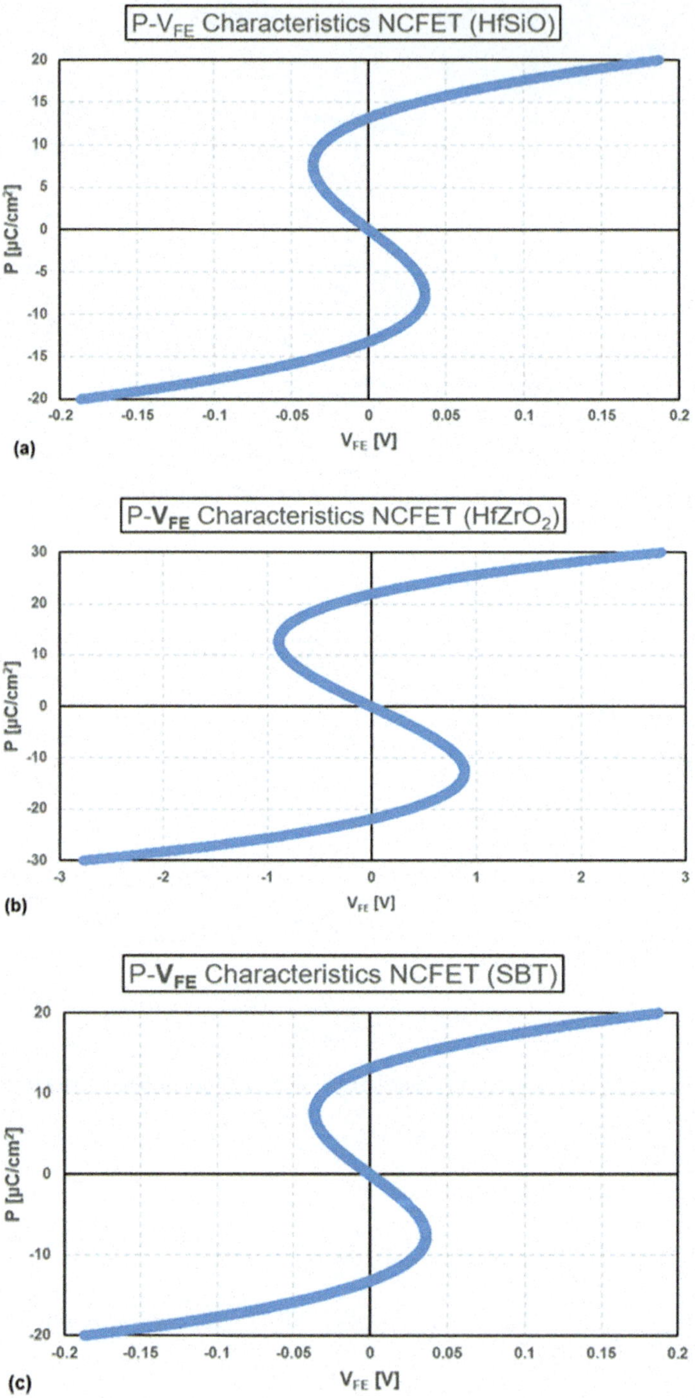

Fig. 14.4 **a–c** Polarization versus Voltage (P–V_{FE}) characteristics obtained

14.5 Conclusion

This work presents a model for NCFET that calculates Current versus Voltage (I–V), Capacitance versus Voltage (C–V), Polarization versus Ferroelectric Voltage (P - V_{fe}) characteristics. The parasitic capacitances are also taken into account for better capacitance matching of NCFET. Ferroelectric materials like Strontium Bismuth Tantalate (SBT), Hafnium Silicon Oxide (HfSiO) and Hafnium Zirconate ($HfZrO_2$) are compared in terms of IV, CV characteristics. The results show superior ON current for Hafnium silicon oxide (HfSiO). The present work considers the impact of different ferroelectric materials on the characteristics of the device which has been ignored in the literature where focus is mainly on impact of variation of ferroelectric material properties like coercive field, remnant polarization. Thus, newer materials need to be explored which will be compatible with the present CMOS technology. Hafnium based oxides can therefore be chosen to be used for a futuristic NCFET design. The comparison of the different ferroelectric material shows which material gives better device performance and can be used for a particular application.

References

1. D. Pushkar, K. Nayak, *Modeling and Simulation of Negative Capacitance MOSFETs*. Dissertation, Indian Institute of Technology Hyderabad (2018)
2. S. Salahuddin, S. Datta, Use of negative capacitance to provide voltage amplification for low power nanoscale devices. Nano Lett. **8**(2)
3. M. Hoffmann, P.V. Ravindran, A.I. Khan, Why do ferroelectrics exhibit negative capacitance? Materials **12**(22), 3743 (2019)
4. G. Pahwa, T. Dutta et al., Analysis and compact modeling of negative capacitance transistor with high on-current and negative output differential resistance-part I. IEEE Trans. Electron Devices **63**(12) (2016)
5. G. Pahwa, T. Dutta et al., Analysis and compact modeling of negative capacitance transistor with high on-current and negative output differential resistance-part II. IEEE Trans. Electron Devices **63**(12), 4986–4992 (2016)
6. G. Pahwa et al., Compact model for ferroelectric negative capacitance transistor with MFIS structure. IEEE Trans. Electron Devices **64**(3), 1366–1374 (2017)
7. A.D. Gaidhane, G. Pahwa, A. Verma, Y.S. Chauhan, Compact modeling of drain current, charges, and capacitances in long-channel gate-all-around negative capacitance MFIS transistor. IEEE Trans. Electron Devices **65**(5), 2024–2032 (2018)
8. A.D. Gaidhane, G. Pahwa, A. Verma, Y. Chauhan, in *2018 4th IEEE International Conference on Emerging Electronics (ICEE)*. Compact Modeling of Drain Current in Double Gate Negative Capacitance MFIS Transistor (IEEE, 2018), pp. 1–5
9. A. Khakifirooz, O.M. Nayfeh, D. Antoniadis, A simple semiempirical short-channel MOSFET current–voltage model continuous across all regions of operation and employing only physical parameters. IEEE Trans. Electron Devices **56**(8), 1674–1680 (2009)
10. R. F. Pierret, *Semiconductor Device Fundamentals* (Addison-Wesley, Reading, MA, 1996)
11. M.A. Wahab, *Design Considerations and Emerging Challenges for Nanotube-, Nanowire-, and Negative Capacitor-Field Effect Transistors*. Dissertation. Purdue University (2016)
12. L.D. Landau, I.M. Khalatnikov,On the anomalous absorption of sound near a second order phase transition point. Dokl. Akad. Nauk SSSR **96**(469) (1954)

13. W. Zhang, K. Bhattacharya, A computational model of ferroelectric domains. Part I: model formulation and domain switching. Acta Mater. **53**(1), 185–198 (2005)
14. H. Mehta, H. Kaur, Modeling and simulation study of novel double gate ferroelectric junctionless (DGFJL) transistor. Superlatt. Microstruct. **97**, 536–547 (2016)
15. T.S. Böscke et al., in *2011 International Electron Devices Meeting*. Ferroelectricity in Hafnium Oxide: CMOS Compatible Ferroelectric Field Effect Transistors (IEEE, 2011)
16. T.S. Böscke et al., Ferroelectricity in hafnium oxide thin films. Appl. Phys. Lett. **99**(10), 102903 (2011)
17. A. Jain, M.A. Alam, Stability constraints define the minimum subthreshold swing of a negative capacitance field-effect transistor. IEEE Trans. Electron Devices **61**(7), 2235–2242 (2014)
18. K. Rabe, C. Ahn, J. Triscone, *Physics of Ferroelectrics: A Modern Per-spective* (Springer-Verlag, New York, NY, USA, 2007)
19. K. Jang et al., Experimental observation and simulation model for transient characteristics of negative-capacitance in ferroelectric $HfZrO_2$ capacitor. IEEE J. Electron Devices Soc. **6**, 346–353 (2018)
20. P. Bidenko et al., Simulation study on the design of sub- kT/q non-hysteretic negative capacitance FET using capacitance matching. IEEE J. Electron Devices Soc. **6**, 910–921 (2018)
21. M.A. Alam, P. Dak, M.A. Wahab, X. Sun, in *2015 IEEE Custom Integrated Circuits Conference (CICC)*. Physics-Based Compact Models for Insulated-Gate Field-Effect Biosensors, Landau-Transistors, and Thin-Film Solar Cells (IEEE, 2015), pp. 1–8

Chapter 15
Comparative Investigation of Single and Double Channel AlGaN/GaN HEMTs for LNAs

Shreyasi Das, Khushwant Sehra, Vandana Kumari, Mridula Gupta, and Manoj Saxena

15.1 Introduction

Gallium Nitride (GaN) has long been an ideal candidate for high power high-frequency applications due to its distinctive features like high breakdown voltage, high electron saturation velocity, and low switching losses [1, 2]. The 2-DEG created is due to the piezoelectric and spontaneous polarization at the interface between AlGaN and GaN layers, making the conventional AlGaN/GaN HEMTs normally-ON devices. AlGaN/GaN heterostructure-based devices have mostly been leading the technology with high current gain and power gain cut-off frequencies with a reduced gate length of about 50–300 nm [3–5]. Further improvement in operational frequency has been experimentally achieved by scaling the channel length below 300 nm, where the short channel effects also dominate [6]. However, the high-power requirement at higher frequency further leads to the search for more compact device architecture that can also provide high current at shorter dimensions [7–10]. Double Channel (DC) has been experimentally demonstrated to deliver high current drive, due to its advantageous architecture consisting of two parallel channels [9, 11–13] which can be a potential advantage in terms of RF circuit designs. DC HEMT also

S. Das
Institute of Radio Physics and Electronics, University of Calcutta, Kolkata, India

K. Sehra · M. Gupta
Department of Electronic Science, University of Delhi, South Campus, New Delhi, India

V. Kumari
Department of Electronics, Maharaja Agrasen College, University of Delhi, New Delhi, India

M. Saxena (✉)
Department of Electronics, Deen Dayal Upadhyaya College, University of Delhi, New Delhi, India
e-mail: msaxena@ddu.du.ac.in

R. Singh et al. (eds.), *The Physics of Semiconductor Devices*, Springer Proceedings in Physics 306, https://doi.org/10.1007/978-981-97-1571-8_15

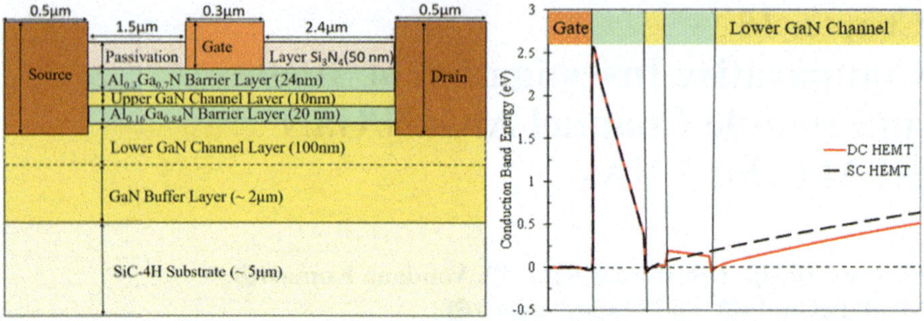

Fig. 15.1 Schematic diagram of double channel (DC) AlGaN/GaN HEMT, with dimensions, used for calibrating the ATLAS TCAD model and conduction band diagram comparison with single channel (SC) AlGaN/GaN HEMT

shows considerably low sheet resistance, besides improved mobility and 2-DEG sheet charge density [14–16].

A low noise amplifier (LNA) with a wide operating frequency range of microwave is extensively used in applications of measurement instrumentations, multimode trans-receivers, 5G millimeter wave communication [17–21]. Two high-performance X-band GaN-based LNAs has been reported by M. Vittori et al. in [17], achieving gain as high as 25 dB with a minimum noise figure (NF_{min}) of 1.74 dB. Cha et al. explored the possibility of using GaN HEMT LNA for survival receiver electronics, given its high power-handling capabilities [18]. A wideband LNA, with operating frequency ranging from 2 to 18 GHz has been discussed by S. Colangeli et al. in [19] using 0.25 μm GaN HEMT. A robust X-band LNA using 0.15 μm AlGaN/GaN HEMT is reported in [21] having gain of 19 dB with a minimum noise figure of 2 dB. LNA for 5G applications has also been reported using GaN HEMTs. A 0.15 μm GaN HEMT-based LNA with an operating range of 26–30 GHz, having a high gain of 20.2 dB with NF_{min} of 2.5 dB is designed in [21].

Owing to the application demands, broadband LNA has drawn much attention from researchers in recent times. Thus, in this paper, a comparative study of dc and RF metrics of DC HEMT with a conventional SC HEMT has been presented. Comparisons are drawn on the basis of linearity and potential application of DC HEMT in LNA circuits.

15.2 Atlas Simulation Results and Discussions

The device currents for SC and DC HEMT were simulated in Silvaco ATLAS [23], and calibrated with previously published work [24]. The details and results of device calibrations are presented in [14]. The same calibrated device has been used in this work to further explore the linearity performance of DC and SC HEMT for LNAs.

15.2.1 Comparison of Linearity Performance of SC HEMTs and DC HEMTs

For modern wireless communication, multichannel transmissions are extensively used. During signal transmission, the closely located neighborhood frequency bands cause intermodulation distortion thus degrading the signal-to-noise ratio (SNR) of the transmitted signal. In this section, the comparison of higher-order transconductance terms, third-order intercept input power (IIP3), VIP3, and third-order intermodulation distortion (IMD3) were investigated at different drain biases. To improve linearity, the change in transconductance with gate voltage should be minimized over the operating range of gate voltage. Flat transconductance over the large range of gate bias results in the lower IMD3 and higher IIP3, thus, improving device linearity. In 2006, Palacios et al. discusses the superior linearity of DC HEMT using a flatter G_m response as compared to SC HEMT [12].

The linearity of a device is ensured when the I_{DS} increases linearly with V_{GS}. To inspect the linearity performance of SC and DC HEMT, similar approach as reported in [25] has been used. It is observed from Fig. 15.2, the maximum transconductance $G_{m,max}$ for the DC HEMT (0.245 S/mm) is more than that of SC HEMT (0.236 S/mm) at the drain bias of 10 V. The higher trans-conductance of DC HEMT is resulted from the significantly enhanced drain current of DC HEMT than that of SC HEMT. In Fig. 15.2, the Gate Voltage Swing (GVS) is defined as 10% drop in transconductance from $G_{m,max}$. A gradual change is also observed in transconductance and GVS for DC HEMT which is significantly higher than that of SC HEMT suggesting improved linearity.

The transfer characteristics of HEMT is represented as a function of V_{GS} as follows,

$$I_{DS}(V_{GS}) = a_0 + a_1 V_{GS} + a_2 V_{GS}^2 + a_3 V_{GS}^3 + a_4 V_{GS}^4 + a_5 V_{GS}^5 + \ldots \quad (15.1)$$

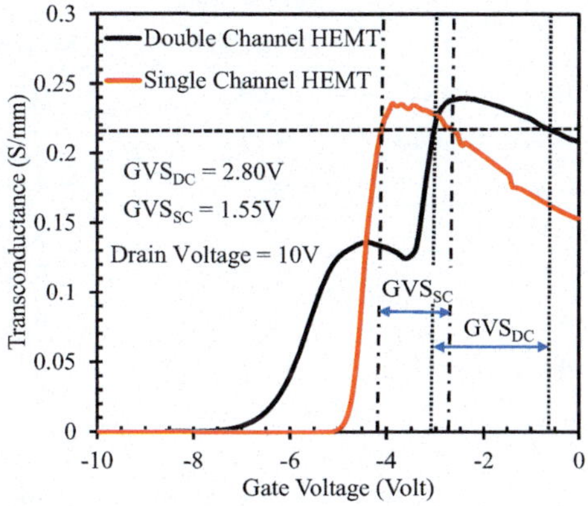

Fig. 15.2 Transconductance curves with GVS range for SC HEMT and DC HEMT

From (15.1), we can calculate the transconductance (G_m) coefficients as dictated by Eq. (15.2),

$$G_m(V_{GS}) = \frac{\partial I_{DS}(V_{GS})}{\partial V_{GS}} = a_1 + 2a_2 V_{GS} + 3a_3 V_{GS}^2 + 4a_4 V_{GS}^3 + \cdots \qquad (15.2)$$

where $a_1, a_2, a_3 \ldots$ are the coefficients of the $G_m - V_{GS}$ curve.

A polynomial based curve fitting for transconductance ($G_m - V_{GS}$ curve) is performed at 5 and 10 V drain biases, in accordance to [25]. We use these coefficients of Eq. (15.2) to compare the linearity and IMD3 performance of the SC and DC HEMT devices.

For better linearity of the device, a_1, as given in Eq. (15.1), should be larger and other higher order coefficients should be smaller. The coefficients for both the devices are extracted from the polynomial fitted to the simulated data of transconductance in MATLAB [26], at 5 and 10 V of drain voltage, and the data is presented in Table 15.1. It is observed from Table 15.1 that the DC HEMT has a larger a_1 value and smaller values of all the ratios of odd terms, thus indicating better linearity as compared to SC HEMT. It can also be inferred from Table 15.1 (by comparing them at 5 and 10 V drain bias individually for both the devices) that, the linearity of DC HEMT has increased considerably at higher drain bias. As for the SC HEMT, though the value of a_1 has increased at $V_{DS} = 10$ V, but the change in odd terms ratio is marginal. Thus, not much change in linearity is attained for SC HEMT with an increase in drain bias.

Figure 15.3a presents the second order derivative of transconductance, which is calculated as $G''_m = \frac{\partial^2 G_m(V_{GS})}{\partial {V_{GS}}^2}$, , compared for DC HEMT with SC HEMT. It is further used to express VIP3 (as given by Eq. 15.3) and is plotted in Fig. 15.3b. VIP3 is the

Table 15.1 The comparison of first few coefficients of DC characteristics for single channel HEMT and double channel HEMT, at drain bias 5 and 10 V are as follows

	Drain voltage 5 V		Drain voltage 10 V	
	DC HEMT	SC HEMT	DC HEMT	SC HEMT
I_{DSS} (A/mm)	1.09	0.927	1.08	0.895
$G_{m,max}$ (S/mm)	0.269	0.253	0.245	0.238
V_{Th} (V)	– 6.5	– 4.5	– 7.25	– 4.8
GVS(V)	1.75	2.4	2.8	1.55
a_1	0.16605	0.10902	0.19548	0.15174
a_2	– 0.15774	0.021825	– 0.06119	– 0.05531
a_3	– 0.58644	0.4159167	– 0.30317	– 0.33352
a_4	– 1.19019	1.042305	– 0.69274	– 1.0099
a_5	– 1.333142	1.441196	– 0.82704	– 1.66893
a_3/a_1	– 3.53170732	3.8150492	– 1.55088	– 2.19797
a_5/a_1	– 8.02855766	13.219556	– 4.23084	– 10.9986

extrapolated gate voltage amplitude, at which the third harmonic becomes equal to the fundamental tone in the drain current of the device [27].

$$VIP3 = \sqrt{24\frac{G_m}{G''_m}} \tag{15.3}$$

The third order intercept input power, IIP3 can be calculated from Eq. 15.4 [27]

$$IIP3 = \frac{2G_m}{3G''_m.R_L} \tag{15.4}$$

In Eq. (15.4), R_L is the resistive load impedance. For Radio Frequency Integrated Circuit (RFIC), R_L is considered as 50 Ω [28]. For better linearity, VIP3 and the third order intercept input power (IIP3) should be high which is directly proportional to G_m and inversely proportional to G''_m. Thus, lower the G''_m, higher are the VIP3 and IIP3. From the above formulations and as observed from Fig. 15.3b, c, higher VIP3 and IIP3 metrics are recorded for DC HEMT, as compared to that of SC HEMT.

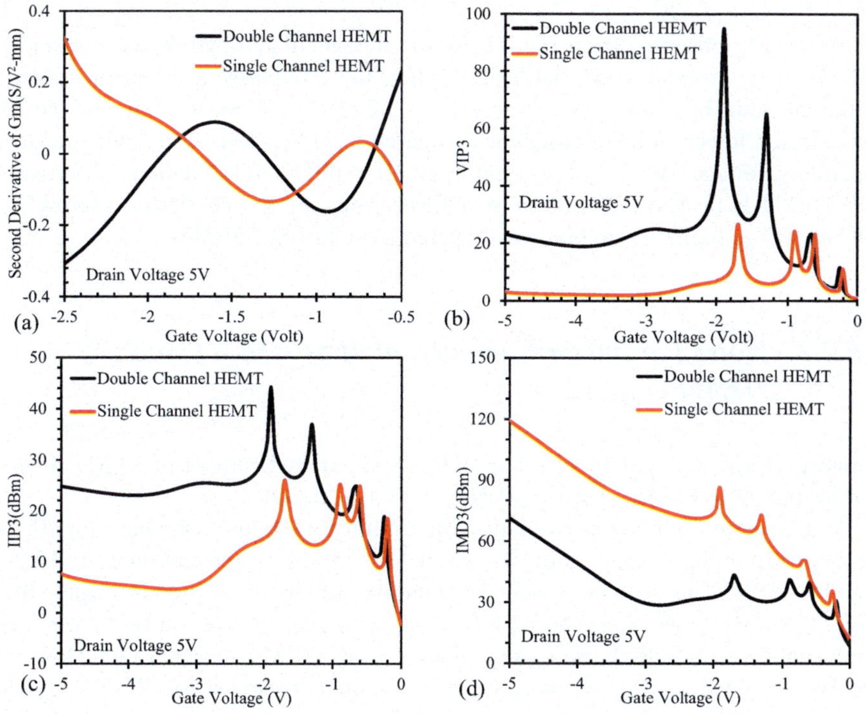

Fig. 15.3 Trends of different linearity parameters for SC HEMT and DC HEMT: **a** second order derivative of transconductance, **b** VIP3, **c** IIP3, and **d** IMD3; all measured at 5 V drain bias for a range of gate voltage

Table 15.2 The comparison of IMD3 values for SC HEMT and DC HEMT, at drain bias 5 and 10 V, and gate voltage corresponding 50% of I_{DSS}

	Drain voltage 5 V		Drain voltage 10 V	
	DC	SC	DC	SC
50% I_{DSS} (mA/mm)	543	463	538	448
Corresponding V_{GS} (V)	– 2.3	– 2.3	– 2.45	– 2.5
G_m (mS/mm)	268	243	270	207
G''_m (S/V^2-mm)	– 0.225	0.181	– 0.627	– 1.98
IMD3 (dB)	6.532	40.17	10.1	47.09

Therefore, better linearity performance can be achieved using DC HEMT structure. Being directly proportional to the Intermodulation Distortion (IMD3), a lower G''_m also results in a lower IMD3, as shown in Fig. 15.3d, which calculated according to the Eq. 15.5 [27].

$$IMD3 = \left(\frac{9}{2}G''_m.(VIP3)^3\right)^2 R_L \tag{15.5}$$

As clearly observed from Fig. 15.3d that the calculated IMD3 is lower for DC HEMT as compared to SC HEMT. Further, the comparison has been drawn at different drain bias voltages in Table 15.2. Table 15.2 shows that the DC HEMT has a much lower value of IMD3 as compared to SC HEMT over a wide range of operating voltage. At a drain bias supply of 10 V, the intermodulation distortion for DC HEMT is far less than that of SC HEMT. Thus, more linearity is ensured from DC HEMT at higher drain bias as compared to that of SC HEMT.

15.2.2 Noise Parameter Comparison of SC HEMT and DC HEMT

Noise performances parameters is used to adjudged the behavior of HEMT devices for designing low noise amplifier (LNA) circuits [29, 30].

In this section, a comparative study of minimum noise figures (shown in Fig. 15.4a) and corresponding current gain (shown in Fig. 15.4b) of conventional SC HEMT and DC HEMT has been presented. The method used in this section is adapted from [31]. For the voltage range – 3.2 to 0 V, A significantly lower noise figure could be visualized from DC HEMT as compared to SC HEMT in Fig. 15.4a. However, below – 3.2 V, the recorded noise figure is significantly lower for SC HEMT. This is because the observed threshold voltage of SC HEMT is ~ – 4 V and ~ – 6 V for DC HEMT. From – 3.2 to – 4 V, kink in the drain current of DC HEMT further translate the kink in the noise figure as seen from the Fig. 15.4a. Thus, as we traverse from – 8 to 0 V, a peak in the noise parameter could be observed when the upper

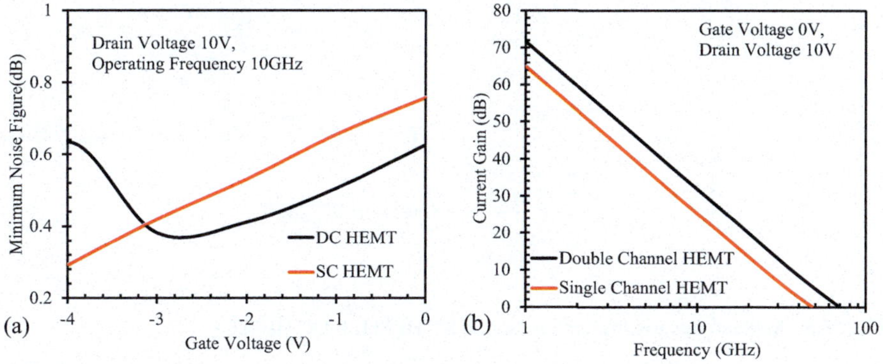

Fig. 15.4 **a** Minimum noise figure at 10 GHz operating frequency, and **b** short circuit current gain at $V_{gs} = 0$ V, for SC HEMT and DC HEMT, $V_{ds} = 10$ V for all cases

channel starts conducting. This peak corresponds to the kink in the transconductance curve of the device, i.e., near the threshold voltage of the upper channel, which is −4 V. The increase in minimum noise figure is proportional to the piezoelectric and spontaneous polarization density, as revealed in analysis. The short-circuit current gain $|h_{21}|^2$ is plotted against the frequency, as shown in Fig. 15.4b reveals higher cut-off frequency for double channel HEMT (69 GHz) as compared to single channel HEMT (48 GHz). Also, the current gain for DC HEMT is considerably higher as compared to SC HEMT, implying more output power.

15.3 ADS Simulation Method, Results, and Discussion

In this section, the simulation models of both the HEMTs have been calibrated in Keysight ADS software [32] for circuit-based RF analysis. The simulation and geometrical parameters generated from the Atlas simulation were utilized to develop the ADS simulation model. The ASM HEMT model [31], ADS library [33], was calibrated against the TCAD data (both DC and RF) to capture the behavior of Single-Channel and Double Channel HEMTs.

The circuits used for calibration of SC HEMT and DC HEMT are shown in Fig. 15.5a, b respectively. For simulating the equivalent of DC HEMT, two ASM HEMTs are considered in parallel, with their source, drain and gate terminals shorted, to model the two simultaneously conducting parallel channels in the device (as shown in Fig. 15.5b). The mobility degradation components, parasitic capacitances, source and drain contact resistance, and thermal resistance and capacitance were also tuned to approximate the ADS model, to be in better coherence with the ATLAS TCAD model in terms of frequency performance as well.

The calibrated DC and S-Parameters of the resulting models are shown in Figs. 15.6a, b and Fig. 15.7a, b, respectively. The presented ADS-simulated drain

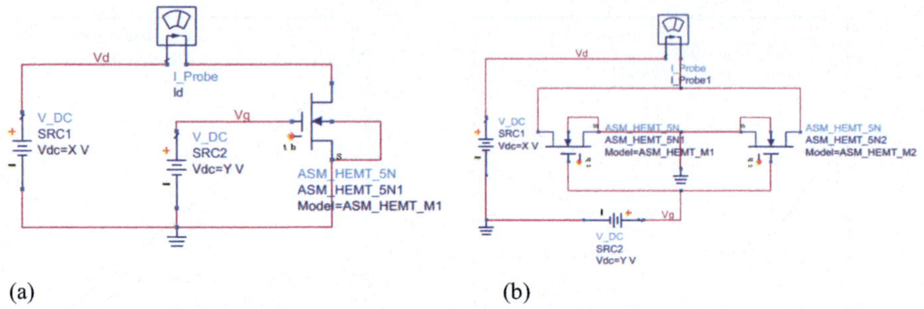

Fig. 15.5 ADS circuits used for calibration: **a** SC HEMT, **b** DC HEMT

current (Fig. 15.6a), transfer characteristics in linear scale (Fig. 15.6b), and S-parameters of SC and DC HEMT (Fig. 15.7a, b) are in close agreement with the extracted results from ATLAS TCAD. The drain current corresponding to 0 V gate bias is much higher for DC HEMT compared to that of SC HEMT as shown in Fig. 15.6a. For the DC HEMT, the lower channel governs the transfer characteristic for lower values of gate voltage, starting from the threshold voltage (V_{Th}) till the kink (as shown in Fig. 15.6b). The kink can be considered the threshold voltage corresponding to the upper channel. After the kink (towards less negative), device current is controlled primarily by the upper channel. Due to the combined effects of both the channels, the device current increases sharply in comparison to SC HEMTs. Due to the contribution of this second channel, a prominent kink is observed in the transfer characteristic and correspondingly in the transconductance curve of double channel HEMT (as shown in Figs. 15.2 and 15.6b).

The reflection coefficient (S11) and transmission coefficient (S21) for SC and DC HEMT, as plotted in Fig. 15.8a, b respectively show a higher output gain and

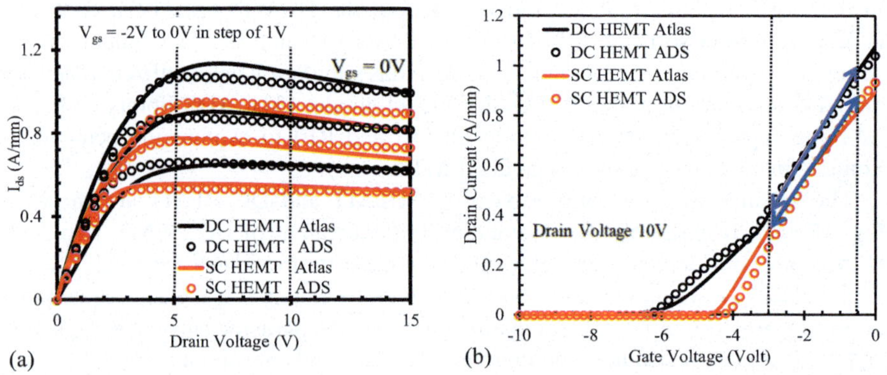

Fig. 15.6 Validation of the ADS simulation model Atlas simulated data **a** output characteristic and **b** transfer characteristic of SC HEMT and DC HEMT having the same specifications as in the schematic diagram (Fig. 15.1) and [14]

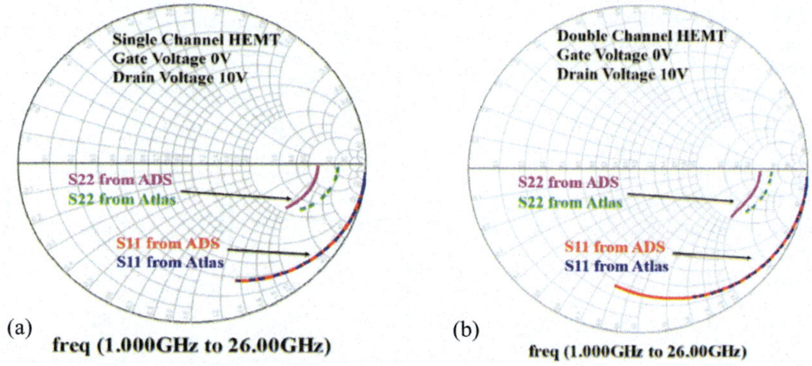

Fig. 15.7 Calibration of frequency performance of **a** SC HEMT and **b** DC HEMT, with the ADS model in terms of reflection coefficients at the input and output ports at $V_{ds} = 10$ V and $V_{gs} = 0$ V

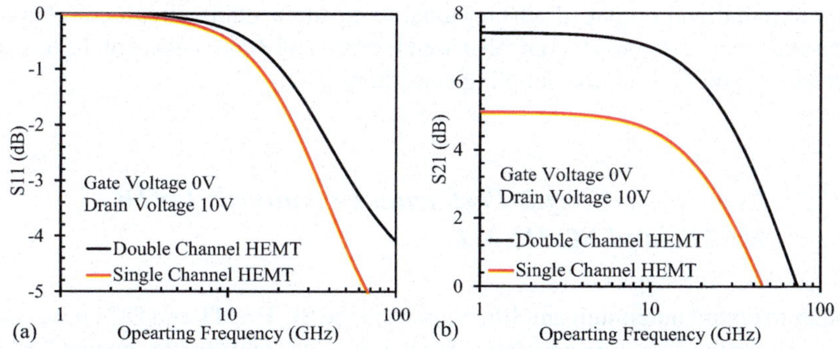

Fig. 15.8 Frequency parameter comparison of SC and DC HEMT in terms of **a** reflection coefficient (S11) and **b** transmission coefficient (S21) at $V_{ds} = 10$ V and $V_{gs} = 0$ V

cut-off frequency for DC HEMT device as compared to that of SC HEMT device. This implies, the LNA designed using DC HEMT should have a wider operating frequency compared to the LNA designed using SC HEMT.

15.3.1 Low Noise Amplifier Circuit in ADS

In this section, a detailed comparative study of SC and DC HEMT-based LNA design has been presented for operation over a X-band of microwave frequency. The basic circuit design was adapted from [19] and necessary changes were made in the circuit, as shown in Fig. 15.9. The component values used for analyzing the circuit is provided in the caption of Fig. 15.9. For better amplification, the operating gate voltage should be close to the threshold voltage such that there is a significant change in drain current

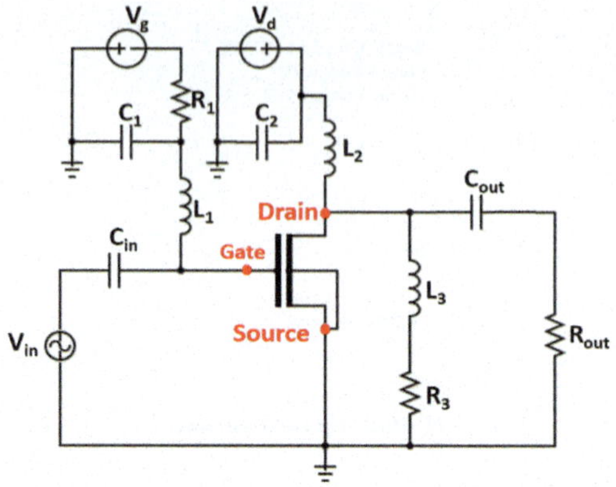

Fig. 15.9 Circuit design for first stage LNA with HEMT used in ADS for implementation, where $C_{in} = C_{out} = 1$ pF, $C_1 = C_2 = 0.1$ pF, $L_1 = L_2 = 0.01$ nH, $L_3 = 2.85$ μH, $R_1 = R_2 = 800\ \Omega$, $R_{out} = 1\ k\Omega$, $V_{in} = 0.2$ V, $V_d = 3$ V and V_g is tuned close to the V_{Th} of the respective HEMTs

with the applied gate voltage. Parallel resistances were connected across the input (gate) terminal and output (drain) terminal to ensure a stable circuit configuration. Bypass capacitors (C_1 and C_2) are also used for gate and drain biasing of the transistor whereas C_{in} and C_{out} are the coupling capacitors.

15.3.2 Low Noise Circuit Performance Analysis for DC HEMT and SC HEMT

In order to obtain maximum amplification from the SC HEMT and DC HEMT based LNA circuits, the single channel HEMT is biased − 4 V and double channel HEMT is biased − 6 V of gate voltage (i.e. near-threshold voltage). The input voltage applied at the gate terminal is an AC signal of amplitude 0.2 V. The frequency of the input voltage was initially set as 10 GHz.

From Fig. 15.10a (first stage LNA output), it is observed that LNA using SC HEMT has better gain as compared to LNA using DC HEMT at the operating frequency of 10 GHz. Similar conclusions can be inferred from Table 15.3, which lists the range of operating frequencies and corresponding gain and minimum noise figures. The phase shift obtained between the output and the input (Fig. 15.10a) is dependent on the considered gate biasing point for the LNA circuit in Fig. 15.9. It is observed that for the X-band of Microwave, i.e., the operating frequency of 8–12 GHz, the LNA using SC HEMT has a comparable gain with DC HEMT (slightly lower in DC HEMT) based LNA. However, the minimum noise figure of DC-based LNA is much lower than its SC counterpart for the entire frequency range as visible from Fig. 15.10b. This gives DC HEMT based LNA a major advantage in RF-based circuits where suppression of noise power plays an essential role while signal transmission. Also, it is evident in Fig. 15.10c that the gain for SC-HEMT based LNA drops rapidly with increase in frequency, which is over 1 dB from 8 to 12 GHz. The gain obtained in lower frequency components is much higher as compared to higher frequency components. In case of wideband applications or multimode communication over the X-band, this will result in non-uniform amplification of different frequency components. However, for the LNA circuit using DC HEMT, marginal reduction (0.06 dB change) in the gain has been observed over the X-band range i.e. 8–12 GHz. This makes DC HEMT-based LNA a more suitable candidate for wideband applications. The gain corresponding to frequencies are listed in Table 15.3 as well.

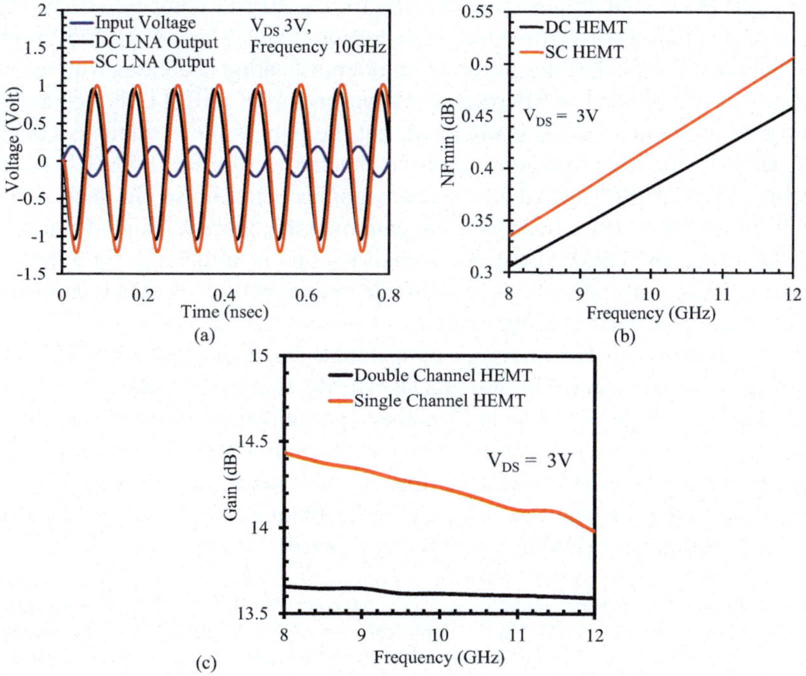

Fig. 15.10 **a** First stage LNA output from SC and DC HEMTs for 10 GHz applied input signal, **b** Minimum noise figure of LNA, and **c** Gain of LNA using a single channel and a double channel HEMT

Table 15.3 Performance comparison of SC and DC HEMT-based LNA circuits at drain voltage 3 V and gate biased near threshold of respective HEMT, for X-band frequencies (8–12 GHz)

	DC HEMT	SC HEMT
Gain (dB) at 8 GHz	13.65	14.98
Gain (dB) at 12 GHz	13.59	13.97
Minimum noise figure (NF_{min}) (dB)	< 0.459	< 0.507

15.4 Conclusion

Results presented in this paper show that Double Channel HEMT exhibits a better performance in terms of current drive, linearity performance, minimum Noise Figure (NF), and current gain, making it a suitable candidate for RF-based circuit applications for Low Noise Amplification (LNA). Significantly higher VIP3 and IIP3, and notably lower NF could be visualized from DC HEMT as compared to SC HEMT. A lower value of G_m'' and higher value of VIP3 for DC HEMT compared to SC HEMT further results in lower intermodulation distortion (IMD3) for DC HEMT, implying more linearity. For SC HEMT, the amount of amplification decreases with operating frequency in the X-band of Microwave. However, for DC HEMT, though a slightly lower gain of about 13.6 dB is obtained, but the gain is more stable as compared to SC HEMT. The minimum noise figure is much lower for DC HEMT based LNA compared to SC HEMT based LNA. This signifies better noise elimination by DC HEMT based LNA. Thus, due to steady gain over the entire X-band of microwave, the LNA using DC HEMT have an extensive scope of utilization for applications of measurement instrumentations, multimode trans-receivers, RADAR applications, and terrestrial and space communications.

From the transfer characteristics of a double channel (DC) GaN HEMT, we observe that the gate control in case of a DC HEMT is affected as the second channel is depleted around $V_{GS} = -6$ V. To meet the requirements for future RF applications, the scaling of the device geometry can impact the ability of the Gate in depleting the second channel. Therefore, in order to improve the overall confinement of the 2DEG channel and to improve the gate action, other techniques such as the back-barrier or the Dual Material Gate (DMG) [34, 35] can be used.

Acknowledgements Authors would like to acknowledge DRDO—SSPL CARS Project No.: 1115/TS/SPL/CARS-95/2022; DU IoE Grant Ref. No.: IoE-DU/MRP/2022/056; and DST—SERB—Power Grant Ref. No.: SPG/2021/00306 for providing financial assistance. One of the authors (Shreyasi Das) would like to acknowledge "Science Academies' IASc—INSA—NASI Summer Research Fellow working under SRFP 2020 vide registration number ENGS1591, Star College Laboratory at Deen Dayal Upadhyaya College, the University of Delhi funded by Department of Biotechnology, Government of India and Semiconductor Device Research Laboratory at Department of Electronic Science, University of Delhi South Campus.

References

1. U.K. Mishra, L. Shen, T.E. Kazior, Y.F. Wu, GaN-based RF power devices and amplifiers. Proc. IEEE **96**(2), 287–305 (2008). https://doi.org/10.1109/JPROC.2007.911060
2. R.S. Pengelly, S.M. Wood, J.W. Milligan, S.T. Sheppard, W.L. Pribble, A review of GaN on SiC high electron-mobility power transistors and MMICs. IEEE Trans. Microw. Theory Tech. **60**(6), 1764–1783 (2021). https://doi.org/10.1109/TMTT.2012.2187535
3. H. Collins, E. Akso, N. Hatai, C.J. Clymore, C. Wurm, R. Hamway, M. Guidry, S. Keller, U. K. Mishra, N-polar GaN MISHEMT with bias-insensitive linearity at 30 GHz. IEEE Microw. Wirel. Technol. Lett. **34**(3), 287–290 (2024). https://doi.org/10.1109/LMWT.2024.3352612

4. P.H. Lee, Y.C. Lin, H.T. Hsu, C.H. Yu, Y.F. Tsao, P. Su, E.Y. Chang, Noise performance investigation of AlGaN/GaN HEMT with tall gate stem for millimeter-wave LNA application. IEEE J. Electron Dev. Soc. **11**, 744–751 (2023). https://doi.org/10.1109/JEDS.2023.3337780
5. M. Micovic, P. Hashimoto, M. Hu, I. Milosavljevic, J. Duvall, Willadsen, W. Wong, A. Conway, A. Kurdoghlian, P.W. Deelman, A. Schmitz, M. Delaney, GaN double heterojunction field effect transistor for microwave and millimeterwave power applications, IEDM technical digest. IEEE Int. Electron Dev. Meet., 807–810 (2004). https://doi.org/10.1109/IEDM.2004.1419298
6. T. Palacios, A. Chakraborty, S. Heikman, S. Keller, S. DenBaars, U. Mishra, AlGaN/GaN high electron mobility transistors with InGaN back-barriers. IEEE Electron Dev. Lett. **27**(1), 13–15 (2006). https://doi.org/10.1109/LED.2005.860882
7. K.H. Hamza, D. Nirmal, A.S.A. Fletcher, L. Arivazhagan, J. Ajayan, R. Natarajan, Highly scaled graded channel GaN HEMT with peak drain current of 2.48 A/mm. AEU Int. J. Electron. Commun. **136**(153774) (2021). https://doi.org/10.1016/j.aeue.2021.153774
8. S. Das, V. Kumari, M. Gupta, M. Saxena, Gate leakage current assessment of AlGaN/GaN HEMT with AlN cap layer, in *Computers and Devices for Communication. CODEC 2019. Lecture Notes in Networks and Systems*, 147, eds. by N.R. Das, S. Sarkar (Springer, Singapore, 2021), 459–464. https://doi.org/10.1007/978-981-15-8366-7_68
9. S. Das, V. Kumari, K. Sehra, M. Gupta, M. Saxena, On the double channel engineering of dual gate AlGaN/GaN HEMTs for heavy ion sensing applications. Micro Nanostruct. **182**(1), 207653 (2023). https://doi.org/10.1016/j.micrna.2023.207653
10. C.-C. Wu, C.-Y. Liu, S. Anand, W.-H. Chieng, E.-Y. Chang, A. Sarkar, Comparisons on different innovative cascode GaN HEMT E-mode power modules and their efficiencies on the Flyback converter. Energies **14**(18), 5966 (2021). https://doi.org/10.3390/en14185966
11. A. Kamath, T. Patil, R. Adari, I. Bhattacharya, S. Ganguly, R.W. Aldhaheri, M.A. Hussain, D. Saha, Double-channel AlGaN/GaN high electron mobility transistor with back barriers. IEEE Electrons Dev. Lett. **33**(12), 1690–1692 (2012). https://doi.org/10.1109/LED.2012.2218272
12. T. Palacios, A. Chini, D. Buttari, S. Heikman, A. Chakraborty, S. Keller, S. DenBaars, U. Mishra, Use of double-channel heterostructures to improve the access resistance and linearity in GaN-based HEMTs. IEEE Trans. Electron Dev. **53**(3), 562–565 (2006). https://doi.org/10.1109/TED.2005.863767
13. H.-Y. Lee, Y.-H. Ju, J.-I. Chyi, C.-T. Lee, Performance comparison of lattice-matched AlInN/GaN/AlGaN/GaN double-channel metal–oxide–semiconductor high-electron mobility transistors with planar channel and multiple-mesa-fin-channel array. Materials **15**(1), 42 (2022). https://doi.org/10.3390/ma15010042
14. S. Das, V. Kumari, K. Sehra, M. Gupta, M. Saxena, TCAD based investigation of single event transient effect in double channel AlGaN/GaN HEMT. IEEE Trans. Device Mater. Reliab. **21**(3), 416–423 (2021). https://doi.org/10.1109/TDMR.2021.3103238
15. R. Chu, Y. Zhou, J. Liu, D. Wang, K. Chen, K. Lau, AlGaN–GaN double-channel HEMTs. IEEE Trans. Electron Dev. **52**(4), 438–446 (2005). https://doi.org/10.1109/TED.2005.844791
16. R. Lingaparthi, N. Dharmarasu, K. Radhakrishnan, A. Ranjan, T.L.A. Seah, L. Huo, Source of two-dimensional electron gas in unintentionally doped AlGaN/GaN multichannel high-electron-mobility transistor heterostructures. Lied Phys. Lett. **118**, 122105 (2021). https://doi.org/10.1063/5.0045910
17. M. Vittori, S. Colangeli, W. Ciccognani, A. Salvucci, G. Polli, E. Limiti, High performance X-band LNAs using a 0.25 μm GaN technology, in *2017 13th Conference on Ph.D. Research in Microelectronics and Electronics PRIME*, 157–160 (2017). https://doi.org/10.1109/PRIME.2017.7974131
18. S. Cha, Y.H. Chung, M. Wojtowwicz, I. Smorchkova, B.R. Allen, J.M. Yang, R. Kagiwada, Wideband AlGaN/GaN HEMT low noise amplifier for highly survivable receiver electronics, in *2004 IEEE MTT-S International Microwave Symposium Digest (IEEE Cat. No.04CH37535)*, vol. 2, pp. 829–831. https://doi.org/10.1109/MWSYM.2004.1339093
19. S. Colangeli, A. Bentini, W. Ciccognani, E. Limiti, A. Nanni, GaN-based robust low-noise amplifiers. IEEE Trans. Electron Dev. **60**(10), 3238–3248 (2013). https://doi.org/10.1109/TED.2013.2265718

20. S. Zafar, E. Aras, B.C. Akoglu, G. Tendurus, M.I. Nawaz, A. Kashif, E. Ozbay, Design of GaN-based X-band LNAs to achieve sub-1.2 dB noise figure. Int. J. RF Microw. Comput.-Aided Eng. (2022). https://doi.org/10.1002/mmce.23379
21. H. Ahn, H. Ji, D. Kang, S.-M. Son, S. Lee, J. Han, A 26–30 GHz GaN HEMT low-noise amplifier employing a series inductor-based stability enhancement technique. Electronics **11**(17), 2716 (2022). https://doi.org/10.3390/electronics11172716
22. C. Gupta, A. Gupta, A.K. Bansal, A. Dixit, Suppression of short-channel effects by double-gate double-channel device design in normally off AlGaN/GaN MIS-HEMTs. IETE J. Res. (2018). https://doi.org/10.1080/03772063.2018.1541764
23. Silvaco Inc., *ATLAS TCAD Tool Version 5.30.0.R* (Santa Clara, CA, USA)
24. T. Zhu, X.F. Zheng, Y.R. Cao, C. Wang, W. Mao, Y.Z. Wang, M.H. Mi, M. Wu, J.H. Mo, X.H. Ma, Study on effect of diamond layer on the performance of double-channel AlGaN/GaN HEMTs. Semicond. Sci. Technol. **35**(5) (2020). https://doi.org/10.1088/1361-6641/ab7773
25. H.-C. Wang, H.-F. Su, Q.-H. Luc, C.-T. Lee, H.-T. Hsu, E.Y. Chang, Improved linearity in AlGaN/GaN HEMTs for millimeter-wave applications by using dual-gate fabrication. ECS J. Sol.-State Sci. Technol. **6**(11) (2017). https://doi.org/10.1149/2.0251711jss
26. MathWorks MATLAB R2021a Academic License. Available https://in.mathworks.com. Accessed 13 Mar 2022
27. N. Garg, Y. Pratap, M. Gupta, S. Kabra, Reliability assessment of GaAs/Al_2O_3 junctionless FinFET in the presence of Interfacial layer defects and radiations. IEEE Trans. Dev. Mater. Reliab. **20**(2), 452–458 (2020). https://doi.org/10.1109/TDMR.2020.2991662
28. A.S. Rai, Reliability analysis of junction-less double gate (JLDG) MOSFET for analog/RF circuits for high linearity applications. Microelectron. J. **64**(1), 60–68 (2017). https://doi.org/10.1016/j.mejo.2017.04.009
29. A. Cappy, Noise modeling and measurement techniques (HEMTs). IEEE Trans. Microw. Theory Tech. **36**(1), 1–10 (1988). https://doi.org/10.1109/22.3475
30. S. Weinreb, M. Pospieszalski, X-band noise parameters of HEMT devices at 300 K and 12.5 K, in *1985 IEEE MTT-S International Microwave Symposium Digest*, 539–542 (1985). https://doi.org/10.1109/MWSYM.1985.1132033
31. J.S. Moon, R. Gabar, D. Brown, I.A. Rodriguez, D. Wong, A. Schmitz, H. Fung, P. Chen, J.-C. Kang, S. Kim, T. Oh, C. Mcguire, 70% power-added-efficiency dual-gate, cascode GaN HEMTs without harmonic tuning. IEEE Electron Dev. Lett. **37**(3), 272–275 (2016). https://doi.org/10.1109/LED.2016.2520488
32. Keysight Technologies, *ADS 2021 Tool* (Santa Rosa, CA, USA)
33. S. Khandelwal, Y.S. Chauhan, *Advanced Spice Model for HEMTs, Technical Manual* (2019). http://iitk.ac.in/asm/index.html
34. J. Pinchbeck, K.B. Lee, S. Jiang, P. Houston, Dual metal gate AlGaN/GaN high electron mobility transistors with improved transconductance and reduced short channel effects. J. Phys. D Appl. Phys. **54**(10), 105104 (2020). https://doi.org/10.1088/1361-6463/abcb34
35. T. Han, H. Zhao, X. Peng, Enhancement of electrons confinement in AlGaN/AlN/GaN heterostructure using BGaN buffer with a small B-content. Superlatt. Microstruct. **126**, 57–62 (2019). https://doi.org/10.1016/j.spmi.2018.12.018

Chapter 16
ZnO Nanorods Orientation Control with Seedless Hydrothermal Growth and Finding Suitable Morphology with Simulation for Tactile Sensors

Rehan Ahmed and Pramod Kumar

16.1 Introduction

Self-powered tactile sensors with a biocompatible material and haptic nano devices can be applied in the artificial electronic skin to mimetic the human skin [1, 2]. The vertical *c*-oriented ZnO nanorods (NRs) are being widely researched for the application of self-powered pressure sensors for monitoring the vital signs [3], transistors at the nanoscale [4], piezoelectric nanogenerators [5], and energy harvesting devices [6]. ZnO NRs are also widely used in electronics, photonics, and optoelectronics and play an important role in the fabrication of such devices. The self-powered devices are usually based on piezoelectric materials such as aluminum nitride [7], lead zirconate titanate (PZT) [8], quartz [9], and zinc oxide (ZnO) [10]. Among these materials, the ZnO nanostructure is the most promising candidate and has been getting attention due to its some favorable properties such that it have a wide and direct band gap of 3.4 eV with a large exciton binding energy of 60 meV [11]. There are many chemical-based approaches for the synthesis of one-dimensional (*c*-oriented) ZnO NRs such as sputter deposition method [12], vapor deposition method [13], galvanic-based cell [14], and hydrothermal method [15]. Among these growth methods, the hydrothermal method is more popular and has been getting great attention due to its advantages of simplicity and low-cost synthesis. Hydrothermal is a chemical-based approach and is carried out in an aqueous solution. The most common method of growth of ZnO NRs is by deposition of an additional seed layer of ZnO (called seeded hydrothermal method) on the substrates for nucleation assistance of vertical growth of ZnO NRs [16]. The addition seed layer requires some other method of deposition which makes

R. Ahmed (✉) · P. Kumar
Department of Physics, Indian Institute of Technology Bombay, Mumbai, Maharashtra 400076, India
e-mail: rehan.phy@iitb.ac.in

R. Singh et al. (eds.), *The Physics of Semiconductor Devices*, Springer Proceedings in Physics 306, https://doi.org/10.1007/978-981-97-1571-8_16

the growth more complex and costly. There have been great amount of research for the development of ZnO NRs by removing the deposition of any ZnO seed layer before the growth process, this method is called the seedless hydrothermal method. The seedless hydrothermal method is done by adjusting the growth parameters (growth conditions) such as growth temperatures, time, and the precursor concentration over the seeded hydrothermal method to achieve vertical ZnO nanorods. In the seedless hydrothermal process, controlling the aspect ratio (lengths and diameters of ZnO NRs), density, growth rate, and verticality of ZnO NRs are very challenging. In this research paper, we have focused on the growth conditions which are responsible for growth morphology of ZnO nanorods in a systematic way. The obtained morphologies are then simulated for the piezoelectric output using COMSOL Multiphysics software which shows the results contrary to the previous research where vertical growth is more preferred for high piezoelectric output.

16.2 Experimental and Simulation Sections

16.2.1 Materials

The ZnO NRs are grown on Gold (Au) coated silicon (Si) wafers by seedless hydrothermal process. Before starting the growth process the Si substrates were cleaned by the standard procedure. Au thin film (90 nm) was deposited on Si substrates by the sputtering method and a 10 nm layer of titanium (Ti) was deposited in between Au and Si for good adhesion purpose by the same method. Thereafter, the Au/Si substrates were vacuum annealed (~ 10^{-5} mbar) at temperature 250–300 °C for 20–25 min. Hexamethylenetetramine (HMT) ($C_6H_{12}N_4$, Sigma Aldrich) and zinc nitrate hexahydrate ($Zn(NO_3)_2.6H_2O$, Sigma Aldrich) were employed as the precursor concentration in a 1:1 ratio in the deionized water (DI) water. All the used chemicals were of the highest purity grade and without any further purifications. High purity DI water (HPLC grade) was used for all the chemical processes.

16.2.2 Growth Method of ZnO Nanorods (NRs)

Vacuum annealed Au/Si substrates were used to grow ZnO NRs by low-cost seedless hydrothermal method, the HMT and zinc nitrate were separately magnetic stirred (5 min) in a 50 ml DI water to get a homogenous solution, and then they are mixed in a beaker. The substrates were then gently kept face down i.e. floating on the growth solution in the beaker, and then the beakers were placed in a pre-heated oven at a fixed temperature and time. After synthesis, the ZnO NRs grown substrates were taken out and rinsed thoroughly in hot DI water (60–70 °C), dried using a nitrogen gun, and baked at 100–120 °C for 3–4 min for removing the absorbed moisture from the samples.

16.2.3 Characterization and Simulation Techniques

The surface morphology and the structural identification of grown ZnO NWs were characterized by Scanning Electron Microscope (SEM, JSM-6390) and X-ray diffraction, respectively. The piezoelectricity of the single ZnO nanorod (NR) was simulated through the finite element method (FEM) in COMSOL Multiphysics software (version 5.5, COMSOL Inc., Palo Alto, CA USA).

16.3 Results and Discussion

16.3.1 Effect of Growth Temperature

Temperature is an important factor (parameter) in controlling the length and diameter of the ZnO NRs in the seedless hydrothermal method. In our experiments, we have investigated the effect of growth temperature at a fixed optimized time of 16 h and others fixed parameters like concentration and substrate treatment. Therefore, to evaluate the effect of growth temperature on the morphology of ZnO NRs, the nanorods were grown at temperatures 60, 70, 80, and 90 °C on vacuum annealed Au/Si substrates while keeping all the other parameters fixed (concentration (100 mM), substrate treatment, and growth time). The results can be seen in Fig. 16.1, which shows SEM images and the XRD spectrum. The SEM images show a higher probability of vertical growth at 90 °C growth temperature which can also be corroborated with the 002 peak in XRD spectra relative to other growth temperatures. 90 °C is turned out to be a good optimum temperature for vertical alignment and a well-defined ideal hexagonal shape of ZnO NRs, which is a similar result as reported in [15, 17]. When the temperature is reduced to 60 and 70 °C, the density and aspect ratio become smaller as compared to 90 °C temperature growth that is clearly shown in SEM and X-ray spectrum in Fig. 16.1, which means that the surface coverage ratio is smaller in 60–70 °C temperature. By increasing the temperature to 90 °C the verticality, density, and surface coverage ratio are enhanced. From the XRD spectrum, it has been clearly shown that the ZnO NRs grown at 90 °C temperature have more verticality as compared to other temperatures and the highest density (220 ± 10 nanorods per 50 μm^2) was reached at 90 °C temperature. In reference [17], the researcher has reported similar trends and investigated the similar temperatures dependent variation in the growth of ZnO NRs by seedless hydrothermal process.

The temperature-dependent growth behavior of ZnO NRs on Au/Si substrates can be understood in the terms of ion mobility and diffusion length from the basic theory

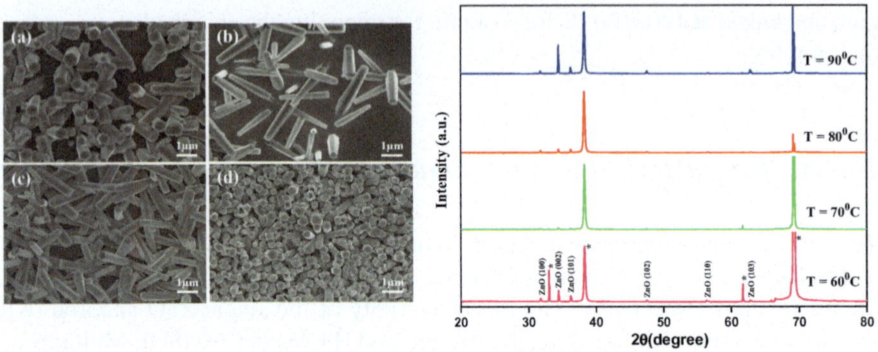

Fig. 16.1 SEM images (left) at temperatures **a** 60 °C, **b** 70 °C, **c** 80 °C, and **d** 90 °C, and XRD spectrum (right) of grown ZnO NRs at respective temperatures, the asterisks (*) show the substrates peaks

of growth and nucleation process of the crystal [15]. Therefore, at low temperature, the diffusion length and mobility are limited on the growth surface, which interdicts the precursor ions to diffuse around the nucleation sites, which results in the large size of nucleation. On the other hand at high temperatures, the nucleation sizes are small because the diffusion length and mobility are large enough. Hence development of high density and verticality are more favorable. Therefore, in our experiments, 90 °C temperature has the optimum value for the dense and vertical aligned of ZnO NRs with length and diameter distribution of about 4–5 μm and 400–500 nm, respectively. Overall temperature can provide control over the ZnO NR dimensions and the growth morphology from mostly tilted to mostly vertical.

16.3.2 Influence of Precursor Concentration

In this section, we have investigated the effect of precursor concentration on the growth of the ZnO NRs, only two materials HMT and zinc nitrate in 1:1 ratio were used in the synthesis process. From our experimental results, it has been confirmed that the density of grown ZnO NRs is directly correlated with the concentration, and it has been also seen that the highest density leads to the verticality of ZnO NRs. The mechanism of growth of ZnO NRs in the growth solution are given by the following chemical equations [18].

$$\text{Decomposition reaction: } C_6H_{12}N_4 + 6H_2O \rightarrow 6HCHO + 4NH_3 \tag{16.1}$$

$$\text{Hydroxyl supply reaction: } NH_3 + H_2O \leftrightarrow NH_4^+ + OH^- \tag{16.2}$$

$$\text{Supersaturation reaction: } Zn^{2+} + 2OH^- \leftrightarrow Zn(OH)_2 \leftrightarrow ZnO + H_2O \tag{16.3}$$

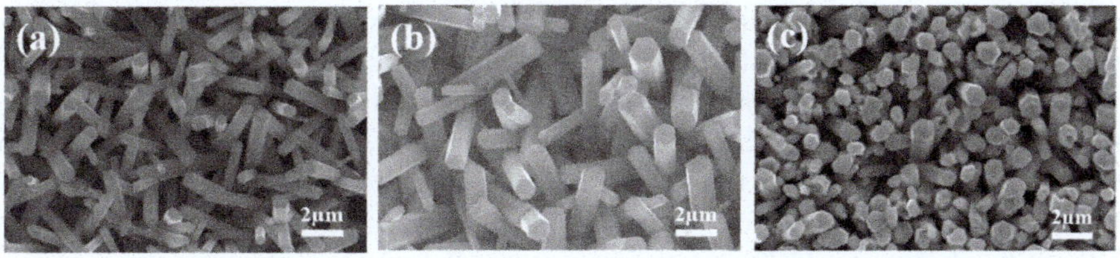

Fig. 16.2 SEM images of ZnO NRs at concentration **a** 50 mM, **b** 70 mM, and **c** 100 mM at 90 °C for 16 h

As in reaction (16.1), first, the HMT ($C_6H_{12}N_4$) hydrolyze into ammonia and HCHO, and then the NH_3 hydrolyze into NH_4^+ and OH^- as given in Eq. (16.2). Finally, the $Zn(NO_3)_2.6H_2O$ provides Zn^{2+} ions, and thereafter from the formation of Zn^{2+} and OH^- results in ZnO (s) as given in Eq. (16.3).

The concentration precursor has an important factor in controlling the density and verticality of the ZnO NRs, and increasing or decreasing the concentration of the precursor solution does influence the resultant products directly. From the above section of growth temperature variation, it has been confirmed that the 90 °C temperature is the optimized value for the better density and verticality of ZnO NRs. Figure 16.2 shows the different density of ZnO NRs, in which at lower concentration Fig. 16.2a, b has lower density and the verticality was also poor compared to 100 mM concentration in Fig. 16.2c. On increasing the concentration, the aspect ratio (length and diameter) increases, similar results were also described in references [19, 20]. The concentration dependence on an optimized temperature shows the growth morphology from higher tilted to higher vertical density.

16.3.3 Piezoelectric Simulation of ZnO Nanorod

Three basic growth morphologies have been seen in the hydrothermal growth process which can be controlled using optimized growth conditions. Most of the research till now focus only on the vertical growth morphology due to the expected high piezoelectric output. To confirm it finite element simulation was carried out for the piezoelectric output of ZnO NRs in COMSOL software. A 3D model of a ZnO material was created in COMSOL to study the piezoelectric behavior. COMSOL software finds the value of generated piezoelectric voltage by applying a pressure (force) on the piezo material.

First, a single *c*-axis oriented vertical regular hexagonal-shaped ZnO NR of length (l) 4 μm and diameter 500 nm is modeled in the simulation geometry. A boundary load of 50 kPa is applied on the top of the NR and the bottom (base) part was fixed on the substrate. The base of the NR is electrically grounded and a piezoelectric potential is calculated on the top of the NR, all the computational simulations were performed in the stationary condition. Figure 16.3 shows the 3D modeling and simulation results of a single ZnO NR in which the generated output potential and stress are presented. The general mathematical formula for theoretical calculations of the piezoelectric voltage (V) along the *c*-axis of single ZnO NR is given by $V = \frac{d_{33}Fl}{A\epsilon_r\varepsilon_o}$, where d_{33} is the piezoelectric coupling coefficient along the *c*-axis, ε_r and ε_o are the relative and vacuum permittivity respectively, and *F* is the pressure (force) applied on the NR. Furthermore, we have performed the simulation by varying the length of the ZnO NR at a fixed diameter (500 nm), the aspect ratio (length/diameter) and the respective generated output voltage is given in Table 16.1.

The results suggest that the output piezoelectric voltage (piezo voltage) increases on increasing the aspect ratio (at constant diameter) of the ZnO NR. Since, in seedless hydrothermal growth, the ZnO NRs grow in mainly three different growth morphology as seen in our experimental results which are mostly tilted, tilted and vertical mixed, and mostly vertical. In Fig. 16.2, it can be clearly shown that the NRs are grown vertical, tilted, and semi-vertical/tilted. So, simulations based on these three different morphologies are conducted. These three morphologies are modeled by a 3 × 3 matrix of ZnO NRs in COMSOL as shown in Fig. 16.4. It shows the simulation results of all the different morphologies. In the 3 × 3 array matrix, ZnO NRs are kept of the same size which is 5 μm long and having a diameter of 500 nm

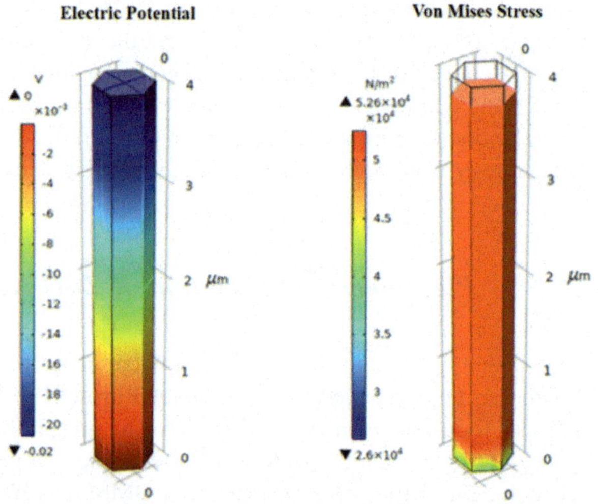

Fig. 16.3 COMSOL simulation of single ZnO NR of length 4 μm and diameter 500 nm at the fixed pressure of 50 kPa

Table 16.1 A comparative study of the output piezoelectric voltage by varying the length (at fixed diameter 500 nm, and fixed pressure 50 kPa) of ZnO NRs in COMSOL

Length (*l*) μm	Aspect ratio	Output voltage (mV)
2	4	10.32
4	8	20.71
6	12	31.06
8	16	41.48

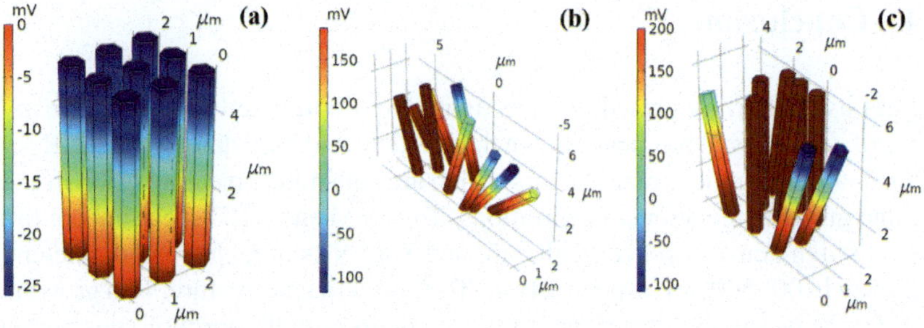

Fig. 16.4 Simulated piezoelectric voltages of 3 × 3 array of ZnO NRs, each 5 μm long and having a diameter of 500 nm, and a 50 kPa pressure was applied on the array. **a** Presents the pure vertical, **b** presents the pure tilted, and **c** presents the semi-vertical or tilted geometries of nanorods

Table 16.2 A comparative study of piezoelectric voltage in various geometry arrays of ZnO NRs

Array geometry	Piezoelectric voltage (mV)
Pure vertical	25.8
Semi-tilted or vertical	79.1
Pure tilted	101.6

and a similarly same pressure of 50 kPa is applied on all the three growth morphologies. The tilted NRs are kept at an angle range from 30 to 80°. Table 16.2 shows comparatively the effect of the growth morphologies of ZnO NRs on the generated piezoelectric voltage. The vertical morphology ZnO NR array produces a minimum voltage of 25.8 mV, while the mostly tilted geometries of NRs produces the highest voltages. The enhancement in the piezoelectric voltage can be seen from the mixed tilted and vertical morphology which is approximately 3 times higher than the mostly vertical morphology arrangement.

The higher piezoelectric voltage could be due to the presence of other orientations of the piezoelectric coefficient (d_{ij}) during the application of pressure on the array, unlike only d_{33} in the vertical array. Another reason could be higher stress levels experienced in some regions by tilted NRs at the same applied force, resulting in higher piezoelectric voltage generation. Unlike in the vertical NRs, the applied pressure acts not only on the top but also on the tilted faces, and hence the NRs faces much higher levels of stress values. These results are contrary to the general view where vertical morphology is more preferred for higher piezoelectric generation due to the higher value of d_{33} coefficient.

16.4 Conclusion

Seedless hydrothermal growth is investigated and optimized for better density and quality of ZnO NRs. It has been shown that vertical and high density ZnO NRs can be grown by adjusting the growth factors which are substrate annealing, growth temperature and growth precursor concentrations. The SEM and XRD spectrum confirm the growth of high density and vertically aligned ZnO NRs of length and diameter about 4–5 μm and 400–500 nm respectively at 90 °C growth temperature, 16 h growth time, and 100 mM precursor concentration. Three main growth morphologies are mostly visible in these growth condition which are mostly vertical NRs, tilted and vertical mixed NRs, and mostly tilted NRs. COMSOL simulation is performed to understand the piezoelectric output response in ZnO NRs, and also for the three different growth morphologies. It is found that the piezo potential increases by increasing the aspect ratio of ZnO NRs, and the tilted array of ZnO NRs shows the enhanced piezo potential over the vertical array due to higher stress values.

Acknowledgements The authors would like to thank Indian Institute of Technology Bombay (IITB) for seed grant and Science and Engineering Research Board (SERB) Government of India for funding. R. Ahmed would like also to thank the Council of Scientific and Industrial Research (CSIR) Government of India for providing the financial fellowship.

References

1. M.-S. Suen, Y.-C. Lin, R. Chen, A flexible multifunctional tactile sensor using interlocked ZnO nanorod arrays for artificial electronic skin. Procedia Eng. **168**, 1044–1047 (2016)
2. W. Deng et al., A flexible field-limited ordered ZnO nanorod-based self-powered tactile sensor array for electronic skin. Nanoscale **8**(36), 16302–16306 (2016)
3. X. Chen et al., Self-powered flexible pressure sensors with vertically well-aligned piezoelectric nanowire arrays for monitoring vital signs. J. Mater. Chem. C **3**(45), 11806–11814 (2015)
4. C. Opoku, et al., Fabrication of field-effect transistors and functional nanogenerators using hydrothermally grown ZnO nanowires. RSC Adv. **5**(86), 69925–69931 (2015)
5. M.S. Al-Ruqeishi, et al., Piezoelectric nanogenerator based on ZnO nanorods. Arab. J. Chem. **12**(8), 5173–5179 (2019)
6. Q. He, et al., P–N junction-based ZnO wearable textile nanogenerator for biomechanical energy harvesting. Nano Energy **85**, 105938 (2021)
7. M.-A. Dubois, P. Muralt, Properties of aluminum nitride thin films for piezoelectric transducers and microwave filter applications. Appl. Phys. Lett. **74**(20), 3032–3034 (1999)
8. T. Morita, Piezoelectric materials synthesized by the hydrothermal method and their applications. Materials **3**(12), 5236–5245 (2010)
9. S. Zhang, F. Yu, Piezoelectric materials for high temperature sensors. J. Am. Ceram. Soc. **94**(10), 3153–3170 (2011)
10. S. Xu, et al., Self-powered nanowire devices. Nat. Nanotechnol. **5**(5), 366–373 (2010)
11. Ü. Özgür, et al., A comprehensive review of ZnO materials and devices. J. Appl. Phys. **98**(4), 11 (2005)
12. W.-T. Chiou, W.-Y. Wu, J.-M. Ting, Growth of single crystal ZnO nanowires using sputter deposition. Diam. Relat. Mater. **12**(10–11), 1841–1844 (2003)

13. S.C. Lyu, et al., Low-temperature growth of ZnO nanowire array by a simple physical vapor-deposition method. Chem. Mater. **15**(17), 3294–3299 (2003)
14. Z. Zheng, et al., General route to ZnO nanorod arrays on conducting substrates via galvanic-cell-based approach. Sci. Rep. **3**(1), 1–5 (2013)
15. S. Xu, et al., Density-controlled growth of aligned ZnO nanowire arrays by seedless chemical approach on smooth surfaces. J. Mater. Res. **23**(8), 2072–2077 (2008)
16. J. Song, S. Lim, Effect of seed layer on the growth of ZnO nanorods. J. Phys. Chem. C **111**(2), 596–600 (2007)
17. I. Cesini, et al., Seedless hydrothermal growth of ZnO nanorods as a promising route for flexible tactile sensors. Nanomaterials **10**(5), 977 (2020)
18. D. Phromyothin, et al., Growth of ZnO nanorods via low temperature hydrothermal method and their application for hydrogen production. Mater. Today Proc. **4**(5), 6326–6330 (2017)
19. L. Vayssieres, Growth of arrayed nanorods and nanowires of ZnO from aqueous solutions. Adv. Mater. **15**(5), 464–466 (2003)
20. L. Vayssieres, et al., Purpose-built anisotropic metal oxide material: 3D highly oriented microrod array of ZnO. J. Phys. Chem. B **105**(17), 3350–3352 (2001)

Chapter 17
A Comparative Demonstration of a 2D MXene-Based Asymmetric Memristor Using Pristine and Nanohybrid $Ti_3C_2T_x$

Soumi Saha, Vivek Adepu, Khush Gohel, Parikshit Sahatiya, and Surya Shankar Dan

17.1 Introduction

Memristors are attracting significant importance across the world owing to their exceptional and amazing characteristics, which are exceptionally suited for non-volatile memory storage devices [1], booting-free computers [2], programmable logic and signal processing, analogic filters [2], remote sensing and low-power applications [3]. Memristors are an efficient solution for the miniaturization of circuits, as one memristor can perform the function of 100 s of transistors [4]. The whole world is looking for a solution for minimizing the power consumption of electronic devices. Replacing the conventional material used to fabricate electronic devices with two-dimensional (2D) materials can be a solution [5]. In the 2D material family, particularly the transition metal carbides/nitrides (MXenes) have been widely studied due to their exciting electronic and optoelectronic properties [5] and unique layered structure [6]. Due to the high conductivity of MXenes, it is expected that the MXene-based memristor will exhibit a poor R_{OFF}/R_{ON} ratio [7]. To alleviate this fundamental issue, recently, nanohybrids of MXene are replacing pristine MXene [8].

The nano-hybrids of MXene are nothing but to tune the electronic properties of the pristine MXene. The electronic properties like bandgap can be modulated by adding nanohybrids to create local Schottky junctions, increasing resistivity and enhancing the R_{OFF}/R_{ON} ratio. Most reports on MXene-based memristors discuss

S. Saha · V. Adepu · K. Gohel · P. Sahatiya · S. S. Dan (✉)
Department of Electrical and Electronics Engineering, Birla Institute of Technology and Science Pilani, Hyderabad Campus, Hyderabad 500078, India
e-mail: ssdan@hyderabad.bits-pilani.ac.in

P. Sahatiya
e-mail: parikshit@hyderabad.bits-pilani.ac.in

R. Singh et al. (eds.), *The Physics of Semiconductor Devices*, Springer Proceedings in Physics 306, https://doi.org/10.1007/978-981-97-1571-8_17

their symmetric current-voltage characteristics. In analog circuit applications, asymmetric memristors [9] are beneficial, but the challenge lies in the fabrication of asymmetric memristors. A comparative study of asymmetric MXene and MXene nanohybrid-based memristors with resistive switching will be highly beneficial for the designers in choosing the correct material combination for the memristive circuit applications.

The detailed process of fabrication and electrical characterization of pristine $Ti_3C_2T_x$ and $SnS/Ti_3C_2T_x$ nanohybrid-based memristor with proper switching mechanisms have been reported in this paper. The active layers of these two devices are made with $Ti_3C_2T_x$ and $SnS/Ti_3C_2T_x$ embedded between the top and bottom copper electrodes on a silicon substrate. The measured resistivity of the pristine MXene is ~ 26 Ω. Inclusion of SnS with pristine MXene, the resistivity of the $SnS/Ti_3C_2T_x$ has been increased to 2.3 kΩ, i.e., 100 times which results in almost 10 times improvement in R_{OFF}/R_{ON} ratio. As the resistivity is significantly less for the pristine MXene, the current response at 5 V is 50 mA. But the measured current for the $SnS/Ti_3C_2T_x$ at 5 V is 0.65 mA. This comparative work undoubtedly would be beneficial to showcase MXene-based memristor for high-density, low-power, and low-cost circuit applications. In Sect. 17.2 deliberates the experimental methods and the materials used in the proposed work. Sections 17.3 and 17.4 deliberate the SnS and MXene characterization and the current-voltage characterization of the device, respectively, and explain the device operation using physical insights into the device.

17.2 Experiment

17.2.1 Material Characterization

The XRD patterns of both Ti_3AlC_2 and fully exfoliated $Ti_3C_2T_x$ are recorded in an X-ray diffractometer to examine the material structure, shown in Fig. 17.1a. The realized diffraction peaks of the Ti_3AlC_2 were akin to the JCPDS card number 52-0875. The resultant $Ti_3C_2T_x$ pattern depicts the exclusion of crystal plane (1 0 4) at $2\theta = 38.890$, and the characteristic peak (0 0 2) has shifted from $2\theta = 9.420$ to 6.810, which implies the exfoliation of Al from Ti_3AlC_2 [10]. The XPS survey spectrum of $Ti_3C_2T_x$ is shown in Fig. 17.1b between 0 and 800 eV binding energy. The high-resolution de- convoluted XPS spectra of the element Ti 2p are shown in Fig. 17.1c, wherein core energy level fitting found that pair of two characteristic doublet peaks, i.e., 2p1/2, 2p3/2 attributed to Ti–C(SP3), Ti^{3+}(SP1), Ti–O(SP1) oxide at appropriate binding energies (B.E.) respectively [11, 12]. Correspondingly, Ti^{2+}(sp3) (2p3/2) and Ti–O–F(sp1) (2p3/2) were also achieved at corresponding B.E.'s for element Ti 2p. The deconvoluted XPS spectrum for element C 1s is shown in Fig. 17.1d. Two characteristic peaks imputed to C–Ti–Tx, graphitic C–C, were observed at B.E.'s 281 and 285 eV, and a new peak attributed to CHx/C–O was realized at 288 eV [13]. Fig. 17.1e displays the core energy level fitted XPS spectrum for element O 1s, in which the contributions ascribed to C–Ti–Ox, C–Ti– (OH)x and Ti–O were obtained at appropriate B.E's depicted below. The core energy level fitted XPS spectrum for element F 1s is shown in Fig 17.1f, wherein the decisive peak realized at B.E. 685 eV imputed to C–Ti–Tx [14]. Furthermore, the peak realized at 689 eV assigned to CF^{2+} due to the grouping of polymeric chains with CF^2 or CF^3 groups [15], and the peak observed at 687 eV assigned to unidentified fluorinated secondary phase or F atoms completely bonded to $Ti_3C_2T_x$ carbon atoms revealed due to the rearrangement of certain Ti atoms [15]. The deconvoluted core-level fitted XPS spectrum realized above for all elements related to $Ti_3C_2T_x$ by the Shirley algorithm was found in correspondence with the existing and reported literature. The synthesized $Ti_3C_2T_x$ material was also examined for molecular interactions by Raman spectroscopy, as illustrated in Fig 17.1g. The Raman shift at 204 cm^{-1} is assigned to the first-order out-of-plane vibrations corresponding to Ti and C elements [16]. The Raman peaks indicated at 421 and 632 cm^{-1} signify the vibrational band of Eg, A1g and were assigned to the in-plane modes of vibration of Ti and C, which belong to the surface functional groups of $Ti_3C_2T_x$ [17]. Figure 17.1h illustrates the high-resolution surface morphology image obtained in FESEM for the delaminated nano-layered structure of $Ti_3C_2T_x$ synthesized using the MILD method.

17.2.2 Device Fabrication

$Ti_3C_2T_x$-based memristor was fabricated on a p-type Silicon wafer with $\prec$100$\succ$ crystal orientation as substrate with the process steps summarized below (1) The silicon wafer was thoroughly cleaned for organic contamination removal using isopropyl alcohol (IPA) and acetone and then dried (100 °C for 20 min) in the high-temperature oven. (2) Copper was deposited on the silicon wafer by thermal evaporation technique using shadow masking with the dimension of 3 × 3 cm with a 0.5 cm mask hole. (3) The MXene ($Ti_3C_2T_x$) was deposited using spin coating for 60 sec at 1000 rpm, followed by drying (100 °C for 20 min). (4) The top metal contact was made of copper on the SnS/$Ti_3C_2T_x$ layer using the thermal evaporation technique. (5) The fabricated device was kept in the high-temperature oven for drying at 100 °C for 1 h. In total, the resistive memory device was fabricated via a simple, cost-effective method, i.e., sandwiching an active layer made of $Ti_3C_2T_x$ between the two Cu contacts.

17.3 Characterization and Device Operation

Figure 17.2a, b depict the schematic representation of the proposed $Ti_3C_2T_x$ and SnS/$Ti_3C_2T_x$-based memristor de- vices. The current vs. voltage characteristics of both proposed devices were performed using a DC probe station and a Keithley 2450 source meter by applying a ± 5 V DC sweep. The ± 5 V DC sweep voltage signal

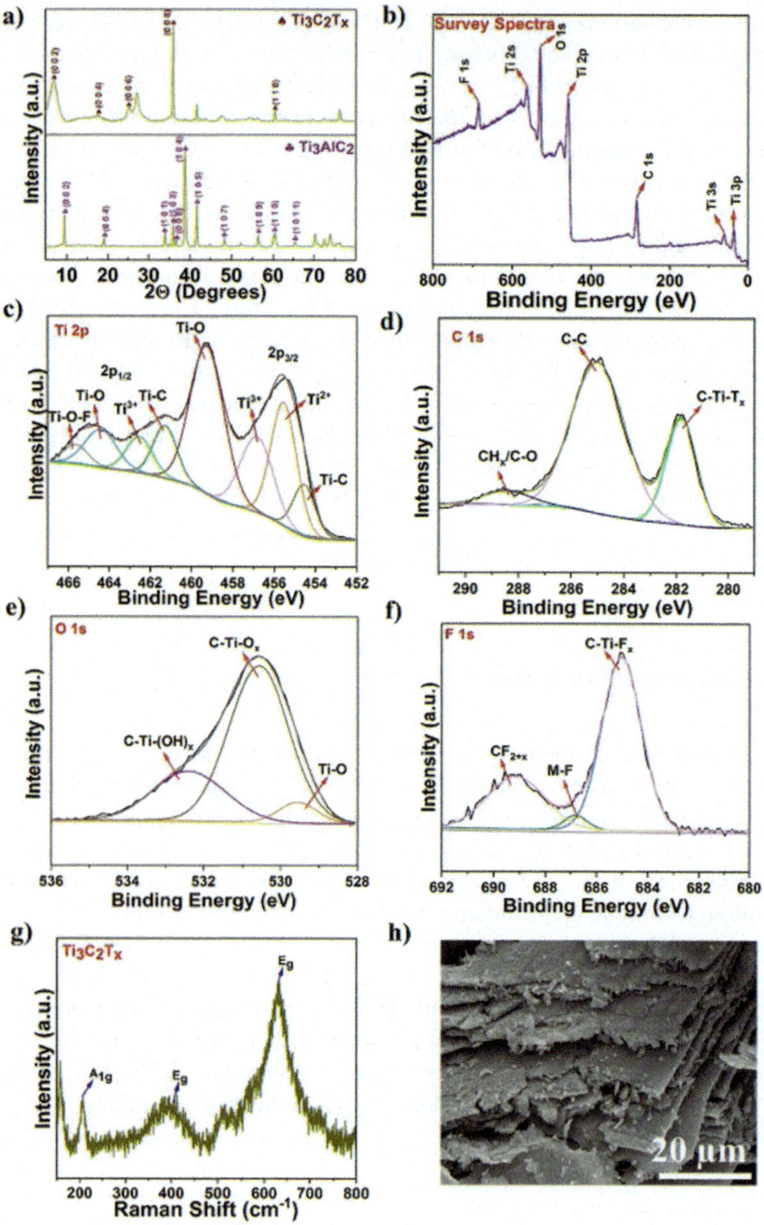

Fig. 17.1 **a** XRD Spectrum of Ti_3AlC_2 and pristine $Ti_3C_2T_x$, **b** XPS survey spectrum of $Ti_3C_2T_x$ synthesized by MILD method, **c–f** high-resolution core level fitted XPS spectra for elements Ti 2p, C 1s, O 1s, F 1s, **g** Raman spectrum for $Ti_3C_2T_x$, **h** FESEM image of delaminated $Ti_3C_2T_x$ synthesized by the MILD method

was applied through the top Cu electrode, and the ground connection was applied through the bottom Cu electrode. Figure 17.2c, d show the measured current response with the applied voltage of the proposed memristor devices both in the linear scale and logarithmic scale. The threshold voltage of the proposed pristine MXene-based memristor is almost 50 mV, and the saturation current is 50 mA. In contrast, the threshold voltage of the proposed SnS/MXene ($Ti_3C_2T_x$)-based memristor is almost 1.5 V, and the saturation current is 0.65 mA. SnS/MXene-based memristor enters into a high resistance state (HRS) at 0–5 V sweep. It can be assumed that 0.2–1.5 V is required to form the conduction bridge in the active layer by observing the current response. The SnS/MXene-based memristor's nonlinear response is evidence of different conduction processes at different applied voltages. For pristine $Ti_3C_2T_x$-based memristor, due to the layer structure with the application of very low voltage like 0.05 V, the device starts to conduct.

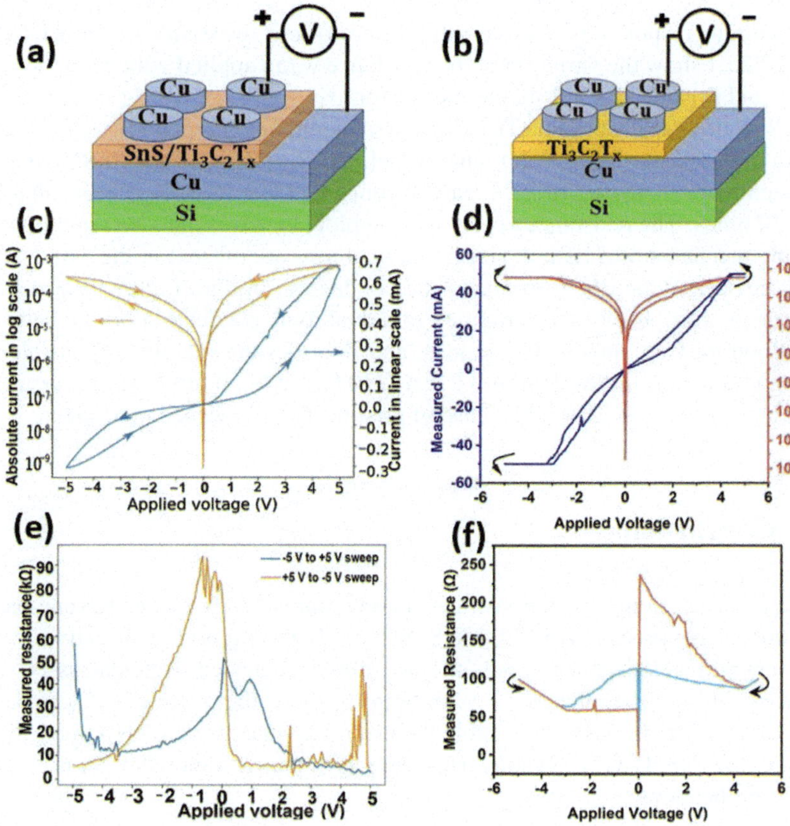

Fig. 17.2 Proposed Cu/SnS/$Ti_3C_2T_x$/Cu and Cu/$Ti_3C_2T_x$/Cu nanohybrid memristor: **a**, b schematic representation of the device, **c**, **d** measured I-V characteristics and **e**, **f** measured R-V characteristics respectively

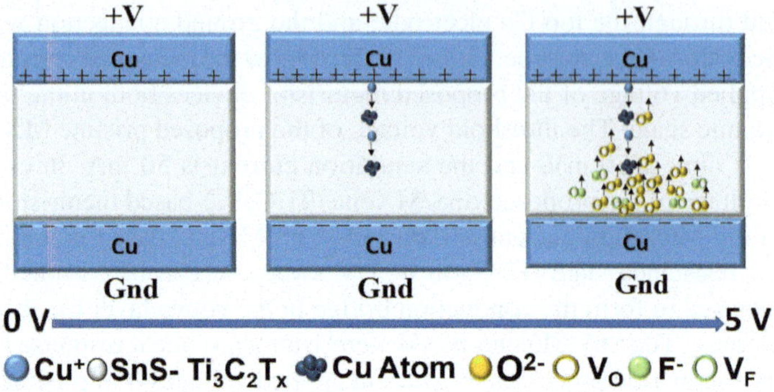

Fig. 17.3 Schematic illustrating dual-ionic channel formation inside the active region. The exact mechanism can be used to explain the switching mechanism of the proposed nanohybrid memristor

The current response is also very high due to less resistivity of the $Ti_3C_2T_x$. Figure 17.2e, f show the variation of the resistance w.r.to applied voltage of the SnS/ $Ti_3C_2T_x$ and pristine $Ti_3C_2T_x$-based memristor.

The R_{OFF}/R_{ON} of the pristine $Ti_3C_2T_x$-based memristor is 2.7 and $I_{ON}/I_{OFF} \approx 250$, but with the inclusion of SnS 2D nanohybrid, the R_{OFF}/R_{ON} of the SnS/ $Ti_3C_2T_x$-based memristor increased to 22. It can be concluded that R_{OFF}/R_{ON} was enhanced almost 10 times. The resistance versus voltage plot also gives the idea regarding the switching mechanism of both devices. The dual ions are present in the active layer of both devices, which help to create the conduction bridge. Figure 17.3 depicts the schematic response of the formation of a dual ionic conduction bridge between top and bottom Cu contacts. Cu, oxides, and fluorides are mainly responsible for shooting up the current level from 0.2 mA (@0.05 V) to 50 mA (@5 V), resulting in very high $I_{ON}/I_{OFF} \approx 250$ in the case of pristine $Ti_3C_2T_x$-based memristor.

17.4 Conclusion

This paper experimentally demonstrates a comparative study of the MXene-based memristors, using pristine $Ti_3C_2T_x$ and SnS/$Ti_3C_2T_x$ nanohybrid as the active layers. The advantages and disadvantages of the two fabricated devices were discussed with reference to different applications. With high I_{ON}/I_{OFF}, the pristine $Ti_3C_2T_x$-based asymmetric memristor will be suitable for analog applications. With a good R_{OFF}/R_{ON} ratio, the SnS/$Ti_3C_2T_x$ nanohybrid-based asymmetric memristor are useful for memory applications.

References

1. S. Ali, S. Khan, A. Khan, A. Bermak, Memristor fabrication through printing technologies: a review. IEEE Access **9**, 95970–95985 (2021). https://doi.org/10.1109/ACCESS.2021.3094027
2. Dissipativity analysis of memristor-based complex-valued neural networks with time-varying delays (ScienceDirect, 2022). https://www.sciencedirect.com/science/article/pii/S00200255 14007518. Accessed 11 Mar 2022
3. V.P.S. Kirar, Memristor: the missing circuit element and its application. Int. J. Electron. Commun. Eng. **6**(12), 1395–1397 (2012)
4. Li, Hardwareimplementation of neuromorphic computing using large-scale memristor crossbar arrays. Adv. Intell. Syst. Wiley Online Library. https://doi.org/10.1002/aisy.202000137. Accessed 11 Mar 2022
5. B. Yalagala, P. Sahatiya, V. Mattela, S. Badhulika, Ultra-low cost, large area graphene/MoS2-based piezotronic memristor on paper: a systematic study for both direct current and alternating current inputs. ACS Appl. Electron. Mater. (2019). https://doi.org/10.1021/acsaelm.9b00086
6. Q.H. Wang, K. Kalantar-Zadeh, A. Kis, J.N. Coleman, M.S. Strano, Electronics and optoelectronics of two-dimensional transition metal dichalcogenides. Nat. Nanotech. **7**(11) (2012). https://doi.org/10.1038/nnano.2012.193
7. X. Chen et al., First-principles calculation and experimental investigation of a three-atoms-type MXene V2C and its effects on memristive devices. IEEE Trans. Nanotechnol.Nanotechnol. **20**, 512–516 (2021). https://doi.org/10.1109/TNANO.2021.3089211
8. V_2C-based memristor for applications of low power electronic synapse. IEEE J. Mag. IEEE Xplore. https://ieeexplore.ieee.org/abstract/document/9316287. Accessed 26 Feb 2022
9. M. Hua, S. Yang, Q. Xu, M. Chen, H. Wu, B. Bao, Forward and reverse asymmetric memristor-based jerk circuits. AEU-Int. J. Electron. C. **123**, 153294 (2020). https://doi.org/10.1016/j.aeue.2020.153294
10. Z. Yu, MXenes with tunable work functions and their application as electron- and hole-transport materials in non-fullerene organic solar cells. J. Mater. Chem. A. https://doi.org/10.1039/C9TA01195A. https://pubs.rsc.org/en/content/articlehtml/2019/ta/c9ta01195a. Accessed 26 Feb 2022
11. Elumalai, *Simultaneous Delamination and Rutile Formation on the Surface of* $Ti_3C_2T_x$ *MXene for Copper Adsorption* (2020)
12. Chem. Asian J. Wiley Online Library. https://onlinelibrary.wiley.com/doi/abs/10.1002/asia.202000090. Accessed 13 Mar 2022
13. *Composites of* TiO_2 *Nanoparticles Deposited on* Ti_3C_2 *MXene Nanosheets with Enhanced Electrochemical Performance* (IOPscience, 2022). https://doi.org/10.1149/2.0981605jes/meta. Accessed 13 Mar 2022
14. *Ultra-Sensitive Electrocatalytic Detection of Bromate in Drinking Water Based on Nafion/* Ti_3C_2Tx *(MXene) Modified Glassy Carbon Electrode* (ScienceDirect, 2022). https://www.sciencedirect.com/science/article/pii/S0925400518306014. Accessed 13 Mar 2022
15. J. Ma et al., Lithium intercalation in anatase titanium vacancies and the role of local anionic environment. Chem. Mater. **30**(9), 3078–3089 (2018). https://doi.org/10.1021/acs.chemmater.8b00925
16. *XPS Analysis of Laser Transmission Micro-Joint Between Poly (Vinylidene Fluoride) and Titanium* (ScienceDirect, 2022). https://www.sciencedirect.com/science/article/pii/S016943320801773X. Accessed 13 Mar 2022
17. First-order Raman scattering of the MAX phases: Ti_2AlN, $Ti_2AlC_{0.5}N_{0.5}$, Ti_2AlC, $(Ti_{0.5}V_{0.5})$2AlC, V2AlC, Ti_3AlC_2, and Ti_3GeC_2—presser. J. Raman Spectrosc. Wiley Online Library. https://doi.org/10.1002/jrs.3036. Accessed 13 Mar 2022
18. High-capacitance mechanism for $Ti_3C_2T_x$ MXene by in situ electrochemical Raman spectroscopy investigation. ACS Nano. https://doi.org/10.1021/acsnano.6b06597. Accessed 13 Mar 2022

Chapter 18
Modelling and Simulation of a Thin-Film Bulk Acoustic Resonator (FBAR) Based Gas Sensor

Ajay Kumar, Rakesh Kumar Saini, Vinay Goyal, and Rahul Prajesh

18.1 Introduction

In the field of gas sensing, Gas Chromatography is one of the most standard available proven methods [1]. It is widely used to identify different components of gas mixtures and has applications in environmental monitoring [2] and the food industry [3]. It enables precise quantitative and qualitative gas analysis; however, the cost of equipment has been creating a barrier to conducting research. Gas sensing by the optical method is usually straightforward and achieves good stability and sensitivity. Most of the optical methods used for gas sensing applications are based on spectroscopy [4, 5]. Their application in gas sensing is seriously narrowed due to relatively high cost and miniaturization issues.

High-frequency acoustic wave devices such as Quartz crystal microbalance (QCM), Surface acoustic wave (SAW), and FBAR are widely used technologies for gas sensing due to their high sensitivity, low power consumption, short response time, wireless sensing, and long-term stability [6]. Amongst them, FBAR offers miniaturized size, high operating frequency, and compatibility to integrate with the circuit on a single chip. Hence, FBAR unquestionably could be a promising candidate as a miniature MEMS gas detector in terms of size, energy consumption, and sensing performance [7]. Volatile organic compounds (VOCs) are generated by vehicles and fuel combustion as well as decorative furniture and building materials. VOCs are cause of nausea, damage to the kidneys, and central nervous system, headaches, and loss of coordination. So, it is of utmost importance that suitable gas sensors should be deployed for the monitoring of such VOCs. Acoustic gas sensors not only offer characteristics such as small size, convenient fabrication, and high sensitivity but

A. Kumar (✉) · R. K. Saini · V. Goyal · R. Prajesh
CSIR-Central Electronics Engineering Research Institute, Pilani 333031, India
e-mail: ajayceeri1024@gmail.com

Academy of Scientific and Innovative Research (AcSIR), Ghaziabad 201002, India

R. Singh et al. (eds.), *The Physics of Semiconductor Devices*, Springer Proceedings in Physics 306, https://doi.org/10.1007/978-981-97-1571-8_18

also provide a facility for wireless monitoring of VOCs. Among acoustic sensors, FBAR has a high resonance frequency, high-quality factor, and limit of detection as low as part per billion (ppb) [8].

In this work, we demonstrated the design & simulation of film bulk acoustic resonator as a TCE gas sensor with finite element modeling (FEM) in COMSOL Multiphysics. The work in the paper is structured as follows, Sect. 18.2 explains the FBAR sensor principle and sensing layer used and Sect. 18.3 explains the geometry and simulation of the FBAR gas sensor. The results and discussion are summarized in Sect. 18.4. Finally, Sect. 18.5 concludes the study of this research paper.

18.2 FBAR Sensor

Film bulk acoustic resonators (FBAR) are the advancement in fabricating bulk acoustic devices that were initially invented by Lakin and Wang in the 80s [9]. Sensors based on FBAR are based on the effects of mass, temperature, gas concentration, light, and pressure that produce shifts in resonant frequency. These sensors are the most preferable ones for biochemical and gas detections with ultra-high sensitivity even at ppb levels as compared to other contemporary sensors [7].

18.2.1 Principle

FBAR structure consists of a piezoelectric layer placed between metal electrodes, where the acoustic wave propagates in the bulk of the piezoelectric thin film. The acoustic wave velocity and thickness of piezoelectric film define the resonance frequency in FBAR structures as shown in Eq. 18.1.

$$f_o = v/2t \tag{18.1}$$

where v and f_o are the acoustic velocity and resonance frequency of the propagating wave and t is the thickness of the piezoelectric thin film. The resonance frequency also depends on other parameters such as the density of piezoelectric material, mass loading on the electrodes, etc. [10]. So, we can observe a change in resonance frequency corresponding to mass loading or density of piezoelectric material and use FBAR as a sensing device. Change in the resonant frequency of an FBAR device is explained by Sauerbrey effect [11] which explains how the resonant frequency decreases when additional mass is observed on the resonator's active surface (Eq. 18.2).

$$\Delta f = -\frac{2f_o^2}{A\sqrt{\rho\mu}}\Delta m \tag{18.2}$$

$$S_m = \Delta f / \Delta m \qquad (18.3)$$

where Δf, Δm, ρ and μ are frequency change, mass change, density (g/cm^3), and shear modulus (gcm^{-1} s^{-2}) respectively. In this work, the mass loading sensing mechanism is utilized to calculate the ratio of change in the resonance frequency (Δf) with change in mass (Δm) and which is the sensitivity of the sensor as given by Eq. 18.3.

18.2.2 Sensing Layer

A highly gas selective and sensitive layer is commonly used for absorbing the gas molecules. We can use various sensing layers on the FBAR devices such as metal oxides (MO) [12, 13], metal–organic framework (MOF) [14], polymers [15], carbonaceous compounds like graphene and CNT [16], and combination of these materials to improve the sensitivity of the sensor [17].

The polymers which may be the sensing materials are polydimethylsiloxane (PDMS), polyisobutylene (PIB), polyepichlorohydrin (PECH), polyethyleneimine (PEI), and polyaminosiloxane [18]. A polymer such as a polyisobutylene shows high affinity toward volatile organic gases. In this work, PIB as a sensing layer is used and tested against various gases such as Trichloroethelene (TCE), Chloromehane (CM), Trichloromethane (TCM), and Dichloromethane (DCM).

18.3 Geometry and Simulation of FBAR Gas Sensor

Finite element method (FEM) simulations of the FBAR gas sensor are performed using COMSOL Multiphysics tool. The study investigates the 2D simulation of an FBAR gas sensor with ZnO as the piezoelectric material along with PIB as the sensing layer.

FBAR is simulated by the FEM analysis, and it is based on dividing the analyzed structure into finite sized elements forming a mesh. Quadrilateral mesh is mapped for each domain of FBAR with size and distribution nodes. Elements sizes are calibrated for general physics and a fixed number of elements in the distribution node. This FBAR model is designed to resonate freely in the z direction since there is a fixed boundary condition at all four sides in the x and y directions. A sinusoidal voltage with 1 V amplitude is applied across the top and bottom Al electrodes. Frequency domain analysis of FBAR has culminated by varying the frequency of the applied voltage (1 V) from 600 to 1250 MHz with the step of 0.1 MHz. FBAR device produces two resonance frequencies based on its materials and dimensions as depicted in Fig. 18.3.

For FBAR applications, ZnO is preferably used as piezoelectric material because it requires low deposition temperature and a high piezoelectric constant [19].

Commonly used electrode materials are platinum (Pt), aluminum (Al), and tungsten (W) and in this work, we used Al as electrode material due to its high acoustic velocity and low density [20].

Here we have proposed an FBAR structure consisting of a stack of Si/Al/ZnO/Al as shown in Fig. 18.1. Device dimensions comprise of width 1000 μm and thickness for Si substrate, bottom electrode, and piezoelectric material are 5, 0.2, and 2 μm respectively. The width for top electrode and PIB layer is 500 nm with thickness 0.1 μm as depicted in Fig. 18.2. The density of the PIB layer is changed due to exposure of gas on the surface of FBAR sensor.

$$\rho_{PIB+gas} = \rho_{PIB} + \Delta\rho \tag{18.4}$$

$$\Delta\rho = k.C_v.M_{gas} \tag{18.5}$$

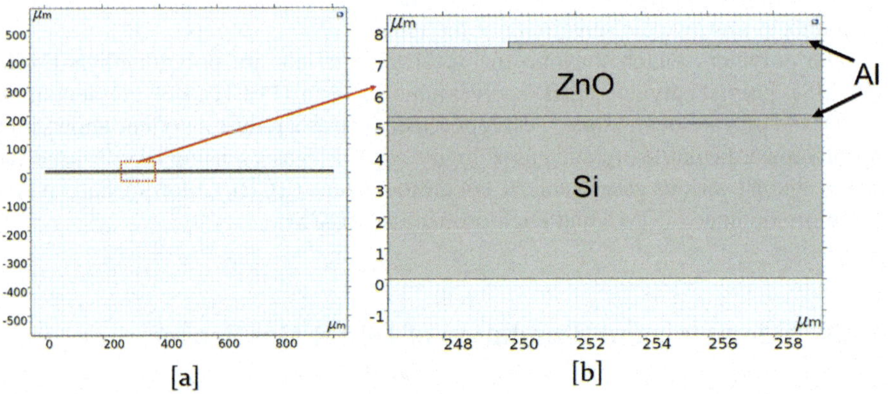

Fig. 18.1 Proposed FBAR schematic design **a** 2D FBAR geometry, **b** zoomed-in image of FBAR to represent the materials in the multilayer structure

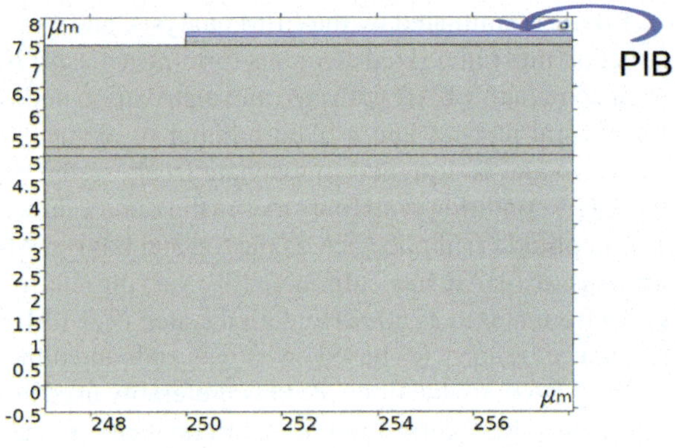

Fig. 18.2 Proposed FBAR sensor schematic with PIB as gas sensing layer

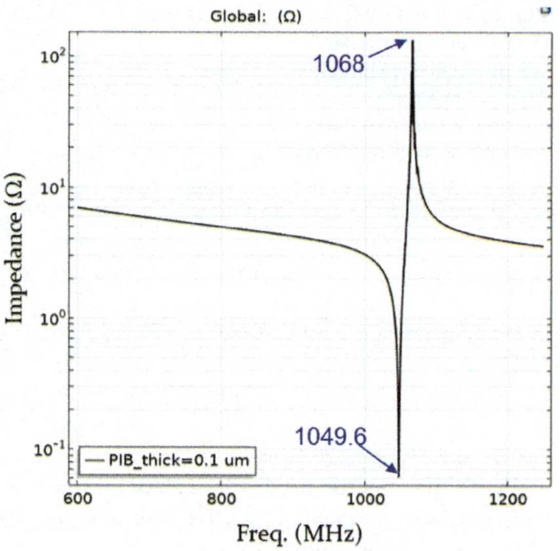

Fig. 18.3 PIB-FBAR frequency versus impedance before gas exposure

where C_v, M_{gas} and k is the volumetric concentration, molar mass and partition coefficient of target gas respectively. Equation 18.4 represents the resultant density of PIB layer where, ρ_{PIB} is the density and $\Delta\rho$ is the shift in the density of the PIB layer due to gas absorption. Equation 18.5 represents the change in density at atmospheric pressure and temperature due to exposure to target gas [21, 22].

18.4 Result and Discussion

The electrical performance of FBAR sensors has been evaluated in the frequency domain analysis before and after gas exposure. Air is taken as reference gas and the resonant frequency of the FBAR gas sensor is evaluated before gas exposure. PIB-FBAR performance in Fig. 18.3 shows the variation of impedance with the frequency curve (Z_{11}) of FBAR. Here we used PIB layer thickness of 0.1 μm because the effect of an additional layer such as the sensing layer cause medium nonlinearity, which leads to acoustic losses and spurious modes.

The resonant frequency f_r is obtained at the minimum impedance of 0.062 Ω which is 1.049 GHz and the anti-resonant frequency f_a is obtained at a maximum impedance of 140 Ω which is 1.068 GHz. The total stored electro-mechanical energy ($k^2{}_{eff}$ (%)) 4.178 is calculated from the resonance and antiresonance frequency with the following relation [22].

$$k_{eff}^2(\%) = \frac{\pi^2}{4}\left(\frac{f_a - f_r}{f_a}\right) \times 100 \tag{18.6}$$

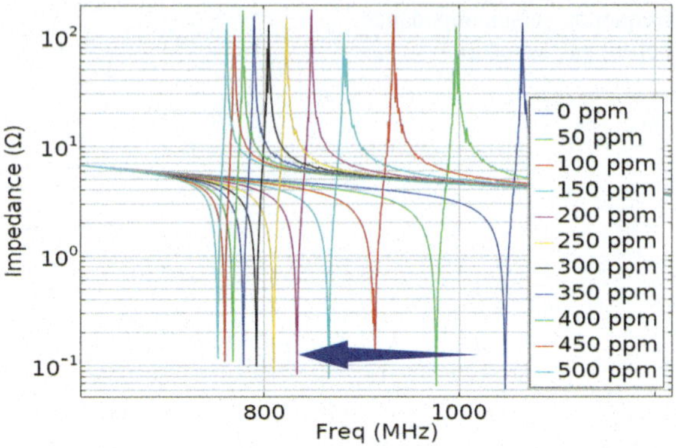

Fig. 18.4 PIB-FBAR (FBAR sensor) impedance versus frequency shift after TCE gas exposure at concentrations from 0 to 500 ppm at steps of 50 ppm

A multilayer structured (PIB-FBAR) is used for the detection of Trichloroethylene (TCE), Chloromethane (CM), Trichloroethane (TCM), and Dichloromethane (DCM) at 1 atm pressure and 25 °C.

The variation in the fundamental resonance frequency of the FBAR sensor is observed when exposed to 50–500 ppm of Trichloroethylene (TCE) is shown in Fig. 18.4. A negative shift trend (depicted in Fig. 18.4 as a blue arrow) in the resonant frequency because of the continuous increase in the density of PIB film after gas adsorption is evident from Fig. 18.4.

Each gas has a different partition coefficient (k), volumetric concentrations (C_v), and molar mass (M_{gas}) as mentioned in [21], which basically decides the change in density of PIB film after gas exposure, and finally FBAR device experience a negative shift in resonant frequency with respect to mass loading on the surface. Figure 18.5 represents the shift in resonant frequency due to even a small change in the concentration of various volatile gases. It is clear from the graph that maximum shift in resonance frequency is observed for TCE, whereas, the resonant frequency shift is negligible for CM. Table 18.1. shows the variation in resonant frequency when the FBAR Sensor is exposed to 100 ppm concentration of various gases namely TCE, CM, TCM, and DCM. The FBAR sensor with 0.1 μm PIB layer is highly sensitive to TCE while it has a fairly good linear response for DCM gas. Gas sensitivity of the FBAR gas sensor can be estimated by the slope of frequency versus gas concentration curve. Maximum gas sensitivity is observed for TCE is (2.619 MHz/ppm) and minimum for CM (0.023 MHz/ppm).

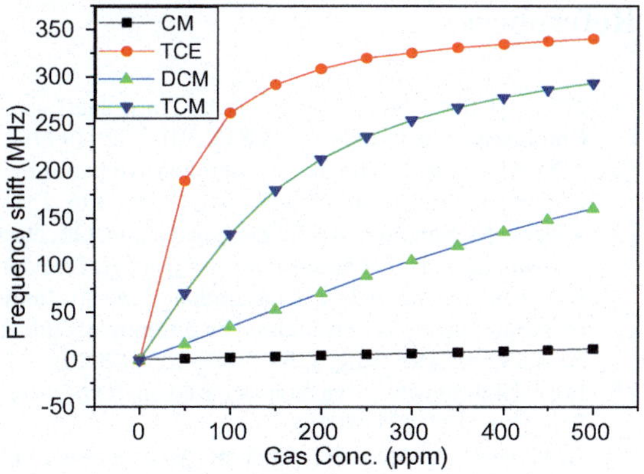

Fig. 18.5 Resonance frequency shift versus different gas concentrations from 0 to 500 ppm

Table 18.1 Comparison of resonance frequency shift and sensitivity of volatile organic gas detection

Target gas	Resonance frequency (MHz)	Resonance frequency shift (MHz)	Sensitivity (MHz/ppm)
Trichloroethylene (TCE)	784.8	261.9	2.619
Chloromethane (CM)	1044.4	2.3	0.023
Trichloroethane (TCM)	976	113.6	1.336
Dichloromethane (DCM)	1029.4	35.4	0.354

18.5 Conclusion

In this work FBAR gas sensor is designed with PIB as a sensing layer and the same was evaluated for various volatile gases such as TCE, CM, TCM, and DCM using COMSOL MULTIPHYSICS 5.4. The performance of the sensor is evaluated by the absolute impedance versus frequency relationship. The proposed PIB-FBAR structure shows a good coupling coefficient of 4.178%, high resonance frequency (1.046 GHz), and a high-quality factor. The sensor exhibits high sensitivity of 2.619 MHz/ppm against TCE gas (100 ppm). On the other hand, sensor exhibits excellent linearity from 100 to 500 ppm for DCM gas however, a low sensitivity as compared to that of TCE and TCM.

Acknowledgements The authors would like to thank Director CSIR-CEERI, Pilani for providing the research facilities. The authors also thank all the scientific and technical staff for providing their valuable feedback and suggestions.

References

1. G.R. van der Hoff, P. van Zoonen, Trace analysis of pesticides by gas chromatography. J. Chromatogr. AChromatogr. A **843**, 301–322 (1999)
2. A.M. Muscalu, T. Górecki, Comprehensive two-dimensional gas chromatography in environmental analysis. TrAC Trends Anal. Chem. **106**, 225–245 (2018)
3. T. Feng, M. Sun, S. Song, H. Zhuang, L. Yao, Gas chromatography for food quality evaluation, in *Evaluation Technologies for Food Quality*, Elsevier, 2019, pp. 219–265
4. B.K. Paul, K. Ahmed, V. Dhasarathan, F.A. Al-Zahrani, M.N. Aktar, M.S. Uddin, A.H. Aly, Investigation of gas sensor based on differential optical absorption spectroscopy using photonic crystal fiber. Alex. Eng. J. **59**, 5045–5052 (2020)
5. H. Li, High sensitivity gas sensor based on IR spectroscopy technology and application. Photon. Sensors **6**, 127–131 (2016)
6. A. Mujahid, A. Afzal, F.L. Dickert, An overview of high frequency acoustic sensors—QCMs, SAWs and FBARs—chemical and biochemical applications. Sensors **19**, 4395 (2019)
7. K. Länge, Bulk and surface acoustic wave sensor arrays for multi-analyte detection: a review. Sensors **19**, 5382 (2019)
8. A. Kumar, R. Prajesh, The potential of acoustic wave devices for gas sensing applications. Sensors Act. A Phys., 113498 (2022)
9. K.M. Lakin, Thin film resonators and filters, in *1999 IEEE Ultrasonics Symposium. Proceedings. International Symposium (Cat. No. 99CH37027)* (1999)
10. Z. Sabani, A.A.M. Ralib, J. Karim, N.B. Saidin, N.F.B.M. Razali, Design and simulation of film bulk acoustic wave resonator (FBAR) Gas sensor based on ZnO thin film, in *2019 IEEE Regional Symposium on Micro and Nanoelectronics (RSM)* (2019)
11. G.J.Z.P. Sauerbrey, The use of quarts oscillators for weighing thin layers and for microweighing. Z. Phys. **155**, 206–222 (1959)
12. L. García-Gancedo, J. Pedrós, E. Iborra, M. Clement, X.B. Zhao, J. Olivares, J. Capilla, J.K. Luo, J.R. Lu, W.I. Milne, et al., Direct comparison of the gravimetric responsivities of ZnO-based FBARs and SMRs. Sens. Act. B Chem. **183**, 136–143 (2013)
13. N. Abraham, R.R. Krishnakumar, C. Unni, D. Philip, Simulation studies on the responses of ZnO-CuO/CNT nanocomposite based SAW sensor to various volatile organic chemicals. J. Sci. Adv. Mater. Dev. **4**, 125–131 (2019)
14. Y.W. Wang, C.Y. Ao, Z.P. Hui, X. Yan, Z. Hu, Y. Chang, H.M. Qu, X.X. Duan, W. Pang, Film bulk acoustic resonator based gas sensor: A sensitive detector for gas chromatography, in *2017 19th International Conference on Solid-State Sensors, Actuators and Microsystems (TRANSDUCERS)* (2017)
15. S. H. Lee, Y. Jung, T. Kim, T. Kim, Y. Kim, S. Jung, Polymer coated film bulk acoustic resonator (FBAR) arrays for indoor air quality (IAQ) monitoring, in *2015 IEEE Sensors* (2015)
16. S.S.B. Hashwan, M.M. Khir, Y. Al-Douri, A.Y. Ahmed, A.S. Algamili, S.S. Alabsi, M.M. Junaid, Analytical modeling of AlN-based film bulk acoustic wave resonator for hydrogen sulfide gas detection based on PiezoMUMPs. J. Phys. Conf. Ser. (2021)
17. Y. Zhao, Q. Yang, Y. Chang, H. Qu, W. Pang, H. Zhang, X. Duan, Detection and discrimination of volatile organic compounds using a single multi-resonance mode piezotransduced silicon bulk acoustic wave resonator (PSBAR) as virtual sensor array. Sens. Act. B Chem. **254**, 1191–1199 (2018)
18. R. Hoffmann, M. Schreiter, J. Heitmann, The concept of thin film bulk acoustic resonators as selective CO_2 gas sensors. J. Sens. Sens. Syst. **6**, 87–96 (2017)
19. X. Zhao, Z. Zhao, B. Wang, Z. Qian, T. Ma, The design of a frame-like ZnO FBAR sensor for achieving uniform mass sensitivity distributions. Sensors **20**, 2408 (2020)
20. G.F. Iriarte, J. Bjurstrom, J. Westlinder, F. Engelmark, I.V. Katardjiev, Synthesis of c-axis-oriented AlN thin films on high-conducting layers: Al, Mo, Ti, TiN, and Ni. IEEE Trans. Ultrason. Ferroelectr. Freq. ControlUltrason. Ferroelectr. Freq. Control **52**, 1170–1174 (2005)

21. S. Johnson, T. Shanmuganantham, Design and analysis of SAW based MEMS gas sensor for the detection of volatile organic gases. Carbon **119**, 0–041316 (2014)
22. A.K. Johar, R. Patel, C. Periasamy, A. Agarwal, D. Boolchandani, FEM modeling and simulation of SMFBAR sensor with PIB as sensing layer for tetrachloroethane (PCE) gas detection. Mater. Res. Express **6**, 015033 (2018)

Chapter 19
Design and Simulation of an Open EWOD Based Digital Microfluidic Device for Droplet Actuation Using COMSOL

Debopam Bhattacharya, Subhadip Chakraborty, Anuraag Mukherjee, Anupam Karmakar, and Sanatan Chattopadhyay

19.1 Introduction

The necessity for rapid, point-of-care devices has given a boost to the development of Lab-on-a-chip (LOC) technologies. In recent times, digital microfluidics (DMF) has emerged as the superior technique for the development of LOC devices due to several advantages like reduced reagent volume (μl, nl, pl), enhanced sensitivity and selectivity, decreased risk of toxic sample operations, user-friendliness, and the ease of assimilating several processes on an individual chip [1]. In DMF, individual droplets can be independently manipulated employing voltages through an array of patterned electrodes. The basic DMF operations consist of transporting, merging, splitting of droplets and droplet dispensation from reservoirs [2]. DMF is typically implemented in two distinct configurations: (i) closed or two plate format and (ii) open or single plate format [3]. In the closed setup DMF, the droplets are sandwiched between the two substrates patterned with electrodes. In this configuration, the top plate consists of the ground electrode and the bottom plate comprises the voltage-controlled electrodes. [3, 4]. On the other hand, an open system DMF is made of a hydrophobic-dielectric layer structure on top of an electrode patterned substrate, with the droplet positioned on the top of the hydrophobic layer. The open setup DMF has both the ground and the activated electrodes in the same plane [3, 4]. Microfluidic systems, both closed and open, have distinct advantages. The two-plate closed DMF systems are capable of basic microfluidic operations however, these

D. Bhattacharya · S. Chakraborty · A. Mukherjee · A. Karmakar · S. Chattopadhyay (✉)
Department of Electronic Science, University of Calcutta, 92, A.P.C. Road, Kolkata 700009, India
e-mail: scelc@caluniv.ac.in

S. Chakraborty
Department of Chemical Biological and Macro-Molecular Sciences, S.N. Bose National Centre for Basic Sciences, Salt Lake, JD Block, Sector-III, Bidhannagar, Kolkata 700106, India

R. Singh et al. (eds.), *The Physics of Semiconductor Devices*, Springer Proceedings in Physics 306, https://doi.org/10.1007/978-981-97-1571-8_19

are not suitable for processes involving mixing and evaporation of liquid droplets as well as its integration with different analytical tools [5]. Further, two-plate closed setup DMF devices are often filled with supplementary media like silicone oil, for the reduction of the required voltage for droplet actuation [6, 7] however, the oil-filled closed DMF systems encompass disadvantages like solute partitioning in the oil, oil leakage, incompatibility for oil-mixable liquids, and inappropriate for applications involving droplet drying [7].

Although droplet splitting is a challenge for the open setup DMF devices [7], they are capable of easy droplet manipulation, ease of assimilation of analytical tools, and are useful for the processes involving mixing and evaporation of droplets [8]. Therefore, integration of both closed and open system DMF is sought for the development of LOC devices. Several design schemes have been proposed for the development of single-plate open DMF devices nevertheless, the challenge exists in the miniaturization of the dielectric-hydrophobic layer thicknesses and the reduction of required actuating voltage [5, 9, 10].

In this work, a novel open Digital Microfluidic device is designed and droplet actuation is simulated employing the Finite Element Analysis (FEA) method. The pressure difference between the leading and trailing side of the droplet, contact angle due to voltage (θ_v), the extent of change in the contact angle ($\Delta\theta = \theta_0 - \theta_v$) and the droplet velocity for three different dielectric materials (HfO_2, Al_2O_3, and Si_3N_4) are evaluated. It is observed that of the three different dielectric layers, HfO_2 provides superior droplet actuation. Further, relevant parameters are investigated for different Teflon (20–200 nm) and HfO_2 (20–260 nm) thickness combinations for the optimized droplet actuation. Actuating voltage is increased from 0.5 to 5 V to study the velocity of the droplet and the contact angle due to voltage (θ_v). Additionally, the variation of the voltage at the leading edge of the droplet and the contact angle due to voltage (θ_v) are investigated for different permittivity of the droplet. The proposed device scheme can be employed in the fabrication of a digital microfluidic device for the electrical droplet actuation of a range of biological samples.

19.2 Simulation Details

19.2.1 Model Description

The FEA of the open EWOD based Digital Microfluidic Device is performed using the COMSOL Multiphysics 5.3a. The schematic representation of the designed device is illustrated in Fig. 19.1 and the Young-Lippmann equation is employed to visualize the droplet actuation [11].

$$\cos\theta_v = \cos\theta_0 + \frac{\varepsilon_r * \varepsilon_0 * V_{app}^2}{2 * \gamma * d} \tag{19.1}$$

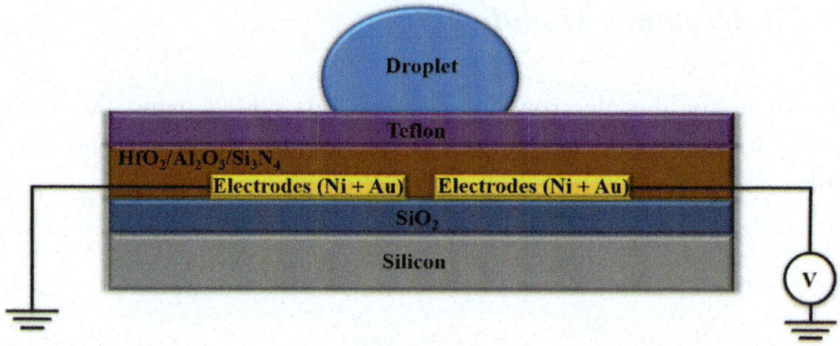

Fig. 19.1 Schematic representation of the designed open EWOD based Digital Microfluidic device

Here θ_v is the modified contact angle due to an applied voltage, θ_0 is the initial contact angle of the droplet, ε_r is the equivalent dielectric constant of the dielectric and the Teflon layer, ε_0 is the permittivity of free space, V_{app} is the applied voltage, γ is the surface tension of water–gas interface and d is the sum of dielectric and Teflon layer thickness. The contact angle of water droplet on a Teflon surface has been recorded to vary from 108° to 114°, thus an average value is taken for the simulation purpose [12, 13]. Three distinct dielectric materials (HfO_2, Al_2O_3, and Si_3N_4) are compared, and the thicknesses of the HfO_2 and Teflon layers are modified for optimal droplet actuation. The simulations are carried out employing the laminar two-phase flow, moving mesh (tpfmm) module, and the voltage drop at the leading side of the droplet for various droplet dielectric constants (5–100) is investigated using the Electrostatics (es) module. The simulation parameters are summarized in Table 19.1. A time-dependent investigation is conducted.

Table 19.1 Simulation parameters for droplet actuation

Name	Value	Description
θ_0	108.85°	The initial contact angle of the water droplet [12, 13]
Γ	0.072 N/m	The surface tension of the water–air interface [14]
ε_r	Variable	The equivalent dielectric constant of the dielectric and the Teflon layer
$d_{dielectric\ layer}$	20–260 nm	Dielectric layer thickness
$d_{Teflon\ layer}$	20–200 nm	Teflon layer thickness
μ_{air}	18.6 μPa-s	Air viscosity
$\mu_{water\ droplet}$	0.89 mPa-s	Water viscosity
V_{app}	3, 10 V	Applied voltage

19.2.2 Electrostatics Module

Applied voltage for the electrowetting purpose is added using the electrostatics module. The module is based on the equations for electric field and the electric displacement field.

$$\nabla \cdot \vec{\mathrm{D}} = \rho_v \tag{19.2}$$

$$E = -\nabla V \tag{19.3}$$

$$D = \varepsilon_0 \varepsilon_r E \tag{19.4}$$

Here, $\vec{D}$ (C/m^2) is the electric displacement field, $\vec{E}$ (V/m) is the electric field strength, ε_0 (F/m) is the permittivity of free space, and ε_r is the relative permittivity and V (V) is the electric potential.

19.2.3 Laminar Two-Phase Flow, Moving Mesh Module

Droplet actuation on an open EWOD device in an air environment is simulated using the laminar two-phase flow moving mesh module. An incompressible flow is considered and the inertial term is neglected in the Navier–Stokes equation to setup the model.

$$\rho \frac{\partial \vec{u}}{\partial t} = \nabla \cdot \left[-pI + \mu \left(\nabla \vec{u} + (\nabla \vec{u})^T \right) \right] + F \tag{19.5}$$

$$\nabla \cdot \vec{\mathrm{u}} = 0 \tag{19.6}$$

Here, ρ (kg/m^3) is the density, $\vec{u}$ (mm/s) is the velocity, p (Pa) is the pressure, I is the unit matrix, μ (Pa * s) is the dynamic viscosity, and F (N/m^3) is any external force acting on the volume of fluid.

No slip boundary condition is applied to all the boundaries except those separating the liquid droplet from the solid. This is expressed as:

$$\vec{u} = 0 \tag{19.7}$$

Navier slip, boundary condition is applied to the interface separating the liquid droplet and solid. Mathematically, this is expressed as:

$$\vec{u} \cdot \vec{n} = 0 \tag{19.8}$$

$$\vec{n} \cdot \mathrm{T} = \frac{-\mu}{\beta}\vec{u} \tag{19.9}$$

β is slip length and $\vec{n}$ is normal to the wall. Navier slip boundary condition permits to define contact angle at the solid wall-fluid interface. The fluid–fluid interface is used to add the surface tension of the water droplet in an air environment. The Young-Lippmann equation links the electrostatics module with the laminar two-phase flow moving mesh module. Finally, a time-dependent study is observed to investigate the droplet actuation on the modelled EWOD device.

19.3 Results and Discussions

19.3.1 Material Comparison

Pressure difference inside the droplet. Figure 19.2 illustrates the pressure difference between the leading and trailing sides of the droplet as a function of time for an applied voltage of 10 V. The pressure difference in the actuated droplets after 2 ms of the applied voltage is 32.17 Pa, 32.05 Pa, and 30.03 Pa for the devices modelled with dielectric layers HfO_2, Al_2O_3, and Si_3N_4, respectively. The pressure difference between the leading and trailing edges of the droplet reduces as it approaches the active electrode.

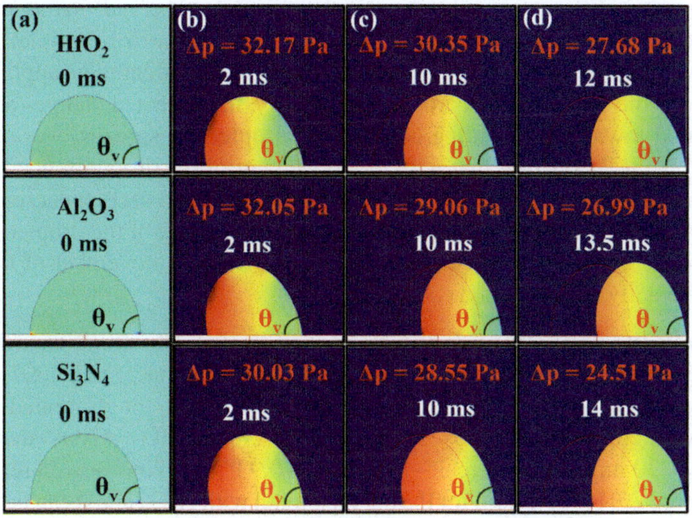

Fig. 19.2 The pressure difference between the droplet's leading and trailing edges as a function of time

To reduce surface tension, a liquid profile prefers to adopt a small surface area, but due to the capillary restrictions of the hydrophobic surface, it assumes a spherical profile [15]. The pressure difference between the inside and outside the droplet profile is given by [15]:

$$\Delta p = p_{inside} - p_{outside} = \frac{2\gamma}{r} \quad (19.10)$$

Here γ is the surface tension and r is the radius of curvature. When one electrode is triggered with 10 V and the other is grounded, opposing charges in the droplet in relation to the applied potential are drawn towards the activated electrode. The surface tension of the liquid–solid interface above the activated electrode reduces and generates an alteration of the contact angle [16]. Thus, an asymmetrical interfacial tension of the liquid droplet is created leading to asymmetric deformation of the droplet. Thereby, a pressure difference inside the droplet is generated [17] leading to its movement. Here r_1 is the radius of the curvature of the droplet arc above the ground electrode whereas r_2 is the radius of the curvature of the droplet arc above the activated electrode.

$$\Delta p = (p_{inside_1} - p_{outside}) - (p_{inside_2} - p_{outside}) \quad (19.11)$$

$$\Delta p = \frac{2\gamma}{r_1} - \frac{2\gamma}{r_2} \quad (19.12)$$

It is observed from Fig. 19.2 that amongst all the dielectrics, HfO_2 provides the largest pressure difference in the droplet. It is further observed that as the droplet moves towards the activated electrode the pressure difference in the droplet decreases. As the droplet moves towards the activated electrode, the effect of an applied voltage spreads over the entire droplet profile. Apparently, the attracted charges are not just accumulated towards the leading edge of the droplet but spreads over the droplet profile. Thus, it is apparent that with the movement of the droplet towards the activated electrode, the asymmetricity of surface tension of the droplet decreases which consequently reduces the asymmetricity of the deformation of the droplet. Thus, the pressure difference inside the droplet decreases with the movement of the droplet towards the activated electrode.

Droplet velocity and voltage-driven contact angle (θ_v). The alteration in the contact angle due to voltage (θ_v) and the amount of change of contact angle ($\Delta\theta = \theta_0 - \theta_v$) for three different dielectric layers HfO_2, Al_2O_3, and Si_3N_4, for an applied voltage of 3 V is illustrated in Fig. 19.3a, b respectively. The $\Delta\theta$ for HfO_2, Al_2O_3, and Si_3N_4 for thicknesses 20 nm–260 nm are 3°–1.6°, 2.77°–0.93°, and 2.62°–0.75° respectively. It is observed $\Delta\theta$ is maximum for the device model with HfO_2 and it can be attributed to the maximum capacitive coupling provided by the HfO_2 dielectric layer. Further, it is observed that for an increase in the dielectric layer thicknesses, the $\Delta\theta$ decreases. Due to the reduced amplitude of the electric field created across the droplet for an

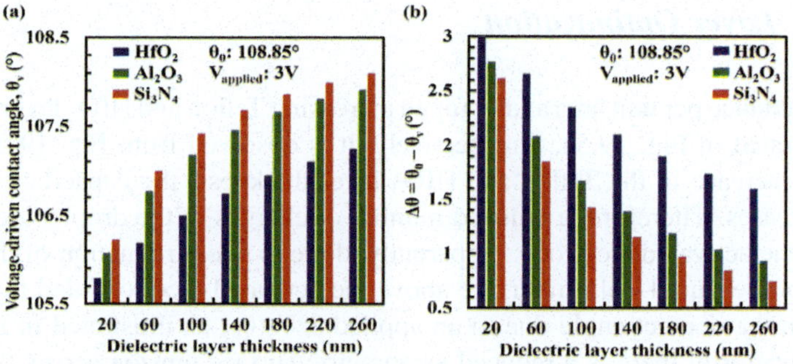

Fig. 19.3 **a** The voltage-driven contact angle (θ_V) for an increasing dielectric layer thickness; **b** the amount of change of contact angle ($\Delta\theta = \theta_0 - \theta_V$) for an increasing dielectric layer thickness

increase in the dielectric layer thickness, it is evident that fewer charges are attracted towards the active electrode. Hence, the droplet's capacitive coupling reduces. Thus, there is less change in the initial contact angle due to an applied voltage for an increasing dielectric layer thickness ($\Delta\theta$).

The nature of velocity acquired by the droplets to move towards the activated electrode is illustrated in Fig. 19.4. There is a lesser magnitude of change in the contact angle due to an applied 3 V ($\Delta\theta$) for Al_2O_3, and Si_3N_4 compared to HfO_2, accordingly, the maximum asymmetricity of the droplet profile occurs for the device modelled with HfO_2. Thus, maximum pressure difference inside the droplet is created for the device modelled with HfO_2 as illustrated in Fig. 19.2, consequently, the droplets acquire the maximum velocity to traverse towards the activated electrode. Further, as the $\Delta\theta$ decreases with an increase in dielectric thickness, the droplet profile acquires reduced asymmetric deformation. Thereby, the pressure difference in the droplet decreases and consequently, the velocity of the droplet decreases with increasing dielectric layer thickness [16, 17]. Therefore, it is observed from Figs. 19.2, 19.3a, b and 19.4 that out of three different dielectric layers, HfO_2 provides superior droplet actuation.

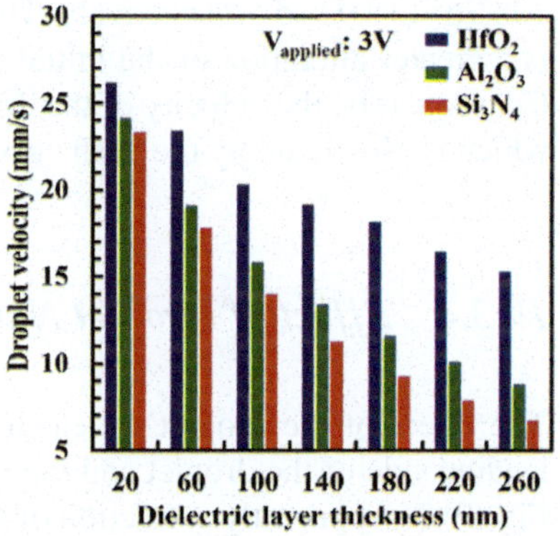

Fig. 19.4 The nature of droplet velocity for three different dielectric layers HfO_2, Al_2O_3, and Si_3N_4 for an increasing dielectric layer thickness

19.3.2 Layer Optimization

The capacitance per unit area and θ_v for an increasing Teflon and HfO_2 layer thickness are illustrated in Fig. 19.5a, b respectively. It is observed from Fig. 19.5a that for both, an increase in the Teflon and HfO_2 layer thickness, the capacitance per unit area decreases. Therefore, a reduced number of charges in the droplet are attracted towards the activated electrode. Apparently, there is a less reduction of the surface tension of the liquid–solid interface above the activated electrode and there is less change in the contact angle due to an applied 3 V (θ_v) as illustrated in Fig. 19.5b [16]. Henceforth, there is a reduced asymmetric surface tension across the droplet and the asymmetric deformation of the droplet decreases for an increase in the Teflon and HfO_2 layer thickness. The asymmetric distortion of the droplet profile induces a pressure difference inside the droplet, leading to its movement [17]. Apparently, due to the reduced asymmetric distortion of the droplet profile for an increase in the Teflon and HfO_2 layer thicknesses, the induced pressure difference in the droplet decreases for an increase in the above-mentioned layer thicknesses. Thus, the droplet velocity decreases for an increase in the Teflon and HfO_2 layer thickness as illustrated in Fig. 19.5c. Further, it can be noticed from Fig. 19.5b that there is a much greater variation of the contact angle due to an applied voltage for the Teflon layer thickness of 20 nm, whereas the variation of the contact angle due to voltage is less for an increase in Teflon layer thickness. Hence, Teflon and HfO_2 layer both of thickness 20 nm provides superior droplet actuation.

19.3.3 Effect of Voltage on Droplet Actuation

The effect of an increasing applied voltage (0.5–5 V) on the contact angle and velocity of the droplet is illustrated in Fig. 19.6. For an increasing applied voltage, an increased number of charges in the droplet are attracted towards the activated electrode causing more change in contact angle. For a droplet of $\theta_0 = 108.85°$, an increase in the applied voltage from 0.5 V to 5 V, the contact angle due to voltage changes from 108.76° to 100.45° respectively. Therefore, it is observed from Fig. 19.6 that there is a greater alteration of the initial contact angle for an increasing applied voltage. Consequently, the velocity of the droplet increases for an increasing applied voltage which is illustrated by the dash curve in Fig. 19.6.

19.3.4 Effect of Droplet Nature on Droplet Actuation

The effect of the droplet's dielectric constant (5–100) on the voltage drop at the leading side of the droplet and the voltage-driven contact angle (θ_v) is illustrated in Fig. 19.7. Apparently, a fraction of the applied voltage gets dropped at the dielectric

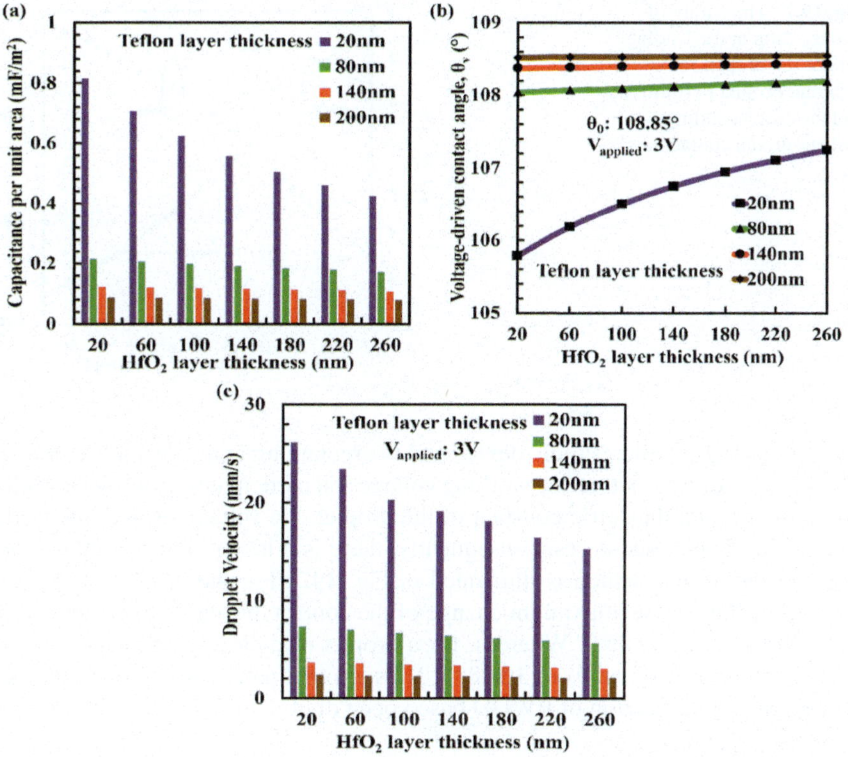

Fig. 19.5 **a** The nature of capacitance per unit area of the designed device for an applied voltage of 3 V for an increasing Teflon and HfO_2 layer thickness, **b** the nature of contact angle due to an applied 3 V (θ_v°) for an increasing Teflon and HfO_2 layer thickness, **c** the variation of droplet velocity for an increase in Teflon and HfO_2 layer thicknesses for an applied voltage of 3 V

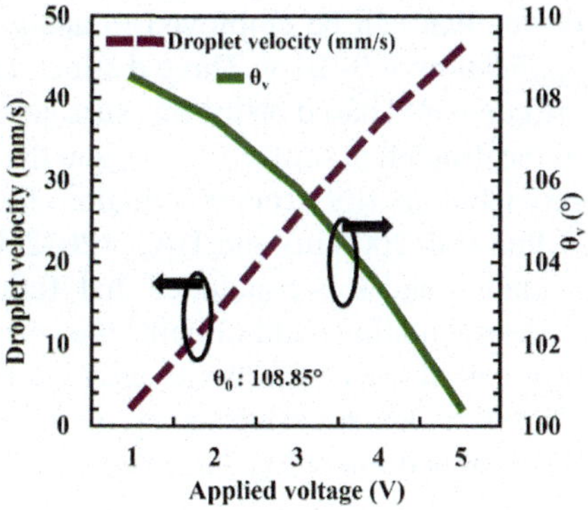

Fig. 19.6 The variation of the contact angle due to voltage (θ_V) and the corresponding droplet velocity for an increasing applied voltage

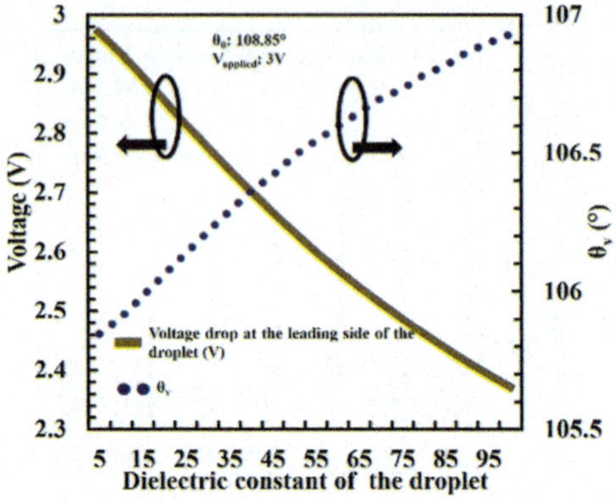

Fig. 19.7 The nature of voltage drop at the leading side of the droplet for different droplet permittivity and the corresponding voltage-driven contact angle (θ_V)

layer of the digital microfluidic device and the remaining at the droplet profile. The voltage at the droplet acts as the working voltage and participates in droplet actuation. For an increasing dielectric constant of the droplet, the voltage drop at the leading side of the droplet decreases, consequently, there is a lesser change in the contact angle of the droplet which is illustrated in Fig. 19.7. For the droplet of dielectric constant 5, there is the maximum change of the contact angle due to an applied 3 V (θ_0: 108.85°, θ_v: 105.85°). Whereas, for a droplet of dielectric constant 100, θ_v is 106.93° reduced from an θ_0 of 108.85°. Thus, droplet nature is an important factor for design and fabrication of EWOD based open digital microfluidic devices.

19.4 Conclusion

The present work sought to design a novel open EWOD based Digital Microfluidic device that can be employed to analyze a wide variety of biological fluids at the application of 3–10 V. Three distinct dielectric materials (HfO_2, Al_2O_3, and Si_3N_4) are compared based on the pressure difference between the leading and trailing edges of the droplet, θ_v, $\Delta\theta = \theta_0 - \theta_v$, and the droplet velocity. It is observed that the HfO_2 provides superior droplet actuation. Further, relevant studies are done for different Teflon (20–200 nm) and HfO_2 (20–260 nm) combinations for the optimal droplet actuation and it is concluded that Teflon and HfO_2 both of 20 nm provides better droplet actuation. Additionally, the voltage drop at the leading edge of the droplet for different droplet permittivity are investigated and the corresponding θ_v is observed. Therefore, an extensive variety of biological fluids can be analyzed by employing the designed open DMF device.

Acknowledgements Mr. Debopam Bhattacharya likes to acknowledge SERB, Department of Science and Technology (DST, GoI) for funding the project titled "Point-of-care Electronic Diagnosis of Anemic Diseases by Employing Impedimetric Techniques", CRG/2018/001922. The authors would also like to acknowledge the funding support from DST PURSE (Ref. no.

DST-PURSE/FUND ALLOCATION/PH I/005/7502; dated 04.08.2012), Government of India for developing the Electrical Characterization Laboratory.

References

1. J. Berthier, Chapter 1—introduction: digital microfluidics in today's microfluidics, 2nd edn., in *Micro and Nano Technologies, Micro-Drops and Digital Microfluidics*, ed. by J. Berthier (William Andrew Publishing, 2013), pp. 1–6. ISBN 9781455725502. https://doi.org/10.1016/B978-1-4557-2550-2.00001-8
2. H. Geng, J. Feng, L.M. Stabrylac, S.K. Cho, Dielectrowetting manipulation for digital microfluidics: creating, transporting, splitting, and merging of droplets. Lab Chip **17**, 1060 (2017). https://doi.org/10.1039/c7lc00006e
3. M. Abdelgawad, A. Wheeler, The digital revolution: a new paradigm for microfluidics. Adv. Mater., **21** (2009). https://doi.org/10.1002/adma.200802244
4. W. Wang, T.B. Jones, Moving droplets between closed and open microfluidic systems. Lab Chip **15**, 2201 (2015). https://doi.org/10.1039/c5lc00014a
5. U.-C. Yi, C.-J. Kim, Characterization of electrowetting actuation on addressable single-side coplanar electrodes. J. Micromech. Microeng.Micromech. Microeng. **16**, 2053–2059 (2006). https://doi.org/10.1088/0960-1317/16/10/018
6. M.G. Pollack, A.D. Shenderov, R.B. Fair, Electrowetting-based actuation of droplets for integrated microfluidics. Lab Chip **2**(2), 96–101 (2002). https://doi.org/10.1039/b110474h. Epub 11 Mar 2002. PMID: 15100841
7. K. Choi, A.H. Ng, R. Fobel, A.R. Wheeler, Digital microfluidics. Annu. Rev. Anal. Chem. **5**, 413–440 (2012). https://doi.org/10.1146/annurev-anchem-062011-143028. Epub 9 Apr 2012. PMID: 22524226
8. M. Abdelgawad, P. Park, A.R. Wheeler, Optimization of device geometry in single-plate digital microfluidics. J. Appl. Phys. **105**, 094506 (2009). https://doi.org/10.1063/1.3117216
9. M. Abdelgawad, A. Wheeler, Rapid prototyping in copper substrates for digital microfluidics. Adv. Mater. **19**, 133–137 (2006). https://doi.org/10.1002/adma.200601818
10. C.G. Cooney, C.-Y. Chen, M.R. Emerling, A. Nadim, J. Sterling, Electrowetting droplet microfluidics on a single planar surface. Microfluid. Nanofluid. **2**, 435–446 (2006). https://doi.org/10.1007/s10404-006-0085-8
11. F. Mugele, J.-C. Baret, J. Phys. Condens. Matter **17**, R70 (2005). https://doi.org/10.1088/0953-8984/17/28/R01
12. M.K. Bernett, W.A. Zisman, Relation of wettability by aqueous solutions to the surface constitution of low-energy solids. J. Phys. Chem. **63**, 1241–1246 (1959). https://doi.org/10.1021/j150578a006
13. D.K. Owens, R.C. Wendt, Estimation of the surface free energy of polymers. J. Appl. Polym. Sci.Polym. Sci. **13**, 1741–1747 (1969). https://doi.org/10.1002/app.1969.070130815
14. N.B. Vargaftik, B.N. Volkov, L.D. Voljak, International tables of the surface tension of water. J. Phys. Chem. Ref. Data **12**, 817–820 (1983). https://doi.org/10.1063/1.555688
15. J. Berthier, Chapter 2—Theory of Wetting **This chapter was written with the collaboration of Kenneth A. Brakke (Mathematics Department, Susquehanna University, Selinsgrove, PA), 2nd edn., in *Micro and Nano Technologies, Micro-Drops and Digital Microfluidics*, ed. by J. Berthier (William Andrew Publishing, 2013), pp. 7–73. ISBN 9781455725502. https://doi.org/10.1016/B978-1-4557-2550-2.00002-X
16. H. Moon, S.K. Cho, R. Garrell, C.-J. Kim, Low Voltage electrowetting-on-dielectric. J. Appl. Phys. **92** (2002). https://doi.org/10.1063/1.1504171
17. S.K. Cho, H. Moon, C.-J. Kim, Creating, transporting, cutting, and merging liquid droplets by electrowetting-based actuation for digital microfluidic circuits. Microelectromech. Syst. J. **12**, 70–80 (2003). https://doi.org/10.1109/JMEMS.2002.807467

Chapter 20
A Capacitive Transducer for Film Thickness Measurement

Imran Ahmad, Zaheeruddin, and Tarikul Islam

20.1 Introduction

The thickness measurement of the thin film is an important problem in the semiconductor process and optimization. Various research has been done on thickness measurement tools and techniques, which shows sensors based on capacitance play a vital role in the development of a device that is inexpensive and may be used for industrial and laboratory applications [1]. The conventional optical method of thickness measurement is costly and requires skilled manpower. The principles of the capacitive measurement of film thickness are well known. The film is placed over an arrangement of measuring electrodes that contact one side of the film surface. In this measurement, highly stable capacitance is needed, but parasitic earth capacitance degrades the measurement accuracy. The metal electrodes are enclosed within a grounded cylindrical shape metal surrounding called a shield to minimize the error caused by parasitic earth capacitance [2, 3].

This study has attempted to design and fabricate a parallel plate capacitive transducer with the fixed lower potential electrode and movable higher potential electrode. The higher potential electrode is connected to a micro screw gauge [4]. Only the vertical distance is precisely varied to get a capacitance for the measurement of the thin film thickness. Another micro screw gauge is connected to a higher potential electrode of one parallel plate capacitive transducer for horizontal movement, which is used to check the film uniformity [5].

I. Ahmad · Zaheeruddin · T. Islam (✉)
Department of Electrical Engineering, Jamia Millia Islamia, New Delhi, India
e-mail: tislam@jmi.ac.in

R. Singh et al. (eds.), *The Physics of Semiconductor Devices*, Springer Proceedings in Physics 306, https://doi.org/10.1007/978-981-97-1571-8_20

20.2 Measurement Technique

A capacitive transducer has been increasingly used for transduction because of its simple design, low cost, low energy consumption, and small size. The capacitance of a parallel capacitor is given by Eq. 20.2.1

$$C = \frac{A \in_0 \in_r}{t} \tag{20.2.1}$$

where C is the capacitance, $\in_0$ is the permittivity of free space, $\in_r$ is the relative permittivity of the medium, and t is the thickness of the dielectric medium present between plates. These variables are used in the capacitive transduction technique [6].

20.3 Design of a Novel Capacitive Transducer

The design consists of two parallel plate capacitive transducers connected with three different micro screw gauges, as shown in Fig. 20.1.

The upper plate of the one capacitive transducer is connected to a micro screw gauge with only vertical movement. The upper plate of another capacitive transducer is connected to two different micro screw gauges with both vertical and horizontal movement. The vertical movement is used to get the capacitance value, which gives the thickness of the part under test. In contrast, the horizontal movement gives capacitance change, which directly indicates film uniformity of part under test. The disc shape capacitor plates are made up of Brass. The diameter of the higher and lower potential electrodes is 18 mm and 32 mm, respectively. The higher and lower potential electrode thickness is 6 mm and 2 mm, respectively. Metal shielding is provided to minimize parasitic earth capacitance. Teflon is used to provide insulation between the electrode and metal shield [7].

20.4 Fabricated Transducer

The capacitance depends on the electric field confined between two plates. The electric field strength decreases with an increase in the distance between the plates. Therefore, the gap between two electrodes is kept lower than the diameter of the electrode to make the field uniform. The conducting plate with a larger surface area can store more electrical charge. Due to this, the area of the lower potential electrode is kept larger than the higher potential electrode. Therefore, a larger capacitance value is obtained with a greater surface area. The plate of the transducer is mounted so that only the gap between plates can be precisely varied to give an accurate capacitance. The selection of electrodes depends on application needs and cost. Brass is selected for electrodes at a reasonably low price in this work [8, 9]. Teflon has been used

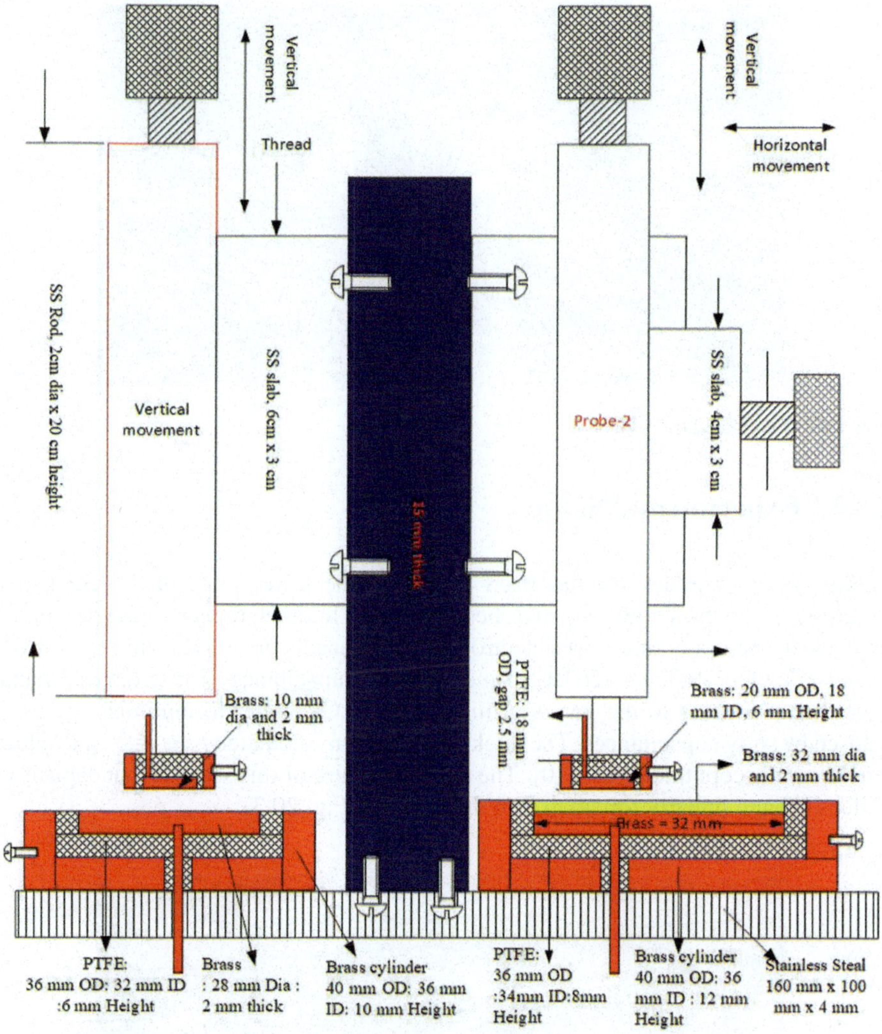

Fig. 20.1 Design of transducer

to provide the insulation between the metal electrode and metal shield, as shown in Fig. 20.2.

Fig. 20.2 Fabricated transducer

20.5 Experimental Setup

Firstly, the part under test has been placed on the lower plate of the capacitive transducer, and the screw gauge has been tightened to touch the top of the part under test. Next, the upper and lower electrode of the transducer is connected to an LCR meter (Hewlett Packard 4263B) to measure the capacitance. The grounded metal shield is connected to the guard terminal of the LCR meter to minimize the error caused by stray capacitances. The thickness has been calculated $t = \frac{A\epsilon_0\epsilon_r}{C}$ (all values are known except thickness) [10]. The desired feature of this setup is that capacitive values change linearly for small film thicknesses (Fig. 20.3).

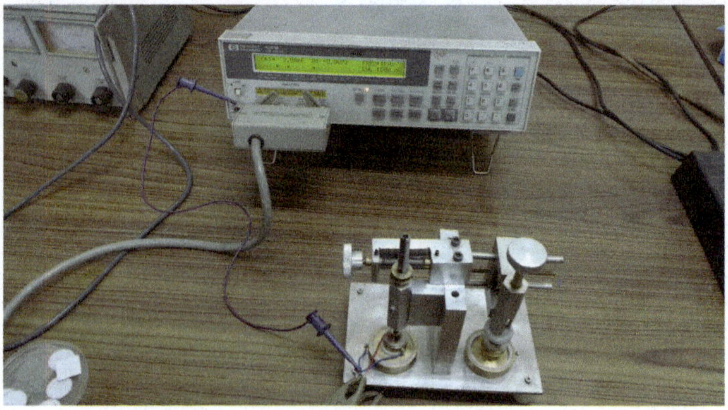

Fig. 20.3 Experimental setup

20.6 Experimental Results

Teflon of different known thicknesses, the dielectric constant is used to calibrate this setup, and the corresponding capacitance value is plotted at three different frequencies as shown in the following curve Fig. 20.4 (Table 20.1).

A further experiment has been performed to calculate the thickness of unknown samples of Teflon by the following Eq. 20.6.1 [11].

$$t = \frac{A\epsilon_0\epsilon_r}{C} \tag{20.6.1}$$

$A = \pi d^2/4$ where d = 18 mm as per design (254.4690049 mm^2).

$\epsilon_0 = 8.85x10^{-12}$F/m (Permittivity of free space), $\epsilon_r = 2.1$ (Fig. 20.5; Table 20.2).

The experimental data have been fitted with a linear curve with a correlation factor (r^2) of 0.909. Experiments were also conducted to measure the capacitance value of film thickness of different insulating materials at different frequencies. Results are shown in Table 20.3.

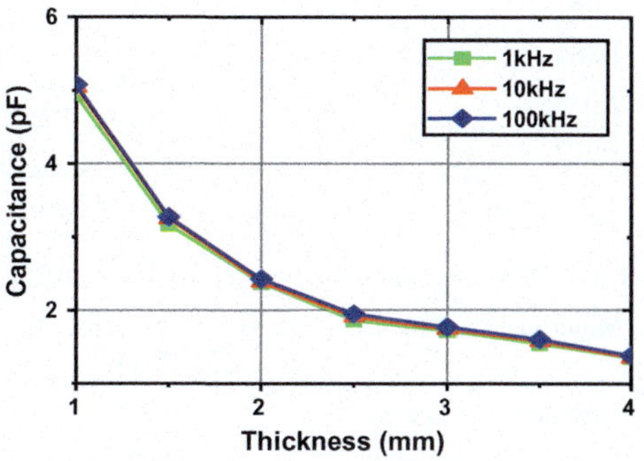

Fig. 20.4 The variation of the capacitance value with change in film thickness

Table 20.1 Capacitance values at different thicknesses and frequency

Thickness (mm)	Capacitance (pF at 1 kHz)	Capacitance (pF at 10 kHz)	Capacitance (pF at 100 kHz)
1.0	4.950	5.055	5.090
1.5	3.180	3.260	3.280
2.0	2.390	2.401	2.440
2.5	1.870	1.918	1.960
3.0	1.73	1.751	1.780
3.5	1.550	1.585	1.610
4.0	1.360	1.364	1.390

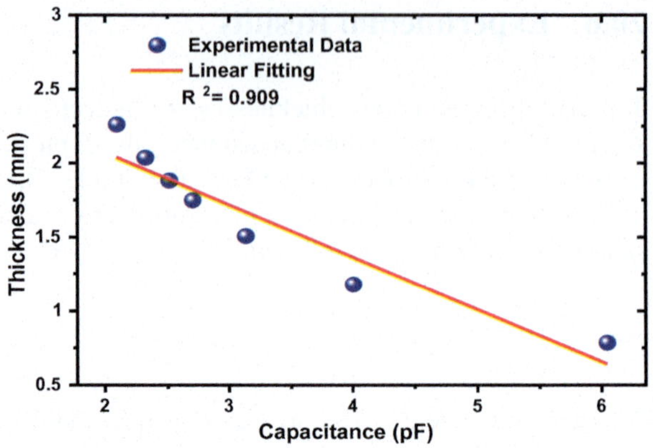

Fig. 20.5 Film thickness (mm) with the variation of the capacitance value (pF)

Table 20.2 Thickness values at different capacitance values

Thickness (mm)	Capacitance (pF at 1 kHz)	Capacitance (pF at 10 kHz)	Capacitance (pF at 100 kHz)
0.788	6.04	6.062	6.07
1.182	4.00	3.965	4.03
1.510	3.13	3.107	3.16
1.751	2.70	2.692	2.75
1.884	2.51	2.520	2.57
2.038	2.32	2.342	2.39
2.262	2.09	2.135	2.25

Table 20.3 Thickness and capacitance for different sample

Material	Thickness (mm)	Capacitance (pF at 1 kHz)	Capacitance (pF at 10 kHz)	Capacitance (pF at 100 kHz)
Alumina	0.372	12.70	12.76	12.79
Glass	0.602	7.850	7.740	7.690
Hard Paper	0.247	19.15	15.267	13.90

20.7 Conclusion

In this study, an effort has been made to demonstrate a novel technique for film thickness measurement using a capacitive transducer. An experimental setup for thickness measurement has been designed and fabricated. Initial experiments were performed to measure the thickness of the micrometer order to the millimeters of different insulating substrates like Teflon film, alumina film, and glass substrates. The thickness measurement results are encouraging, and the proposed method may be used for industrial and laboratory applications. The results presented here can certainly be improved, but they are sufficient to demonstrate the viability of the technique.

References

1. F. Castelli, The thin dielectric film capacitive displacement transducer to nanometer. IEEE Trans. Instrum. Meas.Instrum. Meas. **50**(1), 106–110 (2001)
2. X. Hu, W. Yang, Planar capacitive sensors-designs and applications. Sensor Rev. (2010)
3. J.G. Webster, H. Eren, *Measurement, Instrumentation, and Sensors Handbook: Two-Volume Set* (CRC Press, 2018)
4. B. Hou, B. Zhou, M. Song, Z. Lin, R. Zhang, A novel single-excitation capacitive angular position sensor design. Sensors **16**(8), 1196 (2016)
5. A. Guadarrama-Santana, A. Garcia-Valenzuela, Principles and methodology for the simultaneous determination of thickness and dielectric constant of coatings with capacitance measurements. IEEE Trans. Instrum. Meas.Instrum. Meas. **56**(1), 107–112 (2007)
6. M. Kim, W. Moon, E. Yoon, K.R. Lee, A new capacitive displacement sensor with high accuracy and long-range. Sens. Act. A **130**, 135–141 (2006)
7. M. Kim, W. Moon, E. Yoon, K.R. Lee, A new capacitive displacement sensor with high accuracy and long range, in *The 13th International Conference on Solid-State Sensors, Actuators and Microsystem. Digest of Technical Papers. TRANSDUCERS'05*, vol. 1 (IEEE, 2005), pp. 559–562
8. E.O. Doebelin, D.N. Manik, *Measurement Systems: Application and Design*, 4th edn. (McGraw-Hill, New York, 2007)
9. J. Fraden, J. Fraden, *Handbook of Modern Sensors: Physics, Designs, and Applications*, vol. 3 (Springer, New York, 2004)
10. A. Fertner, A. Sjolund, Analysis of the performance of the capacitive displacement transducer. IEEE Trans. Instrum. Meas.Instrum. Meas. **38**(4), 870–875 (1989)
11. A.A.A. Amir, A.S. Wael, H.A. Mohd, A.Z. Basem, Design of differential cylindrical capacitive displacement sensor. J. Electr. Electron. Eng. **13**(1), 5–8 (2020)

Chapter 21
Performance Projection of Stacked Silicon Nanosheet-FET Architectures for Future Technology Node

Anjali Goel, Akhilesh Rawat, and Brajesh Rawat

21.1 Introduction

To overcome challenges associated with Fin-FET, gate all around nanowire (NW)-FET and nanosheet (NS)-FET have been introduced for 5 nm technology node and beyond [1, 2]. The gate wrapping around the channel in NW-FET enhances the drive current, but a larger spacing between nanowire enhances the aspect ratio of channel stacks [3]. Importantly, NS-FETs have been demonstrated considerable performance improvement over NW-FET and Fin-FET because larger effective channel width enhances the current drivability under the same area footprint, while maintaining the superior gate electrostatic control [4, 5]. Several recent efforts have been demonstrated the excellent switching performance with ON/OFF in between 10^4 and 10^6 and sub-threshold slope (SS) approaching towards around 65 mV/dec for 3 nm technology node on NS-FET [6–8]. So far, stacked nanosheet architectures have been widely adopted by stacking nanosheets with oxide along channels and gate metal is deposited on the top and bottom sides [3, 9, 10]. However, performance of stacked nanosheet-FETs is limited by large parasitic capacitance between sheets when technology node scales down below 2 nm [6, 11, 12]. To overcome this issue, sub-stacked and tree-type stacking architectures have been recently reported [11, 13]. Despite extensive efforts have been devoted to understand the switching performance of the stacked nanosheet-FETs [14], a few works have been explored the switching performance of sub-stacked and tree-type stacking NS-FET architectures [11, 13]. Moreover, no work have been performed to uniformly benchmark the stacked, sub-stacked and tree-type nanosheet-FET architectures in the proper design space. Thus, it becomes worth to understand and compare these device architectures with digital performance figure-of-merits in future technology node.

A. Goel · A. Rawat · B. Rawat (✉)
Department of Electrical Engineering,Nanoelectronic Devices Laboratory (NANOLAB), Indian Institute of Technology, Rupnagar, Ropar 140001, India
e-mail: r.brajesh@iitrpr.ac.in

R. Singh et al. (eds.), *The Physics of Semiconductor Devices*, Springer Proceedings in Physics 306, https://doi.org/10.1007/978-981-97-1571-8_21

In this work, we investigate and benchmark the digital performance metrics of 3-channel stacked, sub-stacked, and tree-type stacking architectures with NS-FETs at 1 nm IRDS technology node. The performance analysis of these devices is done using a fully calibrated technology computer-aided design (TCAD) platform, which accurately essentials the fundamental quantum and short channel device effects. We theoretically compare the short channel performance metrics, such as SS, drain induced barrier lowering (DIBL), and switching performance metrics, such as device delay (τ), power delay product (PDP) of stacked, sub-stacked, and tree-type architectures as a function of ON current. The organization of the paper is as follows. Section 21.2 presents the geometry of the devices along with the modeling methodology. In the Sect. 21.3, I-V characteristics with short channel and digital performance metrics of tree, sub-stacked, and stacked NS-FET are addressed. Finally, Sect. 21.4 lists the key findings.

21.2 Device Design and Simulation Methodology

Figure 21.1 shows the simulated three-channel device schematic and top view of three channels architectures of stacked, sub-stacked, and, tree-type stacking architectures using nanosheets. The device design parameters, such as the gate spacer and the gate length, are chosen from the International Roadmap for Devices and Systems (IRDS) 2020 projection for 1 nm technology node [15]. The gate length of NS-FET is considered to be around 12 nm, and the source-to-gate and gate-to-drain overlap is

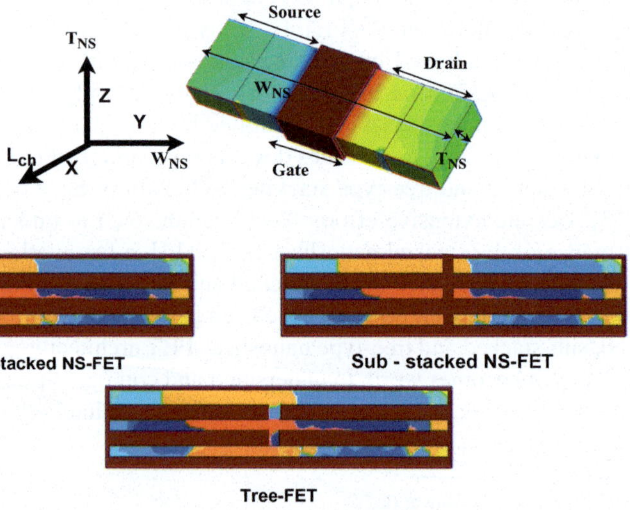

Fig. 21.1 Cross-section view of simulated 3-D NS-FET geometry (top), and channel view of stacked, substacked, and tree-type stacking architectures using nanosheets (middle and bottom)

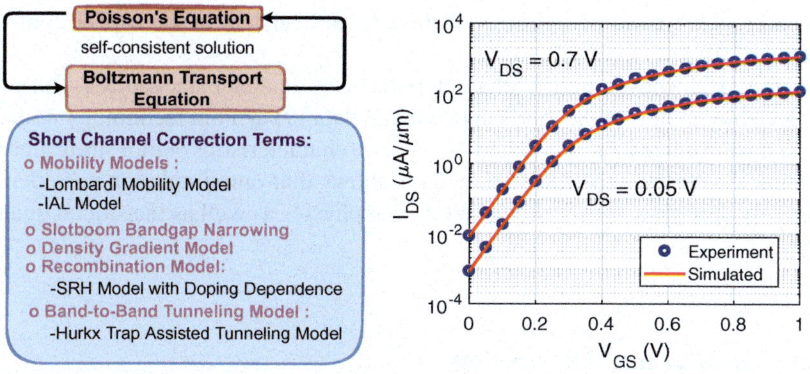

Fig. 21.2 Modeling methodology and simulation results verification: **a** summary of the self-consistent loop with correction terms, and, **b** transfer characteristics of 3 channel stacked NS-FET from our 3-D TCAD simulation and experimental results [9] at $V_{DS} = 0.7V$ and $V_{DS} = 0.05V$ for 1 nm technology node

taken around 4 nm. Further, the stacking of 0.6 nm SiO_2 and 3.7 nm HfO_2 is used as the gate oxide in the three devices corresponding to the effective gate oxide of 1 nm. Furthermore, the source (S) and drain (D) regions of length $L_{S/D} = 10$ nm is assumed to be doped to n-type with the doping concentration of $N_{S/D} = 1 \times 10^{20}$ cm^{-3}. The three uniform sheet with the width (W_{NS}) of around 45 nm , thickness (T_{NS}) for 5 nm and suspension thickness (T_{sus}) of 10 nm are considered in our simulation. These dimensions are in accordance with the experimental data for 1 nm technology node [9]. For uniform performance benchmarking, the width and thickness of channel in stacked, sub-stacked and tree-type stacking architectures are kept same. In particular, the tree-type stacking architecture is constructed by adding interbridges with width of 5 nm and height of 20 nm between the nanosheets. Three NS-FET architectures are therefore compared in the same area footprint.

The novel NS-FET architectures are examined using a fully calibrated three dimensional Sentaurus (3-D) TCAD simulation, based on self-consistent solutions of 3-D Poisson's equation and the Boltzmann transport equation with correction terms. Figure 21.2a shows the self-consistent simulation methodology with quantum and correction model to account scattering-limited carrier mobility, quantum confinement, and band-to-band tunnelling effects in Boltzmann transport equation [16]. The Density-Gradient model is used to account for quantum confinement effects [17]. To account for quasi-ballistic transport, a low field ballistic model is also used. Furthermore, the Slotboom band gap narrowing model is used to account for the narrowing of the band gap caused by significant doping in the source and drain regions [18]. The mobility loss at the silicon and SiO_2 interface due to remote phonon surface and Coulomb scatterings is accounted using the Lombardi mobility, as well as the inversion and accumulation layer mobility models [19]. In addition, the Shockley-Read-Hall recombination model is used in our simulation to activate the generation and recombination conditions of the relevant carriers in the carrier continuity equation [20].

Figure 21.2b shows the I-V characteristics of 3-channel stacked NS-FETs from our modeling methodology and experimental data [9] at 1 nm technology node. For all gate bias voltages, the simulated $I_{DS} - V_{GS}$ characteristics demonstrate an excellent match with the experiment data. This shows that our simulation consideration accurately describes the short channel device physics as well as the silicon quantum effects.

21.3 Results and Discussion

Figure 21.3a shows the transfer characteristics (I_{DS}-V_{GS}) of tree-type, sub-stacked, and stacked nanosheet architectures at $V_{DS} = 0.7$ V for 1 nm technology node. The transfer characteristics of these devices are acquired by keeping the fixed OFF current of around 10 nA/μm, which is achieved by adjusting the gate-metal work function difference. It is observed that tree-type stacking offers considerably higher drain current than that for stacked and sub-stacked architectures. The reason is that the presence of interbridge in-between the stacks enhances the gate control over the channel region. It is observed that the drive current of tree-type architecture is around 2.28 × and 1.38 × higher than that of stacked and sub-stacked architectures, respectively, for all V_{GS}. The sub-stacked architecture importantly provides nearly 1.64 × higher drive current over stacked architecture due to presence of effectively double channels. Figure 21.3b shows the output characteristics of the nanosheet devices at $V_{GS} = 0.7$ V. It is observed that tree-type stacking architecture allows better saturation current when compared to sub-stacked and stacked architectures due to improve gate efficiency. It is found that output conductance (g_{ds}) for tree-type, sub-stacked and, stacked nanosheet architectures is around 117 mS/μm, 116 mS/μm and 115 mS/μm, respectively, at $V_{GS} = 0.7$ V. Therefore, tree-type stacking architecture shows the great promise for achieving nearly 2 × higher drive current over sub-stacked and stacked NS-FET architectures under same area footprint.

Figure 21.4 shows the ON-current (I_{ON}), DIBL and SS for stacked, sub-stacked, and tree-type stacking architectures as function of OFF-current (I_{OFF}) at $V_{DS} = 0.7$ V. The ON and OFF-state currents are computed as $I_{DS}(V_{GS} = V_{DS} = V_{DD})$ and $I_{DS}(V_{GS} = 0, V_{DS} = V_{DD})$, respectively. It is observed that tree-type architecture exhibits 2.3× and 1.32× higher ON-current for all the OFF current values, when compared to sub-stacked and stacked NS-FET, respectively. The I_{ON} for three devices is found to be decreasing with increasing OFF-current because the drain field effect becomes more pronounced for wide range of I_{OFF} value.

Figure 21.4b shows that, as the OFF-current increases, DIBL increases for three devices. The tree-type architecture exhibits lower DIBL compared to sub-stacked, and stacked architectures due to enhanced gate efficiency. Further, stacked architectures has nearly same DIBL compared to tree-type stacking architecture. It is observed from Fig. 21.4c that, tree-type stacking architecture exhibits lower SS over

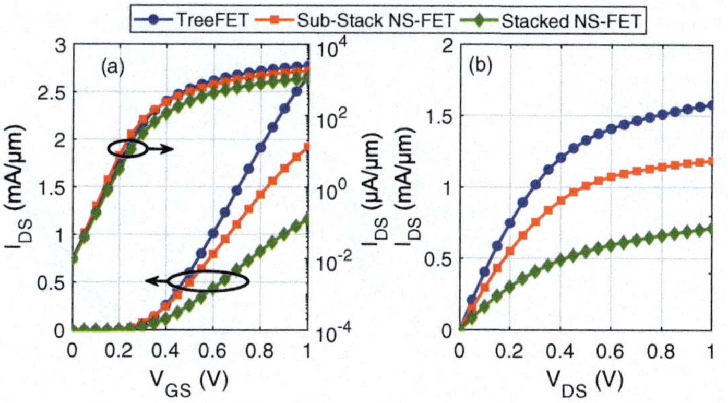

Fig. 21.3 Device I-V characteristics of tree-type, sub-stacked, and stacked NS-FET architectures: **a** transfer characteristics ($I_{DS} - V_{GS}$) at $V_{DS} = 0.7$ V and **b** output characteristics ($I_{DS} - V_{DS}$) at $V_{GS} = 0.7$ V

Fig. 21.4 Short channel performance metrics of tree-type, sub-stacked, and, stacked NS-FET architectures on **a** ON-current (I_{ON}), **b** drain induced barrier lowering (DIBL), and **c** subthreshold slope (SS) as a function of OFF current (I_{OFF})

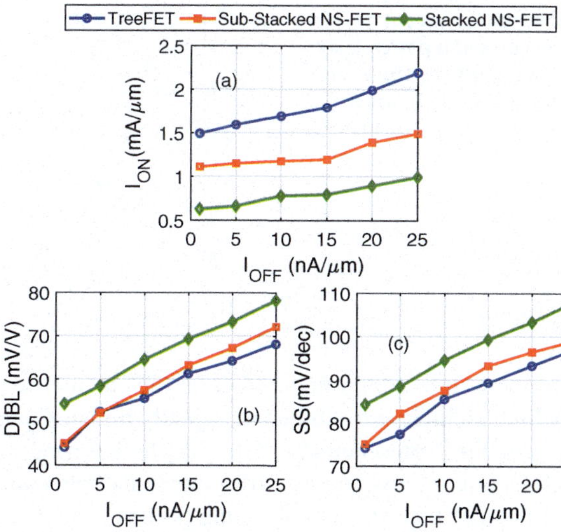

to sub-stacked and stacked architectures. In summary, the short channel performance metrics suggest that tree-type stacking architecture could be more suitable for 1 nm and beyond technology nodes.

Figure 21.5b shows the total gate capacitance (C_g) and gate-drain capacitance (C_{gd}) as a function of V_{GS} at $V_{DS} = 0.7$ V. It is observed that C_g increases abruptly with V_{GS} for three devices, while stacked architecture display a lower C_g over tree and sub-stacked architectures. The C_g represents the combination of C_{gs}+ C_{gd} where C_{gs} is the gate-to-source capacitance. Figure 21.5b shows that, at 1 nm technology node, the C_{gd} contributes significantly in C_g due to higher DIBL values. Figure 21.5 shows that C_{gd} is nearly equal to C_g for three device and linearly increases with V_{GS} due to

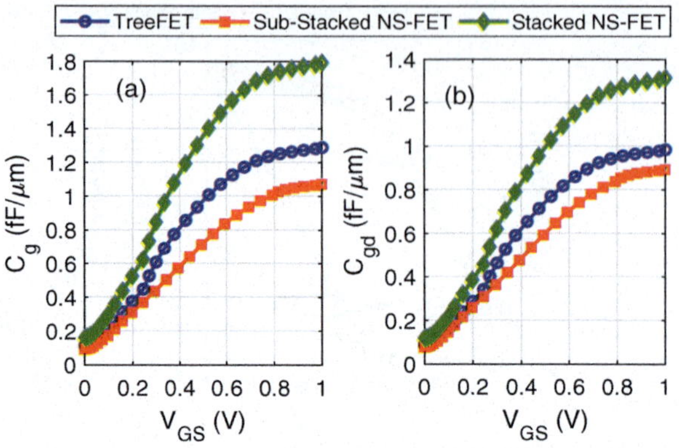

Fig. 21.5 Impact of gate capacitance on tree-type, sub-stacked, and, stacked NS-FET architectures at V_{DS} = 0.7 V: **a** gate capacitance (C_g), and **b** gate-drain capacitance (C_{gd}) as a function of V_{GS}

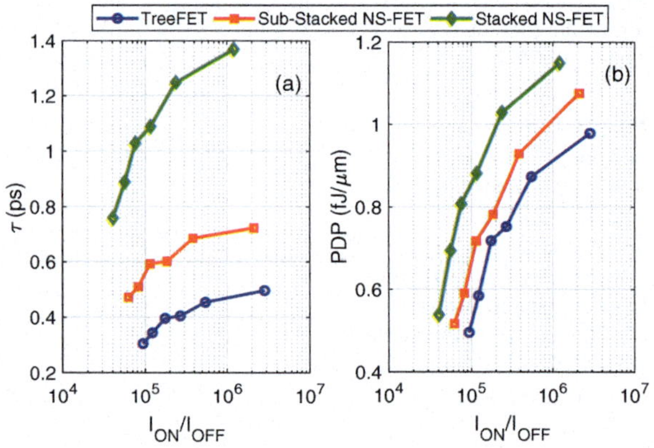

Fig. 21.6 Switching performance metrics of tree-type, sub-stacked, and, stacked NS-FET architectures on **a** device delay (τ), and **b** power delay product (PDP) as a function of I_{ON}/I_{OFF} ratio where I_{OFF} is fixed at 10 nA/μm

increment in the drain field effect on the channel region. Importantly, the tree-type stacking architecture exhibits marginally higher C_g and C_{gd} over sub-stacked architectures due to presence of significantly higher fringing fields. Therefore, sub-stacked architecture promises reduce fringing field effect over tree and stacked architectures.

Figure 21.6 shows the switching performance metrics for tree-type, sub-stacked, and stacked architectures at $V_{DS} = 0.7$ V at different I_{ON}/I_{OFF} ratio. It is observed from Fig. 21.6a that the device delay (τ) for three devices increases considerably with increasing the I_{ON} due to the considerable increment in the gate capacitance (C_g), as observed in Fig. 21.5. Further, the maximum τ of tree-type stacking architecture is found to be 0.498 ps at I_{ON}, which is nearly 1.3 $\times$ and 2.75 $\times$ lower compared to sub-stacked and stacked, respectively. Figure 21.6b shows that the power delay product (PDP) for the three devices increases considerably with increasing the I_{ON} due to the significant increment in the τ and static power. Further, the maximum PDP of tree-type stacking architecture is found to be 0.98 fJ/μm, which is nearly 1.1 $\times$ and 1.2 $\times$ lower than that for sub-stacked and stacked, respectively. Therefore, delay and PDP values suggest that tree-type stacking architecture can be a better choice for smaller channel length with high I_{ON} and high intrinsic speed.

21.4 Conclusion

This work presented a comprehensive digital performance comparison between three-channel tree-type, sub-stack and stacked NS-FET architecture for 1 nm technology node. The tree-type stacking architecture has shown lower subthreshold slope, drain induced barrier lowering, device delay, and power delay product over sub-stack and stacked architecture, making them a promising option for digital applications. Specially, the device delay and power delay product have found to be around 0.498 ps and 0.98 J/μm respectively, which provides an opportunity for low-power and high-speed switching applications at 1 nm technology node.

References

1. P. Feng, S. Song, G. Nallapati, J. Zhu, J. Bao, V. Moroz, M. Choi, X. Lin, Q. Lu, B. Colombeau, N. Breil, M. Chudzik, C. Chidi, Comparative analysis of semiconductor device architectures for 5-nm node and beyond. IEEE Electron Dev. Lett. **38**, 1657–1660 (2017)
2. A. Razavieh, P. Zeitzoff, E.J. Nowak, Challenges and limitations of CMOS scaling for FinFET and beyond architectures. IEEE Trans. Nanotechnol. **18**, 999–1004 (2019)
3. U.K. Das, T.K. Bhattacharyya, Opportunities in device scaling for 3-nm node and beyond: FinFET versus GAA-FET versus UFET. IEEE Trans. Electron Dev. **67**, 2633–2638 (2020)
4. A. Goel, A. Rawat, B. Rawat, Benchmarking of analog/RF performance of fin-FET, NW-FET, and NS-FET in the ultimate scaling limit. IEEE Trans. Electron Dev. **69**(3), 1298–1305 (2022). (Mar.)
5. D. Nagy, G. Espineira, G. Indalecio, A.J. Garcia-Loureiro, K. Kalna, N. Seoane, Benchmarking of FinFET, nanosheet, and nanowire FET architectures for future technology nodes. IEEE Access **8**, 53196–53202 (2020)
6. V. Jegadheesan, K. Sivasankaran, A. Konar, Impact of geometrical parameters and substrate on analog/RF performance of stacked nanosheet field effect transistor. Mater. Sci. Semiconductor Process. **93**, 188–195 (2019)
7. S. Barraud, B. Previtali, C. Vizioz, J.M. Hartmann, J. Sturm, J. Lassarre, C. Perrot, P. Rodriguez, V. Loup, A. Magalhaes-Lucas et al., 7-levels-stacked nanosheet GAA transistors for high performance computing, in *2020 IEEE Symposium on VLSI Technology* (IEEE, 2020), pp. 1–2
8. I. Ferain, C. Colinge, J. Colinge, Multigate transistors as the future of classical metal-oxide-semiconductor field-effect transistors. Nature **479**, 310–316 (2011)
9. N. Loubet, T. Hook, P. Montanini, C.W. Yeung, S. Kanakasabapathy, M. Guillom, T. Yamashita, J. Zhang, X. Miao, J. Wang et al., Stacked nanosheet gate-all-around transistor to enable scaling beyond FinFET, in *2017 Symposium on VLSI Technology* (IEEE, 2017), pp. T230–T231
10. G. Bae, D.I. Bae, M. Kang, S.M. Hwang, S.S. Kim, B. Seo, T.Y. Kwon, T.J. Lee, C. Moon, Y.M. Choi et al., 3 nm GAA technology featuring multi-bridge-channel FET for low power and high performance applications, in *2018 IEEE International Electron Devices Meeting (IEDM)* (IEEE, 2018), pp. 28–7
11. P. Kushwaha, A. Dasgupta, M. Kao, H. Agarwal, S. Salahuddin, C. Hu, Design optimization techniques in nanosheet transistor for RF applications. IEEE Trans. Electron Dev. **67**, 4515–4520 (2020)
12. H.-C. Lin, T. Chou, C.C. Chung, C.J. Tsen, B.W. Huang, C.W. Liu, RF performance of stacked Si nanosheet nFETs. IEEE Trans. Electron Dev. **68**, 5277–5283 (2021)
13. H.Y. Ye, C.W. Liu, On-current enhancement in TreeFET by combining vertically stacked nanosheets and interbridges. IEEE Electron Dev. Lett. **41**, 1292–1295 (2020)

14. C.W. Yeung, J. Zhang, R. Chao, O. Kwon, R. Vega, G. Tsutsui, X. Miao, C. Zhang, C.W. Sohn, B.K. Moon, A. Razavieh, J. Frougier, A. Greene, R. Galatage, J. Li, M. Wang, N. Loubet, R. Robison, V. Basker, T. Yamashita, D. Guo, Channel geometry impact and narrow sheet effect of stacked nanosheet, in *2018 IEEE International Electron Devices Meeting (IEDM)* (IEEE, 2018), pp. 28.6.1–28.6.4
15. International Roadmap for Devices and Systems: MORE MOORE (2020). https://irds.ieee.org/editions/2020/more-moore
16. Synopsys Inc. Mountain View, CA, USA, Version O-2018.06 (2019)
17. S. Rhee, D. Kim, K. Kim, S. Choi, B.G. Park, Y.J. Park, Extension of the density-gradient model to the second-order quantum correction for analysis of the single-charge effect in sub-10-nm MOS devices. IEEE J. Electron Dev. Soc., 1 (2020)
18. D.B.M. Klaassen, J.W. Slotboom, H.C. Graaff, Unified apparent bandgap narrowing in n- and p-type silicon. Microelectron. Reliab. **32**, 125–129 (1992)
19. T. Nishida, C.T. Sah, A physically based mobility model for MOSFET numerical simulation. IEEE Trans. Electron Dev. **34**, 310–320 (1987)
20. J.S. Yoon, J. Jeong, S. Lee, R.H. Baek, Systematic DC/AC performance benchmarking of sub-7-nm node FinFETs and nanosheet FETs. IEEE J. Electron Dev. Soc. **6**, 942–947 (2018)

Chapter 22
Determination of Coupling Coefficient in Large Optical Cavity Distributed Feedback Laser Structure with Metal Surface Grating

Meenakashi Patwal, Deepti Jain, D. S. Rawal, and Ajeet Kumar

22.1 Introduction

Semiconductor laser diodes (LDs) are used in numerous applications. Some of these are pumping of solid-state laser, laser-based gas sensing, free-space communications, fibre amplifiers, medical applications etc. Generally, laser diodes are fabricated in Fabry Perot (FP) mode which has a wide spectral width of around 2 nm. FP lasers have poor output stability due to mode-hopping and multiple mode emissions which is not desirable for applications that require single-mode emissions such as wavelength-division multiplexed high bit rate, long-haul communication systems, and for pumping of Erbium and Ytterbium Doped Fiber Amplifier. Therefore, special structures like DFB LDs [1], Distributed Bragg reflector Laser Diodes (DBR LDs), and Vertical-cavity surface-emitting Laser Diodes (VCSEL LDs) have been developed. DFB laser diode has wavelength selective grating which introduces a periodic change of refractive index or gains along the lasing cavity. Coupling based on perturbation of refractive index and gain is termed index coupling and gain coupling respectively. DFB laser diode with the aid of wavelength-selective grating provides spectral purity as well as temperature stability. The main challenge in DFB fabrication is the formation of a well-defined periodic gratings structure to ensure higher single-mode yield with better wavelength accuracy. So, it is essential to have a fabrication process for gratings that provides low damage, uniform, and reproducible gratings.

M. Patwal (✉) · A. Kumar
Advanced Photonics Simulation Research Laboratory, Delhi Technological University, Shahabad Daulatpur, New Delhi 110042, India
e-mail: meenakshi_phd2k18@dtu.ac.in

M. Patwal · D. Jain · D. S. Rawal
Solid State Physics Laboratory, Lucknow Road, Banarasi Das Estate, Timarpur, New Delhi, Delhi 110054, India

R. Singh et al. (eds.), *The Physics of Semiconductor Devices*, Springer Proceedings in Physics 306, https://doi.org/10.1007/978-981-97-1571-8_22

Fabrication of conventional DFB LD [2, 3] requires an epitaxial regrowth process which introduces defects and damages the gratings. Therefore, the yield is reduced. Thus, making conventional DFB laser diodes less attractive in terms of fabrication as compared to FP laser diodes but DFB LD provides temperature stability, wavelength tunability, spectral purity, and narrow linewidth which is essential in many applications as previously mentioned.

LC DFB laser diode [4, 5] provides the advantages of both Fabry–Perot (single growth) and DFB lasers (wavelength selectivity and temperature stability). In these lasers, the grating is made on both sides of the ridge by periodically etching the top cladding layer. Thus, it eliminates the additional re-growth steps by patterning the grating sideways to the waveguide ridge. The optical mode and grating on both sides of the ridge are evanescently coupled to each other.

LC DFB laser diodes that are index coupled laser diodes suffer from the degeneracy of modes symmetric to the Bragg frequency of the grating [6]. This problem is prominent in perfectly anti-reflecting (AR) coated index-coupled laser diodes. Non-AR coated facets are used to resolve this problem but due to the uncertainty of facets phases [7] yield reduces. The introduction of $\Lambda/4$ phase shift in the middle of the grating breaks the degeneracy of the two lasing modes in the grating stop-band and ensures single-mode lasing at the centre of the stop-band but it causes the high spatial hole-burning effect [8]. Another solution has been the incorporation of gain grating [9] in DFB laser diodes instead of index gratings, which eliminates phase dependency. These are called as Gain-Coupled or complex coupled DFB laser diodes. Due to the small index coupling coefficient in these lasers, there are only minor distortions around the Bragg wavelength [10].

LC DFB lasers with metal surface gratings [10–13] has complex coupled or gain coupled grating incorporated onto both sides of the ridge waveguide (Fig. 22.1). It has great potential for low-cost production because of its relatively simple fabrication. Grating generation is the most critical step in DFB which includes challenges of extremely small feature sizes. Among the LC-DFB lasers reported, grating definition and fabrication have been achieved using e-beam device lithography (EBL) [4, 10, 11], Holographic lithography [12, 14], stepper [13], etc.

Coupling in DFB Lasers is determined by the coupling coefficient (κ) which occurs as a result of the interaction between the grating and the evanescent part of the transverse mode. Therefore, the design and analysis of DFB lasers mainly depends on the (κ) [4]. The coupling coefficient depends on many parameters such as the thickness and composition of the active and cladding layers [4].

Various studies have been done that show its dependence on structural parameters [15]. DFB lasers with lateral surface gratings have sufficiently narrow ridges to enhance coupling between lateral evanescent fields and the grating [15].

Also, various studies have been done so far at different wavelengths to optimize the value of κ by showing its dependence on grating parameters and ridge parameters. Many researchers had shown that κ strongly depends on the grating order [16], the ridge width [15, 17], duty cycle [17], grating width [15], grating thickness [15, 17], ridge height [15], the material for the passivation layer [18], shape of gratings and

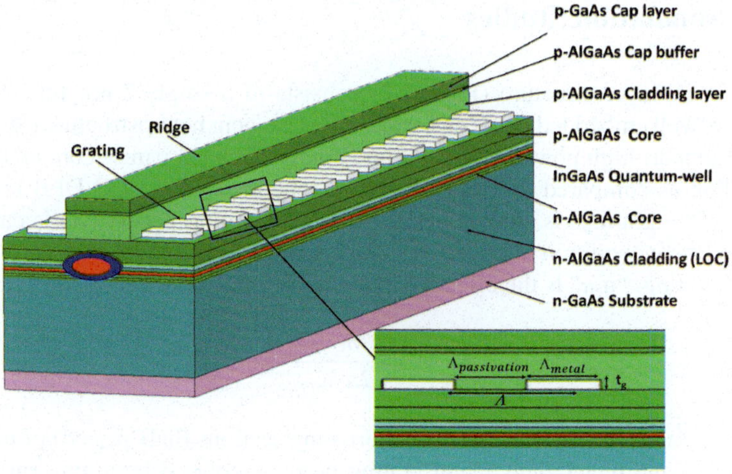

Fig. 22.1 Laterally coupled DFB laser diode with metal surface grating

metal gratings [18] to optimization of maximum-coupling of the evanescent field to the grating in LC-DFB lasers [18].

Generally, the optimum value of κL is around 1 [4]. If $\kappa L \gg 1$ then spatial hole-burning occurs, but on the other hand if $\kappa L \ll 1$ then the feedback is not sufficient for laser operation in single-mode [6]. Therefore, the performance of the DFB laser diode depends predominantly on the precise calculation of the coupling coefficient.

While most of the 9xx nm applications have targeted high-power multimode lasers but in sensing and spectroscopic applications, a very narrow linewidth is required. This necessitates the use of wavelength-selectivity to be incorporated into the structure. The simplest, most straightforward structure that ensures single transverse mode operation is the DFB laser diode by integrating a periodic grating in the device. Most of the research work in DFB Laser is around 1390 and 1550 nm based on InGaAsP quantum well and InP as a substrate because of its wide application in high-speed communications.

With the progress in quantum technologies and spectroscopic applications, ever new wavelengths are emerging. There is only limited research in the field of narrow linewidth DFB lasers based on AlGaAs/InGaAs/GaAs quantum well structures. High power laser diode operation is enabled by using Asymmetric Large Optical Cavity (LOC) epitaxial structures [18] which reduce losses due to free carrier absorption. In this paper, we will design metal surface grating for DFB lasers with asymmetric LOC epitaxial structures and optimize the coupling coefficient by varying structural parameters of the ridge.

22.2 Simulation Studies

The epitaxial-layer structure (Fig. 22.2) consists of a single 7 nm InGaAs/GaAs Quantum-Well embedded between separate confinement heterostructure (SCH) with LOC to support high power. This structure comprises a 0.3 μm N-core and a broad N-cladding as compared to P-cladding. The schematic of the LC DFB laser with metal surface gratings is shown in Fig. 22.1. The grating period (Λ) is calculated from the Bragg condition in Eq. (22.1). "λ_B" is Bragg wavelength and represents the order of grating. "n_{eff}" is the effective index of the optical mode.

$$\Lambda = \frac{m\lambda_B}{2\pi n_{eff}} \tag{22.1}$$

The grating (Fig. 22.3) consists of alternating sections (half-period) of the passivation layer and metal. This metal grating layer is replaced by a uniform layer of 50 nm thickness (t_g), placed on top of the separate-confinement heterostructure layer (SCH layer). The average refractive index of the grating layer was calculated by

$$n_{grating}^2 = \frac{\Lambda_1}{\Lambda} n_{passivation}^2 + \frac{\Lambda_2}{\Lambda} n_{metal}^2 \tag{22.2}$$

The optical confinement layer and single quantum well layer is converted to a single layer with an effective refractive index (n_{eff}). The coupling coefficient derived from the theory of coupled modes is given below

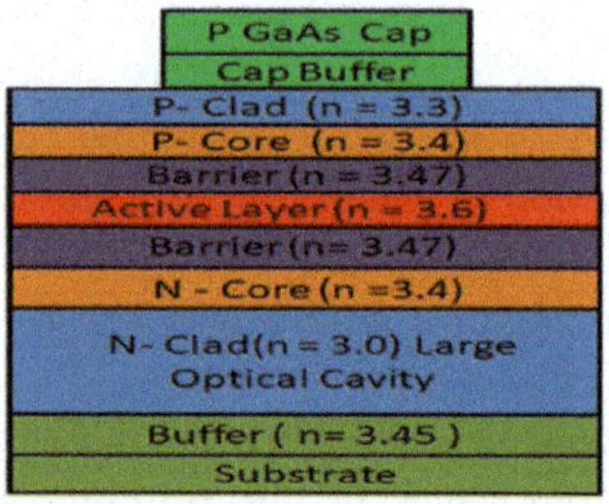

Fig. 22.2 Epitaxial layer structure

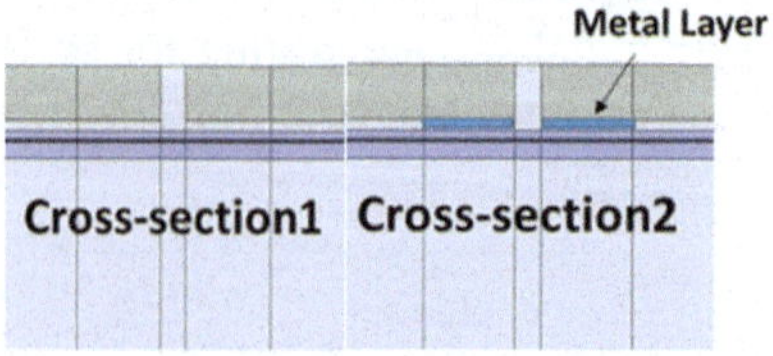

Fig. 22.3 Illustration of one period of the grating

$$\kappa = \frac{n_2^2 - n_1^2}{n_{eff}} \frac{\sin(m\gamma\pi)}{m} \Gamma \tag{22.3}$$

where n_1 is the refractive index of passivation layer Silicon nitride (S_iN_x) and n_2 is refractive index of grating layer, m denotes grating order, γ is (Λ_1/Λ or Λ_2/Λ) which is the duty factor of grating and Γ is the confinement factor which is field overlap integral with the grating region. In order to obtain n_{eff} and Γ, the Finite difference method (FDM) mode solver in Fimmwave was employed.

22.3 Results and Discussion

The optical modal analysis was simulated to obtain fundamental mode profile of the ridge waveguide LC-DFB laser structure as shown in Fig. 22.4.

The dependency of the coupling coefficient on the order of grating for different ridge widths is shown in Fig. 22.5a. It is shown from the simulation studies that the coupling coefficient increases as ridge width decreases and is highest for first order grating. The reason for the increment in coupling coefficient is that the modal pattern of narrower ridge spreads more laterally. Thus, more part of evanescent mode can couple with the grating.

The coupling coefficient (κ) increases with an increase in etch depth (see Fig. 22.5b). This can be explained as etch depth increases n_{eff} decreases. Deep etch ridge is more prone to higher order mode. As etch depth increases residual thickness decreases i.e. nearer to the active layer. Therefore, it can degrade the laser performance significantly.

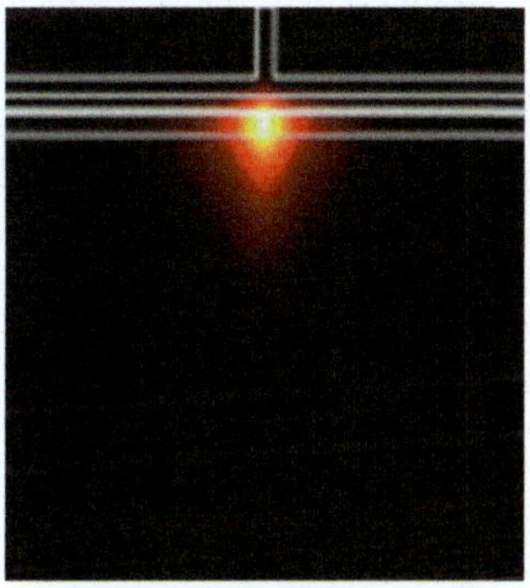

Fig. 22.4 2D optical field distribution of ridge waveguide

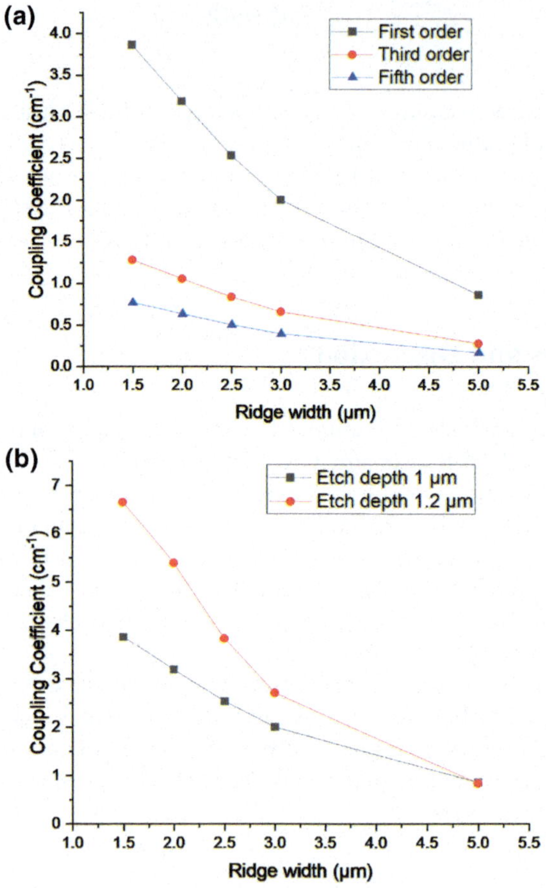

Fig. 22.5 **a** Coupling coefficient versus ridge width at different grating order. **b** Coupling coefficient versus ridge width at different etch depth

22.4 Conclusion

The coefficient of coupling (κ) was calculated by coupled-mode theory and non-uniform FDM. It is obtained from the simulation studies that narrower ridge width and deep etch depth can increase the coupling of electric field and grating but from fabrication point of view such lower feature sizes are difficult to fabricate. Therefore, optimum value of ridge width and etch depth are determined for grating fabrication.

References

1. Y. Mao, Y. Cheng, B. Xu, R. Ji, Y. Li, Record-high power 1.55-μm distributed feedback laser diodes for optical communication, in *2021 Optical Fiber Communication Conference* (Optica Publishing Group, Washington, DC, United States, 2021), pp. W1B-7
2. H. Wenzel, J. Fricke, A. Klehr, A. Knauer, G. Erbert, High-power 980-nm DFB RW lasers with a narrow vertical far field. IEEE Photonics Technol. Lett. **18**, 737–739 (2006). https://doi.org/10.1109/LPT.2006.871127
3. A. Klehr, H. Wenzel, M. Braun, F. Bugge, J. Fricke, A. Knauer, G. Erbert, Five-hundred-milliwatts distributed-feedback diode laser emitting at 940 nm. Opt. Laser Technol. **39**, 333–337 (2007). https://doi.org/10.1016/j.optlastec.2005.07.010
4. L.M. Miller, J.T. Verdeyen, J.J. Coleman, R.P. Bryan, J.J. Alwan, K.J. Beernink, J.S. Hughes, T.M. Cockerill, A distributed feedback ridge waveguide quantum well heterostructure laser. IEEE Photonics Technol. Lett. **3**, 6–8 (1991). https://doi.org/10.1109/68.68030
5. G. Yue, J. Zhang, Z. Wang, X. Zhang, Y. Li, J. Xie, X. Wang, Y. Hao, Research progress of laterally coupled DFB-LD, in *2017 International Conference on Optical Instruments and Technology: Advanced Laser Technology and Applications*, vol. 10619 (SPIE, Beijing, China, 2017), pp. 214–221
6. H. Kogelnik, C.V. Shank, Coupled-wave theory of distributed feedback lasers. J. Appl. Phys. **43**, 2327–2335 (1972). https://doi.org/10.1063/1.1661499
7. J. Buus, Mode selectivity in DFB lasers with cleaved facets. Electron. Lett. **21**, 179–180 (1985). https://doi.org/10.1049/el:19850126
8. H. Soda, Y. Kotaki, H. Sudo, H. Ishikawa, S. Yamakoshi, H. Imai, Stability in single longitudinal mode operation in GaInAsP/InPphase-adjusted DFB lasers. IEEE J. Quantum Electron. **23**, 804–814 (1987). https://doi.org/10.1109/JQE.1987.1073454
9. G.P. Agrawal, N.K. Dutta, *Semiconductor Lasers*, 2nd edn. (Springer, Boston, MA, 1993)
10. M. Kamp, J. Hofmann, A. Forchel, F. Schäfer, J.P. Reithmaier, Low-threshold high-quantum-efficiency laterally gain-coupled InGaAs/AlGaAs distributed feedback lasers. Appl. Phys. Lett. **74**, 483–485 (1999). https://doi.org/10.1063/1.123164
11. R. Schreiner, H. Schweizer, High-frequency operation of 1.57-μm laterally coupled distributed feedback lasers with self-aligned Ohmic contacts and nickel surface gratings. IEEE J. Sel. Top. Quantum Electron. **9**, 1179–1182 (2003). doi: https://doi.org/10.1109/JSTQE.2003.819469
12. S.J. Jang, J.S. Yu, Y.T. Lee, Laterally coupled DFB lasers with self-aligned metal surface grating by holographic lithography. IEEE Photonics Technol. Lett. **20**, 514–516 (2008). https://doi.org/10.1109/LPT.2008.918864
13. A. Assadihaghi, R. Millett, H. Schriemer, K. Dridi, T. Hall, H. Teimoori, K. Hinzer, Threshold current reduction in third-order laterally coupled DFB lasers, in *2009 2nd Microsystems and Nanoelectronics Research Conference* (IEEE, Ottawa, ON, Canada, 2009), pp. 21–24
14. S.J. Jang, Y.T. Lee, J.S. Yu, Coupling coefficient calculation of laterally coupled distributed feedback laser structure with metal surface gratings, in *2008 20th International Conference on Indium Phosphide and Related Materials* (IEEE, Versailles, France, 2008), pp. 1–3
15. H.S. Djie, C. Sookdhis, P. Dowd, Analysis and structure design of distributed-feedback laser (DFB) and distributed Bragg reflector (DBR) laser using regrowth-free surface grating technology, in *Design, Fabrication, and Characterization of Photonic Devices II*, vol. 4594 (SPIE, Singapore, 2001), pp. 250–259
16. A.I. Laakso, J. Karinen, M. Dumitrescu, Modeling and design particularities for distributed feedback lasers with laterally-coupled ridge-waveguide surface gratings, in *Physics and Simulation of Optoelectronic Devices XIX*, vol. 7933 (SPIE, San Francisco, California, United States, 2011), pp. 532–541. https://doi.org/10.1117/12.876135
17. R.R. Millett, K. Hinzer, T.J. Hall, H. Schriemer, Simulation analysis of higher order laterally-coupled distributed feedback lasers. IEEE J. Quantum Electron. **44**, 1145–1151 (2008). https://doi.org/10.1109/JQE.2008.20020

18. W. Zeller, M. Kamp, J. Koeth, L. Worschech, High-power pulsed 976-nm DFB laser diodes, in *Photonic Micro Devices/Microstructures for Sensing II, International Society for Optics and Photonics*, vol. 7682 (SPIE Defense, Security, and Sensing, Orlando, Florida, United States, 2010). https://doi.org/10.1117/12.849981

Chapter 23
Design and Simulation of Stepped Microcantilevers for Energy Harvesting Applications

S. Purohit, V. Swarnalatha, P. Pal, and A. K. Pandey

23.1 Introduction

In recent times, energy harvesting has become very popular field among the researchers especially using piezoelectric elements [1]. The cantilever type structures can produce significant mechanical strain during oscillation and are therefore widely investigated for MEMS-based piezoelectric energy harvesters [2–8]. MEMS piezoelectric energy harvesters have captivated greater attention due to their advantages such as higher energy density, no need of external power, and ease of integration with other micro devices [9, 10]. There are variety of available piezoelectric materials, among them the lead zirconate titanate (PZT) is the most popularly used because of its superior piezoelectricity. However, loss of polarization with continuous usage and poor stability are the major drawbacks of PZT. Piezoelectric properties are affected by the operating temperature and, thus, PZT-based microstructures due to their brittleness cannot be deformed mechanically for longer durations [11–13]. Zinc Oxide (ZnO) is another widely used piezoelectric material because of its CMOS compatibility and eco-friendly nature [14]. ZnO is highly tensile and may undergo mechanical deformation for longer durations without the effect of temperature variation as compared to PZT. Because of its distinctive electrical, optical, and mechanical properties, ZnO has found great potential in fields of solar cells, photo detectors, and it can be integrated with other processes and materials very easily [15, 16]. Fabrication and characterization of ZnO-based thin film microcantilevers using various techniques have also been reported in different research papers [14,

S. Purohit (✉) · V. Swarnalatha · P. Pal
MEMS and Micro/Nano Systems Laboratory, Department of Physics, Indian Institute of Technology Hyderabad, Hyderabad, India
e-mail: ph19resch01005@iith.ac.in

S. Purohit · A. K. Pandey
SENACT Lab, Department of Mechanical and Aerospace Engineering, Indian Institute of Technology Hyderabad, Hyderabad, India

R. Singh et al. (eds.), *The Physics of Semiconductor Devices*, Springer Proceedings in Physics 306, https://doi.org/10.1007/978-981-97-1571-8_23

17]. In this work, different ZnO based microcantilever geometries are designed and simulated using COMSOL Multiphysics and the influence of structural parameters on the performance of cantilever-based energy harvester is analyzed to improve the output voltage and efficiency. The novelty in our study is that it shows the importance and influence of structural changes in a microstructure and the way they affect and modify their mechanical and electrical properties so that a better understanding is developed for improving and enhancing these properties.

23.2 Numerical Analysis

In this section, we briefly describe the procedure for computing the resonance frequency, voltage generated due to the application of mechanical force and the total electrical power generated. Three geometries namely, rectangle, single-stepped and double-stepped rectangular beams are considered. The model geometries of the micro cantilevers are as shown in Fig. 23.1a–c respectively. The length and thickness of all the three beams are kept same. And the width and thickness of the piezoelectric strip (shown in blue color in Fig. 23.1) is 50 μm and 2 μm respectively and rest of its dimensions are same as the cantilever beam. The dimensions of the strip are selected in such a way that it does not hinder the performance of cantilever because there is a possibility of breaking of the cantilever structure due to more thickness. Other dimensions of the beam and the piezoelectric strip are mentioned in Tables 23.1 and 23.2, respectively.

ZnO is selected as a piezoelectric material. Piezoelectric module from Solid Mechanics module in COMSOL is used for the simulation. Total force of 500 μN is applied as a load at the free end of the microcantilever. For observing the amount of voltage generated due to the applied load, a floating potential at top interface and ground at the bottom interface is applied using the Electrostatic module. Resonance frequency is calculated using structural analysis based on Solid Mechanics module in the COMSOL Multiphysics. Silicon is used as a substrate (i.e., material of cantilever beam). Piezoelectric materials produce electrical charge when it

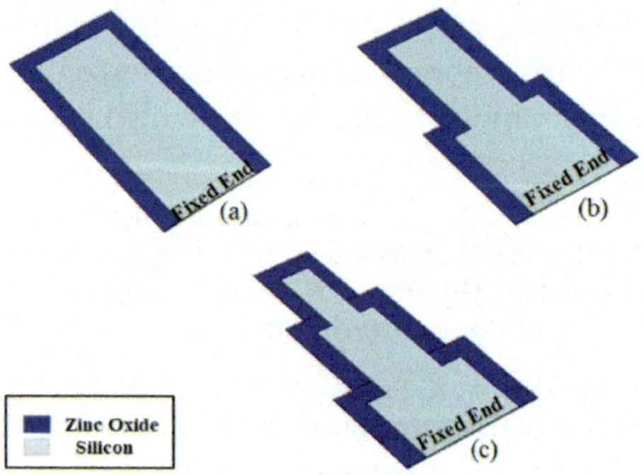

Fig. 23.1 Model geometry of **a** simple rectangular microcantilever, **b** single-stepped rectangular microcantilever, and **c** double-stepped rectangular microcantilever

Table 23.1 Dimensions of the microcantilever

Dimensions	Rectangular (μm)	Single-stepped (μm)	Double-stepped (μm)
Length	700	700	700
Width	350	300	300
Step-length	–	350	250
Thickness	2	2	2

Table 23.2 Dimensions of the piezoelectric strip

Dimensions	Rectangular (μm)	Single-stepped (μm)	Double-stepped (μm)
Length	700	700	700
Width	50	50	50
Step-length	–	350	250
Thickness	2	2	2

is mechanically deformed. The piezoelectric energy harvester is governed by the following equations written in the strain-charge format as Eqs. (23.1) and (23.2) [9]

$$S = s^E T + dE \tag{23.1}$$

$$D = \varepsilon^T E + dT \tag{23.2}$$

where S is the mechanical strain vector, s^E is the elastic compliance tensor (Pa^{-1}), T is the mechanical stress vector ($N\,m^{-2}$), D is the elastic displacement vector ($C\,m^{-2}$), ε^T the dielectric permittivity tensor ($F\,m^{-1}$), E the electric field vector ($V\,m^{-1}$) and d the transverse piezoelectric coefficient tensor ($C\,N^{-1}$). For calculating the output power, electric circuit from the AC/DC Module is used to apply a 10KΩ load resistance across the model geometries.

23.3 Results and Discussions

We have compared the results obtained from the numerical analysis for all the three designs. The comparison of resonance frequency and output voltage are shown in Fig. 23.2a and b, respectively. Rectangular, single-stepped, and double-stepped microcantilever beams are named as Design-1, Design-2 and Design-3 respectively. Total electrical power generated by all the three designs is presented in the Table 23.3. As it can be seen in Fig. 23.2a, single-stepped rectangular beam has the highest frequency of 3358.6 Hz, and the rectangular beam has the least frequency of 2496 Hz among all the designs. And the double-stepped rectangular beam gives the highest value of output voltage generated, i.e., 3.2 V. Rectangular beam gives the least output

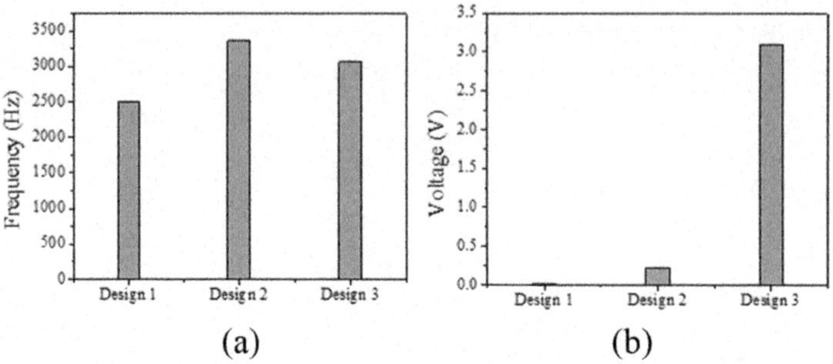

Fig. 23.2 **a** Frequency analysis and **b** output voltage analysis for different microcantilever designs

Table 23.3 Total electrical power generated after applying a load of 10 KΩ

Type of geometry	Electrical power
Rectangular	19.6 nW
Single-stepped	4.85 μW
Double-stepped	1024 μW

voltage of 0.014 V as shown in Fig. 23.2b. Among all the designs, double-stepped rectangular micro cantilever gives an increased output voltage by 21,928.5% and increase in resonant frequency by 22% for the same length.

From the Table 23.3, it is found that the double-stepped microcantilever generates the highest power of 1024 μW. This shows that the structural parameters of the microcantilever beam may affect the amount of output voltage and power generation in piezoelectric energy harvesters. The change in resonance frequency is possibly due to change in mass or stiffness of the beam because of the change in geometries of different microcantilevers. The stress analysis for all the beams can be seen in Figs. 23.3, 23.4, and 23.5. The maximum stresses generated in Design-1, 2 and 3 are 1.49×10^8 N/m^2, 6.81×10^8 N/m^2 and 7.15×10^8 N/m^2 respectively. Due to localization of stress distribution under each step of single or double-stepped microcantilever, higher value of output voltage and power are observed. Also, it can be seen from Figs. 23.4 and 23.5, bumps appear at some point along the length of the cantilever. The possible reason behind these occurrences is the uneven stress distribution along the length due to the steps.

23.4 Conclusions

In this paper, the comparison of three different types of geometries namely, rectangular, single-stepped, and double-stepped micro cantilevers for energy harvesting applications are presented. Any change in geometry of the micro cantilever-based energy harvester can affect and modify the piezoelectric properties. Double-stepped

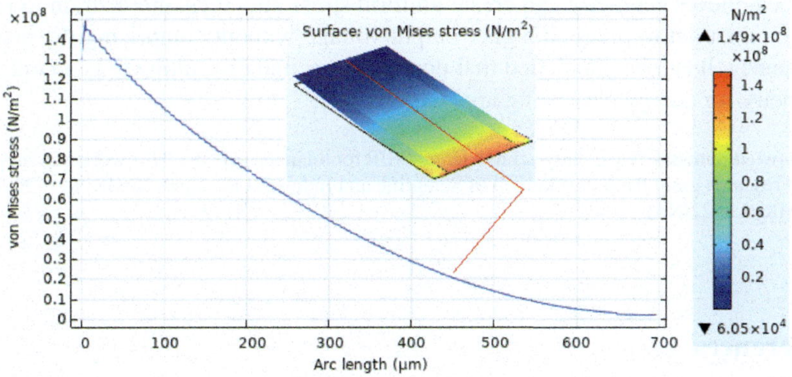

Fig. 23.3 Stress analysis for simple rectangular beam

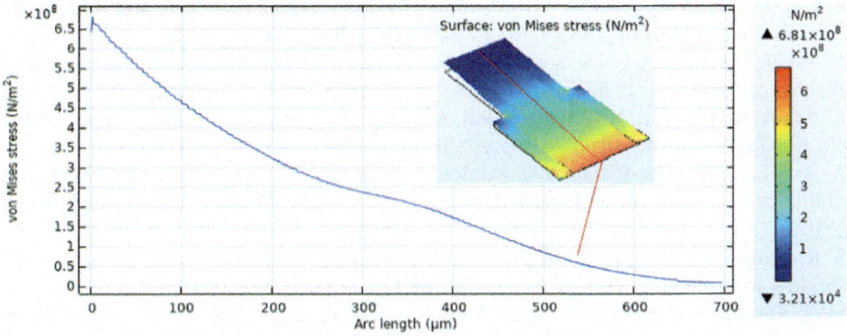

Fig. 23.4 Stress analysis for single-stepped rectangular beam

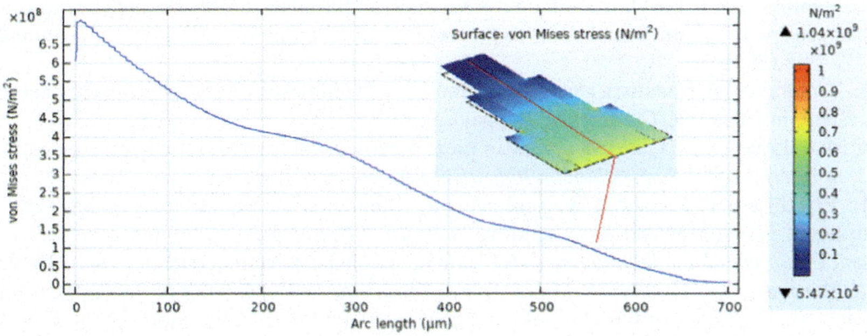

Fig. 23.5 Stress analysis for double-stepped rectangular beam

microcantilever geometry can create uniform stress and, thus, more strain in piezoelectric material to produce higher output voltage without comprising the material occupancy. It can be concluded that double-stepped microcantilever is an optimum geometry for energy harvesting applications.

Acknowledgements We greatly acknowledge CSIR for financial support. This work was supported by the research grant from the Council of Scientific and Industrial Research (CSIR, Ref: 22(0824)/19/EMR-II,0527/NS).

References

1. D.K. Sahoo, A.K. Pandey, Performance of non-uniform cantilever based piezoelectric energy harvester. ISSS J. Micro Smart Syst. **7**, 1–13 (2018)
2. P.M. Kumar, A. Ashok, P. Pal, A.K. Pandey, Frequency tuning of weakly and strongly coupled micromechanical beams. ISSS J. Micro Smart Syst. **9**, 117–130 (2020)
3. A. Ashok, R.P. Nighot, L. Devsoth, M. Yadav, P. Pal, A.K. Pandey, Experimental and theoretical analysis of drag forces in micromechanical beam arrays. Phys. Rev. Appl. **13**(3), 034003 (2020)
4. A. Ashok, R.P. Nighot, N.K. Sahu, P. Pal, A.K. Pandey, Design and analysis of microcantilever beams based on arrow shape. Microsyst. Technol. **25**(11), 4379–4390 (2019)
5. A. Ashok, G. Aparna, P. Pal, A.K. Pandey, An analysis of stepped trapezoidal shaped microcantilever beams for MEMS based devices. J. Micromech. Microeng. **28**(6), 085003–085011 (2018)
6. S. Roundy, P.K. Wright, J. Rabaey, A study of low-level vibrations as a power source for wireless sensor nodes. Comput. Commun. **26**(11), 1131–1144 (2003)
7. M.A. Karami, D.J. Inman, Parametric study of zigzag microstructure for vibrational energy harvesting. J. Microelectromech. Syst. **21**(1), 145–160 (2012)
8. E. Anand, S. Naval, P.K. Sinha, N.K. Das, S. Kundu, Effects of coupling in piezoelectric multi-beam structure. Microsyst. Tech. **26**(4) (2020)
9. S. Sunithamani, P. Lakshmi, Simulation study on performance of MEMS piezoelectric energy harvester with optimized substrate to piezoelectric thickness ratio. Microsyst. Technol. **21**(4), 733–738 (2015)
10. E. Varadrajan, M. Bhanusri, Design and simulation of unimorph piezoelectric energy harvesting system, in *COMSOL Conference*, Bangalore (2013)
11. C.M.A. Lopes, C.A. Gallo, A review of piezo electrical energy harvesting and applications, in *23rd International Symposium on Industrial Electronics* (2014), pp. 1286–1288
12. E. Eba Flora, P. Lakshmi, S. Sunithamani, Simulation of MEMS energy harvester with different geometries and cross section, in *Proceedings of IEEE Conference on Information and Communication Technologies (ICT 2013)*, Casablanca, Morocco (2013), pp. 120–123
13. A. Shashank, S.K. Harisha, A. Mahalakshmi, Modelling and analysis of piezoelectric cantilever energy harvester for different proof mass and material proportion, in *International Conference on Advances in Materials and Manufacturing Applications* (Institute of Physics: Material Science Engineering, Bangalore, India, 2018)
14. T.A. Anderson, A vibration energy harvesting sensor platform for increased industrial efficiency, in *Smart Structures and Materials: Sensors and Smart Structures Technologies for Civil, Mechanical, and Aerospace Systems. Proceedings of SPIE*, vol. 6174, San Diego, California, USA (2006)
15. D. Chaudhuri, S. Kundu, N. Chattoraj, Design and analysis of MEMS based piezoelectric energy harvester for machine monitoring application. Microsyst. Technol. **25**, 1437–1446 (2019)

16. X.D. Xie, A theoretical model for a piezoelectric energy harvester with a tapered shape. Eng. Struct. **144**, 19–25 (2017)
17. Y. Lee, G. Lim, W. Moon, A piezoelectric micro-cantilever bio-sensor using the mass-micro-balancing technique with self-excitation. Microsyst. Technol. **13**, 563–567 (2007)

Chapter 24
Lowering Motional Impedance in Micromachined Frequency-Synthesizer Using Ultra-thin (SiO_2 ~ 30 nm) Internal Dielectric

Satish K. Verma and Bhaskar Mitra

24.1 Introduction

On-chip frequency synthesizer with low operational power requires lower motional resistance (R_m) [1]. The R_m comparable to the characteristics impedance (50 Ω) has potential to eliminate power hungry intermediate matching networks in front-end RF circuits [2, 3]. Electrostatically transduced resonators have high-Q but high motional resistance and low electromechanical efficiency (η_e) [4]. In order to attain low motional resistance with high coupling and high-Q, an extremely thin airgap is required in electrostatically transduced resonators [5–7]. However, this is challenging due to the fabrication process and limited vibration. The above-stated limitations can be eliminated by an electrostatic transducer with narrow airgap which is filled with solid dielectric as internal electrostatic transduction [8, 9]. This work presents design, fabrication, and equivalent circuit analysis of folded flexure beam for optimum motional resistance of resonator.

24.1.1 Design

A schematic of folded flexure structure with internal electrostatic transducer is shown in Fig. 24.1. The structure consists of two beams fixed-free beam. Each flexure forms an electrostatic transducer consisting of a dielectric layer and a metal layer. The two electrodes are electrically isolated but mechanically coupled by coupling beam at the tip. The geometric parameters of the device are listed in Table 24.1. The electrostatic

S. K. Verma (✉) · B. Mitra
Indian Institute of Technology Delhi, New Delhi 110016, India
e-mail: sverma@ee.iitd.ac.in

R. Singh et al. (eds.), *The Physics of Semiconductor Devices*, Springer Proceedings in Physics 306, https://doi.org/10.1007/978-981-97-1571-8_24

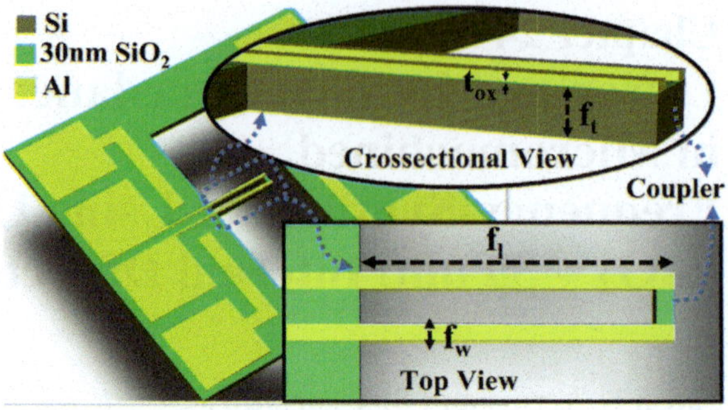

Fig. 24.1 Schematic model of fabricated device depicting 30 nm SiO_2 and metal layer on the drive and sense electrode

Table 24.1 Geometric specification

Parameter	Value	Unit
Flexure length (f_l)	100	μm
Flexure width (f_w)	10	μm
Flexure thickness (f_t)	2	μm
Oxide thickness (t_{ox})	35	nm

force due to the DC bias deflects the flexure out-of-plane while harmonic signal actuates the structure in resonance at resonant frequency.

The resonating MEMS structure can be modelled as a mass-spring-damper system irrespective of its shape which can be a beam, ring, disk or plate. If the body has total kinetic energy U_{Ktot} and velocity at any point is v(x y) then the effective mass, m_{eff} at that point can be determined by [7].

$$m_{eff} = \frac{2U_{Ktot}}{v(x,y)^2} = \frac{2f_t}{v(x,y)^2}\int \rho(x,y)\phi(x,y)^2 dxdy \tag{24.1}$$

where f_t is the thickness, $\rho(x\ y)$ is density and $\phi(x\ y)$ is vibration mode shapes of the resonator. Instead of considering velocity and displacement of each particle in the structure at each point of time, the model of motional behaviour of the system uses the effective mass, m_{eff}, connected with spring having effective stiffness, k_{eff} and damper with damping coefficient, c_{eff}. The simplified mass-spring-damper system with its equivalent electrical circuit is shown in Fig. 24.2. The resonance frequency of the resonator ω_o can be related to effective mass and effective stiffness as,

$$\omega_0^2 = m_{eff}k_{eff} \tag{24.2}$$

$$c_{eff} = \frac{\sqrt{k_{eff}m_{eff}}}{Q} = \frac{\omega_o m_{eff}}{Q} \tag{24.3}$$

where Q is the quality factor of the resonator. If the electrostatic transducer is actuated by an external time-dependent force F(t), then the driving force in the model shown

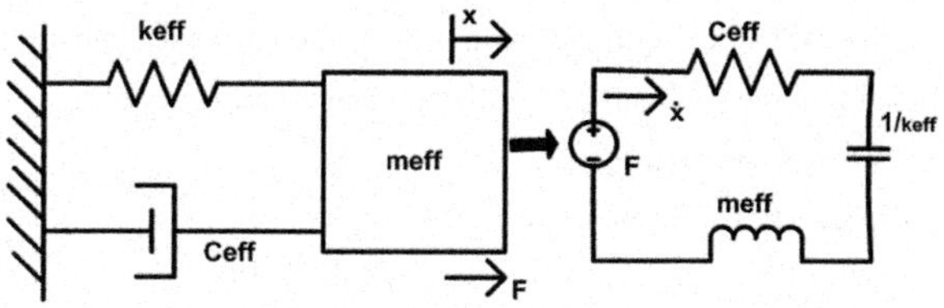

Fig. 24.2 Model of the mass-spring-damper system and its equivalent electrical circuit

in Fig. 24.2, is given by [6]

$$F(t) = m_{eff}\frac{d^2x}{dt^2} + c_{eff}\frac{dx}{dt} + k_{eff}x \tag{24.4}$$

The transduction efficiency, η_e that measures the conversion of an applied input signal into driving mechanical force is given by

$$\eta_e = \frac{F_e}{v_i} = \frac{dC}{dx}V_{bias} \tag{24.5}$$

where C is the transducer capacitance which varies with applied force. The equivalent electrical circuit of the mechanical structure is shown in Fig. 24.6a where Cp represents the input and output capacitances. Likewise, Co is the feedthrough capacitance which includes the fringing field effects and substrate direct parasitic capacitance between input and output. R_m, L_m and C_m are the motional equivalent components of the resonator which relate the physical parameters with the transduction efficiency of the resonator as

$$R_m = \frac{\sqrt{m_{eff}k_{eff}}}{Q\eta_e^2} \tag{24.6}$$

$$C_m = \frac{\eta_e^2}{k_{eff}} \tag{24.7}$$

$$L_m = \frac{m_{eff}}{\eta_e^2} \tag{24.8}$$

The above expressions of motional resistance, motional capacitance and motional inductance provides readily available Butterworth-Von Dyke (BVD) circuit for its resonant behaviour analysis. The physical parameters of resonator are dependent on transduction efficiency, and it improves with high-k dielectric as infer by Eq. 24.6.

To determine the physical parameters and transduction efficiency device's quality factor (Q) is required [10]. Considering viscous fluid, thermos-elastic damping, and acoustic radiation, various losses and hence quality factor has been quantified (Table 24.2) using FEM (COMSOL) analysis. The out-of-plane first mode of resonance (Fig. 24.3b) has resonant frequency of 246.94 kHz. The hormonic response due to resonance is shown in Fig. 24.3b when small signal is applied.

Table 24.2 Quality factor with various damping mechanism

Damping mechanism	Eigen frequency (Hz)	Damping ratio
Thermoelastic damping	2.3e5	5e−7
Squeeze-film damping	2.3e5	1.6e−6
Anchor loss	2.3e5+50i	2.1e−4
Overall	2.29e5+50i	2.1e−4

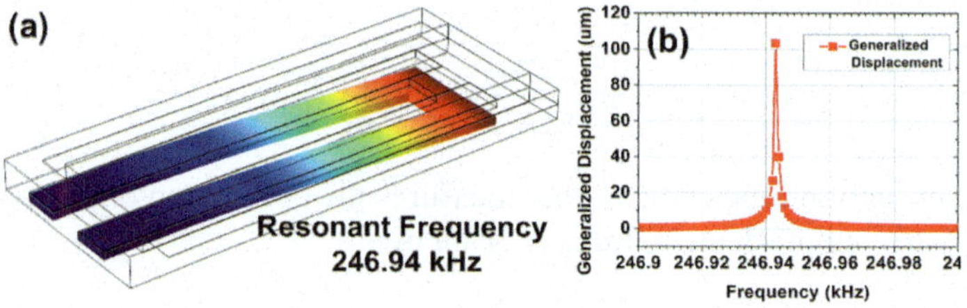

Fig. 24.3 **a** The mode shape of the first resonant frequency at 246.94 kHz. **b** Harmonic perturbation analysis for 1 V DC and 100 mV small signal drive voltage

24.1.2 Fabrication

The fabrication starts with a thermal dry oxidation of Si (100) substrate. The measured SiO_2 layer thickness is found to be 33 nm by ellipsometry as shown in Fig. 24.4. Then a precise through via is opened by patterning and etching the SiO_2 layer to make the bottom silicon contacts in same plane. 200 nm Al is deposited and patterned to define the drive and sense electrodes. The Silicon dioxide and Si is patterned and etched by RIE and wet etching in subsequent processes to define the final metal-insulator-semiconductor device structure. The microscopic and SEM image are of fabricated device is shown in Fig. 24.5.

24.2 Simulation Results and Discussion

To study the effect of motional resistance (R_m) on electrical characteristics of resonator, an equivalent electrical circuit of resonator using BVD model has been developed (Fig. 24.6a). Then we used MATLAB to calculate all the circuit parameters using Q and f_r value from FEM and physical parameters of the device at various bias voltages. Table 24.3 lists the parameter values extracted for two-port resonator at $V_p = 2.5$ V.

The circuit with equivalent parameters values for 10 mV and 2.5 V bias has been simulated using the Cadence virtuoso EDA tool. The device has ~ 38 dB of gain and 5 kHz BW at resonance for 2.5 V bias (Fig. 24.6b). The effect of thin dielectric (transduction gap) and dielectric layer properties on R_m has been studied

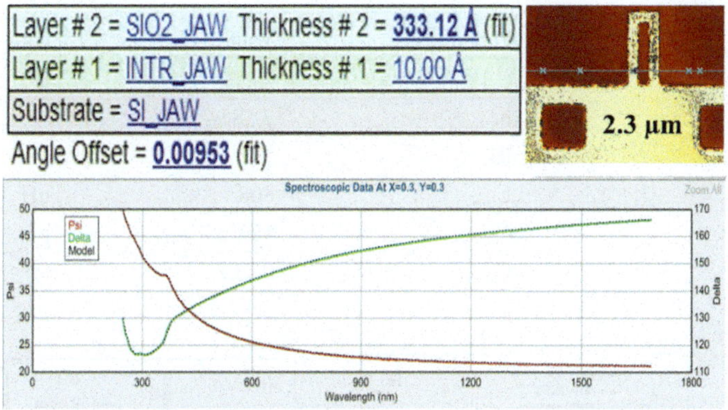

Fig. 24.4 A ellipsometry (fitted curve) and optical profilometric image of the device

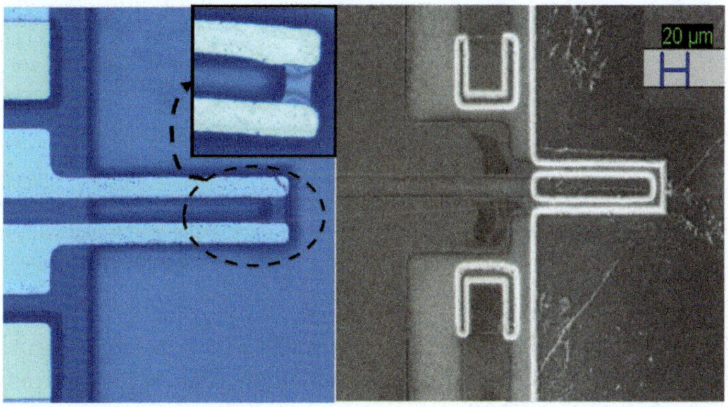

Fig. 24.5 Microscopic (left) and SEM (right) image of fabricated device

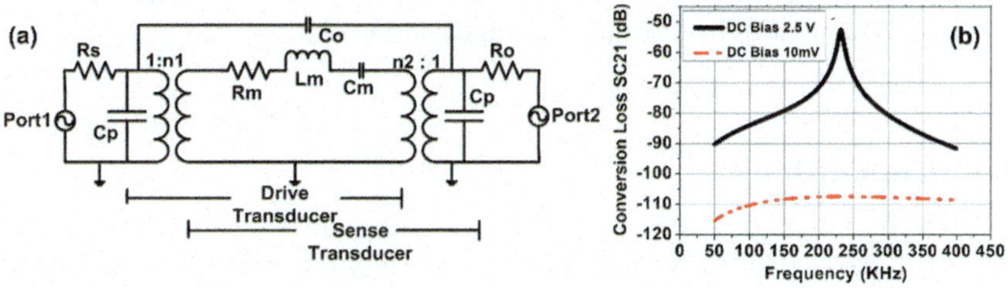

Fig. 24.6 **a** The electrical circuit of two-port MEMS resonator with small signal actuation and resistive load. **b** Conversion gain (38 dB) at 231 kHz resonance frequency at 2.5 V bias voltage

Table 24.3 Extracted parameter of two port resonator

Parameter	Value	Unit
C_p	49.79	fF
n_x	7.58e−5	–
R_m	34	Ω
L_m	808	uH
C_m	583	pF
C_o	0.214	aF

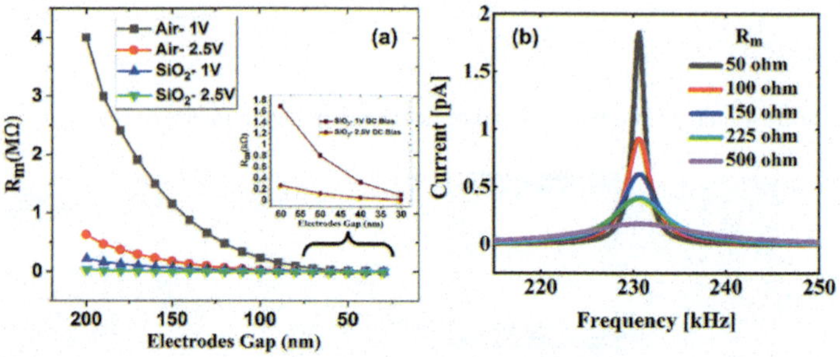

Fig. 24.7 **a** Motional resistance versus electrode gap for air and thin dielectric (SiO_2) with various bias voltages. It illustrates Rm reduces significantly with decreasing gap tuned with bias voltage. **b** Effect of bias dependent motional resistance (R_m) on output sense current of resonator

in Fig. 24.7a. It illustrates R_m reduces exponentially with transduction gap and can be parabolically tuned with bias voltage. Furthermore, R_m can be reduced more towards characteristics impedance (Z_o) by using a high-k dielectric as an internal dielectric transducer. These observations are consistent with analytical model in Eq. 24.6. Figure 24.7b shows the output current at sense port during resonance for various motional resistance Rm by keeping the source and load resistance matched.

24.3 Conclusion

This study presented the effect of thin dielectric on motional resistance (R_m) in micromachined internal dielectric resonators. R_m reduces exponentially with transduction gap and shows parabolic function dependence on bias voltage. A folded cantilever with resonant frequency of 246.94 kHz is designed and its equivalent circuit is simulated. The Q's exceeding 2000 ($Q_{max} = 2282$) in air with f × Q product of 5.64×10^8 is estimated which includes thermoelastic, squeeze film and anchor loss in the model. The frequency response of the equivalent electrical circuit of resonator shows negligible leakage at near 0 V DC bias and as 38 dB conversion gain and 5 kHz BW at resonance for 2.5 V bias with motional resistance as low as of 34 Ω.

Acknowledgements The device fabrication is supported by DST CRG/20191005777. The fabrication and characterization were performed at NRF and CRF IIT Delhi. Satish K. Verma would like to acknowledge Ph.D. fellowship support from Ministry of Education (MoE), Government of India.

References

1. R. Abdolvand, B. Bahreyni, J.E.-Y. Lee, F. Nabki, Micromachined resonators: a review. Micromachines, **7**(9), 160 (2016)
2. K. Wang, A.C. Wong, C.C. Nguyen, VHF free-free beam high-Q micromechanical resonators. J. Microelectromech. Syst. **9**(3), 347–360 (2000)
3. G. Piazza, P.J. Stephanou, A.P. Pisano, Piezoelectric aluminum nitride vibrating contour-mode MEMS resonators. J. Microelectromech. Syst. **15**(6), 1406–1418 (2006)
4. S.A. Bhave, R.T. Howe, Internal electrostatic transduction for bulk MEMS resonators, in *Proceedings of the 2004 Solid State Sensor, Actuator and Microsystems Workshop*, pp. 59–60 (2004)
5. D. Weinstein, S.A. Bhave, Internal dielectric transduction in bulk-mode resonators. J. Microelectromech. Syst. **18**(6), 1401–1408 (2009)
6. M. López, M. de la Candelas, *Micromechanical Resonators with Sub-micron Gaps Filled with High-k Dielectrics* (Infoscience, 2017)
7. L.-W. Hung, High-Q low-impedance MEMS resonators. Technical Report No. UCB/EECS-2012-218, p. 155. http://www.eecs.berkeley.edu/Pubs/TechRpts/2012/EECS-2012-218.html
8. D. Weinstein, S.A. Bhave, Internal dielectric transduction: optimal position and frequency scaling. IEEE Trans. Ultrason. Ferroelectr. Freq. Control **54**(12), 2696–2698 (2007)
9. J.H.T. Ransley, C. Durkan, A.A. Seshia, A depletion layer actuator, in *TRANSDUCERS 2007—2007 International Solid-State Sensors, Actuators and Microsystems Conference* (2007), pp. 1393–1396
10. M. Gologanu, C.G. Bostan, V. Avramescu, O. Buiu, Damping effects in MEMS resonators, in *CAS 2012 (International Semiconductor Conference)*, vol. 1 (2012), pp. 67–76

Chapter 25
Thermally Synthesized Cu/CuO/Cu(Sheet) with Bipolar Resistive Switching

Chandra Prakash and Ambesh Dixit

25.1 Introduction

The continuous demand for fast, low-power, and high-density memory storage devices are compelling to innovative alternative storage devices. Resistive random access memory (RRAM), i.e., the memristor is one such alternative because of its simple two-terminal device structure and utilization of bistable low and high resistance states [1, 2]. Further, RRAM characteristics are also attracting attention due to their potential for artificial neuromorphic computing and resistive switching memory devices. In contrast, the conventional, complementary metal-oxide semiconductor (CMOS) based nanodevices are reaching the scaling limit towards Moore's law [3]. Thus, to reduce the use of CMOS-based nano memory devices because of the relatively higher power consumption, there is a need to explore the alternative non-volatile random-access memory (NV-RAM) with better switching speed, high-density storage capacity, and scalability [4]. Various NV-RAM devices are proposed, such as spin torque transfer magnetoresistive (STT-MRAM), phase change (PC-RAM), ferroelectric, flash-type, and resistive non-volatile random-access memory (RRAM) [5, 6]. The resistive random-access memories are getting attention because of its non-volatile nature, higher endurance, and retention characteristics. Generally, RRAM devices are explored in sandwiched-like device configurations, i.e., an active layer is sandwiched between two metal electrodes [7]. The toggling of ions is attributed to the change between high and low resistance states against the applied electric field, i.e., the switching voltage across the device [8]. That migration of ions, i.e., oxygen vacancy in metal oxides or metal cation like Ag, Cu, and Al from the active electrodes, is the main source of conductive filament formation in RRAM

C. Prakash · A. Dixit (✉)
Advanced Materials and Device (A-MAD) Laboratory, Department of Physics, Indian Institute of Technology Jodhpur, Jodhpur, Rajasthan 342037, India
e-mail: ambesh@iitj.ac.in

R. Singh et al. (eds.), *The Physics of Semiconductor Devices*, Springer Proceedings in Physics 306, https://doi.org/10.1007/978-981-97-1571-8_25

devices, exhibiting a low resistance state while rupturing leads to the high resistance state [9].

Moreover, the switching mechanisms depend on the type of material and device structures. The filamentary switching mechanisms are widely explored to emphasize the resistance change in metal oxide-based RRAM devices [10]. In this mechanism, the formation and rupturing of conductive filament are achieved by applying the bidirectional external field. The filament formation also relies on oxygen vacancies in metal oxide-based RRAM devices, and the type of top contacts also plays a crucial role. The inherent oxygen vacancies in metal oxides make them a suitable candidate for RRAM devices [11]. Numerous metal oxide materials, their composites, and non-oxide materials are explored for RRAM applications, showing both unipolar and bipolar memory characteristics. The metal oxide system includes TiO_2, CuO, ZnO, ZrO_2, Al_2O_3, $BiFeO_3$, HfO_2, etc. [12–14]. These materials offer a wide bandgap and are relatively easy to synthesize, including even low-cost solution processes and pre/post thermal treatment. Further, the process parameters can be engineered to realize bulk to nanostructured thin film structures of these metal oxide materials with a wide range of bandgap engineering possibilities. Copper oxide is an interesting material, showing semiconducting characteristics. Recent studies on monoclinic CuO single crystal also showed the multiferroicity in this system [15]. The copper is also relatively earth-abundant, and its non-toxic nature with fabrication ease makes it a potential system for integration with the existing technologies. However, CuO material is commonly synthesized using vacuum-assisted or solution processes [16]. Hsu et al. [17] and Rehman et al. [18] reported RRAM characteristics on CuO, exhibiting good switching, retention, and endurance properties.

The present work explores the RRAM characteristics of CuO thin films on Cu sheets, synthesized using the single-step thermal treatment of Cu sheets under normal ambient conditions. Further, copper is also used as the top contact using thermal evaporation. Thus, Cu(top electrode)/CuO (active layer)/Cu (bottom electrode) RRAM device configuration is investigated for its current–voltage, retention, and endurance characteristics, showing the potential of thermally treated CuO thin films.

25.2 Materials and Method

The CuO thin film is grown on a commercially available normal-grade copper sheet after cleaning it thoroughly with acetone, isopropanol, and de-ionized water for 10 min each, followed by nitrogen drying. The thoroughly cleaned copper sheet is subjected to controlled thermal treatment at 300 °C for 3 h under normal ambient conditions. A black color thin film is noticed as the conversion of copper into copper oxide using environmental oxygen. The 120–150 nm thick 500 micrometer in diameter circular and 500 μm copper top contacts are deposited using shadow masks in a thermal evaporation system diameter circular shape with the copper source material.

25.3 Result and Discussion

The scanning electron microscopic (SEM-Zeiss make) images and energy-dispersive X-ray (EDX) analysis are carried out on a pristine Cu sheet and CuO thin film on a Cu sheet. The results are shown in Fig. 25.1a–c. The SEM image of a cleaned Cu sheet shows surface imprints and roughness created during cleaning, Fig. 25.1a. The SEM image of the CuO thin film surface is shown in Fig. 25.1b, confirming the dense growth of flower-like morphology in micrometer length scales. The irregular growth structures may be due to the different nucleation sites on the rough surface of the Cu sheet. The EDX elemental analysis, shown in Fig. 25.1c, confirms the stoichiometric ratio of Cu and O in CuO thin film, thus substantiating the formation of CuO material in thin-film geometries with flower-like microstructures. The growth is facilitated on the top surface of the Cu sheet, whereas the bottom surface remains intact and is used as the bottom contact for RRAM devices.

Further to analyze roughness and topography of the thermally grown CuO thin film, AFM images are collected using XE-70 (Park) AFM system on a 1 μm × 1 μm scan area. The topographical features are shown in Fig. 25.1d, substantiating the SEM microscopic results. The large roughness of the material is attributed to the random growth of CuO on copper sheets. The optical micrograph of the copper sheet is shown in Fig. 25.2a together with the thermally treated copper sheet, showing the formation of CuO on top is shown in Fig. 25.2a. The black color of the surface

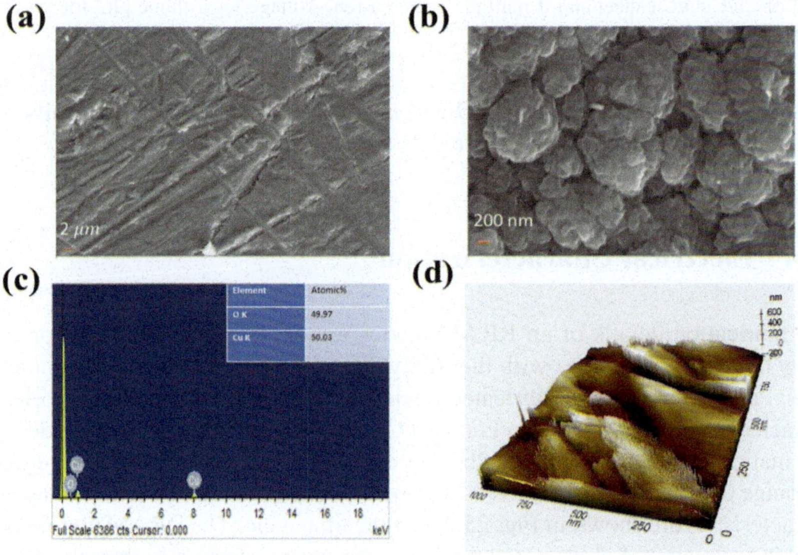

Fig. 25.1 Scanning electron microscopy micrograph of **a** cleaned Cu sheet, **b** thermally treated Cu sheet, i.e., on CuO thin film, **c** EDX spectra of thermally treated Cu sheet, and **d** 3D-AFM image of CuO thin film on Cu sheet

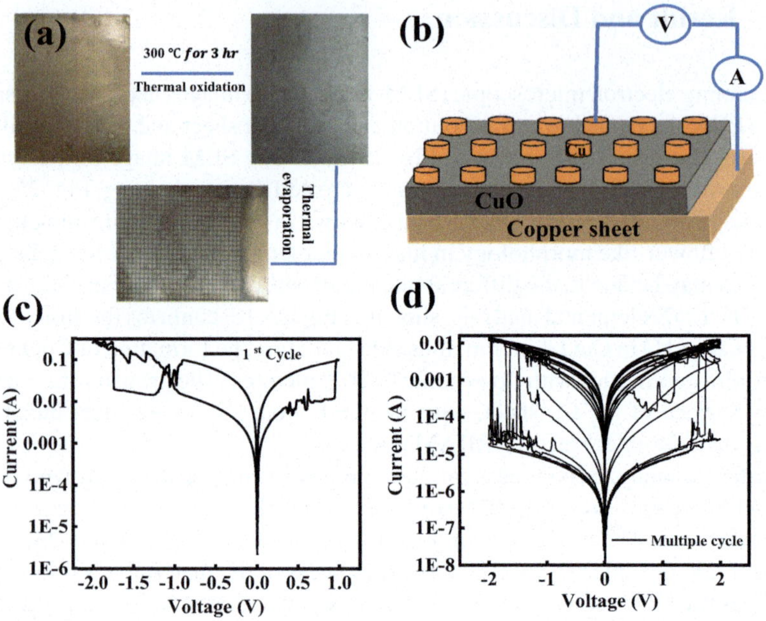

Fig. 25.2 **a** Optical photograph of fabricated device cleaned Cu sheet (left) after thermally treated (middle) and after making electrical contact, **b** schematic diagram of fabricated device Cu (top contact)/CuO (active layer)/Cu sheet (bottom contact), **c** current–voltage logarithmic plot for 1st cycle of Cu/CuO/Cu sheet and **d** multiple cycle current–voltage logarithmic plot for Cu/CuO/Cu sheet

substantiates the formation of CuO and is consistent with the EDX results, where nearly stoichiometric Cu and O elements are noticed.

25.4 Electrical Characterization

The fabrication details of an RRAM device are explained in Fig. 25.2 on a thermally treated copper sheet with thermally deposited copper as the top contact. The optical photograph of the fabricated devices is shown in Fig. 25.2a. The electrical characterization on the Cu(top contact)/CuO/Cu sheet (bottom contact) device, as schematically shown in Fig. 25.2b. The externally applied field is bidirectional in the range $0\ V \rightarrow 2\ V \rightarrow 0\ V \rightarrow -2\ V \rightarrow 0\ V$. The collected current–voltage (I–V) characteristics are shown in Fig. 25.2c for the first cycle. The RRAM characteristics exhibit bidirectional characteristics. Here, the change from a high resistance state (HRS) to a low resistance state (LRS) is considered the set process, whereas the change from LRS to HRS is called the reset process. We also used a compliance current of 20 mA to prevent the breakdown of the RRAM device's inset state. The set process is achieved in the positive polarity, whereas the reset process is achieved in the negative polarity of the applied voltage. The RRAM device showed a high on–off ratio, i.e., $I_{on}/I_{off} \sim 10^3$. The abrupt behavior of the device is attributed to the migration of oxygen vacancy inherently available in CuO after its thermal treatment under the influence of the external electric field [1]. The reproducibility of the device is investigated for different cycles. Figure 25.2d shows the ten cycle data, showing some set and reset voltage variations. The variation in the set and reset voltage is attributed to the rough nature of the CuO surface. Further, after some cycles, the RRAM device starts exhibiting the analog behavior, suggesting the slow migration of oxygen vacancy mediated current conduction competing with Cu ion filament in the presence of an external electric field.

We used Weibull cumulative distribution of Cu/CuO/Cu sheet RRAM devices to further analyze the device in respective HRS and LRS, as shown in Fig. 25.3a. The Weibull cumulative function is defined as $\beta \ln(V) = \ln[-\ln\{1 - f(x)\}]$, where β is the shape parameter, and f(x) is the function for set and reset voltages in the present case [19]. The set and reset voltage of Cu/CuO/Cu sheet RRAM devices are distributed from 0.95 V to 1.52 V and − 1.07 V to − 1.97 V, respectively. This variation is attributed to the non-smooth surface of CuO with probably uneven contacts, leading to such variations. The retention characteristics for Cu/CuO/Cu sheet devices are carried out at read voltage of 0.5 V for 10^3 s under ambient conditions and observed R_{LRS} and R_{HRS} are shown in Fig. 25.3b, substantiating the stability of RRAM devices in this period. We also investigated the endurance for 100 cycles, and the results are summarized in Fig. 25.3c, suggesting that HRS and LRS states are quite robust over these cycles. The current conduction mechanisms are analyzed using logarithmic plots for I–V characteristics of the RRAM device in the set and reset states. The plots are shown in Fig. 25.3d. The current conduction is nearly Ohmic ($I \alpha V^m$, m ~ 1 for both set and reset states, Fig. 25.3d in the applied electric field region for both set and reset states. It is attributed to the thermally generated charge carriers, participating in the current conduction [20, 21]. The abrupt change in the resistance is observed at higher fields, where the device changes its state from HRS to LRS. Here the respective slop is greater than 2, suggesting the trapped charge limited current. Further, Ohmic conduction dominates after switching the device into an LRS state from an HRS state [22].

The microscopic mechanism of this resistive switching behavior can be explained considering the formation of conductive filament and its' rupturing with the external applied electric field in the Cu/CuO/Cu sheet. This process is explained schematically in Fig. 25.4, describing the filament formation during the set and its rupturing in the reset process. The copper ions migrate from the top or bottom contacts, acting as the source in the external field, resulting in filament formation during the set state and bringing the RRAM device to LRS as shown in Fig. 25.4a. Further, after reversing the polarity of the applied electric field, copper metal ions start repelling each other and thus, causing it's' rupturing, exhibiting a transition from LRS to HRS, as shown in Fig. 25.4b [23]. This process is repeated over different field cycles, showing the transition in these two resistive memory states, supported by the formation and rupturing of the filament.

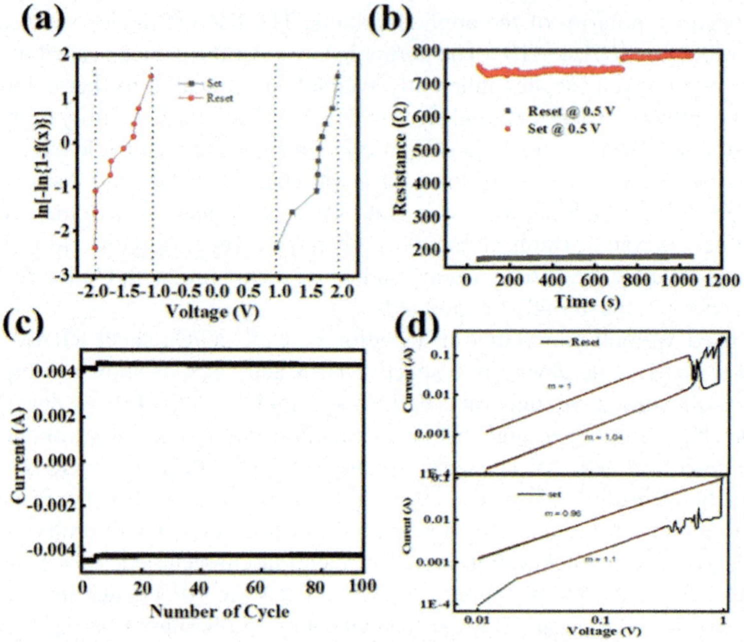

Fig. 25.3 **a** Weibull cumulative distribution for set and reset voltage of Cu/CuO/Cu sheet, **b** retention characteristics of the fabricated device for 1000 s at 0.5 V, **c** endurance characteristics of the fabricated device for 100 cycle at 0.5 and − 0.5 V and **d** log I–log V characteristics of Cu/CuO/Cu sheet in the set and reset state

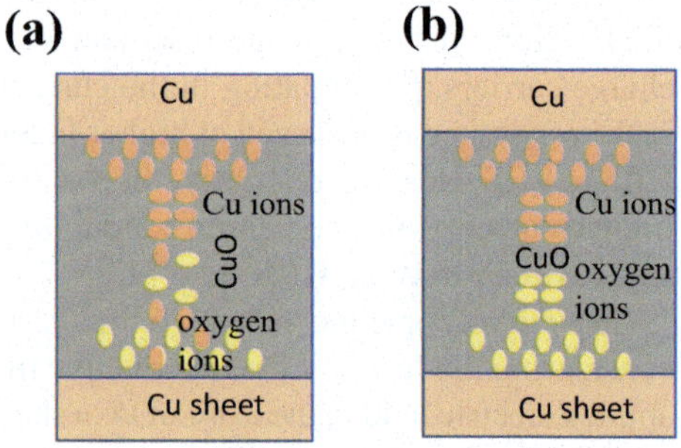

Fig. 25.4 **a** Schematic of the formation of a conductive filament in the set state and **b** rupturing of conductive filament in a reset state in Cu/CuO/Cu

25.5 Conclusion

We successfully demonstrated the synthesis of CuO on a Cu sheet using the thermal treatment of the Cu sheet under normal ambient conditions. It is integrated with top copper contacts to realize Cu/CuO/Cu sheet RRAM device. The bipolar digital resistive switching is demonstrated with a large I_{on}/I_{off} ratio ~ 10^3 with an abrupt change in resistance state, suggesting the digital switching. The variation in switching voltage is also observed in the range of − 1.07 to − 1.97 V for the reset state and 0.95 to 1.52 V in the set state. However, the devices showed high retention (10^3 s), and endurance for 100 cycles. The current conduction is Ohmic mainly in the low field regions for both set and resets states, and the switching is facilitated by the filament formation and its rupturing mediated by the copper metal ions.

Acknowledgements Ambesh Dixit acknowledges SERB, DST, and Gov. of India through project # C.R.G./2020/004023 for carrying out this work. Author Chandra Prakash acknowledges Mr. Sumit Kukreti, Mr. Abhijeet J Kale, Ms. Surbhi Ramawat, Ms. Priyambada Sahoo, Mr. Jitendra K. Yadav, Mr. Piyush Chaudhary, Mr. Ankit K. Yadav from the A-MAD laboratory for the technical discussion and support.

References

1. S. Kwon, H. Seo, H. Lee, K.J. Jeon, J.Y. Park, Reversible bistability of conductance on graphene /CuO_x/Cu nanojunction. Appl. Phys. Lett. **100**, 123101 (2012)
2. J. Fatheema, T. Shahid, A. Mohammad, A. Islam, A comprehensive investigation of MoO_3 based resistive random access memory. RSC Adv. **10**, 19337–19345 (2020). https://doi.org/10.1039/d0ra03415k
3. D. Acharyya, A. Hazra, P. Bhattacharyya, A journey towards reliability improvement of TiO_2 based resistive random access memory: a review. Microelectron. Reliab. **54**, 541–560 (2014). https://doi.org/10.1016/j.microrel.2013.11.013
4. D. Panda, C. Chu, A. Pradhan, S. Chandrasekharan, B. Pattanayak, S.M. Sze, Synaptic behaviour of TiO_x/HfO_2 RRAM enhanced by inserting ultrathin Al_2O_3 layer for neuromorphic computing. Semicond. Sci. Technol. **36**, 045002
5. W. Banerjee, Challenges and applications of emerging nonvolatile memory devices. Electronics **9**(6), 1029 (2020). https://doi.org/10.3390/electronics9061029
6. A. Chen, A review of emerging non-volatile memory (NVM) technologies and applications. Solid State Electron. **125**, 25–38 (2016). https://doi.org/10.1016/j.sse.2016.07.006
7. H.F.R.S. Memory, X. Li, J. Yang, H. Ma, Y. Liu, Atomic layer deposition of Ga_2O_3/ZnO composite films for high performance forming-free resistive memory. ACS Appl. Mater. Interfaces **12**(27), 30538–30547 (2020). https://doi.org/10.1021/acsami.0c06476
8. Y.A. Park, K.D. Sung, C.J. Won, J.H. Jung, N. Hur, Bipolar resistance switching and photocurrent in a $BaTiO_{3-\delta}$ thin film. J. Appl. Phys. **114**, 094101 (2013)
9. J. Lee, W. Schell, X. Zhu, E. Kioupakis, W.D. Lu, Charge transition of oxygen vacancies during resistive switching in oxide-based RRAM. ACS Appl. Mater. Interfaces **11**(12), 11579–11586 (2019). https://doi.org/10.1021/acsami.8b18386
10. X.L. Deng, S. Hong, I. Hwang, J. Kim, H. Jeon, C. Park, Nanoscale confining grains of textured Cu_2O films to single-crystal nanowires and resultant change in resistive switching characteristics. Nanoscale **4**, 2029–2033 (2012). https://doi.org/10.1039/c2nr12100j
11. T. Park, Y.J. Kwon, H.J. Kim, H.C. Woo, G. Seop, C.H. An, Y. Kim, D.E. Kwon, C.S. Hwang, Balancing the source and sink of oxygen vacancies for the resistive switching memory. ACS Appl. Mater. Interfaces **10**(25), 21445–21450 (2018). https://doi.org/10.1021/acsami.8b05031
12. A. Kim, K. Song, Y. Kim, J. Moon, All solution-processed, fully transparent resistive memory devices. ACS Appl. Mater. Interfaces **3**(11), 4525–4530 (2011)
13. C. Kumari, I. Varun, M. Pal, S.P. Tiwari, A. Dixit, Reliable and forming free bipolar resistive switching in solution derived Ag/$BiFe_{0.99}Cr_{0.01}O_3$/FTO device, in *2019 IEEE 14th Nanotechnology Materials and Devices Conference, NMDC 2019* (2019). https://doi.org/10.1109/NMDC47361.2019.9083989

14. K. Chen, M. Kao, S. Huang, C. Li, C. Cheng, Bipolar switching properties and electrical conduction mechanism of manganese oxide RRAM devices. Ceram. Int. **43**, S253–S257 (2017). https://doi.org/10.1016/j.ceramint.2017.05.183
15. Y.S. Hong, J.Y. Chen, C.W. Huang, C.H. Chiu, Y.T. Huang, T.K. Huang, R.S. He, W.W. Wu, Single-crystalline CuO nanowires for resistive random access memory applications. Appl. Phys. Lett. **106** (2015). https://doi.org/10.1063/1.4919102
16. J.W. Han, M. Mayyappan, Copper oxide resistive switching memory for e-textile. AIP Adv. **1**, 032162 (2011). https://doi.org/10.1063/1.3645967
17. C. Hsu, Y. Lin, C. Cheng, W. Jhang, Annealing effect on the performance of copper oxide resistive memory devices. IEEE Trans. Electron. Devices **67**, 3 (2020)
18. S. Rehman, J. Hur, D. Kim, Resistive switching in solution-processed copper oxide (Cu_xO) by stoichiometry tuning. J. Phys. Chem. C **122**(20), 11076–11085 (2018). https://doi.org/10.1021/acs.jpcc.8b00432
19. C. Kumari, I. Varun, S. Prakash Tiwari, A. Dixit, Robust non-volatile bipolar resistive switching in sol-gel derived $BiFeO_3$ thin films. Superlattices Microstruct. **120**, 67–74 (2018). https://doi.org/10.1016/j.spmi.2018.05.008
20. O. Taboubi, M. Radaoui, A. Ben Fredj, S. Romdhane, M.C. Scharber, N.S. Sariciftci, H. Bouchriha, Magnetic field effects on the current of PCPDTBT-based diode. J. Phys. Chem. C **121**, 11727–11732 (2017). https://doi.org/10.1021/acs.jpcc.7b03539
21. D.I. Son, D.H. Park, W.K. Choi, S.H. Cho, W.T. Kim, T.W. Kim, Carrier transport in flexible organic bistable devices of ZnO nanoparticles embedded in an insulating poly(methyl methacrylate) polymer layer. Nanotechnology **20** (2009). https://doi.org/10.1088/0957-4484/20/19/195203
22. L.F. Liu, Unipolar resistive switching and mechanism in Gd-doped-TiO_2-based. Semicond. Sci. Technol. **26**, 115009 (2011). https://doi.org/10.1088/0268-1242/26/11/115009
23. C. Kumari, I. Varun, S.P. Tiwari, A. Dixit, Interfacial layer assisted, forming free, and reliable bipolar resistive switching in solution processed $BiFeO_3$ thin films. AIP Adv. **10** (2020). https://doi.org/10.1063/1.5134972

Chapter 26
Dynamic Tuning of ENZ Region of ITO and Sensing Using a Tapered Optical Fiber

Arun Mambra, Ravi Pant, and J. Mitra

26.1 Introduction

Recently, there has been a surge in research interest in epsilon near zero (ENZ) materials because of their exciting and non-intuitive properties that stem from zero electrical permittivity, including strong field confinement and enhancement, super-coupling and slow light, ultrafast signal modulation, perfect absorption, huge optical nonlinearities, and second and third harmonic generation [1–5]. Bridging the gaps between the theoretical predictions of phenomena associated with near zero index media and their practical realization in real media are challenges that many groups are engaged in. Transparent conducting oxides (TCOs) are one class of systems which exhibit these properties in a relatively simple and straightforward fashion [6]. TCOs such as indium tin oxide (ITO), doped zinc oxide (ZnO), and cadmium oxide (CdO) have already attracted interest due to their ease of fabrication and diverse applications in plasmonics, optoelectronics, and photonics [4, 7]. All these systems exhibit dielectric to metal transition in the near-infrared (NIR) regime, at the ENZ wavelength (λ_{ENZ}). The real part of dielectric function (Re(ϵ)) is positive for $\lambda < \lambda_{ENZ}$ and becomes negative for $\lambda > \lambda_{ENZ}$, thus showing a metal like optical properties and Re(ϵ) $= 0$ for $\lambda = \lambda_{ENZ}$ [6]. Typically, degenerately doped ITO has free electron density $n_e \sim 10^{20}$–10^{21} cm^{-3}, with its $\epsilon(\omega)$ well described by the Drude model. The corresponding plasma frequency (ω_p) provides the dependence of λ_{ENZ} on n_e as $\lambda^2_{ENZ} \propto 1/n_e$ [6]. Here, we have used commercially available ITO coated glass slides with a sheet resistance of 10–100 Ω/sq. Importantly, the origin of doping in ITO i.e. the Sn^{4+}/Sn^{2+} ratio controlled by the oxygen stoichiometry and it's magnitude

A. Mambra (✉) · R. Pant · J. Mitra
Indian Institute of Science Education and Research Thiruvananthapuram, Thiruvananthapuram 695551, India
e-mail: itsmearun18@iisertvm.ac.in

J. Mitra
e-mail: j.mitra@iisertvm.ac.in

R. Singh et al. (eds.), *The Physics of Semiconductor Devices*, Springer Proceedings in Physics 306, https://doi.org/10.1007/978-981-97-1571-8_26

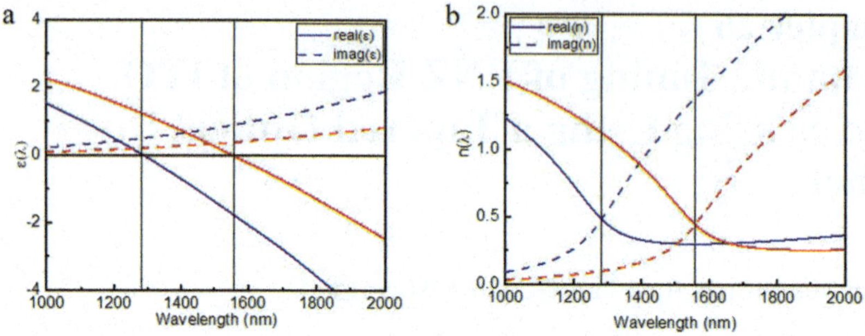

Fig. 26.1 **a** Wavelength dependent dielectric function of ITO with two different λ_{ENZ} 1280 nm (commercial) and 1560 nm (tuned via annealing protocol [6]). **b** Wavelength dependent refractive index for these two systems

allow for both static and dynamic control of n_e, which directly changes λ_{ENZ}. The ability to actively tune λ_{ENZ} opens new avenues to exploit the fascinating properties at desirable wavelengths. Figure 26.1 shows the wavelength-dependent dielectric function and refractive index for the ITO thin films showing the control afforded by the system [6, 8].

In this work, we put forth a dynamic tuning method using electrostatic gating with the help of a solid polymer electrolyte (PEO – $LiClO_4$). We have attempted to showcase an effective method to sense the change in the dispersion of systems by employing the concept of multimodal interference (MMI) within a tapered fiber. Optical fiber based sensing has been explored since the development of modern lossless fibers [9], especially tapered fibers that has expanded the scope of optical sensing, where "leaky" guided modes have been used to sense the dielectric environment via evanescent field coupling, with advances in sensing humidity and biochemical sensors [10]. We have utilized a single-mode fiber (SMF28e) in which the incident energy is confined with the fundamental HE11 mode with a Gaussian profile. Depending on the wavelength of operation, tapering the fiber allows excitation of multiple modes in the tapered region, dominant among which is the non-Gaussian HE12 mode with substantial evanescent field extending beyond the taper boundary. Coupling this mode to a gated ITO film then allows further modulation of the modal dispersion via externally controlled electrostatic gating. It is difficult to realize a dynamic control over the dispersion of an optical device, unless the dispersion of the waveguide, the materials used and the geometry are defined. Recently, dynamic dispersion tuning of silicon nitride ring resonator via electrostatic gating across a graphene layer resulted in generation of tunable frequency comb [11]. Tuning the dispersion of the waveguide using an ENZ material like ITO can be exploited in designing absorption modulators based on ENZ mode absorptions [12].

26.2 Experimental Details and Modeling

Commercially bought ITO thin films, coated on glass (Sigma Aldrich), of thicknesses 20 and 120 nm were used in this project. A narrow strip of ITO (width ~ 5 mm and length ~ 1.3 cm) is isolated to create the active channel with source (S) and drain (D) connections, as shown in Fig. 26.2a. A solid polymer electrolyte solution (SPE) was prepared starting with solutions of 2.5 wt% of *PEO* (100 k molecular weight; Sigma Aldrich) and 2 wt% of $LiClO_4$ (Sigma Aldrich) in methanol and filtered using a PVDF 0.45 μm filter. Equal weights of filtered solutions were mixed and stirred to form SPE solution that was then coated across the etched ITO thin-films. The SPE coating on the active channel allows gating of the channel with the other isolated ITO section acting as the gate electrode. Together, the SPE coated pads realize a three-terminal source-drain-gate electrode device as shown in Fig. 26.2a. Two terminal electrical measurements were performed to characterize the SPE and three terminal gating measurements were performed to evidence the effects of channel carrier density modulation. A source-meter (Keithley Instruments 2400) and a DC power source (Tektronix PWS4305) were used for all the electrical measurements.

Figure 26.2b shows the cross-sectional schematic of a taper in contact with the ITO active channel. The tapered fibers used in the multi-modal interference (MMI) setup were fabricated from 50 cm long single mode fibers (SMF28e, Corning Inc.) with 8 μm core and 125 μm cladding. Splicing and tapering of the fibers were done using a glass processing instrument (Vytran GPX3400) with the dimensions of taper length ~ 1 cm, taper waist ~ 6 μm and taper transition region ~ 3 mm, optimized to yield reproducible and stable MMI in free standing tapers in air. In the final device, the taper-ITO junction is formed prior to coating the active channel with SPE. The MMI was excited using the output noise of a laser line amplifier centered at 1550 nm (Pritel) as a broadband source (BBS) and the IR detector of an optical spectrum analyzer (OSA, Anritsu MS9740A) was employed to detect the spectral transmission. Effective mode index study for a tapered fiber was performed using the electromagnetic wave module of the COMSOL® Multiphysics simulation software. The modal dispersion of the free-standing taper and in contact with ITO

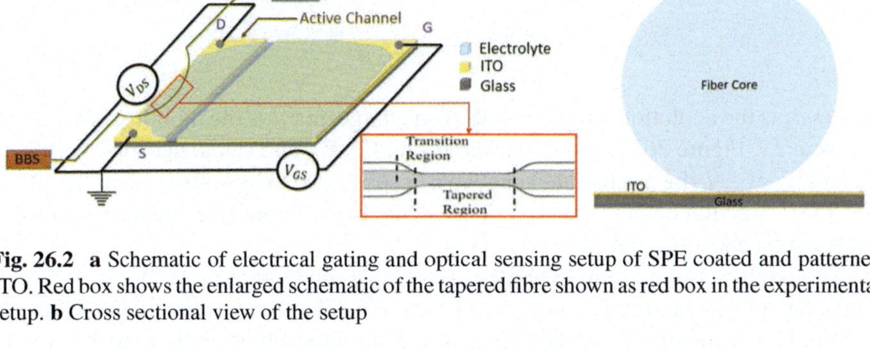

Fig. 26.2 **a** Schematic of electrical gating and optical sensing setup of SPE coated and patterned ITO. Red box shows the enlarged schematic of the tapered fibre shown as red box in the experimental setup. **b** Cross sectional view of the setup

films of various λ_{ENZ}, for the $HE11$ and $HE12$ eigen modes are simulated to yield the effective refractive index corresponding to each mode. The semiconductor module is used to simulate the carrier modulation for ITO with different gate voltage.

26.3 Results and Discussion

Figure 26.3a shows the change in the ITO active channel resistance (R_{ch}) with dc gate voltage (V_{GS}) applied to the SPE. The inset shows the *IV* characteristics measured at each V_{GS}. For the 20 nm ITO thin-film, the non-gated channel resistance was found to be 340 Ω. While positive V_{GS} decreases the R_{ch} signifying increase in n_e, negative V_{GS} increased R_{ch} due to electron depletion. The change in R_{ch} is effected by the formation of an electric double layer (EDL) at the ITO-SPE interface. In this case, the ether O-atoms in the PEO chain coordinates with the Li^+ ions through the Lewis acid–base interaction hence separating the ClO_4^- counter ions [13]. Now, when the SPE is subjected to an electric field the ions move along the electric field towards the surface of the ITO, at the same time electrons in ITO either accumulate or deplete at the surface to form the EDL [14], thereby effectively changing the n_e in the active channel. However, the change in n_e is non-uniform across the depth of the channel due to screening effects of the free carriers. To quantify the depletion width, transport along the channel was simulated in the presence of gate voltage using FEM simulation. Figure 26.4a shows the cross-sectional plot of n_e carrier modulation in ITO with distance from the interface, for a background $n_e = 1 \times 10^{20}\,\mathrm{cm}^{-3}$ and applied $V_{GS} = 2\,\mathrm{V}$, corresponding to an electric field of 2×10^9 V/m. Figure 26.4c shows the variation in local n_e with distance for various electric fields that allows us to conclude that the entire carrier modulation in the ITO films are restricted to within 1 nm of the interface. Evidently 5% of the 20 nm thick ITO film is affected while less than 1% of the 120 nm is affected by the applied V_{GS}. To calculate an effective change in electron density (Δn_e) due to gating, we model the cross-sectional variation in n_e in the ITO film as two lumped resistance elements in parallel (as shown in Fig. 26.4b). Assuming that n_e are constant within each element allows us to derive an analytic expression relating Δn_e to ΔR_{ch} as,

$$\Delta n_e = -\frac{\Delta R_{ch}}{d_1 e \mu R_{ch0}^2} \tag{26.1}$$

where d_1 is the depletion width, *e* is electron charge and μ is the mobility of electrons ~ 30 cm^2/V. Figure 26.3b shows the variation of ΔR_{ch} and calculated Δn_e with V_{GS}. For a V_{GS} of 1 V, the fractional change in n_e is observed to be ~ 40%, which results in a ~ 158% fraction change in λ_{ENZ}. This stems from the inverse square relationship between n_e and λ_{ENZ} i.e., $n_e \propto 1/\lambda_{ENZ}^2$.

Though the pristine fibre (SMF28e) is a single mode fibre in the IR, introduction of the taper allows multiple modes ($HE11$ and $HE12$) to be excited within the tapered region. Consequently, the power transmitted through a fibre with a tapered region

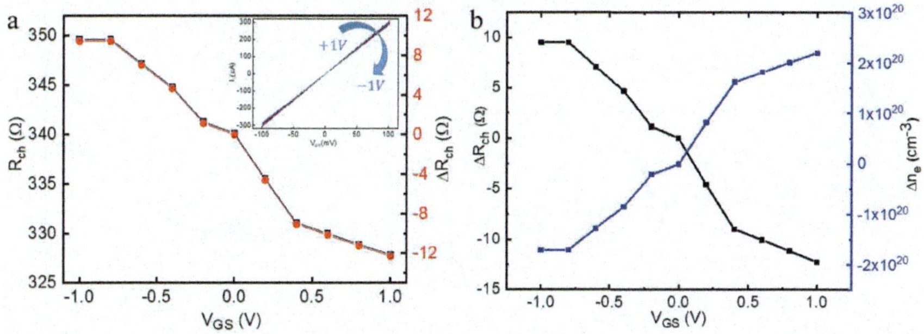

Fig. 26.3 **a** Modulation in the channel resistance (R_{ch}) for change in gate voltage (V_{GS}). Right: ΔR_{ch} with respect to ungated channel resistance. Inset: *IV* characteristic of the active channel for different V_{GS}. **b** Change in number density calculated from the modulation of channel resistance

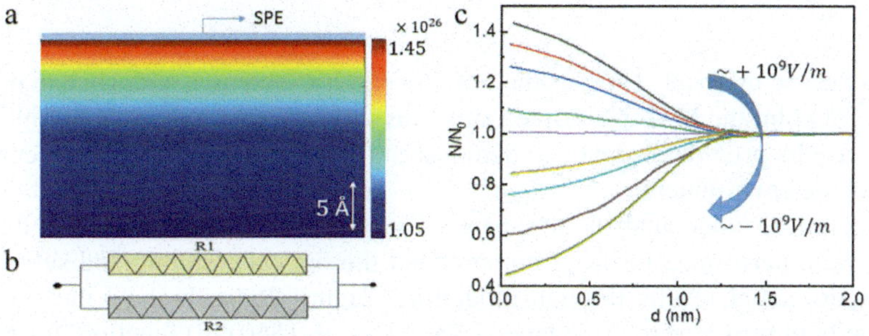

Fig. 26.4 **a** Cross-sectional accumulated number density (m^{-3}) simulated using FEM. Showing an accumulation region of ~ 1 nm. **b** Effective circuit with two parallel resistance $R1$ and $R2$, showing the modulated and unmodulated bulk of ITO, respectively. **c** Number density modulation in the depletion region for different gate voltages

becomes an oscillatory function of wavelength. The oscillations originate due to MMI between the multiple modes allowed by the tapered waveguide. Figure 26.5a (top) shows the MMI pattern observed in the transmitted intensity for a standalone tapered fiber in air. The energy distribution of the non-Gaussian $HE12$ mode is annular with a substantial evanescent field extending beyond the fibre boundary. Hence, MMI is highly sensitive to the refractive index of the surrounding medium and has been used in various refractive index sensing designs [15, 16]. As the ITO films of various λ_{ENZ} are brought in contact to the tapered region the evanescent field is coupled differentially and the interference pattern starts shifting according to the refractive index of the ITOs (Fig. 26.5a (bottom)). Figure 26.5b shows the close-up of one interference minima around 1561 nm for the taper in contact with bare glass ($n = 1.5$) along with those from ITO films with λ_{ENZ} tuned to 1280, 1560, 1640 and 1990 nm. While the ITO film with $\lambda_{ENZ} = 1280$ nm behaves like a metal at 1561 nm with $\text{Re}(\epsilon) < 0$ and $n \sim 0.3$, for $\lambda_{ENZ} = 1560$ nm the optical constants are $\text{Re}(\epsilon) \sim 0$ and $n \sim 0.4$. For λ_{ENZ} of 1640 and 1990 nm the $\text{Re}(\epsilon) > 0$ indicative

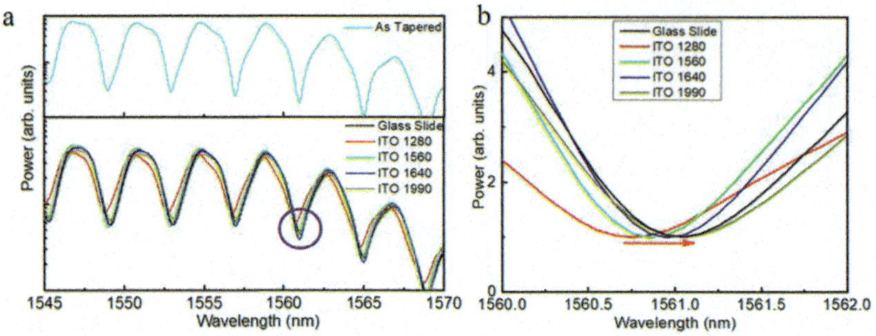

Fig. 26.5 **a** MMI measured for as tapered fiber with a taper waist of 6 μm in air (top), glass slide and ITO with different ENZ wavelength (bottom). **b** Zoomed up view of an interference minima around 1561 nm to understand the shift (violet circle in (**a**)). The red arrow point in the direction of increase in refractive index. *Note* The four-digit number in the legend after the ITO signifies its ENZ wavelength in nanometers

of a dielectric material. Thus, as the refractive index increases with change in λ_{ENZ} of the ITO film the MMI pattern red shifts, as shown by the red arrow in Fig. 26.5b. These results provide evidence to demonstrate that MMI is an efficient method for sensing the environment.

Theoretical mode analysis of tapered fiber in contact with the ITO of different λ_{ENZ} were performed to better comprehend the optical coupling and substantiate the experimental results through modelling. Figure 26.6a, b show the simulated electric field profile across the taper-ITO ($\lambda_{ENZ} = 1560$ nm) junction, for the two modes HE11 and HE12. As discussed, the Gaussian $HE11$ mode is well confined within the core and results in no detectable mode index shift with the changing optical properties of the ITO films for various λ_{ENZ}. By contrast, the HE12 mode (Fig. 26.6b) shows a central maximum with a well-defined lobe towards the periphery of the fiber, which leaks and couples to the environment, thereby being susceptible to any refractive index change in the environment. Figure 26.6c shows a zoomed image of the contact region evidencing coupling of the mode to ITO. Note that the ITO film has a $\lambda_{ENZ} = 1560$ nm, thus $\text{Re}(\epsilon) \sim 0$. Figure 26.7a–c shows the cross-sectional electric field plots for taper-ITO junctions for three films with $\lambda_{ENZ} = 1280$, 1560 and 1990 nm, which are metallic, ENZ and dielectric in nature, respectively. The extent of evanescent wave coupling and the differences between the three cases are self-evident. Thus, either dynamic or static control of the optical properties of ITO, via control of its n_e offers a direct control of the evanescent coupling to the environment, especially the region closest to the taper-ITO junction. The junction platform may be potentially exploited for molecular sensing in a compact geometry.

To quantify the above, Fig. 26.8 plots the variation of the effective mode index (n_{eff}) of the $HE12$ mode across ITO films with various λ_{ENZ}. Co-plotted are the experimentally detected real and imaginary parts of the refractive indices evidencing a qualitative agreement in the trend of calculated n_{eff} and the real part of the corresponding refractive index (n). The effective mode index is seen to increase as the

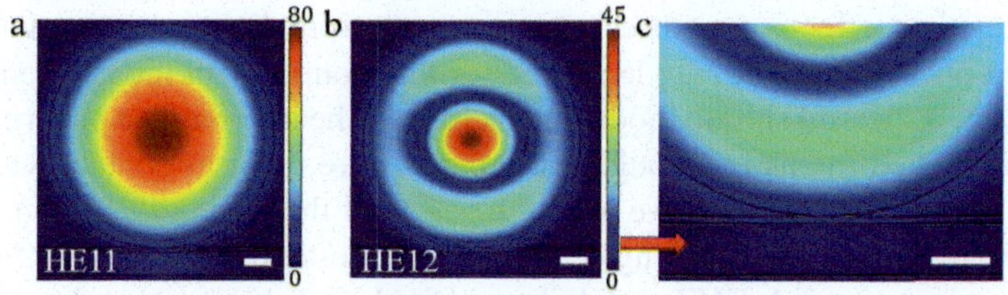

Fig. 26.6 **a** HE11 mode and **b** HE12 mode for a tapered fibre in contact with an ITO coated glass of 1560 nm ENZ. **c** The zoomed up image of the HE12 mode showing the field coupling to the ITO layer. These electric field (V m^{-1}) mode profiles are simulated at $\lambda = 1560$ nm. White line denotes scale bar of 1 μm

Fig. 26.7 HE12 mode profiles zoomed up at the taper-ITO contact region. **a** ITO: metallic with $\lambda_{ENZ} = 1280$ nm. **b** ITO: ENZ with $\lambda_{ENZ} = 1560$ nm. **c** ITO: dielectric with $\lambda_{ENZ} = 1990$ nm. Excitation wavelength is $\lambda = 1560$ nm. White line denotes scale bar of 1 μm

Fig. 26.8 Left: Effective mode index solved using modal analysis, for ITO of different λ_{ENZ} (denoted by the number after ITO in x-axis). Right: Real and imaginary part of refractive index of ITO with different λ_{ENZ}

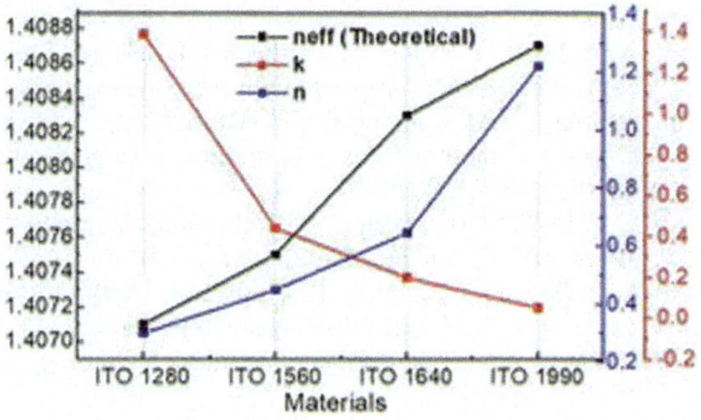

ITO's with increasing refractive index is introduced near the taper. This corroborates with the observed shift in the interference minima being observed as the ITO's optical properties are changed (Fig. 26.5b).

26.4 Conclusion

To conclude we have demonstrated dynamic control of the optical properties of ITO thin films via electrostatic gating that tunes the free electron density of ITO, in turn changing its λ_{ENZ}. Thus as a function of the gate bias, a ITO film can change its optical response from that of a dielectric to a metal, reversibly. We have also demonstrated an effective method to sense the surrounding refractive index using a tapered optical fibre, utilizing leaky modes that couple with the environment and thus produce a spectral shift in modal interference. The interference pattern observed in the transmitted intensity through the tapered fibre, kept in contact with ITO, is detected to systematically evolve as the λ_{ENZ} (and the refractive index) of ITO is varied. Simulations of the coupling between the modes of the fiber and ITO in various configurations i.e. metal, ENZ and dielectric further elucidates the phenomena and showcases the available control parameters that characterize evanescent wave based sensing as explored here.

References

1. S. Campione, D. de Ceglia, M.A. Vincenti, M. Scalora, F. Capolino, Electric field enhancement in ε-near-zero slabs under TM-polarized oblique incidence. Phys. Rev. B **87**(3), 035120 (2013)
2. A. Ciattoni, E. Spinozzi, Efficient second-harmonic generation in micrometer-thick slabs with indefinite permittivity. Phys. Rev. A **85**(4), 043806 (2012)
3. M.G. Silveirinha, N. Engheta, Theory of supercoupling, squeezing wave energy, and field confinement in narrow channels and tight bends using ε near-zero metamaterials. Phys. Rev. B **76**(24), 245109 (2007)
4. N. Kinsey, C. DeVault, J. Kim, M. Ferrera, V.M. Shalaev, A. Boltasseva, Epsilon-near-zero Al-doped ZnO for ultrafast switching at telecom wavelengths. Optica **2**(7), 616–622 (2015)
5. A. Capretti, Y. Wang, N. Engheta, L.D. Negro, Enhanced third-harmonic generation in Si-compatible epsilon-near-zero indium tin oxide nanolayers. Opt. Lett. **40**(7), 1500–1503 (2015)
6. B. Johns, N.M. Puthoor, H. Gopalakrishnan, A. Mishra, R. Pant, J. Mitra, Epsilon-near-zero response in indium tin oxide thin films: octave span tuning and IR plasmonics. J. Appl. Phys. **127**(4), 043102 (2020)
7. G.V. Naik, V.M. Shalaev, A. Boltasseva, Alternative plasmonic materials: beyond gold and silver. Adv. Mater. **25**(24), 3264–3294 (2013)
8. E. Feigenbaum, K. Diest, H.A. Atwater, Unity-order index change in transparent conducting oxides at visible frequencies. Nano Lett. **10**(6), 2111–2116 (2010)
9. K.T.V. Grattan, T. Sun, Fiber optic sensor technology: an overview. Sens. Actuators, A **82**(1–3), 40–61 (2000)
10. P. Wang, H. Zhao, X. Wang, G. Farrell, G. Brambilla, A review of multimode interference in tapered optical fibers and related applications. Sensors **18**(3), 858 (2018)
11. B. Yao, S.-W. Huang, Y. Liu, A.K. Vinod, C. Choi, M. Hoff, Y. Li et al., Gate-tunable frequency combs in graphene–nitride microresonators. Nature **558**(7710), 410–414 (2018)
12. H.W. Lee, G. Papadakis, S.P. Burgos, K. Chander, A. Kriesch, R. Pala, U. Peschel, H.A. Atwater, Nanoscale conducting oxide PlasMOStor. Nano Lett. **14**(11), 6463–6468 (2014)
13. J. Gurusiddappa, W. Madhuri, R.P. Suvarna, K.P. Dasan, Conductivity and dielectric behavior of polyethylene oxide-lithium perchlorate solid polymer electrolyte films. Indian J. Adv. Chem. Sci. **4**, 14–19 (2016)
14. H. Du, X. Lin, Z. Xu, D. Chu, Electric double-layer transistors: a review of recent progress. J. Mater. Sci. **50**(17), 5641–5673 (2015)
15. L.-P. Sun, J. Li, Y. Tan, S. Gao, L. Jin, B.-O. Guan, Bending effect on modal interference in a fiber taper and sensitivity enhancement for refractive index measurement. Opt. Express **21**(22), 26714–26720 (2013)
16. W.B. Ji, H.H. Liu, S.C. Tjin, K.K. Chow, A. Lim, Ultrahigh sensitivity refractive index sensor based on optical microfiber. IEEE Photonics Technol. Lett. **24**(20), 1872–1874 (2012)

Chapter 27
Identifying the Recombination Zone in Perovskite Solar Cells

Spandan Ranpariya and Dhirendra Kumar Sinha

27.1 Introduction

In recent years, the perovskite solar cell has emerged as a prominent candidate among the family of third-generation solar cells [1, 2]. It shows power-conversion efficiency comparable to that of benchmark solar cells in the market [1]. This remarkable performance in terms of power-conversion efficiency is mainly due to the development of high-grade perovskites, advanced processing conditions, and the realization of various device architectures [3]. However, there exist limitations for its large-scale production and usage due to noticeable physical characteristics such as hysteresis in current–voltage [4], loss of charge-carriers due to recombination [5], and ionic effects [6]. Among these, the recombination of photo-generated charge-carriers and their dominant pathways are widely studied [7, 8]. However, identifying the recombination zone within the device architecture helps optimize operating conditions and improve the overall performance of the perovskite solar cells. In this work, we considered methylammonium lead iodide ($CH_3NH_3PbI_3$) as a light-harvesting perovskite material. We specifically considered a *p-i-n* architecture in which the intrinsic perovskite layer is sandwiched between hole-transporting layer (HTL) and electron-transporting layer (ETL). We utilized a one-dimensional drift–diffusion model to simulate the spatial distribution of charge-carriers in the device. The spatial distribution of the charge-carriers is simulated under the following physical conditions (i) direct illumination, i.e., without pre-biasing the device, (ii) illumination with pre-biasing the device at 1.0 V. In the spatial distribution of charge-carrier density, we specify a space as a recombination zone where the density of opposite charge-carriers is observed to be the same orders of magnitude. It is mainly due to a region of high probability for the charge-carriers to recombine. We observe that the recombination zone is located within the perovskite layer when operating without pre-biasing, whereas it is shifted

S. Ranpariya (✉) · D. K. Sinha
Indian Institute of Information Technology Vadodara, Vadodara, Gujarat, India
e-mail: spandanranpariya@iiitvadodara.ac.in

R. Singh et al. (eds.), *The Physics of Semiconductor Devices*, Springer Proceedings in Physics 306, https://doi.org/10.1007/978-981-97-1571-8_27

towards the interface when operating with pre-biasing at 1.0 V. The observation indicates that the pre-biasing helps minimize the recombination in the active layer, thereby improving the overall power-conversion efficiency.

27.2 Recombination Pathways

In perovskite solar cells, the typical recombination pathways, as shown in Fig. 27.1, are: (i) band-to-band recombination and (ii) trap-assisted recombination. In band-to-band recombination, the opposite charge-carriers recombine directly from their respective bands, as depicted schematically in Fig. 27.1a. However, in trap-assisted recombination, the recombination of electron–hole pairs is mediated through the trap states (bulk or interfacial), schematically shown in Fig. 27.1b, c, respectively. These trap-states are located between the valence band and conduction band. These trap-states are due to defects and impurities which are located at the interfaces and the grain boundaries. There are shallow- and deep-energy trap-states. The shallow trap-states are located near the band edge and slow down the movement of the photo-generated charge-carriers. Whereas the deep trap-states are located around the center of the bandgap, which captures the generated charge-carriers before being extracted by the electrodes and lead to the recombination. The recombination via deep trap-states of photo-generated charge-carriers decreases the collection of free charge-carriers, which leads to the decrease in photo-generated current and, thereby, the power conversion efficiency.

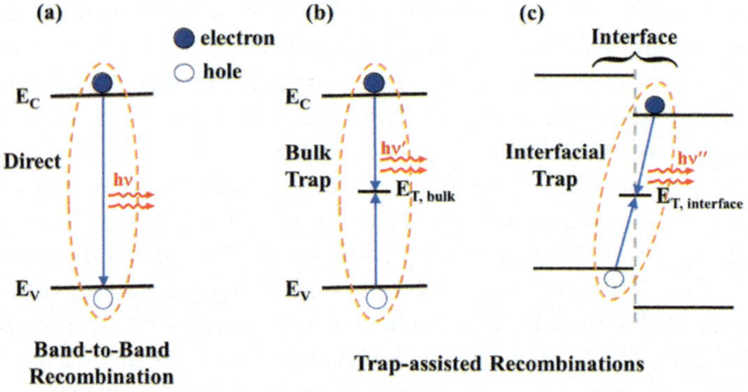

Fig. 27.1 Schematic of the typical recombination pathways in a perovskite solar cell. **a** Band-to-band recombination of the electron–hole pair. **b** Recombination of the electron–hole pairs via trap states in the bulk of the material. **c** Recombination of the electron–hole pairs via interfacial trap states

27.3 Device Structure and Simulation Tool

In this work, we have considered the perovskite material—methylammonium lead iodide ($CH_3NH_3PbI_3$) as an active material which absorbs solar energy and generates electron–hole pairs. The thickness of the perovskite layer is 400 nm. The perovskite layer is sandwiched between spiro-OMeTAD and titanium dioxide (TiO_2), which behave as hole- and electron-transporting layers, respectively. The interface of these transporting layers with the perovskite layer helps the generated electron–hole pairs to dissociate and get collected by the respective electrodes. The thicknesses of the hole-transporting layer and electron-transporting layer are 200 nm and 100 nm, respectively. The gold (Au) and fluorine-doped tin oxide (FTO) are taken as cathode and anode, respectively. The schematic of the device structure is shown in Fig. 27.2. The physical processes such as absorption of solar energy and generation of electron–hole pairs, diffusion of the generated electron–hole pairs towards the interfaces, and dissociation into free charge-carriers are shown schematically in Fig. 27.3.

The device was characterized with MATLAB pdepe solver as a simulation tool. We utilized a one-dimensional drift–diffusion model to simulate the spatial distribution of charge-carriers in the device. The trap-densities in bulk and at the interfaces are independently considered. We used uniform generation of electron–hole pairs during the calculation of the spatial distribution of charge-carrier densities. The spatial distribution of the charge-carriers is simulated under two different physical conditions (i) direct illumination, i.e., without pre-biasing the device, (ii) illumination

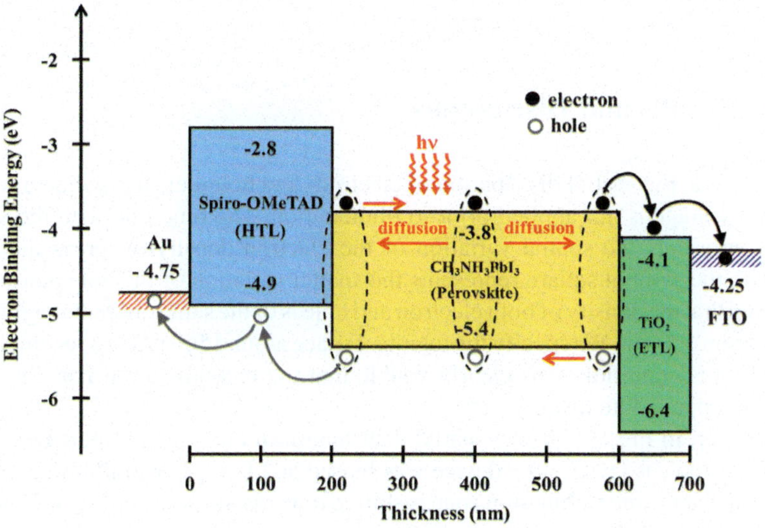

Fig. 27.2 Schematic of the perovskite solar cell. The perovskite methylammonium lead iodide ($CH_3NH_3PbI_3$) is sandwiched between Spiro-OMeTAD as a hole-transporting layer and TiO_2 as an electron-transporting layer. The values of energy level and the thicknesses of each layer are indicated as per the scale shown

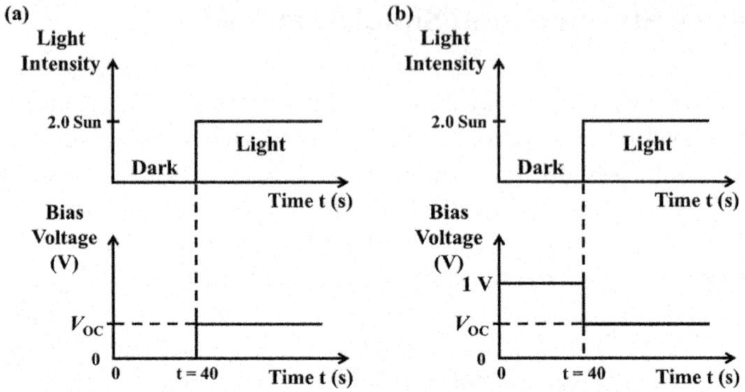

Fig. 27.3 The physical conditions of the simulation are shown schematically. **a** Zero pre-bias is applied to the device in the dark prior to switching ON the light and bias voltage to V_{OC}, and **b** A bias voltage of 1.0 V is applied to the device prior to switching ON the light and biasing at V_{OC}

with pre-biasing the device at 1.0 V. In pre-biasing, the fixed bias is applied to the cell in the dark prior to the measurement. Pre-biasing helps to stabilize the perovskite solar cell and to avoid unnecessary obstacles prior to performing an experiment where both time and applied voltage are covariants [6]. In Fig. 27.3, the physical conditions are shown. Firstly, as shown in Fig. 27.3a, the device is unbiased prior to switching ON the light and biasing at V_{OC}. Secondly, the device is biased at 1.0 V for 40 s prior to turning ON the light and bias voltage to V_{OC}. The second condition is shown in Fig. 27.3b.

27.4 Results and Discussions

In Fig. 27.4a, the spatial distribution of electron- and hole-density is shown wherein pre-bias voltage is not applied prior to illumination. The blue line with the symbol circle represents the spatial variation of the electron density, whereas the brown line with the symbol square represents the spatial variation of the hole density. It is observed that the density of both electron and hole is of the same order of magnitude at the position 500 nm. We specify the region of space around 500 nm as a recombination zone. In this condition, it is clearly visible that the recombination zone is located within the perovskite layer.

However, in Fig. 27.4b, the spatial distribution of electron and hole densities is shown for the case where the device was biased at 1.0 V prior to illumination. It is observed that the recombination zone is shifted towards the perovskite—ETL interfacial layer. As a result, the recombination of charge-carriers within the perovskite layer is minimized. The pre-bias effect may be utilized to condition the device to limit the recombination in the perovskite layer and thereby improve the device performance.

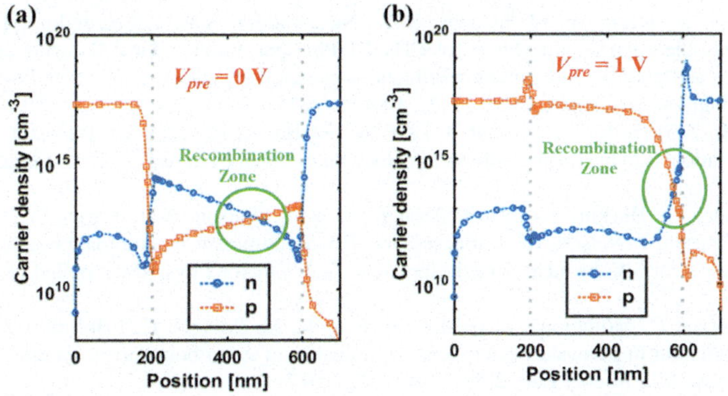

Fig. 27.4 Spatial distribution of charge-carriers, **a** without pre-biasing and **b** with pre-biasing at 1.0 V. The blue line with symbol '○' represents the electron density, whereas the brown line with symbol '□' represents the hole density. The encircle space is a recombination zone

27.5 Conclusions

We have simulated the spatial distribution of electronic charge-carriers in perovskite solar cells biased at V_{OC} and illuminated with a light source. The simulation is performed (i) without applying pre-bias and (ii) with applying a pre-bias of 1.0 V to the device prior to switching ON the light and bias voltage of V_{OC}. Assuming the probability of recombination is relatively high at and around the region of space where the density of opposite charge-carriers is of the same order of magnitude, we specify a recombination zone. We observed that the recombination zone is located within the perovskite layer when simulated without pre-biasing the device. However, the recombination zone shifts towards the perovskite—ETL interfacial layer when it is pre-biased at 1.0 V. The shifting of the recombination zone due to the pre-bias effect may be utilized to limit the recombination within the perovskite layer, thereby optimizing the device to achieve better power-conversion efficiency.

References

1. National Renewable Energy Laboratory (NREL) best research-cell efficiency chart. https://www.nrel.gov/pv/cell-efficiency.html. Last Accessed 16 Mar 2022
2. M.A. Mahmud, T. Duong, J. Peng, Y. Wu, H. Shen, D. Walter, H.T. Nguyen, N. Mozaffari, G.D. Tabi, K.R. Catchpole, K.J. Weber, T.P. White, Origin of efficiency and stability enhancement in high-performing mixed dimensional 2D-3D perovskite solar cells: a review. Adv. Funct. Mater. **32**(3), 2009164 (2021)
3. D.K. Lee, N.G. Park, Materials and methods for high-efficiency perovskite solar modules. Sol. RRL **6**(3), 2100455 (2021)
4. R. Singh, M. Parashar, *Origin of Hysteresis in Perovskite Solar Cells, Book: Soft Matter Thin Film Solar Cells: Physical Processes and Device Simulation* (AIP Publishing, 2020), pp. 1–42.

5. W. Tress, N. Marinova, O. Inganäs, M.K. Nazeeruddin, S.M. Zakeeruddin, M. Graetzel, Predicting the open-circuit voltage of $CH_3NH_3PbI_3$ perovskite solar cells using electroluminescence and photovoltaic quantum efficiency spectra: the role of radiative and non-radiative recombination. Adv. Energy Mater **5**(3), 400812 (2015)
6. P. Calado, A.M. Telford, D. Bryant, X. Li, J. Nelson, B.C. O'Regan, P.R.F. Barnes, Evidence for ion migration in hybrid perovskite solar cells with minimal hysteresis. Nat. Commun. **7**, 13831 (2016)
7. P. Calado, D. Burkitt, J. Yao, J. Troughton, T.M. Watson, M.J. Carnie, A.M. Telford, B.C. O'Regan, J. Nelson, P.R.F. Barnes, Identifying dominant recombination mechanisms in perovskite solar cells by measuring the transient ideality factor. Phys. Rev. Applied **11**(4), 044005 (2019)
8. T.S. Sherkar, C. Momblona, L. Gil-Escrig, J. Ávila, M. Sessolo, H.J. Bolink, L.J.A. Koster, Recombination in perovskite solar cells: significance of grain boundaries, interface traps, and defect ions. ACS Energy Lett. **2**(5), 1214–1222 (2017)

Chapter 28
High Speed Etching of Silicon in NaOH-Based Solution

S. Purohit, V. Swarnalatha, P. Pal, and R. K. Sharma

28.1 Introduction

Silicon is most widely used material in semiconductor industry [1–3]. Wet etching technique is vital for the fabrication of microstructures. It is subdivided into two categories namely isotropic and anisotropic etching. Anisotropic etching is an important process in the manufacturing of MEMS components for numerous applications. Potassium hydroxide (KOH) and tetramethyl ammonium hydroxide (TMAH) are most used etchants in silicon wet anisotropic etching [1, 3–16]. Recently, the effect of NH_2OH on the etching characteristics of KOH and TMAH is investigated to alter the etching characteristics to obtain high etch rate of silicon [6, 7, 17–22]. NaOH is an anisotropic etchant, which can be exploited for silicon wet bulk micromachining to fabricate microstructures. However, limited studies have been conducted pertaining to the etching characteristics of NaOH [23–27]. To find the potential applications of NaOH as an etchant, more study is needed, especially with an additive to improve the etch rate.

In the present work, we have investigated the etching characteristics of Si{100}, Si{110}, and Si{111} in 10 M NaOH without and with the addition of NH_2OH. The main objective of this work is to improve the etch rate, which is inevitable for high productivity. Also, we have studied the etched surface roughness and morphology.

S. Purohit · V. Swarnalatha (✉) · P. Pal
Department of Physics, MEMS and Micro/Nano Systems Laboratory, Hyderabad, India
e-mail: swarnalatha.veerla@phy.iith.ac.in

R. K. Sharma
Department of Chemistry, University of Delhi, Delhi, India

R. Singh et al. (eds.), *The Physics of Semiconductor Devices*, Springer Proceedings in Physics 306, https://doi.org/10.1007/978-981-97-1571-8_28

28.2 Experimental Method

CZ grown, 4-inch, p-type, one side polished Si{100}, Si{110}, and Si{111} wafers with a resistivity of 1–10 Ωcm are used to study the etching characteristics. An oxide layer of 1 μm thick is deposited by thermal oxidation. Oxide layer is pattern using photolithography. Positive photoresist is used as mask for selective etching of oxide layer in buffered hydrofluoric acid (BHF). The photoresist is removed with acetone followed by thorough rinsing in DI water. Now the wafer diced into small chips. Diced samples are then cleaned in a piranha bath (H_2SO_4:H_2O_2::1:1) followed by a DI water rinse to remove any traces of organic material and unwanted particles. A thin layer grown during piranha cleaning may delay etching process. The cleaned samples are therefore dipped in 1% HF for 1 min to remove chemically grown oxide layer. Thereafter the samples are thoroughly rinsed in DI water. Now the etching is performed in pure and 12% NH_2OH-added 10 M sodium hydroxide (NaOH). To keep the temperature constant during the experiment, a constant temperature water bath is used. An etching container made of Teflon is used for the process. A reflex condenser made of thick glass equipped with a double-layered narrow opening is used to circumvent any changes in the etchant concentration caused by continuous heating of the etchant during experiments. The etched samples are characterized using a 3D measuring laser microscope (OLYMPUS OLS4000), optical microscope (OLYMPUS MM6C-PC), scanning electron microscope (SEM). The measurements are taken at different locations of the sample/chip after performing the etching process.

28.3 Results

The etching characteristics including the etch rate and etched surface morphology of Si{100}, Si{110} and Si{111} are measured in pure and 12% NH_2OH added 10 M NaOH solution. The results are presented in following subsections.

Etch Rate

Figure 28.1 shows the etch rates of Si{100}, Si{110} and Si{111} in pure and 12% NH_2OH added 10 M NaOH solutions. It can easily be noticed that the addition of NH_2OH improves the etch rate substantially. In our pervious study, a simple model is proposed to explain the possible etching mechanism in NH_2OH-added alkaline solution (i.e., TMAH/KOH) [22]. In this model, NH_2O^- and OH^- ions are considered as catalysts, while H_2O as the reactive molecule. The active species, which are greatly accessible in NH_2OH-added alkaline solution, are NH_2O^- ions, OH^- ions and H_2O molecules. As the number of reactive species in NH_2OH-added alkaline solution is more in comparison to that in pure alkaline solution, the etch rate significantly increases in the presence of NH_2OH. Hence the addition of NH_2OH into NaOH increase the etch rate significantly.

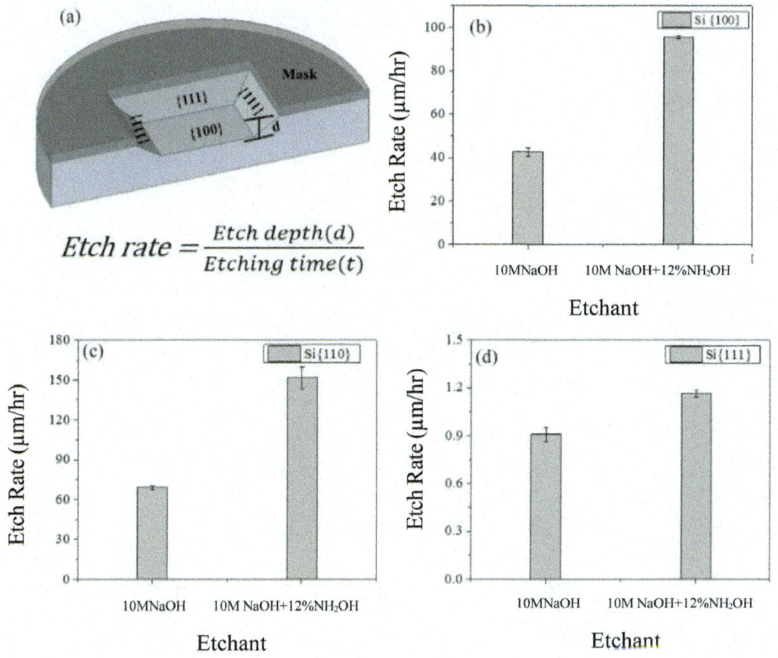

Fig. 28.1 Etch rate of **b** Si{100} **c** Si{110} and **d** Si{111} in pure and 12% NH_2OH-added 10 M NaOH at 70 °C

Surface Roughness

An important etching characteristic for optical MEMS is the etched surface morphology and roughness. It is a result of non-unformal removal of silicon atoms on the surface during etching process due to several reasons such formation of hydrogen bubbles [1, 3, 5, 11, 12, 16, 28]. Figure 28.2 presents the average surface roughness (R_a) and corresponding surface morphologies of the silicon samples etched in pure and 12% NH_2OH-added 10 M NaOH. It is measured using 3D measuring laser microscope and the standard deviation is calculated by scanning, 200 × 200 μm^2 area at six different locations on the same chip. It can easily be noticed from the graphs that surface roughness of Si{100} is approved on the addition of NH_2OH, but it deteriorates in the case of Si{110}. In addition, the fluctuations in the surface roughness is more in the case of Si{110} and Si{111}. Figure 28.3 presents the SEM images of surface morphologies of the Si{100}, Si{110} and Si{111} etched in 12% NH_2OH added 10 M NaOH solution.

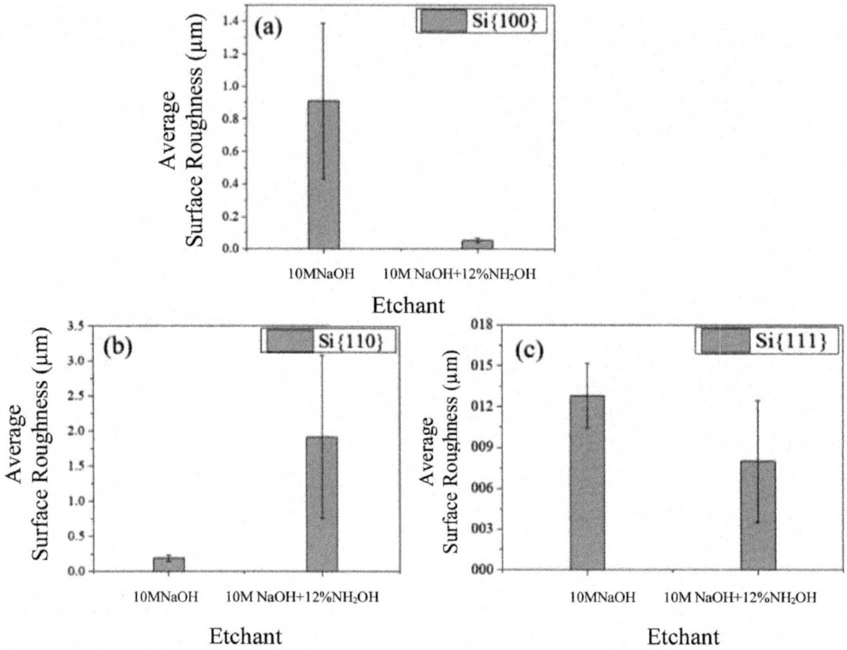

Fig. 28.2 Average surface roughness of **a** Si{100} **b** Si{110} and **c** Si{111} etched in pure and 12% NH2OH added 10 M NaOH at 70 °C

28.4 Conclusions

The etching characteristics (etch rate and etched surface morphology) of Si{100}, Si{110} and Si{111} wafers are investigated in pure 10 M NaOH without and with addition of 12% NH_2OH. The addition of NH_2OH increases the etch rate considerably, which is very useful and desirable to reduce the fabrication time of microstructures that is inevitable to improve the productivity. Moreover, the incorporation of NH_2OH improved etched surface morphology of Si{100}.

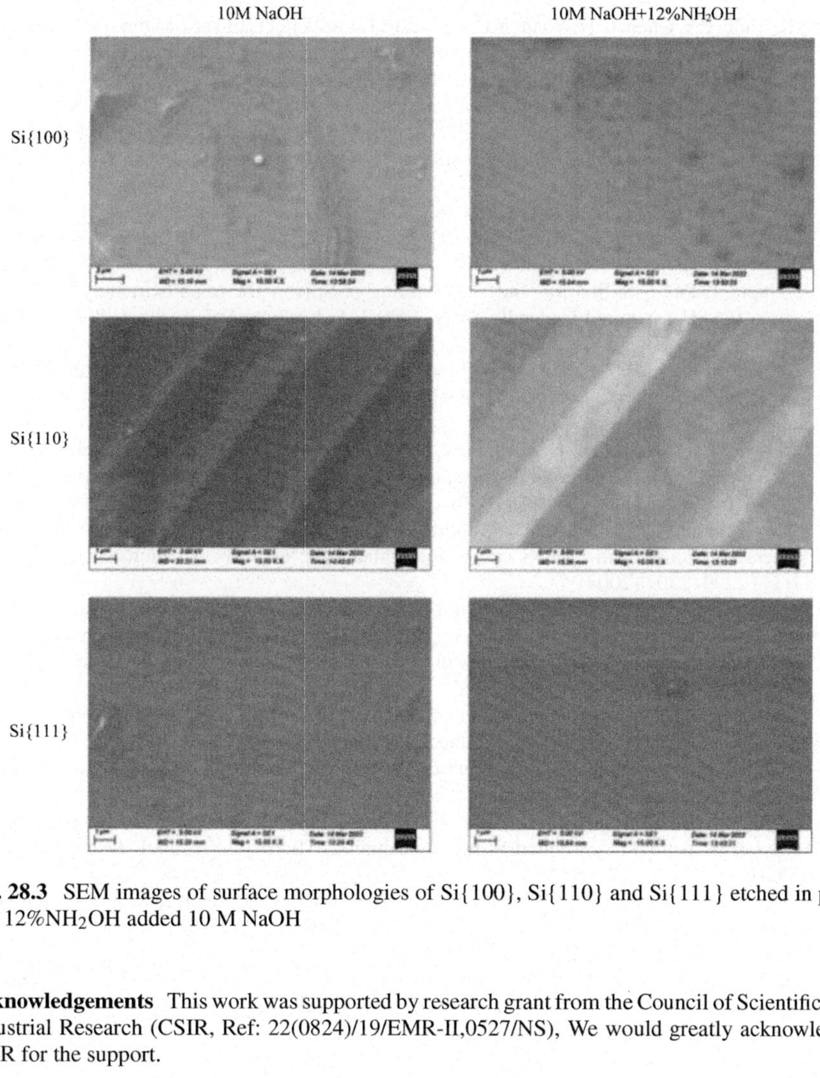

Fig. 28.3 SEM images of surface morphologies of Si{100}, Si{110} and Si{111} etched in pure and 12%NH_2OH added 10 M NaOH

Acknowledgements This work was supported by research grant from the Council of Scientific and Industrial Research (CSIR, Ref: 22(0824)/19/EMR-II,0527/NS), We would greatly acknowledge CSIR for the support.

References

1. P. Pal, K. Sato, *Silicon Wet Bulk Micromachining for MEMS* (Pan Stanford Publishing, Singapore, 2017)
2. S.K. Mishra, Z. Li, S. Beasor, S.M. Pandey, Source/drain profile engineering for enhanced p-MOSFET, US Patent 9,419,082
3. M. Tilli, T. Motooka, V.M. Airaksinen, S. Franssila, M.P. Krockel, V. Lindroos, *Handbook of Silicon Based MEMS Materials and Technologies* (William Andrew Publishing, United States, 2015)

4. C.R. Yang, P.Y. Chen, C.H. Yang, Y.C. Chiou, R.T. Lee, Effects of various ion typed surfactants on silicon anisotropic etching properties in KOH and TMAH solutions. Sens. Actuators A: Physical **119**(1), 271–281 (2005)
5. I. Zubel, Anisotropic etching of Si. J. Micromech. Microeng. **29**(9), 93002–93029 (2019)
6. V. Swarnalatha, A.V. Narasimha Rao, A. Ashok, S.S. Singh, P. Pal, Modified TMAH based etchant for improved etching characteristics on Si{100} wafer. J. Micromech. Microeng. **27**(8) 085003 (2017)
7. A.V. Narasimha Rao, V. Swarnalatha, A. Ashok, S.S. Singh, P. Pal, Effect of NH_2OH on etching characteristics of Si{100} in KOH solution. ECS J. Solid State Sci. Technol. **6**(9) 609–614 (2017)
8. A. Gupta, P. Pal, C.S. Sharma, Surface texturing of Silicon 100 in an extremely low concentration TMAH for minimized reflectivity. ECS J. Solid State Sci. Technol. **8**(10), 622–628 (2019)
9. A.V. Narasimha Rao, V. Swarnalatha, P. Pal, Etching characteristics of Si{110} in 20 wt% KOH with addition of hydroxylamine for the fabrication of bulk micromachined MEMS. Micro and Nano Syst. Letts. **5**(23), 1–9 (2017)
10. K.I. Koo, H. Chung, Y. Yu, J. Seo, J. Park, J.M. Lim, S.J. Paik, S. Park, H.M. Choi, M.J. Jeong, G.S. Kim, Fabrication of pyramid shaped three-dimensional 8 × 8 electrodes for artificial retina. Sens. Actuators A **130**, 609–615 (2006)
11. P. Pal, K. Sato, M.A. Gosalvez, M. Shikida, Study of rounded concave and sharp edge convex corners undercutting in CMOS compatible anisotropic etchants. J. Micromech, Microeng. **17**(11) 2299–2307 (2007)
12. P. Pal, K. Sato, M.A. Gosalvez, Y. Kimura, K. Ishibashi, M. Niwano, H. Hida, B. Tang, S. Itoh, Surfactant adsorption on single crystal silicon surfaces in TMAH solution: orientation-dependent adsorption detected by in-situ infra-red spectroscopy. J. Microelectromech. Syst. **18**, 1345–1356 (2009)
13. I. Zubel, M. Kramkowska, Possibilities of extension of 3D shapes by bulk micromachining of different Si (*h k l*) substrates. J. Micromech. Microeng. **15**(485), 93 (2005)
14. P. Pal, K. Sato, Various shapes of silicon freestanding microfluidic channels and microstructures in one step lithography. J. Micromech. Microeng. **19**, 1–11 (2009)
15. H. Yang, K. Xu, Y. Yang, T. Li, J. Jiao, X. Li, Y. Wang, A novel method to fabricate single crystal nano beams with (111)-oriented Si micromachining. Microsyst. Technol. **14**(8), 1185–1191 (2008)
16. P. Pal, V. Swarnalatha, A.V. Narasimha Rao, A.K. Pandey, H. Tanaka, K. Sato, High speed silicon wet anisotropic etching for applications in bulk micromachining: a review. Micro Nano Syst. Lett. **9**(1), 1–59 (2021)
17. P.K. Menon, A.V. Narasimha Rao, A.L. Murthy, A.K. Pandey, P. Pal, High speed etching of silicon in KOH+ NH_2OH solution at lower temperatures for the fabrication of through holes in silicon wafer. Micro Nano Lett. 1–6 (2020)
18. A.V.N. Rao, P. Pal, A.K. Pandey, V. Swarnalatha, P.K. Menon, H. Tanaka, K. Sato, Aging effects of KOH + NH_2OH solution on the etching characteristics of silicon. ECS J. Solid State Sci. Technol. **8**(11), 685–692 (2019)
19. A.V.N. Rao, V. Swarnalatha, P. Pal, Etching characteristics of Si 110 in 20 wt% KOH with addition of hydroxylamine for the fabrication of bulk micromachined MEMS. Micro Nano Syst. Lett. **5**(1), 1–9 (2017)
20. V. Swarnalatha, A.V.N. Rao, P. Pal, Effective improvement in the etching characteristics of Si{110} in low concentration TMAH solution. Micro Nano Lett. **13**(8), 1085–1089 (2018)
21. V. Swarnalatha, P. Pal, A.K. Pandey, A.V.N. Rao, Y. Xing, H. Tanaka, K. Sato, Systematic study of the etching characteristics of Si{111} in modified TMAH. Micro Nano Lett. **15**(1) 52–57 (2020)
22. V. Swarnalatha, K.T. Vismaya, A.V. Narasimha Rao, P. Pal, A.K. Pandey, H. Tanaka, K. Sato, Etching mechanism behind the high speed etching of silicon in NH_2OH-added alkaline solutions. IEEJ Trans. Sens. Micromach. **140**(1), 24–30 (2020)

23. E. Vazsonyi, Z. Vertesy, A.L. Toth, Anisotropic etching of silicon in a two-component alkaline solution. J. Micromech. Microeng. **13**, 165–169 (2003)
24. U. Gangopadhyay, S.K. Dhungel, K. Kim, Novel low cost chemical texturing for very large area industrial multi-crystalline silicon solar cells. Semicond. Sci. Technol. **20**, 938–946 (2005)
25. P. Akhter, A. Baig, A. Mufti, Dissolution of Si (100) layers in NaOH aqueous solutions. J. Phys. D: App. Phys. **22**(12) (1989)
26. É. Vázsonyi, CS. Dücső, A. Pekker, Characterization of the anisotropic etching of silicon in two-component alkaline solution. J. Micromech. Microeng. **17**(9), 1916 (2007)
27. P. Akhter, A. Baig, A. Mufti, Dissolution of Si (100) layers in NaOH aqueous solutions. J. Phy. D: App. Phy. **22**(12), 1924 (1989)
28. P. Pal, M.A. Gosalvez, K. Sato, Silicon micromachining based on surfactant-added tetramethyl ammonium hydroxide: etching mechanism and advanced application. Jpn. J. Appl. Phys. **49**, 056702 (9 pages) (2010)

Chapter 29
Parylene Membrane Transfer on PDMS Microchannel for Microvalve Fabrication

Imran Ahmad and Bhaskar Mitra

29.1 Introduction

Parylene has been widely adopted for fabricating microvalves and micropumps due to its biocompatibility and chemical inertness. Quake valves [1], which are pneumatically actuated polymer membrane microvalves, are typically used in microfluidic systems. Typically, they use a thin PDMS membrane, but parylene valves are likely to perform better [2]. This paper describes the parylene bonding technology and its application in the transfer of thin parylene membrane, which is a crucial step in parylene microvalve fabrication [3]. This transferred Parylene membrane seals the channels of variable length over the device side of PDMS [4].

29.2 Lithography on Si-Wafer to Fabricate a Mold

Fabrication of SU8 mold is done by optical lithography of negative photoresist SU8 2035 (Fig. 29.1).

Silanization of the SU8 mold was done in a vacuum desiccator using a drop of trichlorosilane. PDMS (10:1) is poured and demolded as per the standard procedure [5].

I. Ahmad (✉) · B. Mitra
Department of Electrical Engineering, Indian Institute of Technology, Delhi, India
e-mail: imran.ahmad@ee.iitd.ac.in

R. Singh et al. (eds.), *The Physics of Semiconductor Devices*, Springer Proceedings in Physics 306, https://doi.org/10.1007/978-981-97-1571-8_29

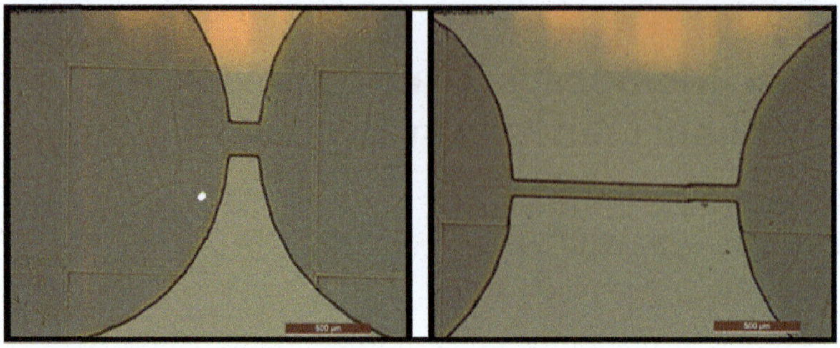

Fig. 29.1 Fluidic channel after lithography

29.3 Membrane Transfer Process Using Silicon

In the first process, the polished side of the silicon wafer is spin-coated with a sacrificial photoresist (S1813). Next, 2 μm parylene-C is uniformly deposited on PR. using a chemical vapour deposition (CVD) system to define the membrane material. Then a 200 nm thick parylene-C is deposited on the device side of PDMS using a parylene CVD system. Finally, the PDMS chip is bonded to the Si-wafer by parylene-parylene bonding at 230 °C [2]. After bonding, the sacrificial PR has been etched in acetone, but the acetone does not fully etch the PR due to strong bonding. So, the carrier Si-wafer has been etched in 30% KOH to release the membrane. The parylene membrane is successfully transferred using this method [6]. However, some fluidic channels with small dimensions get distorted during the etching of the silicon wafer (Fig. 29.2).

29.4 Membrane Transfer Using a Teflon Carrier

In the second process, Teflon is used as a carrier wafer. Teflon sheet is the best candidate since it is flexible to bend, non-sticky, and easily removed from the bonded PDMS chip. The thickness of Teflon is 0.5 mm. To achieve the required smoothness, thermal polishing is performed by sandwiching the Teflon sheet between two polished 4-inch Si-wafers and heated at 300 °C (above the glass transition temperature of Teflon) with enough weight applied (20 kg). Then the membrane transfer process is done as before, with Teflon instead of silicon carrier. The Teflon sheet is brought into contact with PDMS and heated to 230 °C with enough applied pressure to achieve

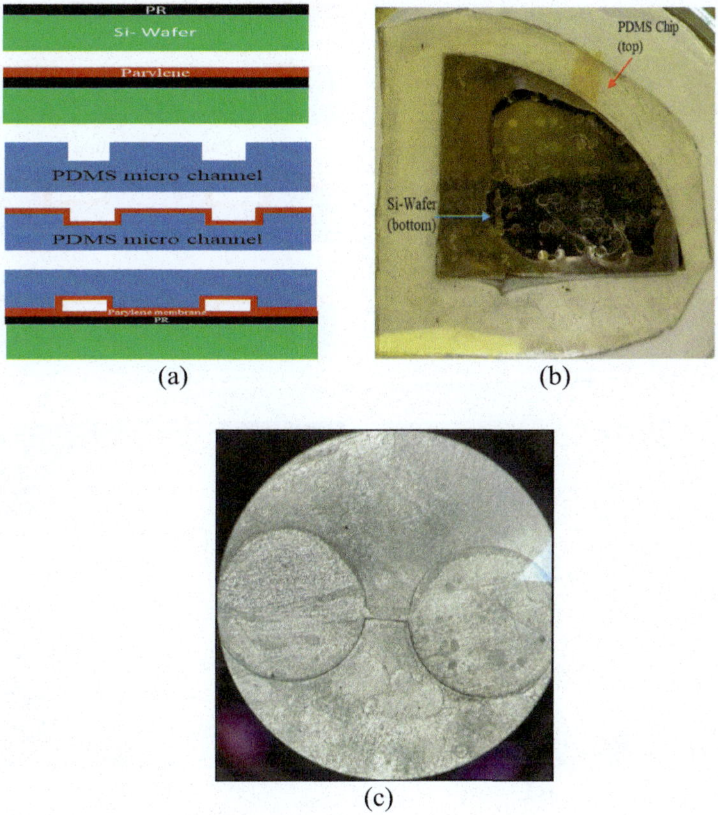

Fig. 29.2 **a** Process flow for membrane transfer using si-wafer, **b** parylene-parylene bonding at 230 °C, **c** transferred Parylene membrane using si wafer

the parylene-parylene bonding [7, 8]. The Teflon is peeled off by application of slight force with a tweezer, leaving the membrane intact (Fig. 29.3).

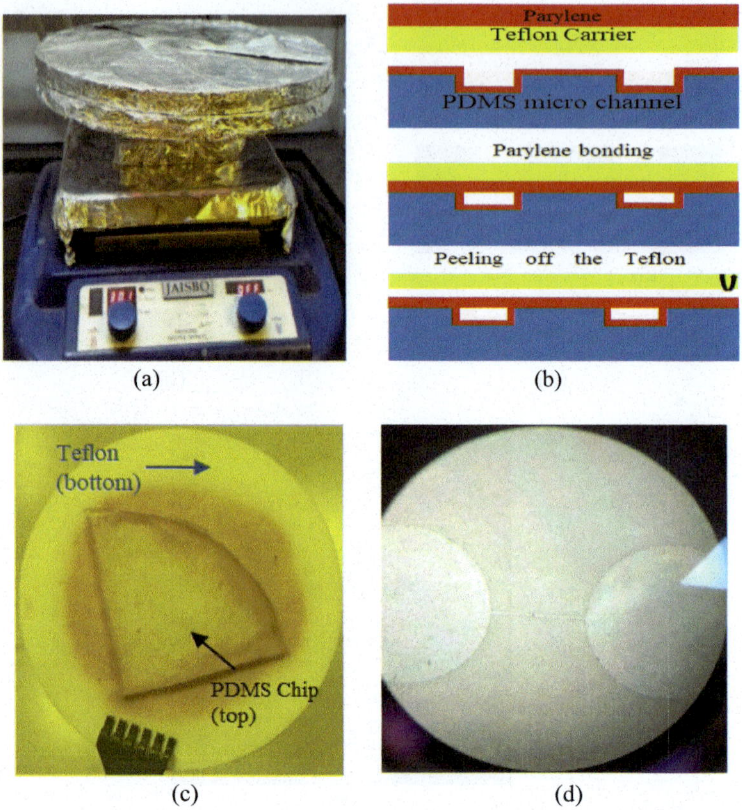

Fig. 29.3 **a** Thermal polishing of Teflon, **b** process flow of membrane transfer using Teflon, **c** parylene-parylene bonding at 230 °C, **d** transferred parylene membrane using Teflon

29.5 Conclusion

In this study, an effort has been made to discuss two methods for the transfer of thin parylene membrane onto a PDMS chip. By utilizing parylene-parylene bonding, a flat Parylene membrane has been transferred to a PDMS chip consisting of the microchannels. This parylene-compatible technique and simple method to transfer parylene membranes to the PDMS chip has applications in microfluidic device fabrication like microvalves and micropumps.

References

1. M.A. Unger, H.P. Chou, T. Thorsen, A. Scherer, S.R. Quake, Monolithic microfabricated valves and pumps by multilayer soft lithography. Science **288**(5463), 113–116 (2000)
2. H.S. Kim, K. Najafi, Wafer bonding using parylene and wafer-level transfer of free-standing parylene membranes, in *TRANSDUCERS'03. 12th International Conference on Solid-State Sensors, Actuators and Microsystems. Digest of Technical Papers* (Cat. No. 03TH8664), vol. 1. (IEEE, 2003), pp. 790–793
3. Z. Hua, Y. Xia, O. Srivannavit, J.M. Rouillard, X. Zhou, X. Gao, E. Gulari, A versatile microreactor platform featuring a chemical-resistant microvalve array for addressable multiplex syntheses and assays. J. Micromech. Microeng. **16**(8), 1433 (2006)
4. I. Miranda, A. Souza, P. Sousa, J. Ribeiro, E.M. Castanheira, R. Lima, G. Minas, Properties and applications of PDMS for biomedical engineering: a review. J. Funct. Biomater. **13**(1), 2 (2021)
5. J. Kim, A.M. Stockton, E.C. Jensen, R.A. Mathies, Pneumatically actuated microvalve circuits for programmable automation of chemical and biochemical analysis. Lab Chip **16**(5), 812–819 (2016)
6. J. Melin, S.R. Quake, Microfluidic large-scale integration: the evolution of design rules for biological automation. Annu. Rev. Biophys. Biomol. Struct. **36**, 213–231 (2007)
7. P. Rezai, P.R. Selvaganapathy, G.R. Wohl, Plasma enhanced bonding of polydimethylsiloxane with parylene and its optimization. J. Micromech. Microeng. **21**(6), 065024 (2011)
8. I.E. Araci, S.R. Quake, Microfluidic very large-scale integration (mVLSI) with integrated micromechanical valves. Lab Chip **12**(16), 2803–2806 (2012)

Chapter 30
RF Analysis of Tapered Angle Hetero-Junction Dopingless TFET for Low Power Applications

Monika Sharma, Rakhi Narang, Manoj Saxena, and Mridula Gupta

30.1 Introduction

With the ongoing miniaturization in semiconductor industry, power dissipation is one of the major issues that come into consideration because of reduction in device dimensions and their operating supply voltages. Thermionic emission based devices exhibits the 60 mV/decade limit of sub-threshold swing, whereas because of tunneling phenomenon TFET provides an advantage of achieving lowest tunable sub-threshold swing below 60 mV/decade. TFET suffers from a major disadvantage of lower $I_{ON,}$ which is overcome by implementing hetero-junction TFETs [1–3]. Random doping fluctuations (RDFs) [4] is a major stumbling block for formation of abrupt junction at source/drain-channel for sub-90 nm gate length. Hence, for overcoming the fabrication issues dopingless TFETs [5] are considered, In Dopingless TFET [5], the doping concentration in the source and drain region are created with the help of charge plasma concept, in which the work-function of the source side metal electrode is chosen such that $\emptyset m \leq \chi s + \frac{Eg}{2}$ and the work-function of the drain side metal electrode is chosen such that $\emptyset m \geq \chi s + \frac{Eg}{2}$ [6] for enhancing the p-type and n-type doping respectively. For enhancing the I_{ON} current level to few milli-amperes, III–V materials based TFETs' are in research these days because of their lower effective masses in comparison to silicon for electrons and holes due to

M. Sharma · M. Gupta (✉)
Semiconductor Device Research Laboratory, Department of Electronic Science, University of Delhi, South Campus, New Delhi 110021, India
e-mail: mridula@south.du.ac.in

R. Narang
Department of Electronics, Sri Venkateswara College, University of Delhi, New Delhi 110021, India

M. Saxena
Department of Electronics, Deen Dayal Upadhyaya College, University of Delhi, New Delhi 110078, India

R. Singh et al. (eds.), *The Physics of Semiconductor Devices*, Springer Proceedings in Physics 306, https://doi.org/10.1007/978-981-97-1571-8_30

which they assist the tunneling of electrons [7, 8] and ease in fabrication with the reduction in threshold voltage and sub-threshold swing. Hence, by incorporating the effect of device engineering, i.e., tapered angle effect on InGaAs based DL-TFET [9] shows 0.206 mA/μm I_{ON} current, which makes it a powerful device for using it in low power circuit application with high switching speed. By introducing tapered-angle in InGaAs Dopingless TFET [9] is found to be most promising architecture as it increases the I_{ON} to 0.206 mA/μm. In this paper, TCAD based simulation study is carried out on the optimized structure obtained in [9] for examining the RF performance parameters of hetero-angle dopingless TFET using the ATLAS device simulator.

30.2 Device Structure and Simulation Standards

The diagrammatical representation of DL TFET structures are shown in Fig. 30.1. The $In_{1-x}Ga_xAs$ material is chosen for designing the DL-TFET structure with $x = 0.75$ having the band-gap almost similar to that of silicon. The source electrode is deposited at 5.8 eV (i.e., platinum) for developing 'p-type' doping concentration and drain electrode is deposited at 4.0 eV (i.e., zirconium) for enhancing 'n-type' doping. The drain electrode Zr (zirconium) forms ZrO_2 quickly. Hence, Li metal is used as a reluctant in molten LiCl salt at 650 °C, which leads to the formation of Zr metal without forming the Li_2ZrO_3 compound. In particular, Zr metal can be produced without formation of Li_2ZrO_3 (i.e., no ZrO_2 remained) when LiCl salt containing a high concentration of Li metal is used. However, Zr metal can be produced with Li_2ZrO_3 when LiCl salt containing both Li metal and Li_2O was added [10]. This can also be done using the RAPET method [11], i.e., ZrO_2 with Mg powder was carried out in a closed stainless steel cell, at 750 °C which leads to the formation of Zr nano-particles. The thickness of HfO_2 is taken to be 2 nm under source/drain electrode region and 4 nm between source-gate region/gate-drain region as shown in Fig. 30.1a. For creating the elevated oxide layer, the conventional photolithography technique is used with silicon nitride mask and then selective etching is done for patterning the oxide layer [12]. The physical models used in simulation are: concentration dependent mobility, field dependent mobility, Shockley-read hall, Fermi–Dirac statistics, band gap narrowing, band-to-band non local tunneling models with their default values.

The device simulation study is carried out on TCAD ATLAS device simulator [13].

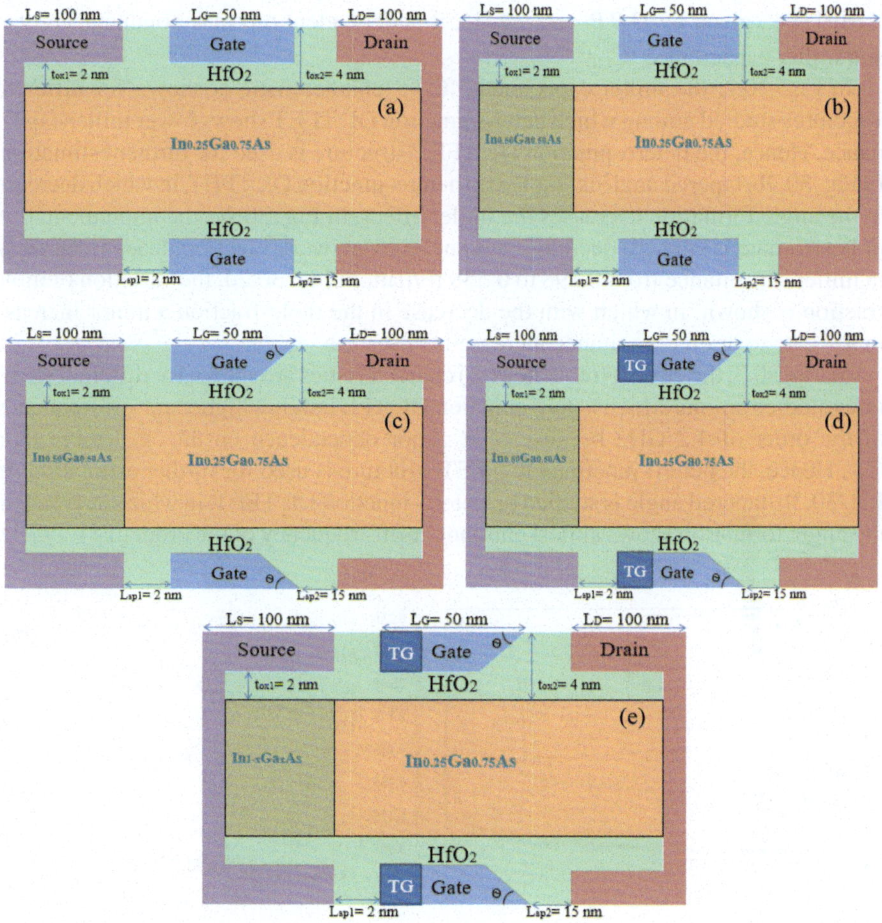

Fig. 30.1 Diagrammatical representation of DL-TFET, **a** basic DL TFET, **b** hetero junction DL TFET, **c** tapered angle hetero-junction InGaAs DL-TFET, **d** hetero-gate hetero-junction InGaAs DL-TFET with tapered oxide layer and **e** different composition of 'x' for source in hetero-gate hetero-junction InGaAs DL-TFET with tapered oxide layer

30.3 Results and Discussion

The RF study is carried out on the device structures shown above for validating them for the analog performance figure of merit. The characteristic of gate-drain capacitance (C_{GD}) plays a vital role in estimating the device performance speed, cut-off frequency (F_T) (i.e., $F_t = \frac{g_m}{2\pi C_{gg}}$), it depicts that cut-off frequency is dependent on trans-conductance 'g_m' and gate-drain capacitance 'C_{GD}'.

Miller-capacitances 'C_{GS} & C_{GD}', decrease in the miller-capacitances and increase in trans- conductance will increase the cut off frequency, gain-bandwidth product

(GBP) (i.e., given by $GBP = \frac{g_m}{2\pi 10 C_{gd}}$), it is dependent on trans-conductance 'g_m' and miller capacitance 'C_{GD}'.

In Fig. 30.2, the miller capacitance 'C_{GD}' characteristic is shown for different structures studied among which hetero-junction DL TFET shows lower miller capacitance. Hence, the hetero-junction DL TFET structure is used for further estimation. In Fig. 30.2b, tapered angle is studied on hetero-junction DL TFET, in which decrease in the angle formation decreases the miller effect. In Fig. 30.2c, the implementation of hetero-gate is studied, decrease in tunneling gate work function, leads to increase in miller capacitance from 0.398 to 0.558 fF/μm. In Fig. 30.2d, the variation of mole fraction is shown, in which with the decrease in the mole-fraction a minor increase in the gate to drain capacitance increase is observed.

In Fig. 30.3, the cut-off frequency (F_t) characteristics are shown for different structures studied among which hetero-junction DL TFET shows higher cut-off frequency of the order of 4.2 GHz because of it major dependence on the 'g_m' more than C_{gg}. Hence, the hetero-junction DL TFET structure is used for further estimation. In Fig. 30.3b, tapered angle is studied on hetero-junction DL TFET, in which decrease in the angle formation shows almost similar cut-off frequency of the order of 11.1 GHz.

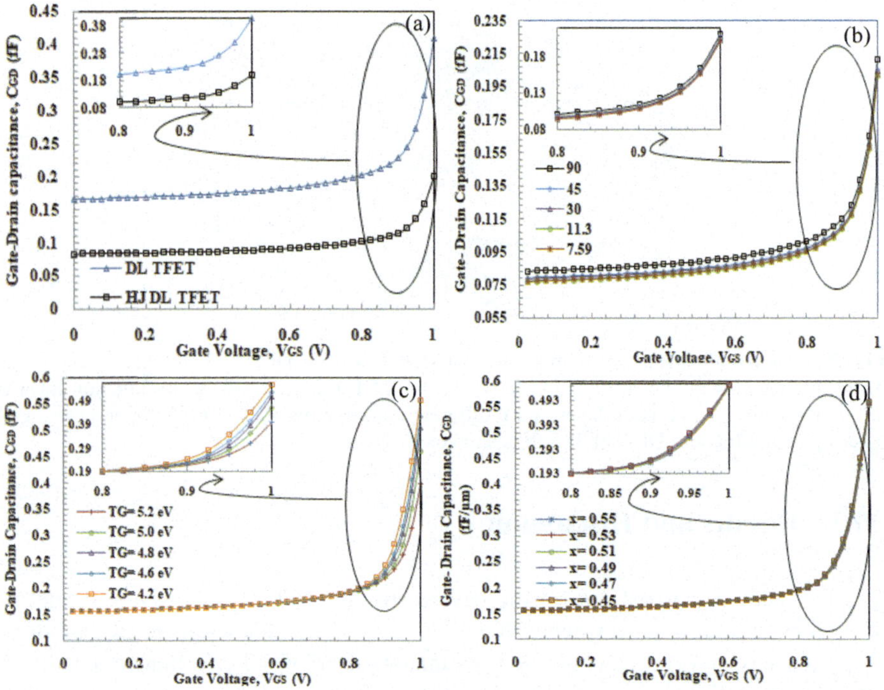

Fig. 30.2 Comparison of gate to drain capacitance characteristics of DL-TFET, **a** basic DL TFET with Hetero junction DL TFET, **b** tapered angle hetero-junction InGaAs DL-TFET, **c** hetero-gate hetero-junction InGaAs DL-TFET with tapered oxide layer and **d** different composition of 'x' for source in hetero-gate hetero-junction InGaAs DL-TFET with tapered oxide layer

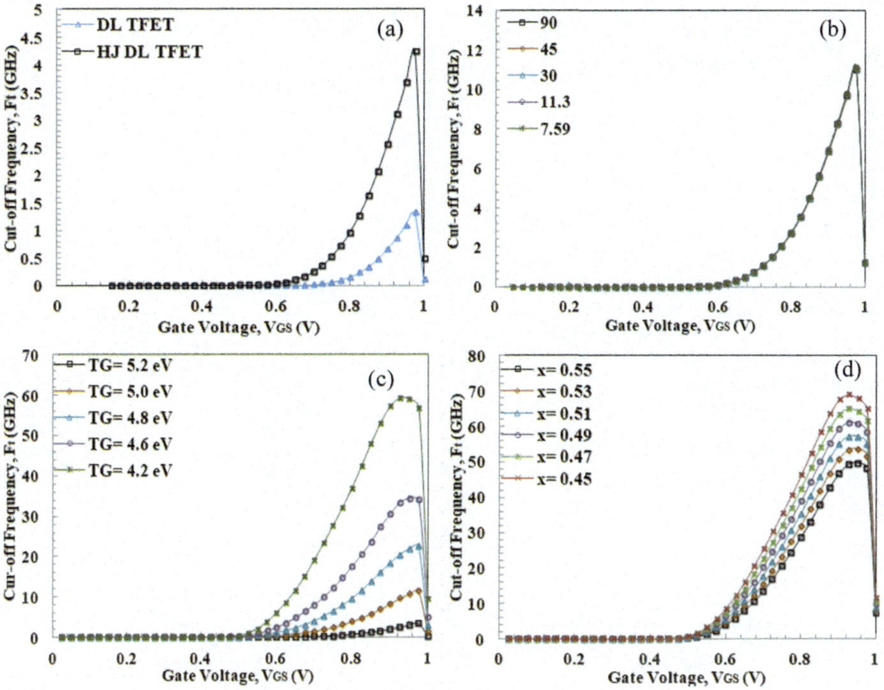

Fig. 30.3 Comparison of cut-off frequency characteristics of DL-TFET, **a** basic DL TFET with hetero junction DL TFET, **b** tapered angle hetero-junction InGaAs DL-TFET, **c** hetero-gate hetero-junction InGaAs DL-TFET with tapered oxide layer and **d** different composition of 'x' for source in hetero-gate hetero-junction InGaAs DL-TFET with tapered oxide layer

In Fig. 30.2c, the implementation of hetero-gate is studied, decrease in tunneling gate work function, leads to increase in the cut-off frequency from 33.8 to 59.3 GHz. In Fig. 30.2d, the variation of mole fraction is shown, in which with the decrease in the mole-fraction increase in the cut-off frequency upto 69 GHz.

In Fig. 30.4, the gain-bandwidth product (GBP) characteristics are shown for different structures. Gain-bandwidth is majorly dependent on the trans-conductance 'g_m' and 'C_{GD}', in Fig. 30.4a the hetero-junction DL TFET shows increase in the GBP. In Fig. 30.4b, it shows minor increase in the characteristics with the reduction in the tapered angle.

In Fig. 30.4c, the gain-bandwidth product is majorly dependent on the 'g_m' rather than C_{GD}, because of which the GBP increases to the level of 14.5 GHz even after reduction in the miller capacitance. In Fig. 30.4d, the reduction in the $In_{1-x}Ga_xAs$ mole- fraction results in more staggered hetero-junction and hence, increase in the gain-bandwidth product. For higher range of V_{GS} supply, the GBP reduces with the reduction in miller capacitance (C_{GD}) and it shows the variation 16.9–1.93 GHz.

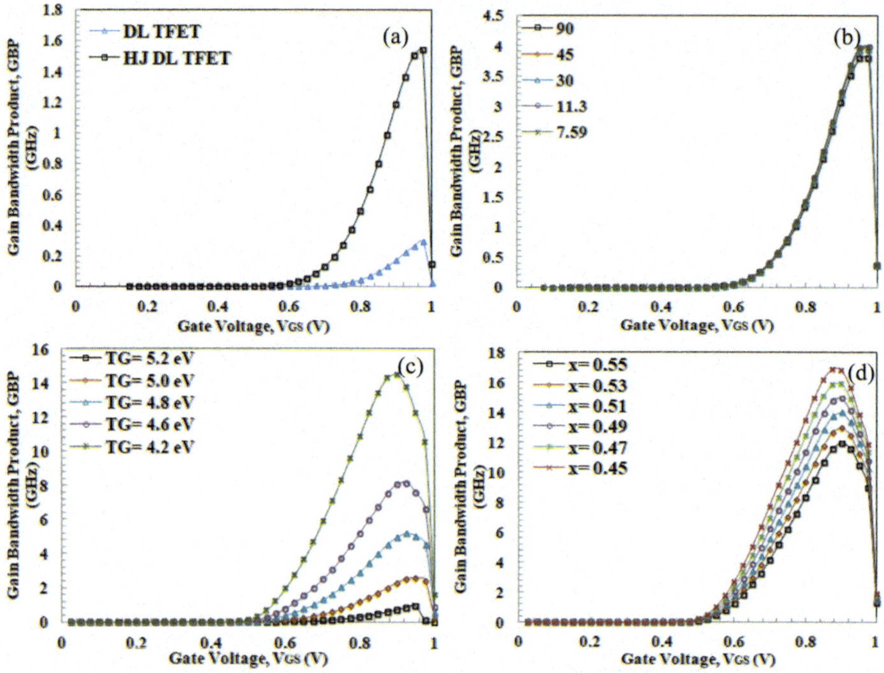

Fig. 30.4 Comparison of gain-bandwidth characteristics of DL-TFET, **a** basic DL TFET with Hetero junction DL TFET, **b** tapered angle hetero-junction InGaAs DL-TFET, **c** hetero-gate hetero-junction InGaAs DL-TFET with tapered oxide layer and **d** different composition of 'x' for source in hetero-gate hetero-junction InGaAs DL-TFET with tapered oxide layer

30.4 Conclusion

In this article, the RF analysis of tapered angle dopingless TFET is presented for different gate engineered structures. The InGaAs dopingless TFET at tapered angle 'ɵ' at 7.59°, TG (tunnelling gate work-function) at 4.2 eV with source side $In_{1-x}Ga_xAs$ composition at 0.45 is found to be most optimized structure showing highest cut-off frequency of the order of 69 GHz and gain-bandwidth product of the order of 16.9 GHz. which makes it useful for digital applications and can be used for faster switching circuits.

Acknowledgements Authors would like to acknowledge University Grants Commission (Award letter no. 3713/NET-JULY 2018) for providing necessary financial assistance during the course of this work and Monika Sharma would like to acknowledge Department of Electronic Science, University of Delhi, South Campus, New Delhi-110021.

References

1. Narain Arora, World Scientific (1993)
2. A.M. Ionescu, H. Riel, Nature **479**, 329–337. https://doi.org/10.1038/nature10679
3. K. Boucart, A.M. Ionescu, IEEE Trans. Electron Devices **54**(7), 1725–1733 (2007). https://doi.org/10.1109/TED.2007.899389
4. S.M. Turkane, G.K. Kharate, A.K. Kureshi, Indian J. Sci. Technol. **10**, 14 (2017). https://doi.org/10.17485/ijst/2017/v10i14/95495
5. M.J. Kumar, S. Janardhanan, IEEE Trans. Electron Devices **3285–91**, 6010 (2013)
6. G. Gupta, B. Rajasekharan, R.J.E. Hueting, IEEE Trans. Electron Devices **64**(8) (2017)
7. M. Sharma, R. Narang, M. Saxena, M. Gupta, Adv. Nat. Sci. Nanosci. Nanotechnol. **10**(3), 035009(8pp), (Aug. 2019), Published Online: 29, Online ISSN: 2043-6262 (IOP Publishing). https://doi.org/10.1088/2043-6254/ab38b1
8. X. Duan, J. Zhang, S. Wang, Y. Li, S. Xu, Y. Hao, IEEE Trans. Electron Devices **65**(3), 1223–1229 (2018)
9. M. Sharma, R. Narang, M. Saxena, M. Gupta, IETE Tech. Rev. (2020). https://doi.org/10.1080/02564602.2020.175822
10. E.-Y. Choi, D.H. Heo, Reduction of zirconium oxide compounds by lithium metal as a reductant in molten LiCl salt. J. Nucl. Mater. **512**, 193–198 (2018). https://doi.org/10.1016/j.jnucmat.2018.10.015
11. M. Eshed, S. Pol, A. Gedanken, M. Balasubramanian, Zirconium nanoparticles prepared by the reduction of zirconium oxide using the RAPET method. Beilstein. J. Nanotechnol. **2**, 198–203 (2011). (Epub 2011 Apr 6. PMID: 21977431; PMCID: PMC3148038). https://doi.org/10.3762/bjnano.2.23
12. H.S. Oh, S.J. Cha, Field effect transistors having gate tapered electrodes for providing high breakdown voltage capability and methods of forming same, US Patent, US6046474A, 2000
13. ATLAS Device User Guide, Version 5.16

Chapter 31
Impact of Pocket Doped Mg_2Si/Si Heterojunction Ge Gated TFET for Low Optical Power Detection at 1550 nm

Manisha Khurana, Upasana, Manoj Saxena, and Mridula Gupta

31.1 Introduction

Due to ever increasing demand of highly sensitive and low cost photodetectors, photo transistors have gained popularity due to its internal trans-impedance gain, miniaturization and low power consumption while maintaining the high performance. Various applications of photodetectors include optical interconnects (OIs), image sensing, military applications, biomedical instruments, remote sensing of environment, pollution monitoring and industrial process monitoring [1, 2]. Now a days, OIs [3] have gained an enormous deal of attention for faster data transfer with larger bandwidth both in the infrared and ultraviolet regime. Wavelengths in the near infrared (NIR) regime find significant applications in this direction with majorly three available optical windows 850 nm, 1330 nm and 1550 μm [4]. Silicon based NIR photodetectors has already been explored by researchers and reached a good level of maturity in various forms of photodetectors including MSM photodetectors [5], p-n photodiodes [6], p-i-n architectures [7], avalanche photodiodes [8, 9], Opto-electronic MOSFET [10] etc. Due to the bandgap limitations of silicon, low bandgap material with finitely high absorption in the infrared regime has been explored by many researchers. In this direction germanium phototransistors (Field effect phototransistors) monolithically integrated with the Si technology [4, 11] can pave the way to an effective photodetector with large gain of phototransistor. The limitation of conventional field effect

M. Khurana · Upasana · M. Gupta (✉)
Semiconductor Devices Research Laboratory, Department of Electronic Science, University of Delhi South Campus, New Delhi 110021, India
e-mail: mridula@south.du.ac.in

M. Saxena
Department of Electronics, Deen Dayal Upadhyaya College, University of Delhi, New Delhi 110078, India

R. Singh et al. (eds.), *The Physics of Semiconductor Devices*, Springer Proceedings in Physics 306, https://doi.org/10.1007/978-981-97-1571-8_31

transistors (FETs) like MOSFET include high subthreshold swing and low commutation speed which can be over taken by modified FETs architectures including Junctionless FET (JL-FET) [12], tunnel FET [13] (TFET) etc. Thus, TFET based phototransistor [14, 15] has been explored in the literature with different photoabsorber materials and architectures to find its suitability for radiation detections. In this work, we have presented germanium gated tunnel field effect phototransistor along with narrow bandgap source material and the feasibility of Magnesium silicide (Mg_2Si) and silicon heterojunction [16] has been utilized with experimentally verified extinction coefficient (k) and absorption coefficient (α) for Germanium layer [17]. The simulation based exploration has been done using ATLAS-2D device simulator [18] with LUMINOUS module of SILVACO. Mg_2Si is non toxic, naturally abundant low band gap material which is widely accepted thermoelectric material and helped in resolving major issues like limited energy resources and global warming [16]. The advantage of exploring narrow bandgap Mg_2Si source material is an abrupt discontinuity of energy bands at the source-channel junction which effectively reduce the tunneling barrier at the junction. Also, for further device performance improvement, a heavily n^+ doped pocket [19] has been explored along with the Mg_2Si source to explore its suitability as phototransistor application at 1550 nm.

Section 31.2 represents the device architecture with simulation methodology used in the study with the principle of operation. Section 31.3 presents the outcome of the study including the dark and the optical characterization of the device along with the comparison with the already reported architectures followed by the conclusion in Sect. 31.4. Optical performance evaluation has been done using the figure of merits sensitivity [20], responsivity and detectivity [21].

31.2 Device Architecture and Simulation Methodology

Figure 31.1 represents the proposed germanium gated pocket doped Mg_2Si/Si heterojunction [16] based tunnel field effect phototransistor (Ge-Mg_2Si/n^+ Si TFET). Material properties of germanium and Mg_2Si have been given in Table 31.1. Here Ge photoabsorber layer has been used over the complete channel in between the gate and the oxide layer with finitely large absorption coefficient [17] of 2512 cm^{-1} at the studied wavelength of 1550 nm (near infrared radiations). The absorption of radiations takes places inside the Ge photo layer whereas the conduction (tunneling) region is the Si TFET device with Mg_2Si source. The device has been illuminated vertically over the gate inside the optical window as shown in Fig. 31.1. Upon illumination, the absorption of photons takes place inside the Ge layer and an exponentially reduced optical power penetrates inside the device from the top towards the Si channel. Thus, a thick Ge layer of 200 nm has been chosen so that a minimum optical power reaches the silicon channel that separates the photo absorption region and conduction region in the proposed architecture that eventually reduce the noise in the device. The advantage of low bandgap Mg_2Si in the source region introduce a sharp discontinuity at the source channel junction that effectively results in a low tunneling barrier width as compared to Si source. Also, heavily doped n^+ silicon pocket has been incorporated in between source and channel region that further reduce the tunneling barrier width at the source-channel junction. The channel length (L) of the device is 100 nm, pocket length ($L_n = 10$ nm), Ge thickness (t_{Ge}) = 200 nm, Si channel thickness (t_{Si}) = 20 nm, oxide (HfO_2) thickness = 3 nm, Ge doping ($10^{18}/cm^3$ p-type), source doping ($5 \times 10^{19}/cm^3$ p-type), drain doping ($10^{19}/cm^3$ n-type), channel doping($10^{16}/cm^3$ n-type) and pocket doping ($10^{19}/cm^3$ n-type).

SILVACO ATLAS-2D device simulator has been used to study the electrostatic behavior of transistors. The models include free carrier recombination models: SRH (Shockley–Read–Hall) and AUGER, CVT has been used to evaluate the carrier mobility, doping and temperature and to model the quantum tunneling effect at the source channel junction, Non-local band to band tunneling model (BTBT) has been used. LUMINOUS module has been called to study the device behavior in the illuminated state.

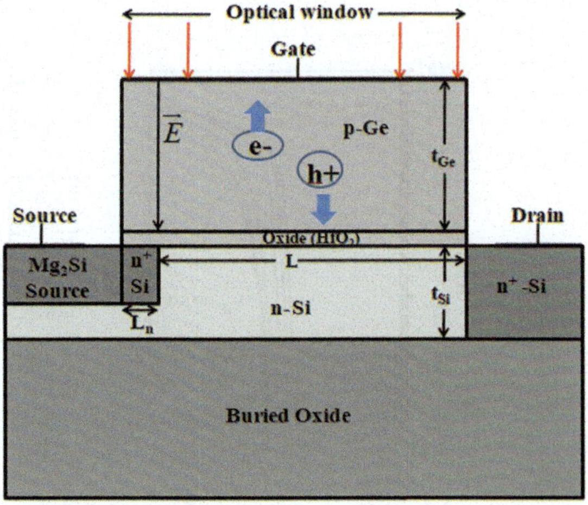

Fig. 31.1 Schematic view of Ge-Mg_2Si/n^+ Si Tunnel field effect phototransistor

Table 31.1 Material property

Magnesium silicide (Mg_2Si) [22]		Germanium (Ge) [17]	
Electrical parameter	Value	Optical parameter	Value
Relative permittivity	20	Wavelength (nm)	1550
Bandgap (eV)	0.77	Imaginary refractive Index (k)	0.031
Electron affinity (eV)	3.59	Absorption coefficient(α) (cm^{-1})	2512
Hole mobility ($cm^2V^{-1} s^{-1}$)	70	Real refractive Index (n)	4.27
Electron mobility ($cm^2V^{-1} s^{-1}$)	370	Quantum efficiency (η_0)	0.6

31.3 Results

Figure 31.2a represents the dark state transfer characteristics of the proposed Ge-Mg_2Si/n^+ Si tunnel field effect phototransistor for different pocket lengths of L_n = 2 nm, 4 nm, 6 nm, 8 nm and 10 nm case where an increment of drain current has been observed with increased pocket length as compared to no pocket case. Also, a left shift in the characteristics has been obtained that indicates reduced gate voltage requirements for the device to turn on (i.e., starting of tunneling). The results have been verified through the energy band profile and electron concentration shown in Figure 31.2b, c respectively. Figure 31.2b clearly depicts the band shifting (inside the tunneling window) at the source channel junction with increased pocket length from 2 to 10 nm with minimum tunneling barrier for L_n = 10 nm. Due to the reduction in the tunnelling barrier, an increase in the electron concentration has been observed (Fig. 31.2c) inside the Si channel that results from the tunneling of electrons from the valence band of Mg_2Si source to the conduction band of Si channel.

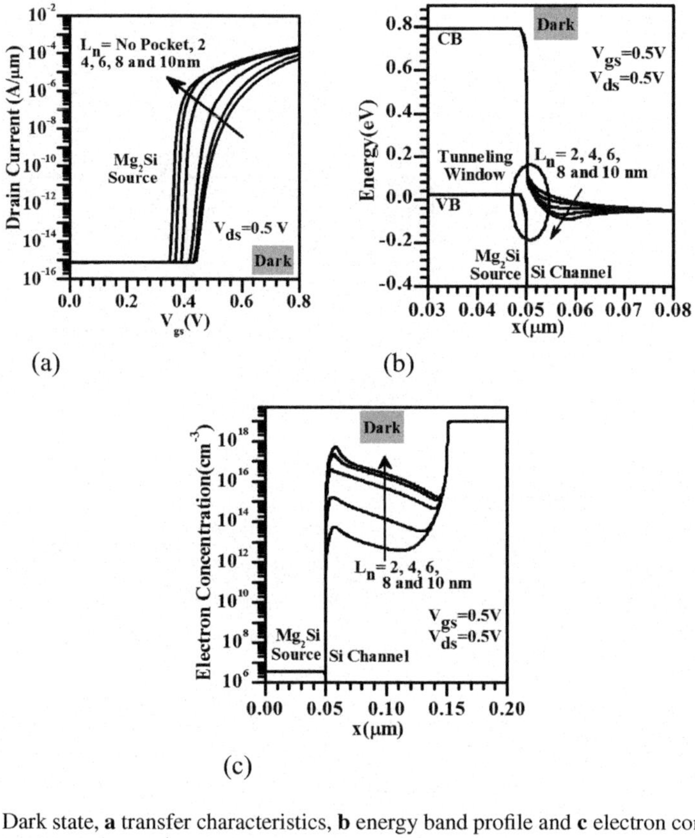

Fig. 31.2 Dark state, **a** transfer characteristics, **b** energy band profile and **c** electron concentration of the proposed Ge-Mg_2Si/n^+ Si TFET at V_{ds} = 0.5 V and L = 100 nm

By studying the advantage of implementing heavily n^+ doped Si pocket at the source channel junction in the dark condition, the proposed architecture has been explored for its application in the illuminated condition at 1550 nm wavelength. The comparison of light and dark condition has been shown in Fig. 31.3a at V_{ds} = 0.5 V and $P_i = 10^{-3}$ μW/μm^2 for without pocket and pocket length of 10 nm (best optimized case) where a left shift in the drain current is noticed under illumination. Left shift of transfer characteristics for both the cases (no pocket and L_n = 10 nm case) represents the generation of optical voltage at the gate terminal that effectively reduce the threshold voltage for the device. The calculated value of V_{opt} is 0.18 V for both the cases representing photo absorption inside the Ge layer only and zero absorption or noise in the device. Also, pocket doped photodetector shows an advantage of steep subthreshold characteristics than the no pocket case. Illumination effect on different pocket lengths device architecture has been shown in Fig. 31.3b for L_n = 2 nm, 4 nm, 6 nm, 8 nm and 10 nm at V_{ds} = 0.5 V and $P_i = 10^{-3}$ μW/μm^2 and a comparison has been done with no pocket case where a large left shift has been observed at larger pocket lengths towards L_n = 10 nm case with increased illuminated drain current.

The effect can be verified through energy band diagram at the source channel junction and shown in Fig. 31.4a for different pocket lengths under illumination marked by tunneling window. It has been observed that increased dip in the energy bands at the source channel junction for larger L_n reduces the tunneling barrier width (barrier between the valence band energy of source region and conduction band energy of the channel region).

The photodetector sensitivity is the ratio of photocurrent to dark current at applied gate and drain bias [23]. Here the sensitivity plot has been shown in Fig. 31.4b for no pocket and pocket doped cases at V_{ds} = 0.5 V and $P_i = 10^{-3}$ μW/μm^2. It has been shown that a peak point of sensitivity has been achieved at specific V_{gs} for each case (shown in Table 31.2). Again it can be verified from the sensitivity plots that

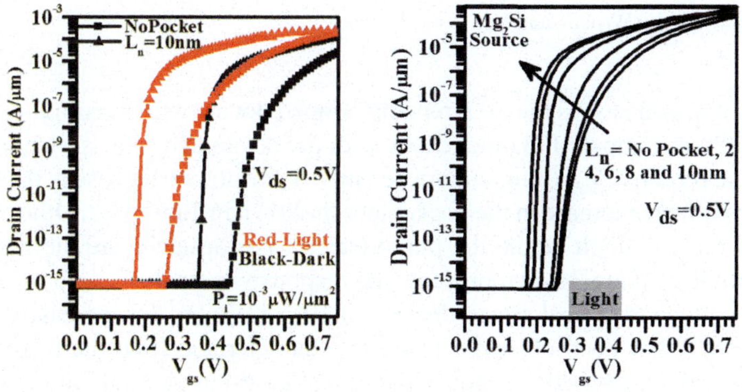

(a) (b)

Fig. 31.3 **a** Transfer characteristics comparison of the proposed device with No pocket and Pocket case (L_n = 10 nm) in the dark and illuminated condition (Light) and **b** improved transfer characteristics (Light) with different pocket length

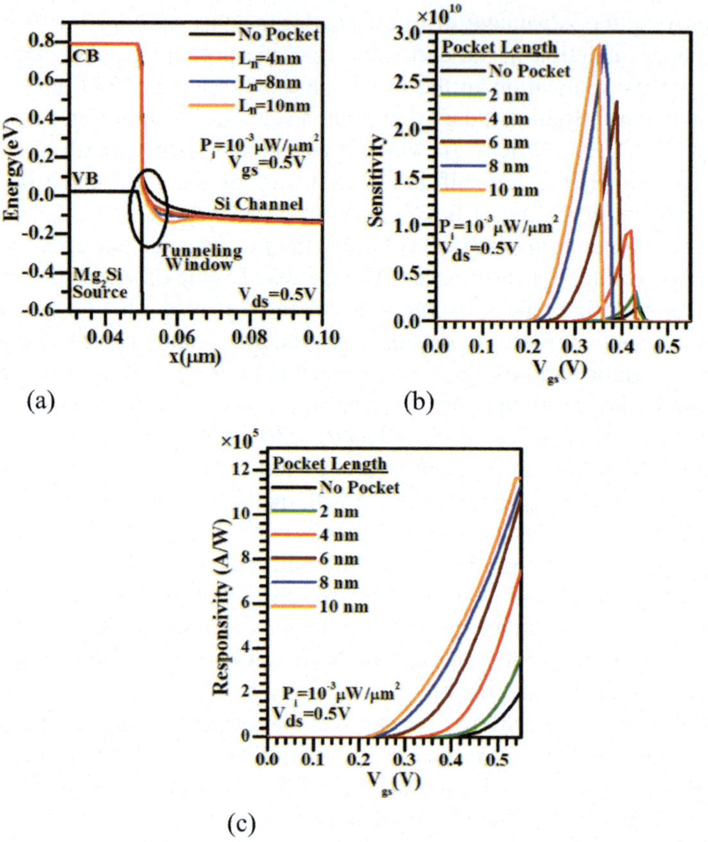

Fig. 31.4 **a** Energy band diagram at the Mg_2Si /Si interface depicting the effect of increasing L_n and **b** sensitivity and **c** responsivity Vs V_{gs} comparison of without pocket and increased pocket length at $P_i = 10^{-3}$ $\mu W/\mu m^2$ and $V_{ds} = 0.5$ V

the best optimized case is the 10 nm pocket doped device with maximum sensitivity (2.9×10^{10}) as compared to other cases with the advantage of least required $V_{gs} =$ 0.35 V. The required V_{gs} of maximum sensitivity for no pocket case is 0.44 V and it decreases with the increase in the pocket length shown in Table 31.2. Other important figures of merits which define the photodetector performance are the responsivity and detectivity. Figure 31.4c represents the responsivity analysis with respect to the gate bias voltage at optical power density of 10^{-3} $\mu W/\mu m^2$ for increased n^+ pocket length. Here it has been observed from the characteristics that the responsivity is increasing with increased V_{gs} for a particular case but the maximum sensitivity gate bias point (obtained from the Fig. 31.4c) is the optimized gate voltage for its application as phototransistor. The reason of increased responsivity beyond the peak sensitivity point is the dominant effect of the gate bias voltage instead of the optical radiations. Thus the optimized gate bias voltage for optical detection of the 1550 nm radiations is the respective peak sensitivity points (Fig. 31.4b). Also, the responsivity of the device increased with increased pocket length from no pocket case and starts to saturates at $L_n = 10$ nm case. The optimized responsivity values for each pocket case at peak sensitivity points have been tabulated in Table 31.2. Also, Signal to noise performance of the photodetector has been done using the specific detectivity (D) given by Eq. (31.1). There exist three types of noise in a semiconductor photodetector i.e. shot noise, Johnson noise (or thermal noise) and flicker noise [24]. The shot noise is related to the statistical fluctuations in the current flow, johnson noise is generated by the thermal motion of the electrons and flicker noise is usually observed at low frequencies. Detectivities values obtained using shot noise assumption are generally over-estimated by an order of magnitude as compared to experimetal noise. Here the noise analysis has been done based on dark current fluctuations only and the thermal fluctuations of the free carriers (Johnson noise factor) has been neglected based on literature study for Ge based photodetctors. The reason of neglecting Johnson noise factor is the applied gate bias voltages which is ~ 0.4 V. At these high voltages, the shot noise dominates over Johnson noise [25] and the detectivity is given by:

$$D = R\sqrt{\frac{A}{2qI_{Dark}}} \tag{31.1}$$

where A = Area of illumination and R is Responsivity [21, 26].

Thus, the novelty of the proposed Mg_2Si based pocket doped photodetector is 20 times increase in the sensitivity with 17 times increase in responsivity and detectivity as compared to without pocket architecture with ~ 0.1 V less required V_{gs}.

Table 31.2 Analysis of optical parameters at at $P_i = 10^{-3}$ μW/μm^2 and $V_{ds} = 0.5$ V

L_n	V_{gs}	Sensitivity	Responsivity (A/W)	Detectivity (Jones)
No Pocket	0.44	1.44×10^9	1.42×10^4	2.81×10^{16}
2 nm	0.43	3.06×10^9	2.43×10^4	4.80×10^{16}
4 nm	0.42	9.39×10^9	1.13×10^5	2.23×10^{16}
6 nm	0.39	2.28×10^{10}	1.82×10^5	3.60×10^{17}
8 nm	0.36	2.86×10^{10}	2.23×10^5	4.41×10^{17}
10 nm	0.35	2.9×10^{10}	2.5×10^5	4.94×10^{17}
OC-TFET[27]	0.4	5.6×10^7	3.3×10^3	4.3×10^{14}
JL-TFET[21]	-0.3	8×10^7	99	1.48×10^{14}

31.4 Conclusion

This work proposes germanium gated pocket doped Mg_2Si/Si heterojunction based tunnel field effect phototransistor (Ge-Mg_2Si/n^+ Si TFET). In this article pocket length optimization has been done to best optimize the photodetector performance i.e., sensitivity, responsivity and detectivity optimization with minimum required gate bias voltage. It is analyzed from the study that the pocket doped architecture shows better performance as compared to no pocket case. Also, different pocket length shows different characteristics in terms of dark as well as illuminated drain current and it is concluded from the observations that the pocket length of 10 nm with Mg_2Si source is the best optimized architecture. The maximum sensitivity (2.9 $\times 10^{10}$) has been achieved for pocket length of 10 nm which is 20 times larger than the no pocket case at least required V_{gs} of 0.35 V. The responsivity of 2.5 $\times$ 10^5 A/W and detectivity of 4.94 $\times$ 10^{17} Jones has been obtained which is 17 times higher than the no pocket case. Thus it is concluded that the pocket doped architecture with Mg_2Si source architecture can be used as an effective NIR photodetector at 1550 nm.

Acknowledgements Authors would like to thank Department of Electronic Science, University of Delhi and Council of Scientific and Industrial Research (letter no. 22(0724)/17/EMR-II) to pursue this research work.

References

1. J.W. Zeller, C. Rouse, H. Efstathiadis, P. Haldar, J.S. Lewis, N.K. Dhar, P. Wijewarnasuriya, Y.R. Puri, A.K. Sood, Development of silicon-germanium visible-near infrared arrays. Image Sens. Technol. Mater. Devices, Syst. Appl. **3**(9854), 985408 (2016). https://doi.org/10.1117/12.2229650
2. R. Gautam, M. Saxena, R.S. Gupta, M. Gupta, Analytical model of double gate MOSFET for high sensitivity low power photosensor. JSTS J. Semicond. Technol. Sci. **13**, 500–510 (2013)
3. J. Bashir, E. Peter, S.R. Sarangi, A survey of on-chip optical interconnects. ACM Comput. Surv. **51**, 1–34 (2019)
4. L. Colace, G. Assanto, Germanium on silicon for near-infrared light sensing. IEEE Photonics J. **1**, 69–79 (2009). https://doi.org/10.1109/JPHOT.2009.2025516
5. Y. Su, S. Li, Z. Wu, Y. Yang, Y. Jiang, J. Jiang, Z. Xiao, P. Zhang, T. Zhang, Materials science in semiconductor processing high responsivity MSM black silicon photodetector. Mater. Sci. Semicond. Process. **16**, 619–624 (2013). https://doi.org/10.1016/j.mssp.2012.11.008
6. A. Ramaswamy, M. Piels, N. Nunoya, T. Yin, J.E. Bowers, High power silicon-germanium photodiodes for microwave photonic applications. IEEE Trans. Microw. Theory Tech. **58**, 3336–3343 (2010)
7. J.M. Fédéli, L. Virot, L. Vivien, J.M. Hartmann, Y. Bogumilowicz, D. Marris-Morini, E. Cassan, C. Baudot, F. Boeuf, High-performance waveguide-integrated germanium PIN photodiodes for optical communication applications, in *2014 7th International Silicon-Germanium Technology and Device Meeting (ISTDM)*, vol. 1 (2014), pp. 131–132. https://doi.org/10.1109/ISTDM.2014.6874690
8. L. Virot, P. Crozat, J.M. Fédéli, J.M. Hartmann, D. Marris-Morini, E. Cassan, F. Boeuf, L. Vivien, Germanium avalanche receiver for low power interconnects. Nat. Commun. **5**, 3–8 (2014). https://doi.org/10.1038/ncomms5957

9. P. Vines, K. Kuzmenko, J. Kirdoda, D.C.S. Dumas, M.M. Mirza, R.W. Millar, D.J. Paul, G.S. Buller, High performance planar germanium-on-silicon single-photon avalanche diode detectors. Nat. Commun. **10** (2019). https://doi.org/10.1038/s41467-019-08830-w
10. K. Ang, M. Yu, G. Lo, D. Kwong, Low-voltage and high-responsivity germanium bipolar phototransistor for optical detections in the near-infrared regime. IEEE Electr. Device Lett. **29**, 1124–1127 (2008)
11. L. Colace, V. Sorianello, S. Rajamani, Investigation of static and dynamic characteristics of optically controlled field effect transistors. J. Light. Technol. **32**, 2233–2239 (2014)
12. S. Singh, M. Khosla, G. Wadhwa, B. Raj, Design and analysis of double—gate junctionless vertical TFET for gas sensing applications. Appl. Phys. A **127**, 1–7 (2021). https://doi.org/10.1007/s00339-020-04156-3
13. S.M. Turkane, A.K. Kureshi, Review of tunnel field effect transistor (TFET). Int. J. Appl. Eng. Res. **11**, 4922–4929 (2016)
14. S. Joshi, P.K. Dubey, B.K. Kaushik, Photosensor based on split gate TMD TFET using photogating effect for visible light detection. IEEE Sens. J. **20**, 6346–6353 (2020)
15. N. Dagtekin, A.M. Ionescu, Partially gated lateral tunnel field effect transistor for optical applications partially gated lateral tunnel field effect transistor for optical applications, 232105 (2014). https://doi.org/10.1063/1.4904026
16. J. Madan, M. Dassi, R. Pandey, R. Chaujar, R. Sharma, Numerical analysis of Mg_2Si/Si heterojunction DG-TFET for low power/high performance applications: Impact of non- idealities. Superlattices Microstruct. **139** (2020). https://doi.org/10.1016/j.spmi.2020.106397
17. V. Sorianello, A. Perna, L. Colace, G. Assanto, H.C. Luan, L.C. Kimerling, Near-infrared absorption of germanium thin films on silicon. Appl. Phys. Lett. **93**, 93–96 (2008). https://doi.org/10.1063/1.2987999
18. ATLAS Users Manual, Device Simulation Software, Silvaco Int., Santa Clara, CA, 2014. Version 5.24.1.R. (n.d.)
19. D.B. Abdi, M.J. Kumar, In-built N^+ Pocket p-n-p-n tunnel field-effect transistor. IEEE Electron Device Lett. **35**, 1170–1172 (2014). https://doi.org/10.1109/LED.2014.2362926
20. H. Xu, J. Li, B.H.K. Leung, C.C.Y. Poon, B.S. Ong, Y. Zhang, N. Zhao, A high-sensitivity near-infrared phototransistor based on an organic bulk heterojunction. Nanoscale **5**, 11850–11855 (2013). https://doi.org/10.1039/c3nr03989g
21. H. Ferhati, F. Djeffal, Planar junctionless phototransistor: a potential high-performance and low-cost device for optical-communications. Opt. Laser Technol. **97**, 29–35 (2017)
22. M.I. Baleva, E. Goranova, M. Marinova, A. Atanasov, Heterojunctions between silicon and the semiconducting metal silicides β-$FeSi_2$ and $MgSi_2$. ECS Trans. **8**, 151–156 (2019). https://doi.org/10.1149/1.2767301
23. W. Kong, G. Wu, K. Wang, T. Zhang, Y. Zou, D. Wang, L. Luo, Graphene-β-Ga_2O_3 heterojunction for highly sensitive deep UV photodetector application. Adv. Mater. **28**, 10725–10731 (2016)
24. K. Wang, X. Qiu, Z. Lv, Z. Song, H. Jiang, Ultrahigh detectivity, high-speed and low-dark current AlGaN solar-blind heterojunction field-effect phototransistors realized using dual-float-photogating effect. Photonics Res. **10**, 111–119 (2022)
25. Z. Wu, Y. Zhai, H. Kim, J.D. Azoulay, T.N. Ng, Emerging design and characterization guidelines for polymer-based infrared photodetectors. Acc. Chem. Res. **51**, 3144–3153 (2018). https://doi.org/10.1021/acs.accounts.8b00446
26. M.I. Pintor-Monroy, M.G. Reyes-Banda, C. Avila-Avendano, M.A. Quevedo-Lopez, Tuning electrical properties of amorphous GaO thin films for deep UV phototransistors. IEEE Sens. J. **21**, 14807–14814 (2021). https://doi.org/10.1109/JSEN.2021.3074623
27. H. Ferhati, F. Djeffal, Boosting the optical performance and commutation speed of phototransistor using SiGe/Si/Ge tunneling structure. Mater. Res. Express. **5**, 65902 (2018)

Chapter 32
MEMS Acoustic Sensor for Low Frequency Applications

Washim Reza Ali, Anshuman Raunak, Ashish Kumar, and Mahanth Prasad

32.1 Introduction

Majority of the acoustic sensors developed so far have focused on acoustic signal detection either for hearing aid applications [1, 2] or aero-acoustic applications [3–5]. However, in recent times, acoustic signal detection in low frequency range (below 200 Hz) has attracted attention due to its application in natural disaster warning [6], medical treatment [7], structural health monitoring [8] and most importantly in the detection of low frequency noise (*LFN*). *LFN* in the range of 10 Hz- 200 Hz has been recognized as a powerful stressor [9]. The most prominent effects on human health are emotional changes such as annoyance, agitation, and distraction, the development of cardiovascular diseases, sleep disorders, high blood pressure, fatigue, dizziness, tinnitus, aural pain and even dental wear [10–13]. Thus, *LFN* can produce significant effect on physical and mental health of human at high *SPL* levels ~ 120 dB [14]. Besides, exposure to moderate levels of *LFN* (70 dB SPL) at the frequency region of 31.5–125 Hz for 2 h has been reported to upset neuroendocrine activity in humans [15]. Common sources of *LFN* include wind turbines, refrigerator, rocket engines, drilling of oil and gas wells etc. [11, 13]. Therefore, detection of *LFN* is utmost necessary to control its emission from these various sources. This is where acoustic sensors are employed. A number fiber optic acoustic sensors have been reported [16, 17] for low frequency measurements due to the inherent advantage of remote

W. R. Ali (✉)
Silicon Austria Labs, Villach 9500, Austria
e-mail: rezali786@gmail.com

A. Raunak
Defence Laboratory Jodhpur, Jodhpur, Rajasthan 342011, India

A. Kumar
National Institute of Technology, Raipur 492010, India

M. Prasad
Central Electronics Engineering Research Institute, Pilani, Rajasthan 333031, India

R. Singh et al. (eds.), *The Physics of Semiconductor Devices*, Springer Proceedings in Physics 306, https://doi.org/10.1007/978-981-97-1571-8_32

measurement and immunity to electromagnetic interference. But, all these sensors require some input power for operation. Also, reflectivity of non-metallic diaphragm sensor is low which results in attenuation and noise. However, adding a metallic layer on the diaphragm can improve reflectivity but it increases the diaphragm thickness and fabrication complexity. That is why in the current work, the design of a piezoelectric acoustic sensor has been proposed. It offers the advantages of requiring no input power and comparatively simpler fabrication process. The sensor is designed for operation in the SPL range of 40–120 dB and in the low frequency band of 5 Hz–2 kHz.

32.2 Design and Simulation

The front side of the device consists of piezoelectric ZnO layer sandwiched between two gold electrodes, placed on an oxidized Si-diaphragm. Figure 32.1 is showing the 3-D model of the design used for simulation.

The *Si*-diaphragm has been created through bulk-micromachining of *Si* from the backside. A microchannel has also been created for providing a connection between the cavity and the outside environment. A *Pyrex* glass wafer was anodically bonded for sealing the cavity after diaphragm formation.

Optimization of Si-diaphragm thickness is required to operate the device in linear region. The generalized guideline for diaphragm design is that the deflection of the diaphragm should be less than 25% of the diaphragm thickness for linear variation of deflection with the applied pressure [18–20]. Based on this guideline, different diaphragm thicknesses from 5 to 20 μm in steps of 5 μm has been considered for simulation. In all these simulations, the device size has been kept fixed at 4 mm × 4 mm. The thicknesses of different layers used for simulation are: Si (5–20 μm) + thermal-oxide (1 μm) + Au (0.2 μm) + ZnO (2 μm) + Au (0.2 μm). The simulation has been carried out using *MEMS-CAD* tool *Coventorware*. The material properties, utilized in the simulation, is given in Table 32.1. The simulation results are shown in Figs. 32.2, 32.3 and 32.4.

The design of the microchannel decides the low cut-off frequency of the sensor. The microchannel is necessary for pressure compensation. The low cut-off frequency is given by [3]:

$$f_c = \frac{1}{2\pi R_t C_c} \tag{32.1}$$

The cavity compliance, C_c is given by:

$$C_c = \frac{V}{\rho_0 C_0^2} \tag{32.2}$$

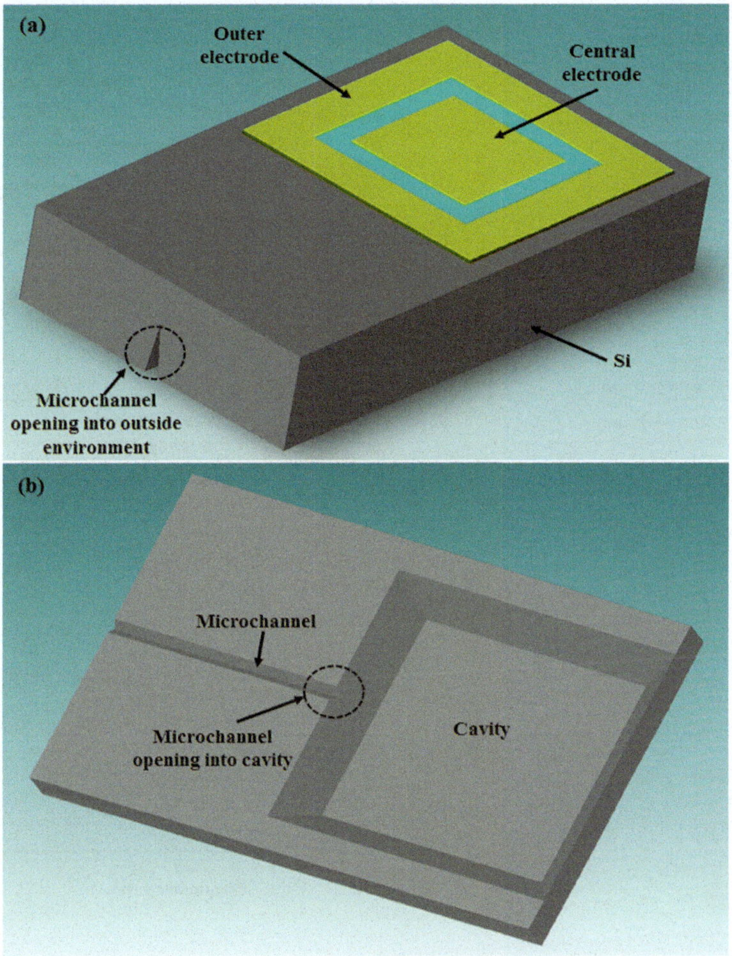

Fig. 32.1 3-D model of the device used for simulation: **a** front view and **b** back view

Table 32.1 Properties of the materials used in the simulation

Parameters	Si	SiO_2	Au	ZnO
Density (kg/m^3)	2331	2300	19,280	5680
Young's Modulus (GPa)	130	69	80	210
Poisson's Ratio	0.278	0.17	0.42	0.33
Dielectric Constant	11.7	3.9	–	8.55
Piezo-electric co-efficient (pC/N)	–	–	–	− 5

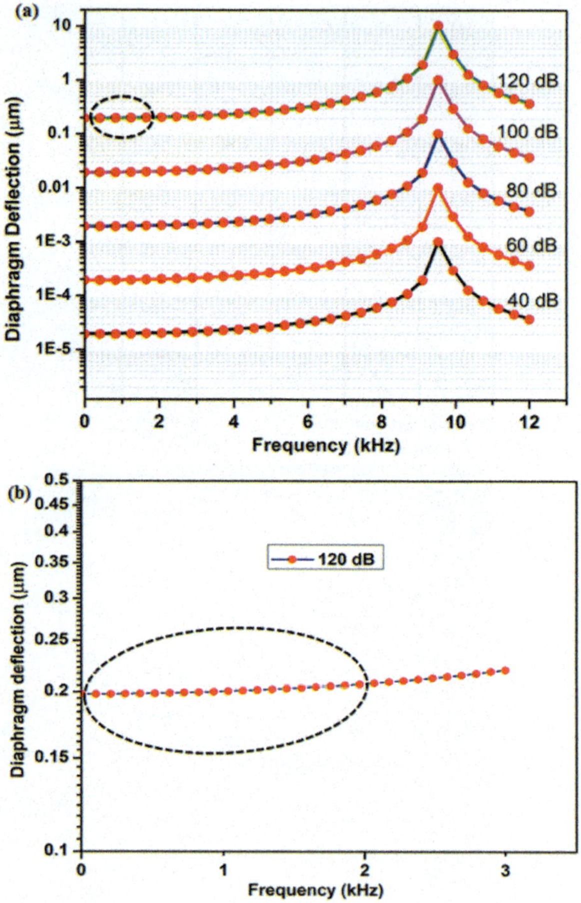

Fig. 32.2 Harmonic analysis of the sensor showing, **a** variation of deflection with frequency for the SPL range of 40–120 dB and **b** enlarged view of the flat portion showing response within the frequency band of 2 kHz at the maximum SPL of 120 dB

Microchannel resistance R_t is given by:

$$R_t = \frac{128\mu L}{\pi D_d^4} \tag{32.3}$$

where, V is the volume of the cavity, ρ_O is the density of air, C_O is the speed of sound in air, μ is the viscosity of air, $L = 3$ mm is the effective length of the microtunnel and D_d is the hydraulic diameter.

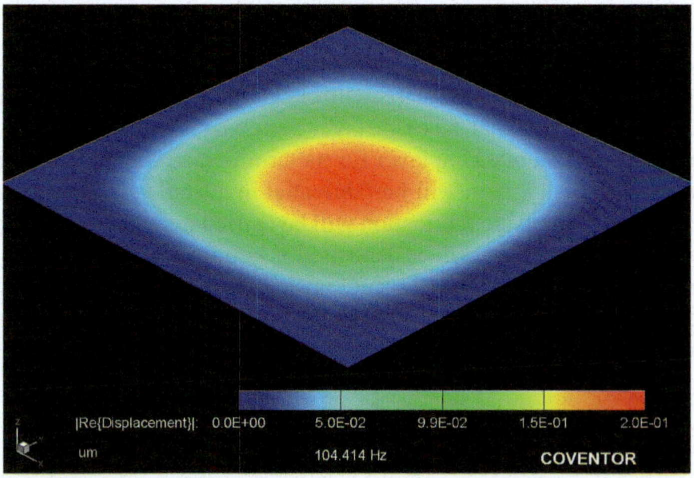

Fig. 32.3 Central deflection of the diaphragm at an applied *SPL* of *120 dB*

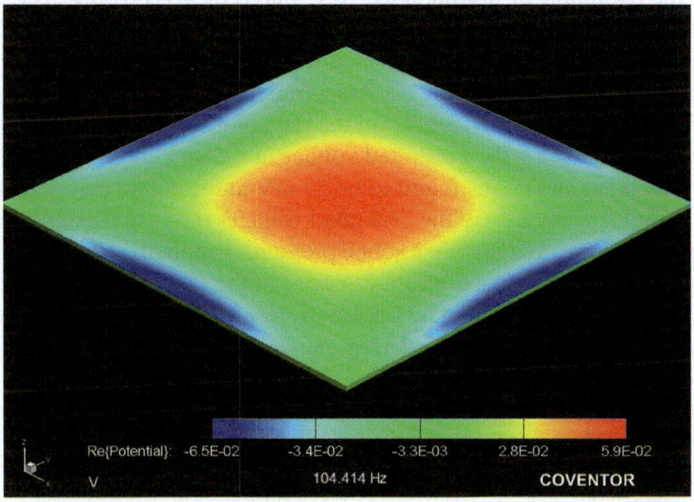

Fig. 32.4 Generated voltage at an applied *SPL* of *120 dB*

The size and thickness of the diaphragm dictate the resonance frequency of the device which is given by [3]:

$$f_r = \frac{1}{2\pi M_d C_d} \tag{32.4}$$

The resonance frequency is dependent on the diaphragm mass (M_d) and compliance (C_d).

32.3 Proposed Fabrication Process Flow

The proposed fabrication process starts with cleaning of Si (100) wafers using RCA1 and RCA2 cleaning. After that, an oxide layer is grown on the wafer. On the backside, the oxide layer acts as a masking layer for patterning microchannel and cavity. The microchannel and cavity, both will be fabricated in a single step using wet TMAH etching. Thereafter, frontside fabrication will be carried out on the oxide layer which acts as an insulating layer. Au will be deposited as bottom electrode on the oxide layer using sputtering. Next, piezoelectric ZnO layer will be deposited using sputtering. Thereafter, another Au layer will be deposited and it will act as top electrode. The whole process flow is summarized in Fig. 32.5.

32.4 Result and Discussion

Optimization of the Si-diaphragm thickness is necessary to develop a reliable device structure for operation in the intended SPL range of 40–120 dB. The optimized diaphragm thickness for a device size of 4 mm $\times$ 4 mm was found to be 10 μm. The deflection obtained for this optimized thickness is 0.2 μm within the flat band at the maximum SPL of 120 dB (Fig. 32.3). Also, maximum deflection is obtained at the centre of the structure as is evident from the red color in Fig. 32.3.

Figure 32.4 is showing the generated voltage at the maximum SPL of 120 dB i.e. 20 Pa (ref. 20 μPa) and it was found to be 59 mV at the centre and $-$ 65 mV at the edges. The opposing polarity of the generated voltages is due to the opposing stresses created at the centre and edges when sound pressure is applied on the diaphragm from the topside. That is why, electrodes are placed at the central and outer regions [as shown in Fig. 32.1a] so as to obtain optimum sensitivity. Therefore, the central and outer sensitivity of the device are determined to be 2.95 mV/Pa and 3.25 mV/Pa, respectively.

Figure 32.2 is showing the result of harmonic analysis which gives the variation of diaphragm displacement when harmonic pressure loads are applied. It has been observed from the analysis that the deflection becomes maximum at a particular frequency at all the SPL levels (Fig. 32.2a) which is nothing but the resonance frequency of the device and it was found to 9.6 kHz. Figure 32.2b is showing the enlarged view of the flat portion showing response within the frequency band of 2 kHz at the maximum SPL of 120 dB. Similar flat response has been observed within the band of 2 kHz for the whole operating SPL range of 40–120 dB. However, for the sake of simplicity, enlarged response is shown for the maximum SPL of 120 dB only. Also, the resonance frequency is obtained analytically from Eq. (32.4) and is found to be 9.2 kHz which is in good agreement with the simulated result.

The lower cut-off frequency was obtained from the microchannel design utilizing Eq. (32.1) and it was found to be 5.7 Hz. Moreover, it has been experimentally

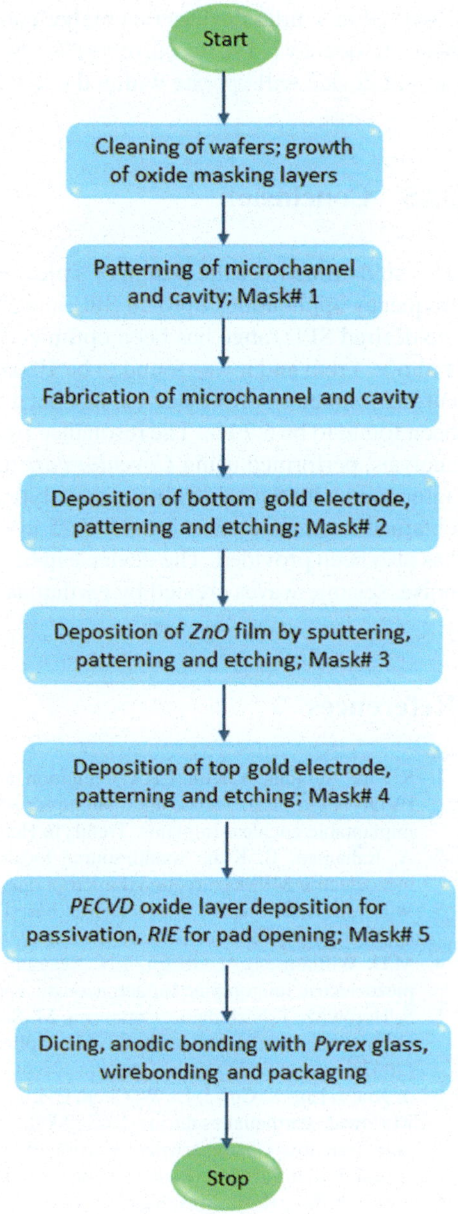

Fig. 32.5 Proposed fabrication process flow of the device

observed in similar devices that higher cut-off frequency is nearly 1/5th of the resonance frequency [3] so the current sensor will operate linearly upto 2 kHz. Therefore, current device will operate within the low frequency band of 5 Hz–2 kHz.

32.5 Conclusion

This study illustrates the design of a piezoelectric-ZnO based MEMS sensor for low frequency applications upto an SPL of 120 dB. The thickness of the diaphragm for the desired SPL range has been optimized using Coventorware for a device size of 4 mm $\times$ 4 mm and it was found to be 10 μm. The microchannel, responsible for low cut-off frequency of the device, has been designed. The low cut-off frequency has been found to be 5.7 Hz. The resonance frequency has been obtained from harmonic analysis, performed using *Coventorware* and it was found to be 9.6 kHz which is in agreement with the value obtained analytically. The average sensitivity of the device obtained from simulation is 2.95–3.25 mV/Pa. A proposed fabrication process flow has also been provided. The device can be employed in the sensing of low frequency noise, seismic waves created by earthquakes, volcanic eruptions etc.

References

1. S. Park, X. Guan, Y. Kim, F.P.X. Creighton, E. Wei, I.J. Kymissis, H.H. Nakajima, E.S. Olson, PVDF-based piezoelectric microphone for sound detection inside the cochlea: toward totally implantable cochlear implants. Trends in Hearing **22**, 1–11 (2018)
2. A. Rahaman, B. Kim, Sound source localization by Ormia ochracea inspired low–noise piezoelectric MEMS directional microphone. Sci. Rep. **10**, 9545 (2020)
3. W.R. Ali, M. Prasad, Piezoelectric based MEMS acoustic sensor for wide frequency applications. IEEE Sens. **21**(24), 27352–27360 (2021)
4. M.D. Williams, B.A. Griffin, T.N. Reagan, J.R. Underbrink, M. Sheplak, An AlN MEMS piezoelectric microphone for aeroacoustic applications. JMEMS **21**(2), 270–283 (2012)
5. S. Horowitz, T. Nishida, L. Cattafesta, M. Sheplak, Development of a micromachined piezoelectric microphone for aeroacoustics applications. J. Acoust. Soc. Am. **122**(6), 3428–3436 (2007)
6. Y. Xia, J. Liu, X. Cui, J. Li, W. Chen, C. Liu, Abnormal infrasound signals before 92 M $\geqq$ 7.0 worldwide earthquakes during 2002–2008. Asian J. Earth Sci. **41**(4–5), 434–441 (2011)
7. G.V. Batanov, Characteristics of etiology of immediate hypersensitivity in conditions of exposure to infrasound. Radiats. Biol. Radioecol. **35**(1), 78–82 (1995)
8. R.D. Whitlow, R. Haskins, S.L. McComas, C.K. Crane, I.L. Howard, M.H. McKenna, Remote bridge monitoring using infrasound. J. Bridge Engg. **24**(5), 1–11 (2019)
9. H.G. Leventhall, Low frequency noise and annoyance. Noise Health **6**, 59–72 (2004)
10. K.P. Waye, J. Bengtsson, A. Kjellberg, S. Benton, Low frequency noise "pollution" interferes with performance. Noise Health **4**, 33–49 (2001)
11. J.A. Alves, F.N. Paiva, L.T. Silva, P. Remoaldo, Low-frequency noise and its main effects on human health—a review of the literature between 2016 and 2019. Appl. Sci. **10**, 1–27 (2020)
12. M.A. Cavacas, V. Tavares, G. Borrecho, M.J. Oliveira, P. Oliveira, A. Águas, J.M. Santos, Industrial noise and tooth wear—experimental study. Int. J. Med. Sci. **12**, 3–8 (2015)

13. R.J. Lubner, N.S. Kondamuri, R.M. Knoll, B.K. Ward, P.D. Littlefield, D. Rodgers, K.G. Abdullah, A.K. Remenschneider, E.D. Kozin, Review of Audio vestibular symptoms following exposure to acoustic and electromagnetic energy outside conventional human hearing. Front. Neurol. **11** (2020)
14. G. Leventhall, Low frequency noise. What we know, what we do not know, and what we would like to know. J. Low Freq. Noise Vibr. Act. Control 79–104 (2009)
15. L. Liu, P. Lu, S. Wang, X. Fu, Y. Sun, D. Liu, J. Zhang, H. Xu, Q. Yao, UV adhesive diaphragm-based FPI sensor for very-low-frequency acoustic sensing. J. IEEE Photonics **8**(1), 1–9 (2016)
16. X. Fu, P. Lu, J. Zhang, Z. Qu, Y. Li, P. Hu, W. Yan, W. Ni, D. Liu, J. Zhang, Micromachined extrinsic fabry-pérot cavity for low-frequency acoustic wave sensing. Opt. Express **27**(17), 24300–24310 (2019)
17. K.P. Waye, J. Bengtsson, R. Rylander, F. Hucklebridge, P. Evans et al., Low frequency noise enhances cortisol among noise sensitive subjects during work performance. Life Sci. **70**, 745–758 (2002)
18. M.D. Giovanni, *Flat and Corrugated Diaphragm Design Handbook*, 1st edn. (Marcel Dekker, New York, 1982)
19. W.R. Ali, M. Prasad, Design and fabrication of piezoelectric MEMS sensor for acoustic measurements. SILICON **14**, 6737–6747 (2022)
20. X. Wang, B. Li, O.L. Russo, H.T. Roman, K.K. Chin, K.R. Farmer, Diaphragm design guidelines and an optical pressure sensor based on MEMS technique. Microelectron. J. **37**, 50–56 (2006)

Chapter 33
Aptamer Functionalized CVD Grown Monolayer WS_2 Based FETs for Real-Time Detection of *E. coli*

Abdul kaium Mia, Abhilasha Bora, and P. K. Giri

33.1 Introduction

A biosensor is an analytical device that is used for the detection and analysis of biomolecular interactions and has a biorecognition element with a transducer element. Real-time and label-free detection of pathogenic bacteria is essential for continuous monitoring of public health and food quality. The current methods of detection such as polymerase chain reaction (PCR) are time consuming and require well-trained personnel for the operations. Electrical biosensors are of a type of device that can produce results in real-time, are easy to use and hence, can be used as point of care devices [1]. In recent years 2D transition metal dichalcogenides (TMDs) such MoS_2, WS_2, $MoSe_2$, WSe_2 have attracted a lot of attention because of their excellent opto-electrical properties such as high on–off ratio, mobility, and tuneable bandgap [2–4]. WS_2 is a layered material and its opto-electrical properties vary with layer number. The indirect bandgap becomes direct when reduced to monolayer [5]. The FET biosensors based on TMDs are highly sensitive for real-time detection of pathogens because of their finite bandgap unlike graphene [6, 7]. Because of the planar structure with a finite bandgap, WS_2 is compatible with well-established fabrication semiconductor technologies [8]. There are different methods of growth of monolayer WS_2 such as hydrothermal, exfoliation, CVD, etc. [9, 10].

Here we have grown high quality, layer controlled, large area of monolayer WS_2 via CVD technique. The 2D WS_2 were studied by Raman and PL spectroscopy. ssDNA aptamer is used as biorecognition element because of its high specificity, stability, and low complexity. The small size of the aptamer makes it an ideal

A. Mia · P. K. Giri (✉)
Centre for Nanotechnology, Indian Institute of Technology Guwahati, Guwahati 781039, India
e-mail: giri@iitg.ac.in

A. Bora · P. K. Giri
Department of Physics, Indian Institute of Technology Guwahati, Guwahati 781039, India

R. Singh et al. (eds.), *The Physics of Semiconductor Devices*, Springer Proceedings in Physics 306, https://doi.org/10.1007/978-981-97-1571-8_33

candidate as a receptor to bind the analytes within the Debye length. Aptamers are commonly used in optical biosensors [11]. Herein we have used it as an electrical biosensor for the first time. We fabricate an FET using the WS_2 monolayer and study its electrical characteristics.

33.2 Materials and Method

WS_2 monolayer was directly grown on highly doped SiO_2/Si substrate by CVD technique. Prior to growth, the substrates were ultrasonically cleaned with acetone, 2-propanol followed by DI water for 10 min each. 10 mg of WO_3 and 200 mg of sulfur powder were used as precursors and kept in two different quartz boats inside a 1-inch diameter smaller tube. The smaller tube was kept inside a 2-inch diameter tube inside a two-zone muffle furnace. During the growth, temperature for the sulfur and WO_3 region was typically 150 and 940 °C, respectively. The substrates were kept upside down above the WO_3 powder and high purity Ar was used as a carrier gas with 100 standard cubic centimeters per minute (sscm) flow rate. The temperature of the substrate region ramps to 940 °C in 47 min and during this time the temperature of the sulfur region was approximately 150 °C. After the growth, the system is allowed to cool down to room temperature naturally. The single standard DNA aptamer was purchased from Integrated DNA Technologies, USA.

33.3 Characterization

The Micro-Raman and photoluminescence (PL) analysis were conducted in high resolution spectrometer (LabRam HR800, Jobin Yvon) with a 532 nm laser excitation at room temperature. UV lithography was carried out by tabletop direct laser writing lithography system (Dilase 250, Kloe-Franc) with 375 nm laser. Electrical characterizations were performed by using, Keithley 4200 Semiconductor Characterization System (Keithley Instruments).

33.3.1 Fabrication and Functionalization

Back gated FET was fabricated directly on as grown WS_2 on highly doped SiO_2/Si substrate. Dilase 250, direct UV laser writing lithography system was used for patterning the source and drain contacts separated by 10 μm on WS_2 film. For the metal contact, Cr/Au (5/60) nm was deposited via e-beam and thermal evaporation respectively followed by lift-off. Finally, the contact pads were passivated using S1813 photoresist and 8 μm channel was made for the interaction of the samples with WS_2 surface. In order to capture *Escherichia coli*, we have used ssDNA 5'-GGC

TGG ATG GGG CGT GTG GGA GAG GAG GGG CCC ACG CGG GCA GAC AGG GTG CGC AAC GGT GCG GAC AGC G-3' aptamer, which were immobilized on WS_2. The different concentration of the aptamer in 1-X PBS solution at 7.0 PH was dropcasted on the substrate and kept overnight at 4 °C.

33.4 Result and Discussion

The optical image of the monolayer WS_2 on SiO_2/Si substrate shows uniform growth over 100 μm area, as shown in Fig. 33.1a. Atomic force microscopy (AFM) shows that the thickness of the flake is around 0.9 nm which corresponds to single layer WS_2 as reported. The Raman spectrum shows two prominent Raman peaks at 356.2 cm^{-1} and 419.2 cm^{-1} corresponding to in-pane ($E^1{}_{2g}$) and out of plane (A_{1g}) modes, respectively. The 532 nm laser excites others phonon modes which are labelled in Fig. 33.1 c. The separation between the A_{1g} and $E^1{}_{2g}$ modes is an indirect method to determine the layer thickness of WS_2. The separation of 62.2 cm^{-1} suggests the growth of monolayer WS_2 [12], which is consistent with the AFM analysis.

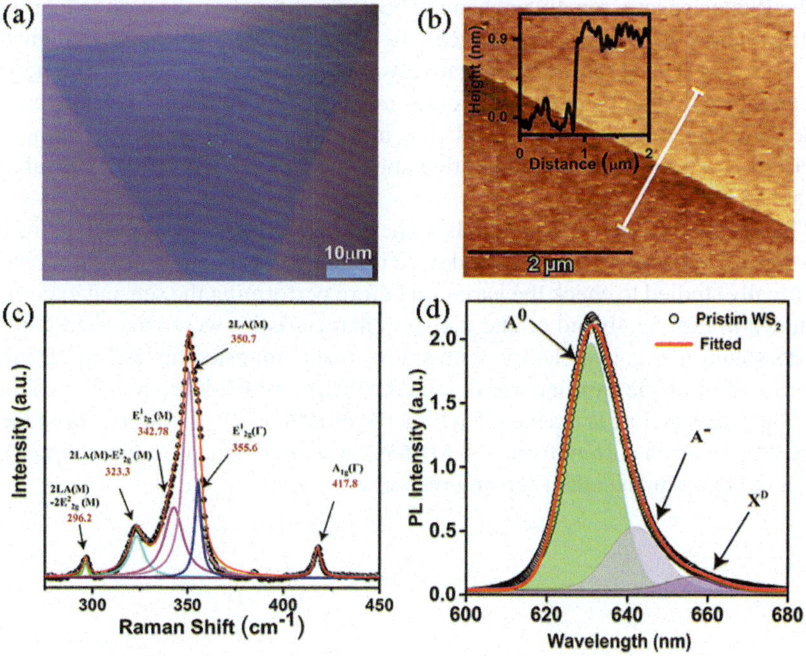

Fig. 33.1 **a** Optical image of the as grown Monolayer (1L) WS_2 on Si/SiO_2 substrate, **b** atomic force microscopy image of the 1L WS_2, the inset shows the height profile, **c** Raman spectra of the 1L WS_2 film fitted with multiple peaks, **d** PL spectra of single layer WS_2, fitted with three peaks

Table 33.1 Excitonic contributions in PL spectra for pristine and functionalized WS_2

Excitons transitions	Pristine 1L WS_2			Functionalized 1L WS_2		
	Peak centre (nm)	Contribution %	Energy (eV)	Peak centre (nm)	Contribution %	Energy (eV)
Neutral	630.8	74.4	1.966	634	67.0	1.956
Trions	642.1	21.3	1.931	643	29.7	1.928
Defect	656.9	4.3	1.887	659	3.3	1.882

PL spectrum of the monolayer WS_2 flake is shown in Fig. 33.1d. We have deconvoluted the spectra in order to find the contributions from different excitons. The PL spectrum consists of a neutral exciton (A^0), a negatively charged trion (A^-), and a defect exciton (X^D) [13]. From the PL data we can see that the defect transition is very weak and hence, this indicates high quality growth. The intensity of monolayer neutral exciton peak shows the nature of direct bandgap of WS_2 complementary of its bulk counterpart. The addition of the aptamers shows a systematic redshift of the PL spectra upto 1 μM concentration but at high concentration, the saturation position shows slight blue shift as compared to 1 μM, as shown in Fig. 33.2a and b. The Raman spectra remain unchanged because of functionalization indicating the quality of the WS_2 film does not decrease because of functionalization, as depicted in Fig. 33.2c [3]. The aptamers are negatively charged, hence as a result of functionalization the trion binding energy decreases as well as the contribution from the trions peak also increases, as expected and shown in Fig. 33.2d. The detailed comparison of PL observations between the pristine and 1 mM functionalized WS_2 is tabulated in Table 33.1 (Fig. 33.2).

The schematic representation of the experimental condition is shown in Fig. 33.3. An optical image of the actual FET device is shown in FIg. 33.4a. The devices were electrically studied to check the variation before performing the sensing experiment, as shown in Fig. 33.4b and c. The transfer characteristics were measured with 0.01 X PBS solution as gate dielectric with a gold coated tungsten tip. Before adding the bacteria solution, the baseline was obtained with a constant drain to source voltage of 1 V. Once the baseline is reached, 5 μL of 10^3 cfu/ml bacteria solution was added. A sharp drop of drain current from 336 to 86 nA was observed, as shown in Fig. 33.4d, which indicates an excellent sensing response.

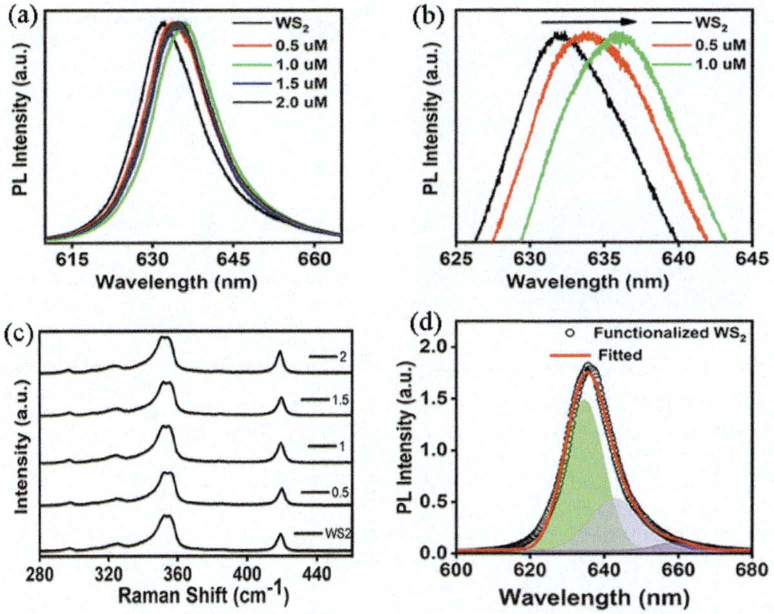

Fig. 33.2 **a** PL spectra of the functionalized 1L WS_2 with different concentrations of aptamers in 1-X PBS, **b** enlarged PL spectra of 1L WS_2 with three different concentrations (0, 0.5, 1 μM) aptamers, **c** stacked Raman spectra of different concentration aptamer functionalized 1L WS_2, **d** Gaussian fitted PL spectra of 1 μM aptamer functionalized 1L WS_2

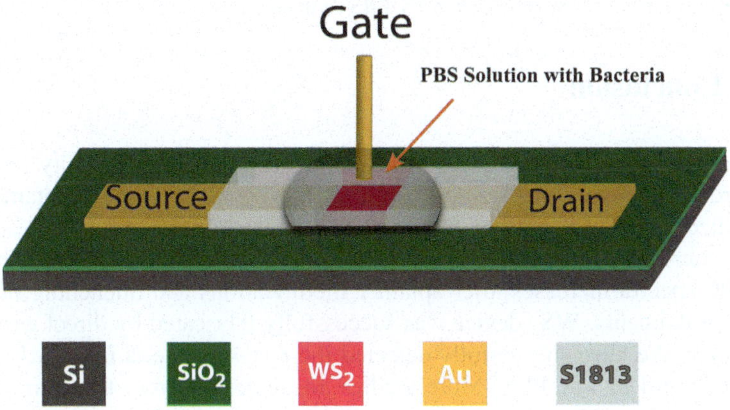

Fig. 33.3 Schematic representation of final sensing device the electrical measurements configurations

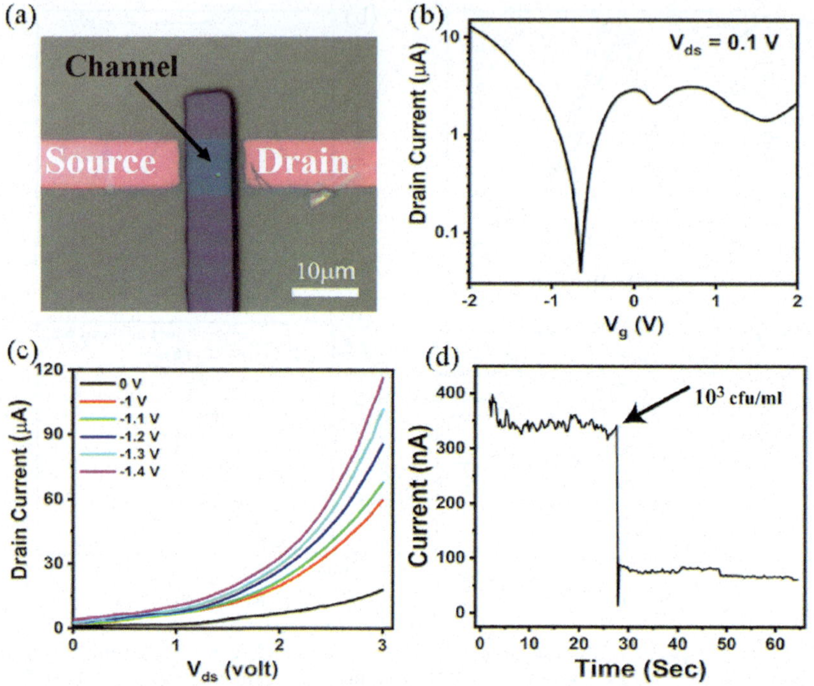

Fig. 33.4 **a** Optical image of the final device with 8 μm micro channel, **b** transfer characteristic of a the WS_2 FET in 1-X PBS solution, **c** output characteristics of 1L WS_2 FET varying the gate voltage, **d** real time response of *E. coli* in 1X-PBS solution

33.5 Conclusion

In summary, we have successfully synthesized large area high quality monolayer WS_2 directly on highly doped SiO_2/Si substrates by CVD technique. Aptamers were immobilized on the WS_2 surface. Because of the negatively charged ssDNA aptamers, electron transfer to WS_2 takes place, a result of which the intensity of the trions peak in the PL spectra increases after aptamer modifications and quenching of neutral exciton peak occurs. WS_2 device was successfully fabricated by direct writing UV lithography. We have successfully detected the *E. coli* bacteria in a 5 μL solution of 10^3 CFU/ml in 1-X PBS. More studies are under progress for optimizing the sensitivity. This result is important for realizing the next generation label free, real-time detection of bio analytes using 2D material-based FETs.

Acknowledgements The authors like to thank BioAptagen Laboratories Private Limited for providing the Bio-lab facilities and aptamers, CIF facility, IIT Guwahati is acknowledged for AFM and Raman characterizations. The financial support from MEITY (Grant No. 5(1)/2021-NANO), Govt. of India, and SERB (Grant number CRG/2021/006397), Govt. of India for carrying out part of this work.

References

1. I. Prattis, E. Hui, P. Gubeljak, G.S. Kaminski Schierle, A. Lombardo, L.G. Occhipinti, Trends Biotechnol. **39**, 1065 (2021)
2. A. Bolotsky, D. Butler, C. Dong, K. Gerace, N.R. Glavin, C. Muratore, J.A. Robinson, A. Ebrahimi, ACS Nano **13**, 9781 (2019)
3. P. Fathi-Hafshejani, N. Azam, L. Wang, M.A. Kuroda, M.C. Hamilton, S. Hasim, M. Mahjouri-Samani, ACS Nano **15**, 11461 (2021)
4. D. Ovchinnikov, A. Allain, Y.-S. Huang, D. Dumcenco, A. Kis, ACS Nano **8**, 8174 (2014)
5. Y. Liu, X. Hu, T. Wang, D. Liu, ACS Nano **13**, 14416 (2019)
6. N. Masurkar, N.K. Thangavel, S. Yurgelevic, S. Varma, G.W. Auner, L.M. Reddy Arava, Biosens. Bioelectron. **172**, 112724 (2021)
7. J. Wei, Z. Zhao, K. Lan, Z. Wang, G. Qin, R. Chen, Talanta **236**, 122839 (2022)
8. A.K. Mia, M. Meyyappan, P.K. Giri, Two-dimensional transition metal dichalcogenide based biosensors: from fundamentals to healthcare applications. Biosensors **13**, 169 (2023)
9. S. Cao, T. Liu, S. Hussain, W. Zeng, X. Peng, F. Pan, Mater. Lett. **129**, 205 (2014)
10. H. Yin, X. Zhang, J. Lu, X. Geng, Y. Wan, M. Wu, P. Yang, J. Mater. Sci. **55**, 990 (2020)
11. A.K. Mia, A. Bora, M.T. Hossain, S. Sinha, P.K. Giri, Fast detection of Staphylococcus aureus using thiol-functionalized WS_2 quantum dots and Bi_2O_2Se nanosheets hybrid through a fluorescence recovery mechanism. J. Mater. Chem. B **11**, 10206–10217 (2023)
12. C. Cong, J. Shang, X. Wu, B. Cao, N. Peimyoo, C. Qiu, L. Sun, T. Yu, Adv. Opt. Mater. **2**, 131 (2014)
13. H.R. Gutiérrez, N. Perea-López, A.L. Elías, A. Berkdemir, B. Wang, R. Lv, F. López-Urías, V.H. Crespi, H. Terrones, M. Terrones, Nano Lett. **13**, 3447 (2013)

Chapter 34
Superconducting Stub Tuner Impedance Matching Circuit for High-Frequency Measurements of Nanoscale Devices

Anusha Shanmugam, Prasanta Kumbhakar, and Madhu Thalakulam

34.1 Introduction

Measuring the state of a gated quantum dot with high fidelity requires sensitive, fast, and noninvasive read-out, which is a necessity to realize semiconductor qubits. The spin-state of the quantum dot can be measured using a charge sensor by exploiting the property of spin-to-charge conversion [1]. Quantum point contacts (QPCs) are sensitive electrometers that can be used to detect single-electron tunnelling events in a capacitively coupled quantum dot [2]. The fast and fragile nature of single-electron transport demands high-frequency measurements as opposed to DC measurements which are time-averaged and poisoned by 1/f noise. Radio-frequency reflectometry measurements in electrostatically gated quantum dots allow time-resolved charge detection using a QPC or a single-electron transistor as a charge detector [3, 4]. Combining sensitivity with fast operation is challenging because of the large RC time-constant constituted by the device resistance (~ > 100 kΩ) and parasitic capacitance of the wires (pF) used in the circuit, limiting the measurement bandwidth to a few kHz.

In order to perform high-frequency measurements in high impedance devices like quantum dots and QPCs, one needs to impedance match the high resistance of these devices to the characteristic impedance Z_0 (50 Ω) of the circuit. Using a lumped element impedance transformer such as an LC circuit limits the resonant frequency and bandwidth (BW) to a few tens of MHz due to the stray capacitances of the wires [5]. In this work, we use a stub tuner, which consists of a pair of terminated transmission lines for impedance matching the QPC to 50 Ω [6–8].

A. Shanmugam · P. Kumbhakar · M. Thalakulam (✉)
Indian Institute of Science Education and Research Thiruvananthapuram, Kerala 695551, India
e-mail: madhu@iisertvm.ac.in

R. Singh et al. (eds.), *The Physics of Semiconductor Devices*, Springer Proceedings in Physics 306, https://doi.org/10.1007/978-981-97-1571-8_34

34.2 Design

A stub tuner is a pair of transmission lines whose electrical length determines the operating frequency while their difference determines the matched load [9]. We have used single open stub shunt tuning, i.e., one line is shunted with the QPC of impedance Z_L while the other one is kept open. For a single-stub shunt tuning circuit, the length of the transmission line is chosen such that the admittance of the combined line and load is of the form, $Y = 1/Z_0 + iB$, where Z_0 is the characteristic impedance of the transmission line and B is the susceptance and shunting the line at D_1 with an open transmission line of length D_2 and admittance $- iB$. The input impedance of the circuit is given by the parallel combination of two arms.

$$Z_{total} = Z_0\left[\tanh(\gamma D_2) + \frac{Z_0 + Z_L\tanh(\gamma D_1)}{Z_L + Z_0\tanh(\gamma D_1)}\right] \tag{34.1}$$

where $\gamma = \alpha + i\beta$, is the propagation constant. For the lossless case, the attenuation constant $\alpha = 0$. β is the phase constant of the transmission line ($\beta = \frac{2\pi f\sqrt{\varepsilon}}{c}$), f is the frequency. The reflection coefficient is given by $\Gamma = \frac{Z_L - Z_0}{Z_L + Z_0}$. We chose the lengths D_1 and D_2 such that the reflection coefficient vanishes at resonant frequency f [10].

$$D_1 = \frac{\lambda_0}{2\pi}\tan^{-1}\sqrt{\frac{R}{Z_0}} \tag{34.2}$$

$$D_2 = \frac{\lambda_0}{2} - \frac{\lambda_0}{2\pi}\tan^{-1}\left(\frac{R - Z_0}{\sqrt{RZ_0}}\right) \tag{34.3}$$

We have calculated the stub lengths for impedance matching the QPC at different points along the pinch-off curve at a resonant frequency of 3 GHz. The lengths of transmission lines D_1 and D_2 are tabulated in Table 34.1.

The variation of the lengths D_1 and D_2 with load resistance at a resonant frequency of 3 GHz is shown in Fig. 34.1. In reflectometry measurements, the reflected power of a radio frequency signal is modulated by the change in the load terminating the waveguide at a resonant frequency. We have designed the stub tuner for impedance matching the QPC at the middle of the last conductance plateau, where the transconductance shows maximum variation at a resonant frequency of 3 GHz. The lengths

Table 34.1 Stub lengths calculated for various load resistances

Conductance	Resistance (kΩ)	D_1 (mm)	D_2 (mm)
$0.25e^2/h$	103.2	9.782	10.059
$0.5e^2/h$	51.6	9.724	10.117
$0.75e^2/h$	34.4	9.680	10.162
e^2/h	25.8	9.643	10.199
$2e^2/h$	12.5	9.528	10.315

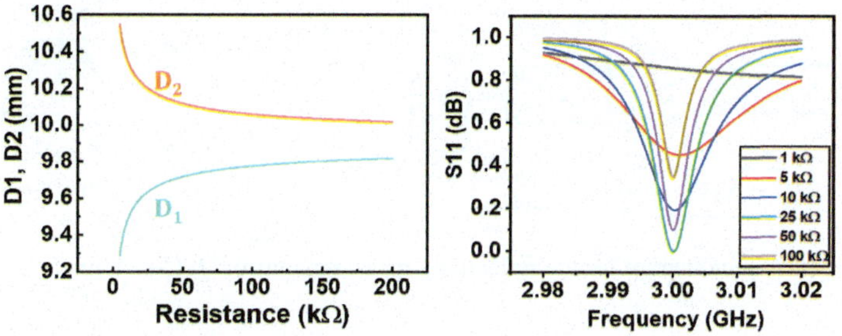

Fig. 34.1 **a** Calculated length of stubs D_1 and D_2 for various load resistances to operate at 3 GHz frequency, **b** reflection coefficient S_{11} for varying load resistances

D_1 and D_2 corresponding to e^2/h at 3 GHz are $D_1 = 0.964$ cm and $D_2 = 1.02$ cm. The reflected signal shows the maximum dip in the S_{11} for a load resistance of 25 kΩ, and the dip changes if the resistance changes on either side of the matched load (Fig. 34.1).

34.3 Device Fabrication

A stub tuner for load resistance $R_L = 25$ kΩ and resonant frequency f = 3 GHz has been fabricated on a sapphire substrate, using photolithography followed by aluminium metallization. QPC has been fabricated on a GaAs/AlGaAs heterostructure by photolithography and electron beam lithography. The QPC is galvanically coupled to the stub tuner for conducting reflectivity measurements (Fig. 34.2).

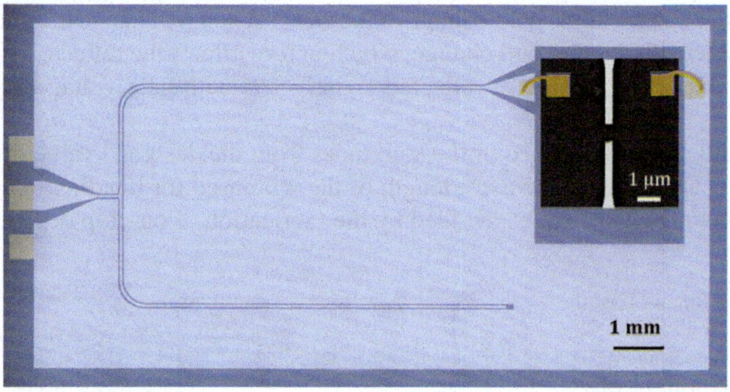

Fig. 34.2 Optical micrograph of the stub tuner similar to the one used for measurements, showing the galvanically coupled QPC. SEM image of the QPC is shown in the inset

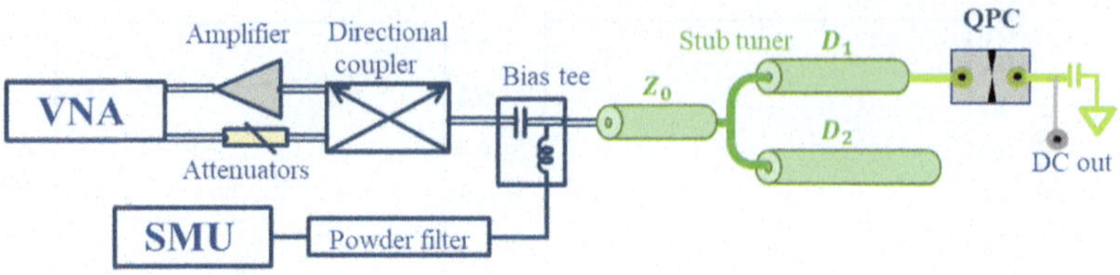

Fig. 34.3 Schematic diagram for the reflectivity measurements of QPC-stub tuner

34.4 Experiment

Reflectivity measurements are taken at 4 K as described elsewhere [11]. The measurement setup consists of a bias-tee for applying DC bias to the source-drain of QPC, followed by a directional coupler to apply a microwave signal and measure the reflection coefficient. A 50 μV AC signal is applied to the source drain of QPC, and the output is measured through a lock-in amplifier. The gate voltage is applied using a source measure unit (SMU). Reflectivity measurements (S_{11}) are performed using a vector network analyzer (VNA) at a source power of 0 dBm. The measurement setup is shown in Fig. 34.3.

34.5 Results

When a negative voltage is applied to the QPC gates, the source-drain current starts decreasing, and the channel pinches off at a voltage of − 3.28 V. The pinch-off curve of the QPC is shown in Fig. 34.4. The measured S_{11} from the QPC-stub tuner exhibits resonance at a frequency of 1.96 GHz. The reflected signal from the QPC-stub tuner varies with changes in the gate voltage of the QPC. When we change the gate voltage, the resistance of the QPC will change, which in turn affects the reflected signal from the stub tuner. The change in reflected signal (S_{11}) with gate voltage is shown in Fig. 34.4.

The shift in the resonance of the stub tuner from the designed value of 3 GHz to 1.96 GHz can be due to the extra length of the wire used for bonding the QPC chip to the stub tuner. This can be avoided by the fabrication of on-chip devices.

Fig. 34.4 **a** Pinch-off of the QPC measured at 4 K. **b** Reflection coefficient S_{11} of stub tuner with varying Gate voltage (conductance) of QPC

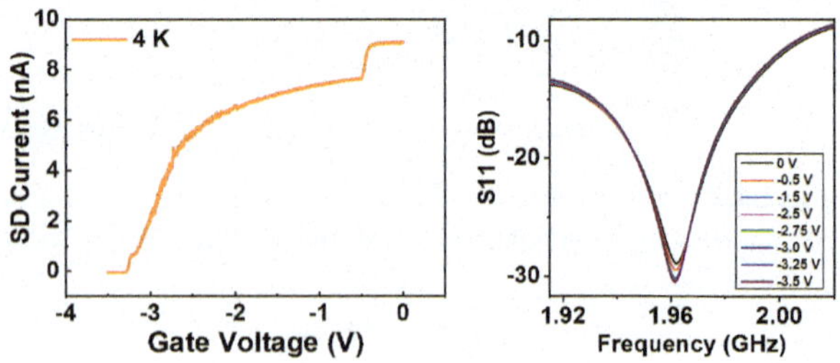

34.6 Conclusion

We have designed a stub tuner impedance matching circuit for matching the QPC impedance to that of the line for performing high-frequency measurements. We have fabricated an aluminium stub tuner for impedance matching 25 kΩ at a resonant frequency of 3 GHz. From the reflectivity measurements of the QPC-stub tuner system at 4 K, it can be seen that the reflection coefficient changes with QPC resistance. The major advantage of using a stub tuner is that resonant frequency and matched load can be easily tuned by varying the lengths of the stubs. Stub tuners can be a good replacement for low-bandwidth lumped element impedance transformers.

Acknowledgements The authors acknowledge the funding received from MoE-STARS under the grand MoE/STARS1/363.

References

1. J.M. Elzerman, R. Hanson, L.H. Willems van Beveren, B. Witkamp, L.M.K. Vandersypen, L.P. Kouwenhoven, Single-shot read-out of an individual electron spin in a quantum dot. Nature **430**, 431 (2004)
2. M. Field et al., Measurements of coulomb blockade with a noninvasive voltage probe. Phys. Rev. Lett. **70**, 1311–1314 (1993)
3. R.J. Schoelkopf, P. Wahlgren, A.A. Kozhevnikov, P. Delsing, D.E. Prober, The radio-frequency single-electron transistor (RF-SET): a fast and ultrasensitive electrometer. Science (80-). **280**, 1238–1242 (1998)
4. D.J. Reilly, C.M. Marcus, M.P. Hanson, A.C. Gossard, Fast single-charge sensing with a rf quantum point contact. Appl. Phys. Lett. **91**, 1–4 (2007)
5. N. Ares et al., Sensitive radio-frequency measurements of a quantum dot by tuning to perfect impedance matching. Phys. Rev. Appl. **5** (2016)
6. G. Puebla-Hellmann, A. Wallraff, Realization of gigahertz-frequency impedance matching circuits for nanoscale devices. Appl. Phys. Lett. **101**, 9–13 (2012)
7. S. Hellmüller et al., Optimization of sample-chip design for stub-matched radio-frequency reflectometry measurements. Appl. Phys. Lett. **101**, 4–7 (2012)
8. T. Hasler et al., Shot noise of a quantum dot measured with gigahertz impedance matching. Phys. Rev. Appl. **4**, 1–7 (2015)
9. M. David, *Pozar, Microwave Engineering*, 3rd edn. (Wiley, USA, 2005)
10. V. Ranjan et al., Clean carbon nanotubes coupled to superconducting impedance-matching circuits. Nat. Commun. **6** (2015)
11. P. Kumbhakar, A. Shanmugam, C.H. Sharma, J.L. Reno, M. Thalakulam, Quantum point contact galvanically coupled to planar superconducting resonator for ultra-sensitive broad-band electrical amplification. Quantum Sci. Technol. **6**, 045006 (2021)

Chapter 35
Spice Simulation of Solution Processed Bottom Gate Bottom Contact Organic Thin Film Transistor

Saiganesh Puttur, Fiheon Imroze, Sankaran Aniruddhan, and Soumya Dutta

35.1 Introduction

Substantial improvement in device performance of solution processed OTFTs, owning to the inherent advantages like low cost and low temperature processing along with mechanical flexibility, has shown the possibility of their potential deployment in radio-frequency identification tags, back-plane circuit of flexible displays, sensors, and biodegradable circuits. A systematic methodology to build up circuit design and circuit simulation from the device characteristics is highly desirable for establishing a self-consistent design rule. In the present work, the electrical characteristics of fabricated OTFTs, exhibiting p-type behaviour in accumulation mode, are used for circuit design.

35.2 Methods

35.2.1 Fabrication of OTFT

The solution based OTFTs consisting of poly[2,5-bis(3-tetradecylthiophen-2-yl) thieno[3,2-b]thiophene] (PBTTT-C14) as semiconductor and cross-linked poly(4-

S. Puttur (✉) · F. Imroze · S. Aniruddhan · S. Dutta
Department of Electrical Engineering, IIT Madras, Chennai 600036, Tamil Nadu, India
e-mail: saiganeshp@smail.iitm.ac.in

F. Imroze
e-mail: ee16d301@ee.iitm.ac.in

S. Aniruddhan
e-mail: ani@ee.iitm.ac.in

S. Dutta
e-mail: s.dutta@ee.iitm.ac.in

R. Singh et al. (eds.), *The Physics of Semiconductor Devices*, Springer Proceedings in Physics 306, https://doi.org/10.1007/978-981-97-1571-8_35

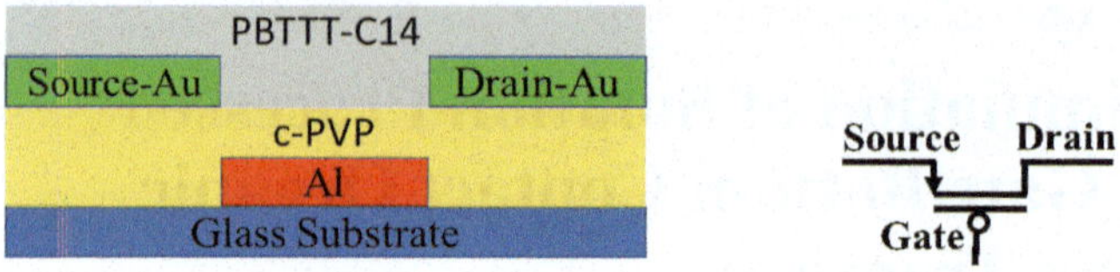

(a) Schematic of bottom-gate bottom-contact OTFT device architecture.

(b) Schematic Symbol for OTFT.

Fig. 35.1 Schematic of bottom-gate bottom-contact OTFT device architecture

vinyl phenol) (c-PVP) as polymer dielectric were fabricated in our laboratory. The output and the the transfer characteristics of the fabricated OTFTs were chosen for circuit design [1, 2]. The schematic of the device structure is shown in Fig. 35.1a. The patterned gate on the glass substrate was fabricated using photolithography process, followed by the thermal evaporation of the aluminium. A layer of c-PVP was spun on top of the gate electrode to realize the gate insulator. The patterning of the bottom contact source-drain were obtained using photolithography process, followed by thermal evaporation of gold with subsequent lift-off process. Finally, the PBTTT-C14, the polymer semiconductor, was spun inside a glove box. The schematic symbol of OTFT is displayed in Fig. 35.1b.

35.2.2 Spice Model and Parameter Extraction

In order to design a circuit, the spice device level-1 model parameters were extracted from the output and transfer characteristics of the fabricated OTFTs [3]. Since the characteristics of the OTFT are very similar to those of metal-oxide-semiconductor FET (MOSFET), Spice Level-1 model of MOSFET [4] was employed to define the characteristics of the OTFT. The level-1 model of MOSFET is described by Eq. (35.1),

$$I_{ds} = \begin{cases} 0 & , |V_{gs}| < |V_T|. \\ \frac{W}{L} K_P[(V_{gs} - V_T) - \frac{V_{gs}}{2}](1 + \lambda V_{ds}) & , |(|V_{gs}| - |V_T|)| > |V_{ds}|, |V_{gs}| \geq |V_T|. \\ \frac{W}{2L} K_P (V_{gs} - V_T)^2 (1 + \lambda V_{ds}) & , |(|V_{gs}| - |V_T|)| \leq |V_{ds}|, |V_{gs}| \geq |V_T|. \end{cases} \tag{35.1}$$

where $V_T = |\ V_{TO}\ | + \gamma(\sqrt{2\psi + |V_{SB}|} - \sqrt{2\psi})$. The parameters such as threshold voltage, process transconductance coefficient, channel length modulation parameters bearing their usual symbols were extracted from the output and the transfer characteristics of the solution processed OTFT, as tabulated in Table 35.1.

Table 35.1 Extracted parameters for level-1 spice model of MOSFET

Parameters	For L = 80 μm
$K_p(=\mu C_{ox})$(Process transconductance coefficient) AV^{-2}	8.9×10^{-11}
V_{TO}(Zero-bias threshold voltage) V	− 2.5
γ(Body-effect parameter) $V^{0.5}$	0.5
ψ(Surface potential) V	0.65
λ(Channel-length modulation parameter) V^{-1}	0.0035
$C_{gso} = C_{gdo}$ (Gate overlap capacitance) Fm^{-1}	6×10^{-10}
Rs=Rd (Source and drain resistance) Ω	1×10^{6}

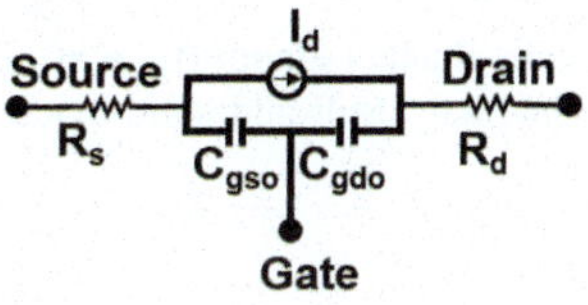

(a) Equivalent circuit model of OTFT representing overlap capacitance C_{gso} and C_{dso}, and the Contact resistance R_s and R_d.

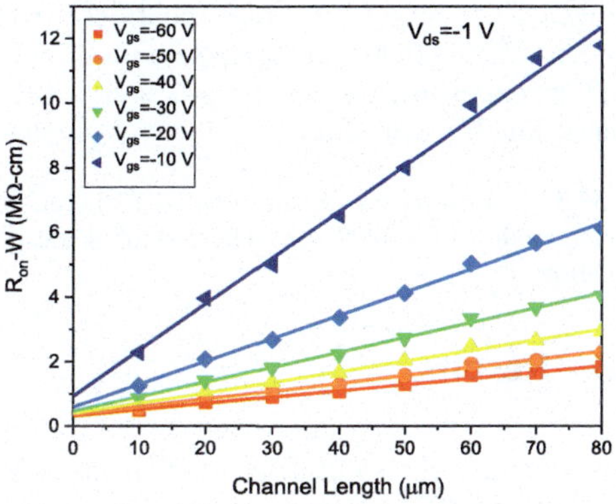

(b) Extraction of contact resistance using TLM.

Fig. 35.2 Equivalent circuit model of OTFT representing overlap capacitance C_{gso} and C_{dso}, and the Contact resistance R_s and R_d

The gate-source capacitance (C_{gso}) and gate-drain capacitance (C_{dso}), representing the gate overlap capacitance with source and drain, respectively (Fig. 35.2a), were extracted using the capacitance versus voltage (C-V) characteristics. The transfer line method (TLM) [5] was adapted to estimate the contact resistance R_s and R_d. The OTFTs with different channel lengths varying from 10 μm to 80 μm were fabricated and characterized to perform the TLM, as depicted in Fig. 35.2b. The contact resistance was extracted from the y-intercept and was divided equally to estimate source (R_s) and drain (R_d) contact resistances. The parameters like surface potential and body effect were taken as fitting parameters. The simulated results replicating the transfer characteristics (Fig. 35.3a) and the output characteristics (Fig. 35.3b) were observed to be within the acceptable range of agreement with the experimental results.

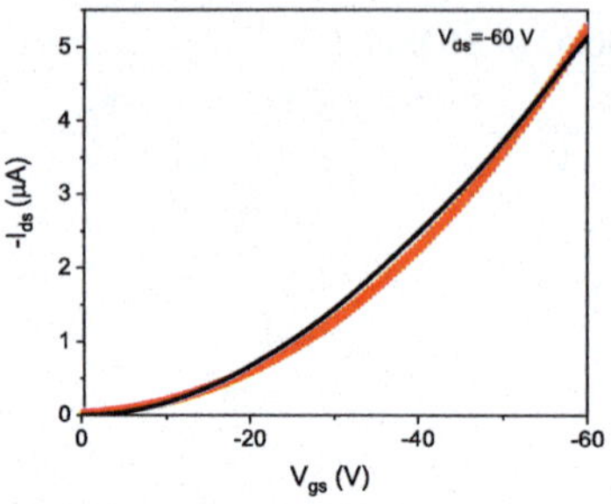

(a) Transfer characteristics of PBTTT-based OTFT with channel length (L) = 80 μm and channel width (W) = 3000 μm, where symbol red is experimental result and solid black line is simulation.

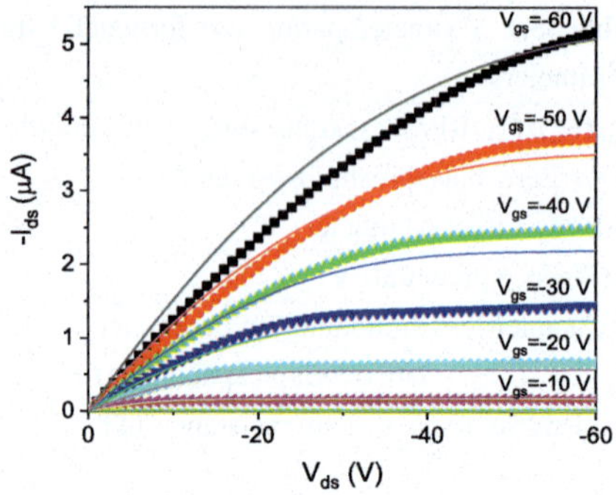

(b) Output characteristics of the same OTFT, where symbols are experimental results and solid lines are simulation.

Fig. 35.3 Transfer characteristics of PBTTT-based OTFT with channel length (L) = 80 μm and channel width (W) = 3000 μm, where symbol red is experimental result and solid black line is simulation

35.3 Results

The schematic for saturation load inverter consisting of inverting transistor with channel width W_0 and saturation load transistor with channel width W_1 are illustrated in Fig. 35.4a. The circuit simulation were performed using LTSpice XVII simulator [6]. Dual bias configuration for saturation load inverter with V_{dd}=80 V and the voltage transfer characteristics (VTC) of the inverter for various saturation load channel

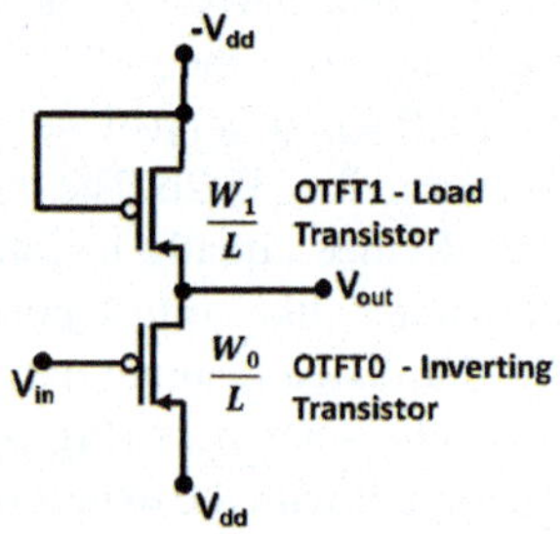

(a) Schematic of dual bias saturation load inverter.

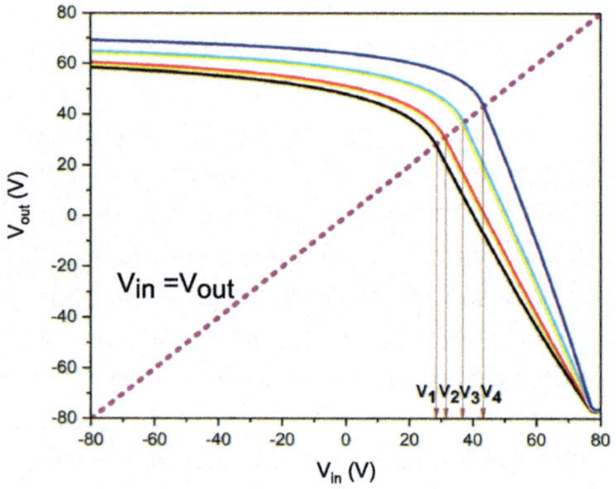

(b) Voltage transfer characteristics of the inverter corresponding to $\frac{W_0}{W_1}$ =2(black line), 3(red line), 6(green line), 12(blue line).

Fig. 35.4 Schematic of dual bias saturation load inverter

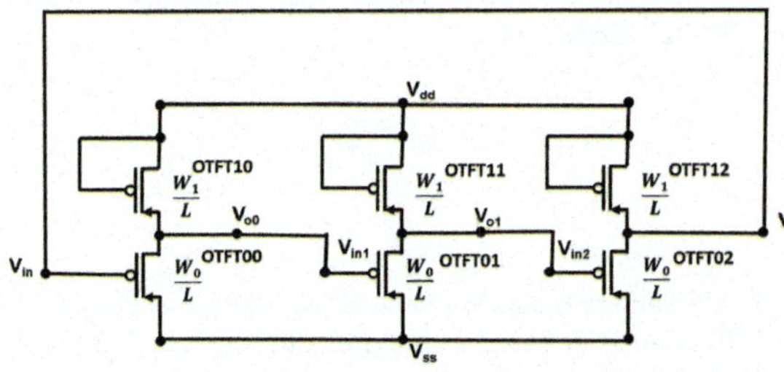

(a) Schematic of three stage ring oscillator

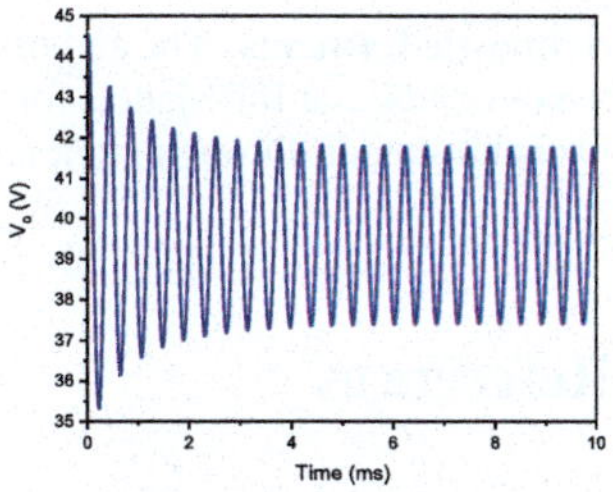

(b) The output of a three-stage ring oscillator having a frequency of 2083 Hz.

Fig. 35.5 Schematic of three stage ring oscillator

widths are shown in Fig. 35.4b. The dotted line V_{in}=V_{out} intersects the VTC curve, providing the intersection co-ordinate value as switching threshold voltage V_1, V_2, V_3, and V_4. The channel width of the inverting transistor W_0= 3000 μm was kept constant while the channel width of the load transistor was varied as W_1=1000 μm (black line), W_1= 500 μm (red line), W_1= 300 μm (green line) and W_1= 250 μm (blue line). The gain of the inverter was observed to increase with the decrease in width W_1 keeping the inverting transistor width W_0 constant. Increase in the $\frac{W_0}{W_1}$ ratio shifts the trip point to the higher voltages ($V_4 > V_3 > V_2 > V_1$), showing the higher slopes of the transition, as expected [7].

The three-stage ring oscillator was simulated using three dual bias configuration saturation load inverters connected back-to-back with $V_{dd} = -80$ V and $V_{ss} = 80$ V, as displayed in Fig. 35.5a. The ring oscillator constructed from an inverter with $\frac{W_0}{W_1} = 2$, $\frac{W_0}{W_1} = 3$ and $\frac{W_0}{W_1} = 6$ resulted in no oscillations. However, the ring oscillator consisting of an inverter with $\frac{W_0}{W_1} = 12$ resulted in oscillations with a frequency of 2083 Hz and a single-stage delay of 160 μs, as shown in Fig. 35.5b. The ring oscillator with a higher $\frac{W_0}{W_1}$ resulted in oscillations corresponding to higher gain of the inverter to generate sustained oscillations [7].

35.4 Conclusions

A systematic approach to extract the spice model level-1 parameters from the experimental data and their deployment in simulating OTFT based circuits, such as saturation load inverter and the three-stage ring oscillator are demonstrated in the present paper. This work offers a foundation for realizing the circuits using solution-based OTFTs. Since the fabrication involves photolithography for patterning gate, source, and drain, a physical layout for these simulated circuits is feasible for fabricating the OTFT based circuits.

Acknowledgements The authors would like to acknowledge the Centre for NEMS and Nanophotonics (CNNP) at IIT Madras for providing fabrication facility and DST, Government of India, for financial support through grant number CRG/2018/004984.

References

1. F. Imroze, C. Mithun, K. Logesh, P. Venkatakrishnan, S. Dutta, Effect of recessed electrodes on contact resistance in organic thin film transistor based on polymer dielectric, in *2020 Device Research Conference (DRC)*. (IEEE, 2020), pp. 1–2
2. F. Imroze, M.C. Ajith, P. Venkatakrishnan, S. Dutta, Organic thin film transistors incorporating recessed electrodes on polymer gate dielectric. IEEE Electron Device Lett. **43**(3), 434–437 (2022). https://doi.org/10.1109/LED.2022.3143075
3. IEC/IEEE test methods for the characterization of organic transistors and materials. IEC 62860:2013(E) IEEE Std 1620-2008, pp. 1–28 (2013). https://doi.org/10.1109/IEEESTD.2013.6617657
4. H. Shichman, D.A. Hodges, Modeling and simulation of insulated-gate field-effect transistor switching circuits. IEEE J. Solid-State Circ. **3**(3), 285–289 (1968)
5. D. Gundlach, L. Zhou, J. Nichols, T.N. Jackson, P. Necliudov, M.S. Shur, An experimental study of contact effects in organic thin film transistors. J. Appl. Phys. **100**(2), 024509 (2006)
6. LTspice Simulation Software. https://www.analog.com/en/design-center/design-tools-and-calculators/ltspice-simulator.html
7. E. Cantatore, E. Meijer, Transistor operation and circuit performance in organic electronics, in *ESSCIRC 2004-29th European Solid-State Circuits Conference (IEEE Cat. No. 03EX705)* (IEEE, 2003), pp. 29–36

Chapter 36
Inducing Phase Transitions in MoS_2 by Ionic Liquid Gating

Sreevidya Narayanan, Anoop Kamalasanan, and Madhu Thalakulam

36.1 Introduction

Electrolyte gating is different from conventional solid dielectric gating in the way that it consists of freely moving ions. Ionic liquids are most commonly used for electrolyte gating owing to their ultra-low vapour pressure and compatibility with high vacuum. When a droplet of ionic liquid is placed between the channel surface and the gate electrode and a voltage is applied to the gate, a strong electric field is established at the channel-liquid interface on account of the formation of an Electric Double Layer (EDL). This EDL is constituted by the rearrangement of ions on the electrolyte/channel surface as a result of the voltage applied to the gate electrode. Figure 36.1 shows the schematic of EDL formation and rearrangement of ions with the application of liquid voltage (V_{IL}).

Ions present in ionic-liquids have a diameter approximately of 2 nm. Close proximity of liquid ions to channel surface creates a huge electric field of the order of 10 MV/cm [1]. This electric field induce high carrier densities of the order of 10^{15}–10^{14} cm^{-2}, an order more than what can usually induced by high-K solid dielectrics. The carrier density can be tuned by changing V_{IL}, also making it superior to chemical doping.

Large carrier densities induced by ionic liquids enable one to explore new physics. In semiconductors of lower dimensions such as few-layered MoS_2, a semiconductor to metallic character can be induced by high-K solid dielectrics, but taking the system to a superconducting state requires even higher carrier densities, by an order, which can only be achieved by ionic liquid gating. In addition, a superconducting transition in potassium tantalite [1], transforming a low-carrier paramagnetic oxide into a high-carrier ferromagnet at room temperature [2] etc. are also possible by EDL gating.

S. Narayanan · A. Kamalasanan · M. Thalakulam (✉)
Indian Institute of Science Education and Research, Thiruvananthapuram, Kerala 695551, India
e-mail: madhu@iisertvm.ac.in

R. Singh et al. (eds.), *The Physics of Semiconductor Devices*, Springer Proceedings in Physics 306, https://doi.org/10.1007/978-981-97-1571-8_36

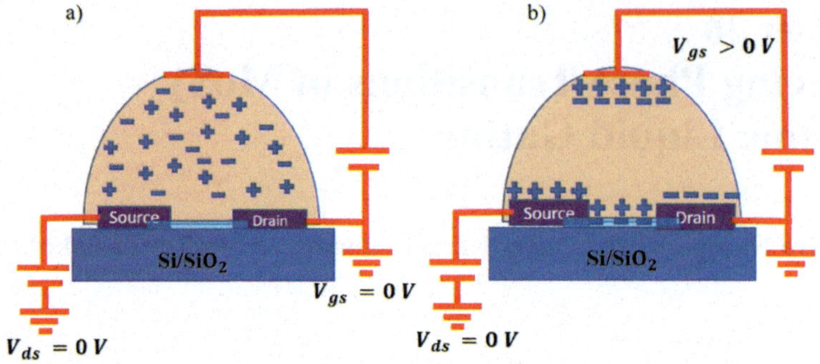

Fig. 36.1 Illustration of ionic liquid gating technique showing rearrangement of cations and anions **a** without any applied voltage and **b** positive voltage applied to gate electrode

Two commonly used ionic liquids for transport studies are EMIM-TFSI (1-Ethyl-3-methylimidazolium bis (trifluoromethylsulfonyl)imide) and DEME-TFSI (*N*,*N*-diethyl-*N*-methyl-*N*-(2-methoxyethyl) ammonium bis (trifluoromethanesulfonyl) imide). The ionic liquid used in our studies is DEME-TFSI which has been popularly used due to its wide electrochemical window which enables a wider range of V_{IL} [3, 4]. Our channel material, Molybdenum Disulphide (MoS_2), belongs to the family of transition metal dichalcogenides (TMDCs) and is a semiconducting layered material. Mechanically exfoliated flakes have a high degree of crystallinity with an atomically flat and dangling bond-free surface. Its mechanical flexibility and transparent nature make it a potential candidate for transparent electronic devices. MoS_2 is an n-type semiconductor, mostly due to defects such as Sulphur vacancies. EDL gating has allowed ambipolar character on MoS_2 by tuning the liquid-gate voltage [2]. The ultra-high electric field due to the EDL gating strongly confines the carriers to the layers adjacent to the ionic liquid. Therefore, the transport on a few-layered MoS_2 approaches the monolayer limit irrespective of the number of layers, under EDL gating [5]. Semiconducting-metal-superconducting transitions are also achieved as a result of carrier density changes induced by ionic liquid voltage [3]. Here we report an insulator to metal transition and the appearance of a re-entrant insulating phase on a few layered MoS_2 under EDL gating. We estimate a thickness corresponding to ~ 4–5 layers using the optical contrast method [6].

36.2 Device Fabrication

Few-layered MoS_2 flakes are mechanically exfoliated from the bulk and further transferred onto Si/SiO_2 wafers using the dry transfer technique. Thickness is estimated using the optical contrast method [6]. Next, the sample is patterned by photolithography followed by Cr (3 nm)/Au (30 nm) metallization. The second lithography is

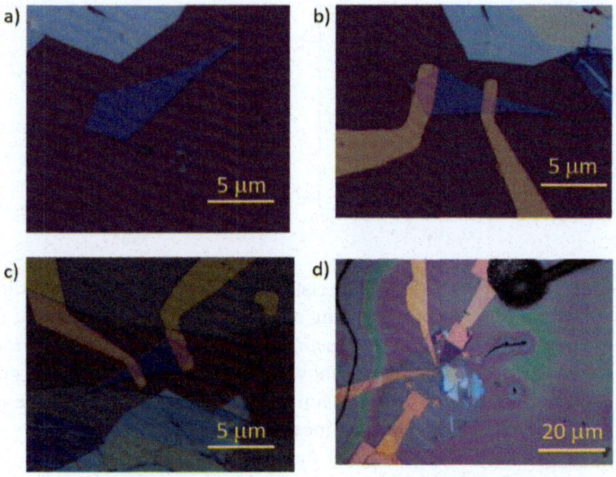

Fig. 36.2 Steps involved in Ionic liquid gating **a** MoS_2 on Si/SiO_2 substrate. **b** Patterning the sample by photolithography. **c** After metallization and lift off. **d** After dropping ionic liquid

done to selectively open windows for the gate electrode and channel surface to minimize the interaction of ions in the liquid with the contact metal surface. The ionic liquid DEME-TFSI is carefully dropped onto the channel and the liquid gate. In order to remove any moisture content, the ionic liquid is baked at 120 °C under vacuum for approximately 12 h. After dropping ionic liquid, the device is quickly mounted onto a 4 K cryocooler. Figure 36.2 shows the optical image of various stages of the device fabrication.

36.3 Experiment and Results

Before applying the V_{IL}, the operational voltage range is characterized by acquiring the electrochemical window. Figure 36.3 shows the electrochemical window of the ionic liquid DEME-TFSI. There is a sudden increase of current after 4.5 V suggesting the safe operating voltage is between 0 and 4 V.

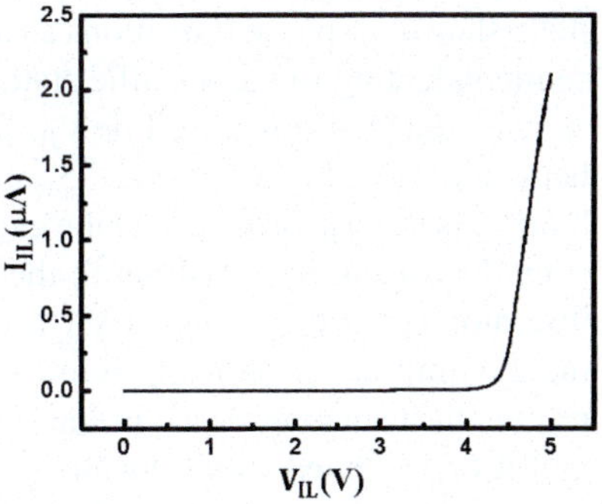

Fig. 36.3 Current–voltage characteristics of the ionic liquid DEME–TFSI at room temperature

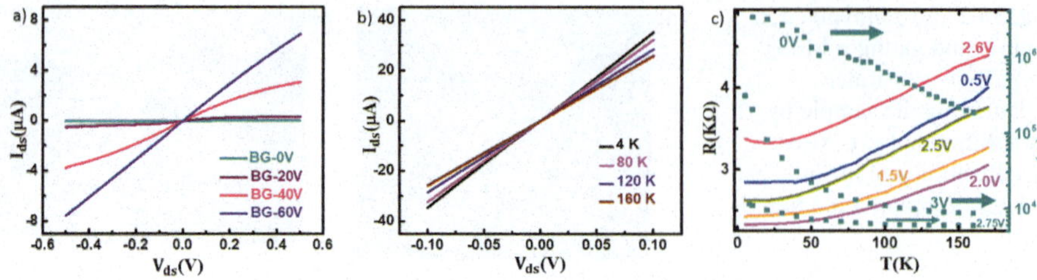

Fig. 36.4 Transport characteristics. **a** I–V characteristics of the device as a function of V_{BG} before dropping ionic liquid, showing an non-linear behavior. **b** I–V characteristics after applying a gate voltage of 0.5 V, showing linear behavior. **c** The R–T behavior of the device starting from 150 K down to 4 K for various V_{IL} showing the re-entrant insulating phase as the V_{IL} is increased. Right—axis (log scale, marked in green colour) shows the resistance in the insulating phase, R–T plot of which is indicated by dotted lines (green colour). Green arrows points to the right (green) axis which is represented in log scale for the insulating regions

Two-probe measurement of the fabricated device is done in order to study the Current–Voltage (I–V) and Resistance–Temperature (R–T) characteristics. The measurements are performed in a 4 K cryo-cooler. Figure 36.4 summarizes the electrical characterizations. Before dropping the ionic liquid, I–V characteristics of MoS_2 shows normal semiconducting behavior with a resistance of 10 MΩ at room temperature. The device shows an n-type characteristics; source–drain current (I_{ds}) increases with back gate voltage (V_{BG}) as shown in Fig. 36.4a. As V_{IL} is increased to 0.5 V, device shows a metallic character with linear I–V with no back-gate response, Fig. 36.4b. Resistance decreased as V_{IL} increases up to 2.0 V. Device shows more metallic behavior as V_{IL} increases, as shown in Fig. 36.4c. When V_{IL} becomes 2.75 V, device starts showing insulating behavior. This re- entrant insulating phase is surprising as the carrier density is expected to increase with V_{IL}.

36.4 Conclusions

Two probe temperature-dependent transport measurements of few-layered MoS_2 flake show two phase transitions as a function of V_{IL}. The first transition is from the semiconducting to the metallic state. Without applying liquid voltage device shows semiconductor behaviour. For V_{IL} 0.5 V through 2.6 V, the device shows metallic behaviour. For higher V_{IL} (2.75 and 3.0 V) the metallic state turns into an insulating. As the V_{IL} is increased the carrier concentration is expected to increase which is evident from the transition to the metallic state. But the reason for the metal-to-insulator transition for higher V_{IL} is unclear. It may be because the system is reaching the 2D limit and is showing some anomalous behaviour at high carrier densities. A re-entrant insulating phase in ionic liquid gated monolayer WS_2 has been reported and it is attributed to the localization of induced carriers due to weak monolayer screening in the high gating limit [7]. The exact model for these observations is open to pursuit. Our experiments are confined down to a temperature of 4 K. At lower temperatures, the system can go from metallic to superconducting behaviour. The two-dimensional (2D) superconductivity is of high academic interest. Superconducting devices operating in the 2D limit are also technologically interesting for applications such as quantum electrical quantum information technologies. Ionic liquids have limited electrochemical window. Ionic gels, a mixture of ionic liquid and a polymer mixed in a suitable medium of solvent, are proposed to overcome the limitations of ionic liquid gating [8, 9]. Ionic gels raises the electrochemical stability of the electrolyte gates without varying any of its properties.

Acknowledgements The authors acknowledge the funding received from SERB, Govt of India under the grant CRG/2018/004213.

References

1. K. Ueno et al., Discovery of superconductivity in $KTaO_3$ by electrostatic carrier doping. Nat. Nanotechnol. **6**, 408–412 (2011)
2. S.W. Stanwyck, Electrolyte gating of nanoscale nanoscale devices. A dissertion submitted to the department of Applied physics and the committee on graduate studies of Stanford university in partial fulfillment of the requirements for the degree of doctor of philosophy (2017), http://purl.stanford.edu/xy317dg0262
3. J. T. Ye et al, Superconducting dome in a gate-tuned band insulator, https://www.science.org.
4. Y. Zhang, J. Ye, Y. Matsuhashi, Y. Iwasa, Ambipolar MoS_2 thin flake transistors. Nano Lett. **12**, 1136–1140 (2012)
5. J.M. Lu et al, Evidence for two-dimensional Ising superconductivity in gated MoS_2, https://www.science.org
6. A. Varghese, C.H. Sharma, M. Thalakulam, Topography preserved microwave plasma etching for top-down layer engineering in MoS_2 and other van der Waals materials. Nanoscale **9**, 3818–3825 (2017)
7. J. Lu et al., Full superconducting dome of strong Ising protection in gated monolayer WS2. Proc Natl Acad Sci USA **115**, 3551–3556 (2018)
8. J. Seidemann, Iontronics-field effect study of different devices, using techniques of ionic liquid gating, https://tel.archives-ouvertes.fr/tel-01759252
9. L. Chu et al., Charge transport in ion-gated mono-, bi-, and trilayer MoS_2 field effect transistors. Sci Rep. (2014). https://doi.org/10.1038/srep07293

Chapter 37
Ultrasensitive Reduced Vanadium Dioxide-Based MEMS Pirani Gauge with Extended Dynamic Range

Manu Garg, Khanjan Joshi, Dhairya S. Arya, Sushil Kumar, Mujeeb Yousuf, Ankur Goswami, and Pushpapraj Singh

37.1 Introduction

Micromachined Pirani gauges are being successfully deployed for in-situ vacuum characterization of rough to medium vacuum. The ability to detect in-package vacuum in real-time with high precision and reliability makes them a prominent choice over their conventional counterparts such as helium leak test and Q-factor extraction techniques [1–5]. Additionally, the heterogeneous integration offers seamless calibration of microsensor output over time.

Existing implementations of micromachined Pirani gauge are focused on scaling laws and increasing the gaseous conduction regime [6]. The scaling laws are implemented by increasing the length of the sensing element while decreasing the width and thickness [6]. The gaseous conduction regime is improved by reducing solid conduction and radiation losses associated with the sensing element [7]. Further, the width should be increased to increase the sensing area for gaseous conduction however, this results in increased solid conduction loss [6, 7]. This trade-off between the gaseous and solid conduction limits the sensitivity and operating range of a typical Pirani gauge. Multiple heat sinks and caping wafer around the sensing element are also introduced to increase the sensitivity of the Pirani gauge [8, 9]. Another crucial parameter linked to solid conduction is the effective thermal conductivity of the sensing element. Integration of lower thermal conductivity materials with the

M. Garg · K. Joshi · D. S. Arya · S. Kumar · M. Yousuf · P. Singh (✉)
Centre for Applied Research in Electronics (CARE), Indian Institute of Technology Delhi, New Delhi 110016, India
e-mail: prsingh@care.iitd.ac.in

A. Goswami
Department of Materials Science and Engineering (DMSE), Indian Institute of Technology Mandi, Kamand, Himachal Pradesh 175075, India

R. Singh et al. (eds.), *The Physics of Semiconductor Devices*, Springer Proceedings in Physics 306, https://doi.org/10.1007/978-981-97-1571-8_37

sensing element can reduce the effective thermal conductivity of the sensing element and hence solid conduction [10].

Metal oxides are prominent materials for electronic applications like memories, transistors, biosensors, and photodetectors [7, 11]. Moreover, metal oxides exhibit high oxidization resistance and are thermally stable. In particular, vanadium dioxide (VO_2) is actively being explored in the aforementioned applications. VO_2 is a transition metal oxide that undergoes a phase transition at a critical temperature (T_c) of 340 K [12, 13]. Below this T_c, VO_2 exhibits an insulating monoclinic phase and exhibits a high resistance region. This region is useful in Pirani gauge application where higher resistance infers lower power consumption to heat the sensing element. Further, the lower thermal conductivity of VO_2 is beneficial in minimizing solid conduction loss and the high-temperature coefficient of resistance offers high sensitivity at lower power consumption.

In this paper, we propose an ultrasensitive VO_2 based MEMS Pirani gauge for *in-situ* vacuum characterization of packaged devices. The device is operated below the critical temperature of VO_2 phase transition i.e., in the insulating monoclinic phase. A post-CMOS compatible fabrication process flow is proposed for fabricating the device. Simulations are carried out in COMSOL Multiphysics and the gauge can be employed for vacuum characterization from 0.1 to 10^5 Pa.

37.2 Design and Description

The schematic illustration of VO_2 based MEMS Pirani gauge is shown in Fig. 37.1. VO_2 is the active element in the sensing element, whereas Si_3N_4 provides mechanical stability to the suspended structure. A constant current is applied between the source and drain terminals which in turn heats the sensing element due to the Joule heating phenomenon. The overheated temperature ($T_{avg} - T_{ref}$) of the sensing element depends on the pressure in the ambient. Mathematically, the average overheat in temperature (ΔT) across the membrane is given as [14]:

$$\Delta T = \frac{\delta}{\varepsilon}\left(1 - \frac{\tanh\left(\frac{\sqrt{\varepsilon(p)}l}{2}\right)}{\left(\frac{\sqrt{\varepsilon(p)}l}{2}\right)}\right) \tag{37.1}$$

where δ is the ohmic heat generation due to Joule heating, $\varepsilon(p)$ is the pressure-dependent heat loss, and l is the length of the sensing element. Further, δ and $\varepsilon(p)$ are given as [14]:

$$\delta = \frac{I_{bias}^2 R_o}{\lambda_s \cdot w \cdot l \cdot z} \tag{37.2}$$

$$\varepsilon = \frac{\eta \lambda_{gas}(P)}{\lambda_s \cdot g \cdot z} \tag{37.3}$$

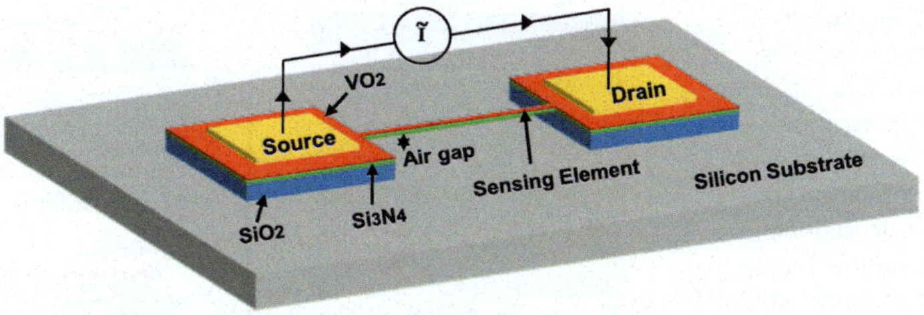

Fig. 37.1 Schematic of VO_2 Pirani gauge biased in constant current mode indicating the associated heat loss mechanisms

where I_{bias} is the applied current bias, R_0 is the nominal resistance of the VO_2 film at $T_{ref} = 293.15$ K, λ_s is the thermal conductivity of VO_2, w and z are the effective width and thickness of the sensing element respectively, η is the correction factor incorporating fringing heat flux, $\lambda_{gas}(P)$ is the pressure-dependent thermal conductivity of gas, and g is the suspended gap. Considering the 1D analytical model presented in [14], the heat balance equation can be given as:

$$\frac{\partial^2 t(x)}{\partial x^2} + \delta - \varepsilon t(x) = 0 \quad (37.4)$$

where, $t(x)$ is the temperature distribution across the sensing element and x is the length coordinate of the sensing element. Corresponding to the overheat temperature as per Eq. (37.1), the output voltage (V_P) is given as:

$$V_P = I_{bias} \cdot R_o(1 + \alpha(\Delta T)) \quad (37.5)$$

where α is the temperature coefficient of resistance.

37.3 Proposed Fabrication Process Flow

The fabrication of reduced vanadium dioxide-based Pirani gauge is a two-step lithography process (Fig. 37.2). Firstly, a Si/SiO_2 wafer with 1 μm of SiO_2 is taken as a starting substrate. This layer of SiO_2 acts as a sacrificial layer for the formation of a suspended filament of reduced VO_2. Now, 200 nm of silicon nitride is deposited using RF sputtering which is then followed by a deposition of 15 nm of vanadium. The Silicon nitride layer provides mechanical strength to the VO_2 layer and prevents it from vapor hydrofluoric (VHF) acid during the releasing process. The sample is then annealed for 20 min to form the vanadium pentoxide in a three-zone (zone 1@450 °C, zone 2@475 °C, zone 3@450 °C) furnace. The oxidation is carried out at standard pressure (1 atm) with an oxygen flow rate of 1 sccm. Vanadium pentoxide is the highest oxidation state of vanadium and therefore further oxidation is not possible. Now, the sample is vacuum annealed at 500 °C for reduction of V_2O_5 to form VO_2

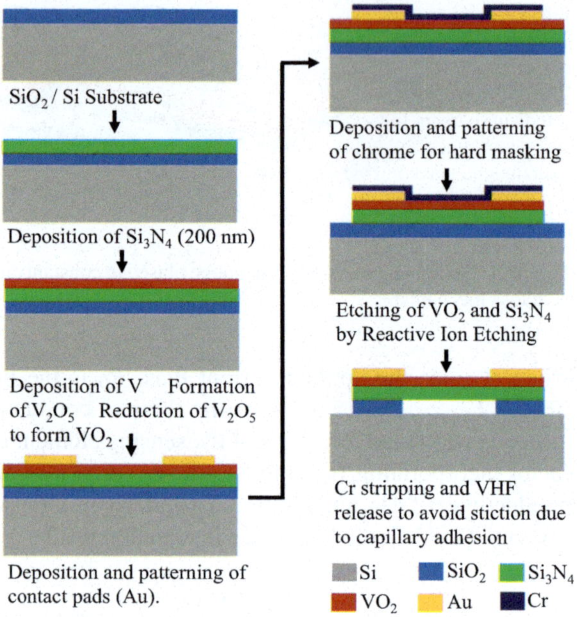

Fig. 37.2 Fabrication process flow of the VO_2 based MEMS Pirani gauge. The overall fabrication process is a two-step lithography process. The structure is released using vapor hydrofluoric acid to avoid stiction due to capillary forces. Cr metal is used as a hard mask for patterning the sensing element and to prevent the film form vapor hydrofluoric acid

for 20 min using rapid thermal annealing. Afterward, Cr/Au (20 nm/100 nm) is lifted off at the anchor areas to form the contact pads. An additional layer of Cr is deposited and patterned to form the beam structure. This layer of Cr serves as a masking layer to etch the unwanted VO_2 and Si_3N_4. Reactive Ion Etching (RIE) is then performed down to SiO_2 to form the beam structure. Finally, the structure is released using VHF to avoid stiction due to capillary forces. It can be noted that the Cr masking layer is still present to protect the VO_2 layer from VHF which is finally removed after suspension using RIE.

37.4 Results and Discussion

The electrothermal FEM analysis of the gauge is carried out using COMSOL Multiphysics®. The base of the substrate is set at T_{ref}, as the boundary condition. Also, the Multiphysics coupling between heat transfer and the AC/DC module is done to demonstrate the effect of Joule heating on the conduction through the sensing element. Figure 37.3 shows the normalized variation of resistivity with the temperature of a typical VO_2 film [12]. The curve is fitted with the sigmoidal Boltzmann function as per the equation given below:

$$\rho = \frac{0.01959}{1 + e^{\left(\frac{T-49.23825}{5.31544}\right)}} \quad (37.6)$$

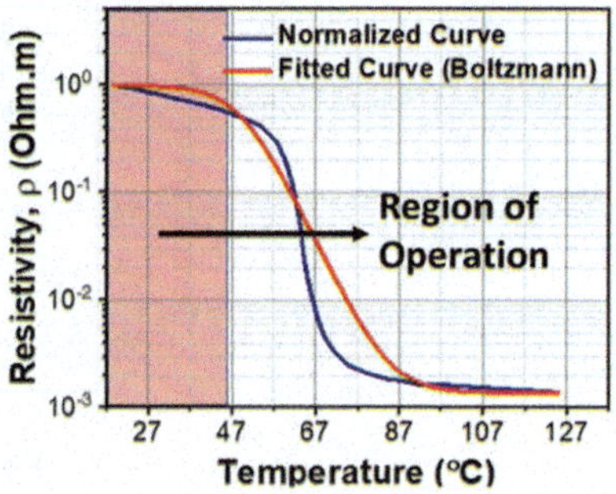

Fig. 37.3 Temperature-dependent resistivity of VO_2 (Blue). The red curve indicates the fitted curve using the sigmoidal function via the Boltzmann curve. Fitting accuracy is 94%

The red curve shown in Fig. 37.3 shows the fitted resistivity variations with the temperature at a fitting accuracy of 94%. Based on the relation given in Eq. 37.6, the resistivity is appended to the material properties of VO_2 in the COMSOL material library.

Further, the pressure-dependent environment is created by sealing the device in the air cavity whose thermal conductivity depends on the pressure inside the cavity. The device is biased at 2 mA between source and drain. This value of biasing current is selected in conjunction with the region of operation given in Fig. 37.3. Figure 37.4 shows the simulated temperature profile over the entire length of the beam for a pressure range of 0.1–10^5 Pa. The temperature profile follows the Gaussian distribution with a maximum temperature of ~ 40 °C at the center of the beam at 0.1 Pa. It can be noted that this temperature is within the specified region of operation for the worst-case scenario. Any increase in cavity pressure will decrease the value of ΔT as per Eq. 37.1 and reflect a lower temperature of the sensing element, thereby limiting the ΔT in the specified region of operation. Eminently, the suspended filament experiences a maximum change in temperature at a lower pressure range due to the lesser number of molecules available for gaseous conduction. Also, the heating temperature is well within the safe limit to avoid thermal interaction with the onboard electronics [15].

Figure 37.5 shows the calibration curve (S-shaped) of the proposed VO_2 based MEMS Pirani gauge for a pressure range of 0.1–10^5 Pa. The change in output voltage variations is plotted from Fig. 37.6 to check the useful detection range of the proposed gauge as per the relation given below:

$$V_L(p) = V_{max} - V_p \tag{37.7}$$

$$V_H(p) = V_p - V_{max} \tag{37.8}$$

where, $V_L(p)$ and $V_H(p)$ are the output voltage variations corresponding to lower and upper detection limits respectively. Considering the 50 μV detection limit of

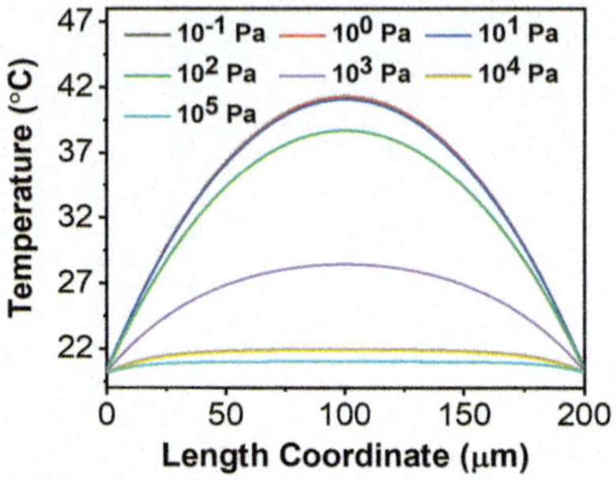

Fig. 37.4 The temperature profile of the sensing element along the length of the sensing element. The maximum temperature in the worst-case scenario (~ 0.1 Pa) is ~ 40 °C. The average temperature at this temperature is 40 °C which follows the safe limit to avoid thermal interaction with onboard electronics

Fig. 37.5 Calibration curve i.e., output voltage versus pressure of the proposed VO_2 based MEMS Pirani gauge for the pressure range of 0.1–10^5 Pa

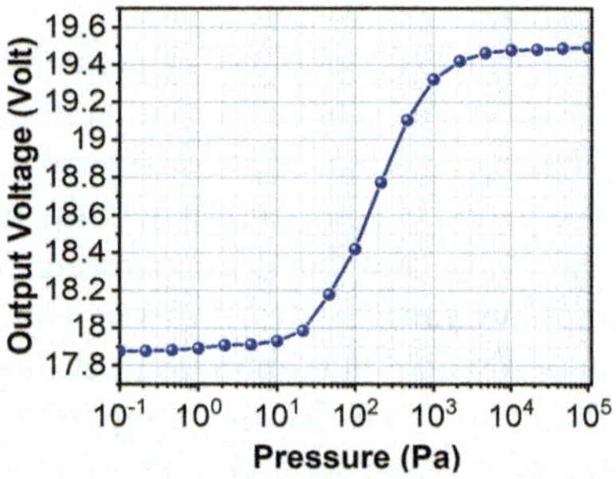

Fig. 37.6 Change in output voltage variations for extracting useful operating range. Considering the 50 μV detection limit of the measuring instrument, the gauge can be employed for vacuum detection of 0.1–10^5 Pa

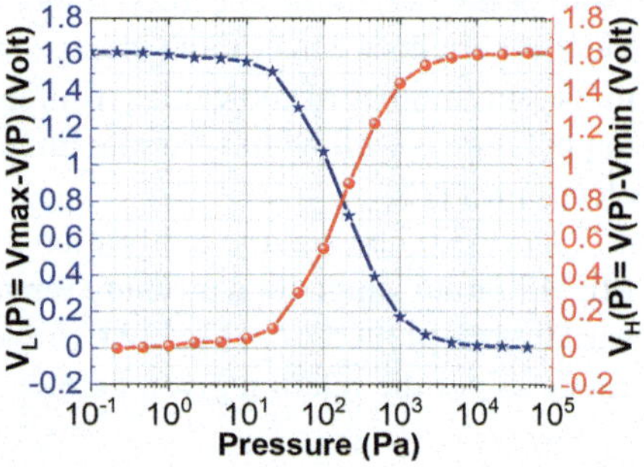

the measuring instrument, the gauge can be employed for vacuum detection of 0.1–10^5 Pa.

37.5 Conclusions

The VO_2 based MEMS Pirani gauge is simulated and a successful application for vacuum detection of packaged electronics devices is presented. The monoclinic phase operation of VO_2 provides an ultrasensitive detection at a biasing current of 2 mA. The post-CMOS compatible two-step lithography fabrication process is proposed for manufacturing the device. Considering the 50 μV detection limit of the measuring instrument, the gauge can be employed for vacuum detection of $0.1–10^5$ Pa.

Acknowledgements The authors would like to acknowledge the Nanoscale Research Facility (NRF) at the Indian Institute of Technology New Delhi, India, 110016 for proving the fabrication and characterization facilities.

References

1. Y.C. Chen, W.C. Lin, H.S. Wang, C.C. Fan, K.C.H. Lin, B.C. Chou, M.C.M. Liu, in *Differential Micro-Pirani gauge for monitoring MEMS wafer-level package*. 2015 28th IEEE International Conference on Micro Electromechanical Systems (MEMS), pp. 89–92. IEEE (2015)
2. S.-W. Cheng, J.-C. Weng, H.-C. Hsu, Y.-C. Sun, Y.-C. Chen, W. Fang, in *Modulate the Chamber Pressure of the Hermetic Sealed MEMS Device by Varying the Cavity Depth of Cap Si*. 2015 IEEE Sensors (IEEE, 2015), pp. 1–4
3. C.-W. Cheng, C.-H. Chu, L.-M. Hung, W. Fang, in *12 Inch MEMS Process for Sensors Implementation and Integration*. 2017 19th International Conference on Solid-State Sensors, Actuators and Microsystems (TRANSDUCERS) (IEEE, 2017), pp. 402–405
4. K.-C. Liang, C.-W. Cheng, C.-H. Lin, W. Fang, in *A Novel Low Pressure Sensor With Fin-Structures*. SENSORS (IEEE, 2012), pp. 1–4
5. X. Wang, C. Liu, Z. Zhang, S. Liu, X. Luo, A micro-machined Pirani gauge for vacuum measurement of ultra-small sized vacuum packaging. Sens. Actuators A **161**(1–2), 108–113 (2010)
6. J. Zhang, W. Jiang, J. Zhou, X. Wang, in *A Simple Micro Pirani Vasuum Gauge Fabricated by Bulk Micromachining Technology*. TRANSDUCERS 2009–2009 International Solid-State Sensors, Actuators and Microsystems Conference (IEEE, 2009), pp. 280–283
7. M. Garg, D.S. Arya, S. Sharma, S. Kumar, W. Uddin, S. Das, Y. Chiu, P. Singh, Highly responsive metal oxide (V_2O_5)-based NEMS Pirani gauge for in-situ hermeticity monitoring. J. Microelectromech. Syst. **30**(3), 340–342 (2021)
8. Y. Kong, B. Jiao, L. Zhang, S. Yun, D. Chen, in *Design and Fabrication of Wafer-Level Packaged MEMS Pirani Gauge With Sorrounded Heat Sinks*. 2017 19th International Conference on Solid-State Sensors, Actuators and Microsystems (TRANSDUCERS) (IEEE, 2017), pp. 950–953
9. L.-M. Zhang, B.-B. Jiao, S.-C. Yun, Y.-M. Kong, D.-P. Chen, Investigation and optimization of Pirani vacuum gauges with monocrystal silicon heaters and heat sinks. J. Microelectromech. Syst. **26**(3), 601–608 (2017)
10. M. Garg, D.S. Arya, S. Kumar, Y. Chiu, P. Singh, Reducing solid conduction in electrothermally driven MEMS Pirani gauge using integrated polymeric thin film. Appl. Phys. Lett. **120**(8), 084101 (2022)
11. R. Santos, J. Loureiro, A. Nogueira, E. Elangovan, J.V. Pinto, J.P. Veiga, T. Busani, E. Fortunato, R. Martins, I. Ferreira, Thermoelectric properties of V_2O_5 thin films deposited by thermal evaporation. Appl. Surf. Sci. **282**, 590–594 (2013)

12. R. McGee, A. Goswami, S.A.M. Bukhari, L. Zhou, K. Shankar, T. Thundat, Fabrication of phase change microstring resonators via top down lithographic techniques: incorporation of VO_2/TiO_2 into conventional processes. J. Microelectromech. Syst. **28**(5), 766–775 (2019)
13. K. Miyazaki, K. Shibuya, M. Suzuki, H. Wado, A. Sawa, High temperature coefficient of resistance of low-temperature-grown VO_2 films on TiO_2-buffered SiO_2/Si (100) substrates. J. Appl. Phys. **118**(5), 055301 (2015)
14. C.H. Mastrangelo, R.S. Muller, Microfabricated thermal absolute-pressure sensor with on-chip digital front-end processor. IEEE J. Solid-State Circuits **26**(12), 1998–2007 (1991)
15. B. Xiao, T. Dong, E. Halvorsen, Z. Yang, Y. Zhang, N. Hoivik, D. Gu, N.M. Tran, H. Jakobsen, Integrated micro Pirani gauge based hermetical package monitoring for uncooled VO $\times$ bolometer FPAs. Microsyst. Technol. **17**(1), 115–125 (2011)

Chapter 38
Lower Drying Temperature Process for Hole Transport Layer PEDOT: PSS in PCDTBT: PCBM Devices

Madhu Rawat and S. Sundar Kumar Iyer

38.1 Introduction

Organic photovoltaic (OPV) devices have made continuous and impressive improvements in power conversion efficiency (PCE) in the last several years by use of appropriate and novel material layers for building the device [1–10]. While these reported results are indicative the potential of OPV, in many practical applications, the choice of the layers in building solar cells are based on their compatibility with manufacturing processes and the device stability requirements for the application. Here devices with poly (3, 4-ethylene dioxythiophene):poly (styrene sulfonate)—popularly PEDOT:PSS—as hole transport layer (HTL); with a blend layer of donor poly[N-9′-heptadecanyl-2,7-carbazole-alt-5,5-(4′,7′-di-2-thienyl-2′,1′,3′-benzothia diazole)] and acceptor phenyl-C61-butyric acid methyl ester—popularly referred to as PCDTBT:PCBM—as the photoactive layer (PAL) are considered.

In literature a number of HTLs are reported including MoO_3, Fe_3O_4, NiO and WO_3 [11–14], which are inorganic metal oxides. Many of these HTLs exhibit excellent stability, dominant hole collection efficiency, and electron-blocking capability. However, these inorganic HTL materials require vacuum deposition process for incorporating them in devices, which is not typically suitable for economical and large-scale fabrication with roll-to-roll printing processes. This is because, the

M. Rawat · S. S. K. Iyer (✉)
National Centre for Flexible Electronics, Indian Institute of Technology Kanpur, Kanpur 208016, India
e-mail: sskiyer@iitk.ac.in

Samtel Centre for Display Technologies, Indian Institute of Technology Kanpur, Kanpur 208016, India

S. S. K. Iyer
Department of Electrical Engineering, Indian Institute of Technology Kanpur, Kanpur 208016, India

R. Singh et al. (eds.), *The Physics of Semiconductor Devices*, Springer Proceedings in Physics 306, https://doi.org/10.1007/978-981-97-1571-8_38

mentioned inorganic oxides are not easily solubilized in common organic solvents for solution-processing. PEDOT:PSS which is soluble in water has emerged as a popular choice for solution processable HTL for organic photovoltaic cell (OPV). The work function of PEDOT:PSS is −5.10 eV which gives better compatibility with the highest occupied molecular orbital level for many donor materials [15–30]. The water soluble and economical PEDOT:PSS as an HTL layer is also attractive due to its low toxicity, high transparency, high conductivity and suitability for roll-to-roll device fabrication [31–33].

One widely studied PAL for OPV devices that could potentially provide reasonable stability and lifetime for practical applications is PCDTBT:PCBM [34, 35]. These PALs could, however degrade when subjected to temperatures above 70 °C [36] during post PAL device processing. Thus, combining PCDTBT:PCBM PAL and water solution based PEDOT:PSS as HTL on top of it can result in sub-optimal PCE as the latter layer is typically dried at 100 °C during its formation.

Inverted OPV device structure glass-substrate/ITO/ZnO/PCDTBT:PCBM/PEDOT:PSS/Ag/glass-encapsulation layers was considered for studying the effect of drying temperature of the PEDOT:PSS HTL layer on the device performance. Three sets of complete devices and stacks (without the final Ag and encapsulation layers) were considered—devices with HTL layers formed at (i) 100 °C drying in lab ambient; (ii) 70 °C drying in lab ambient; and (iii) 70 °C drying in vacuum. The measurements results and parameters extracted from devices or stacks subjected to the three different PEDOT:PSS drying process mentioned above are qualified in the rest of this paper with "100 °C_PEDOT:PSS", "70 °C_PEDOT:PSS" and "70 °C_PEDOT:PSS (vacuum)" respectively. Since PEDOT:PSS layers are typically dried at 100 °C, 100 °C_PEDOT:PSS may be considered the control samples against which the other two splits may be compared.

38.2 Experimental Section

38.2.1 Device Fabrication

For the experiments, PEDOT:PSS (AI4083) was procured from Heraeus Clevios™; PCDTBT (1-Material) and PCBM from NanoC; Zinc acetate dehydrate $Zn(CH_3COO)_2{\cdot}2H_2O$ and Iso propyl alcohol (IPA), ethalonamine, and O-Xylene from Sigma-Aldrich. These were used without further purification. The chemicals H_2O_2 and NH_4OH were procured from a local vendor.

Commercially available glass with indium doped tin oxide (ITO) layers were used as the starting substrate for the fabrication. The sheet-resistance of transparent ITO layer was 10–15 Ω/sq. For every substrate the ITO layer was patterned using photolithography technique to form final shape of the cathode and the external contact point for the devices to be fabricated. The patterned substrate was then cleaned using de-ionized water, hydrogen peroxide, and ammonium hydroxide (ratio 4:1:1) at 70 °C for 15 min. The cleaned substrate was first dried by nitrogen gas, followed by exposure to UV-ozone plasma for 10 min. The zinc oxide (ZnO) layer was then formed on the substrate with the sol–gel technique. The ZnO precursor (0.384 g/5 ml in IPA with 50 μL ethalonamine) was spin coated at 2500 rotation per minute (rpm) for 1 min. Following the coating, the layer was annealed immediately on a preheated hot-plate at 150 °C for 10 min. to form the ZnO layer. The substrate with ZnO layer was then transferred to an N_2 glove box for the deposition of other layers.

The photoactive layer was deposited onto the ZnO layer by spin coating 1500 rpm with a solution containing PCDTBT-PCBM (1:2 wt. ratio) dissolved in o-xylene solvent. A maximum concentration of 18 mg/ml was used in the solution, which was stirred overnight in nitrogen atmosphere at 70 °C. The solution was filtered using a 0.45 μm polyvinylidene difluoride (PVDF) syringe filter. The spin coating was carried out at 1500 rpm. Subsequently, the film stack was annealed at 70 °C for 10 min on the hot plate. This produced a uniform photoactive layer on the substrates of thickness 101 ± 1.73 nm.

DI water solutions of PEDOT:PSS as HTL was then spin coated at 5000 rpm on the PAL. As per the experiment planned, two sets of film stacks were annealed on a hotplate—at 100 °C or at 70 °C. One set of film stacks were annealed in a specially designed vacuum annealing chamber attached to the glove box with a base pressure of ~ 0.2 mbar and an annealing temperature of 70 °C.

All the substrates with the films on them as mentioned above were then transferred to a thermal evaporator integrated with the glove box for depositing silver (Ag) anode. The targeted thickness of 100 nm Ag was deposited at 0.2–0.5 Å/s on the top PEDOT:PSS layers using a shadow mask. The base pressure before the deposition was 3.2×10^{-6} mbar. The final thickness of the Ag electrode was controlled in-situ using a quartz crystal digital thickness monitor.

38.2.1.1 Characterizations

UV–Visible absorption and transmittance spectra were recorded using Perkin Elmer Lambda 35, UV–Visible double beam spectrophotometer with a solid-state sample holder. Photo luminance for the films was recorded by using Perkin Elmer lambda Fluorolog FL3C-21. Asylum MFP-3D, atomic force microscope (AFM) was used for topographic imaging. Current–Voltage characteristics of devices were recorded using Keithley 2400 source meter. Newport Sol2A Class ABA solar simulator was used for the illumination of devices. The devices were characterised under illumination in accordance to recommendations in Solmat [37].

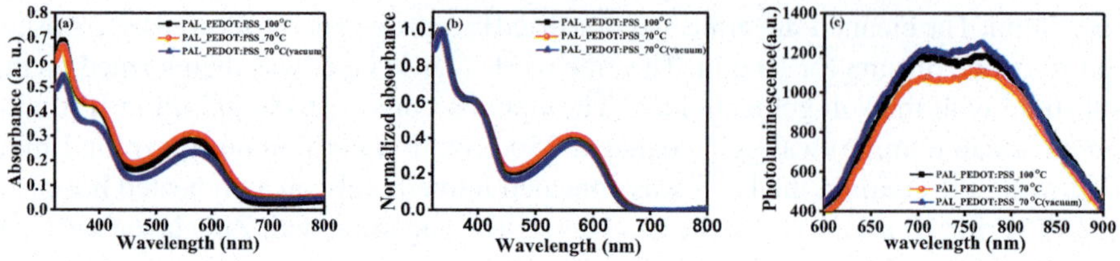

Fig. 38.1 **a** UV–Vis absorption spectra; **b** normalised UV–Vis absorption spectra; and **c** photoluminescence spectra for PEDOT:PSS_100 °C, PEDOT:PSS_70 °C and PEDOT:PSS_70 °C (vacuum). The stack considered consists of glass-substrate/ITO/ZnO/PCDTBT:PCBM/PEDOT:PSS layers

38.3 Result and Discussion

38.3.1 UV–Vis Absorption

The UV–Vis (Ultraviolet Visible) absorption spectra for the film stack glass-substrate/ITO/ZnO/PCDTBT:PCBM/PEDOT:PSS for the three category of PEDOT:PSS drying are shown in Fig. 38.1a. For being better able to compare the spectrum without the effects of thickness variations, the absorbance normalised with respect to (w.r.t.) the PCBM absorption peak occurring in the samples around 335 ± 1 nm and are shown in Fig. 38.1b. From the data, it is observed that the PEDOT:PSS layer drying condition has little effect on the PAL absorption. In the samples observed, the absorption of the PEDOT:PSS_70°C sample shows slightly higher absorption for the PCDTBT absorption peak at 567 nm than the other layers. The PEDOT:PSS_100 °C sample is marginally better than that in the PEDOT:PSS_70 °C (vacuum) stack with peak PCDTBT absorption at 565 nm. The PCDTBT peak absorption in the PEDOT:PSS_70 °C (vacuum) sample is at 569 nm. The red shift for 70 °C temperature process appears to indicate better PCDTBT interactions.

38.3.2 Photoluminescence

The photoluminescence (PL) data for the glass-substrate/ITO/ZnO/PCDTBT:PCBM/PEDOT:PSS stack are shown in Fig. 38.1c. The Stoke's Shift in the spectra with respect to the absorption spectra is evident from the redshift of the PL spectra w.r.t. the absorption spectrum of the samples. It is observed that the PL for PEDOT:PSS_70 °C (vacuum) stack is higher than that for the PEDOT:PSS_100 °C stack which in turn is higher than the PEDOT:PSS_70 °C stack. These differences could be due to the fraction of excitons dissociating into electron–hole pairs or the annihilation/quenching of excitons and thus not being available to luminesce. Formation of electron–hole pair can help with higher photovoltaic (PV) response, while quenching of exciton will lower PV response. The antibatic nature

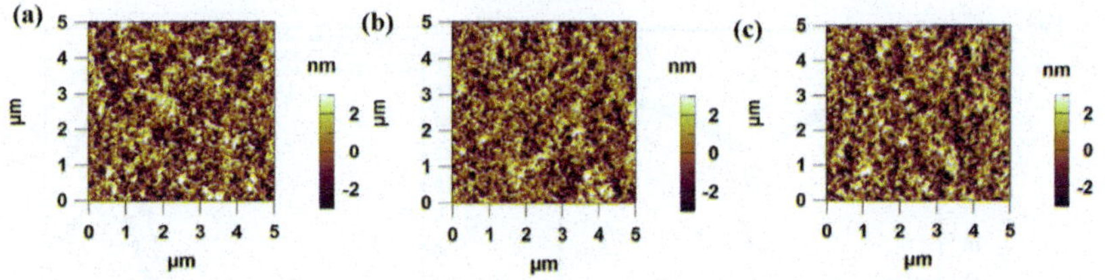

Fig. 38.2 AFM morphology image of PEDOT: PSS layer dried at **a** 100 °C in ambient, **b** 70 °C in ambient and **c** 70 °C in vacuum

of the PL response in all the stacks measured is likely due to higher quenching of excitons close to the ZnO/PCDTBT interface which is where the incident light enters the PAL.

38.3.3 AFM Micrographs

The surface roughness of the PEDOT:PSS layer for the three types of device stacks were characterised with AFM and the images obtained are shown in Fig. 38.2. The PEDOT:PSS_70 °C (vacuum) films shows the highest root mean square (RMS) surface roughness of 1.72 ± 0.08 nm. The PEDOT:PSS_100 °C films shows an RMS roughness of 1.68 ± 0.14 nm. The PEDOT:PSS_70 °C samples shows an RMS roughness of 1.59 ± 0.02 nm.

38.3.4 Electrical Characterisation

The comparison experiment was repeated for two runs. In each run two substrates were used for each experimental split. Each substrate contained four devices of areas 1, 0.81, 0.49 and 0.25 cm^2. All devices were electrically characterised in dark conditions, as well as, under standard illumination (1 sun intensity, AM1.5G spectrum) to obtain the current versus voltage (V) characteristics. The current densities (J) were determined by dividing the measured current by the device area. Representative J–V characteristics obtained for the OPV devices in the dark are shown in Fig. 38.3 and under standard illumination are shown in Fig. 38.4.

The dark J–V characteristics of the representative devices are shown in a semi-log plot in Fig. 38.3a. It is evident that the PEDOT:PSS_70 °C devices show the lowest off current and closer to ideal behaviour compared to the other samples. From the data for Fig. 38.3a, the conductance ($G = dJ/dV$) at every bias point was determined for the curves and plotted w.r.t. corresponding V in Fig. 38.3b. The location of the peak value of the G-V plots is proportional to the built-in potential V_{bi} in the device [38]. Thus, the V_{bi} for PEDOT:PSS_100 °C and PEDOT:PSS_70 °C samples are

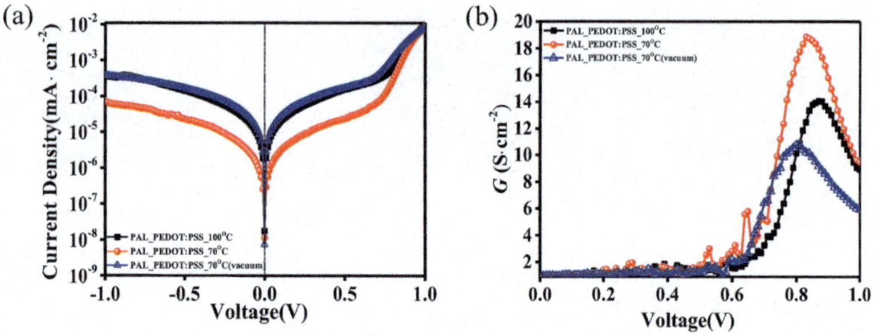

Fig. 38.3 **a** The *J–V* characteristics of the devices in dark; and **b** the *G–V* plot derived from the dark *J–V* characteristics for representative devices from each experimental split

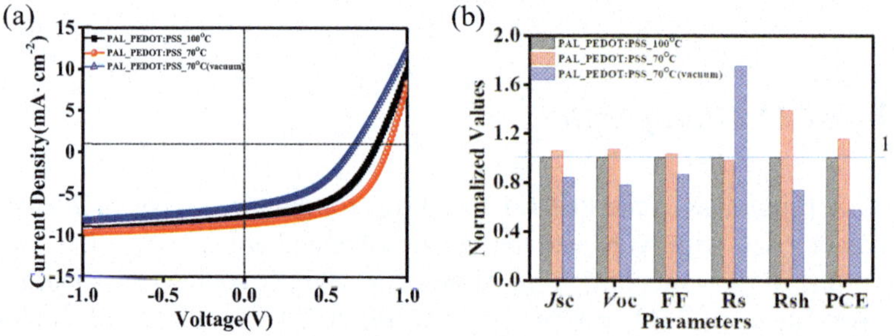

Fig. 38.4 For representative samples for each experimental split **a** the *J–V* characteristics for representative device under standard illumination of 1 sun and AM1.5G spectrum. **b** The solar cell parameters from the representative device normalised w.r.t. the respective PEDOT:PSS_100 °C device parameter values

similar with the former being marginally higher. The PEDOT:PSS_70 °C (vacuum) samples annealed show significantly lower V_{bi}.

The *G* value at the peak and the full width half maximum (FWHM) are indicative of the magnitude of the traps and their spread within the energy gap [39]. The PEDOT:PSS_70 °C devices have the least amount of traps, which are distributed over a smaller energy range in the energy gap. The PEDOT:PSS_100 °C samples have significantly higher traps, but over a similar range in the energy gap. The PEDOT:PSS_70 °C (vacuum) samples have the highest number of traps in the devices studied; and they are also spread wider in energy within the energy gap of the device PAL.

The device characteristics under standard illumination are shown in Fig. 38.4a. The solar cell parameters evaluated from these characteristics are normalised w.r.t. PEDOT:PSS_100°C device values are shown in a bar chart in Fig. 38.4b. The solar cell parameters evaluated for all the devices and the median values are indicated in Table 38.1. The mean and standard deviation values are also indicated within the parenthesis in the table.

Table 38.1 Median values of solar cell parameters for PCDTBT:PC60BM OPV devices from different experimental splits under standard illumination

Experimental split	J_{sc} (mA cm^{-2})	V_{oc} (V)	FF (%)	PCE (%)
PEDOT:PSS (100 °C)	8.22 (8.06 ± 0.85)	0.79 (0.81 ± 0.01)	48.35 (51.00 ± 3.18)	3.14 (3.33 ± 0.47)
PEDOT:PSS (70 °C)	8.55 (8.51 ± 0.32)	0.87 (0.86 ± 0.04)	52.73 (52.69 ± 3.40)	3.92 (3.84 ± 0.42)
PEDOT:PSS (70 °C)	6.59 (6.77 ± 0.61)	0.64 (0.63 ± 0.01)	42.92 (44.0 ± 5.86)	1.81 (1.91 ± 0.42)

The mean and standard deviation values of parameters are given within brackets

From Fig. 38.4 and Table 38.1, it is evident that the PEDOT:PSS_70 °C devices show the best solar cell parameters and are superior to the PEDOT:PSS_100 °C samples. The PEDOT:PSS_70 °C (vacuum) samples show the worst parameters. These appear to be in line with the results from earlier optical and electrical analysis discussed.

The trend in the short circuit current J_{SC} values in the devices from Fig. 38.4 and Table 38.1 confirm that the trend observed between the PL for experimental splits in Fig. 38.1c was due to exciton dissociation and not due to quenching. Thus drying PEDOT:PSS at 70 °C in the ambient results in better harvesting of the excitons formed from the incident light.

While PEDOT:PSS_100 °C appears to have the highest V_{bi} from Fig. 38.3b, the higher trap levels and spread within the energy gap also observed from Fig. 38.3b could explains its lower open circuit voltage V_{OC} w.r.t. PEDOT:PSS_70 °C in Fig. 38.4b. The samples PEDOT:PSS_70 °C (vacuum) has lower V_{bi} as well as lots of traps spread over its energy gap. Thus, devices in that split have the lowest V_{OC} is the lowest.

The trend in the series resistance R_s values could be explained from trend in variation in roughness of the PEDOT:PSS layer in Fig. 38.2. The higher trap values observed from Fig. 38.3b are in line with the trend seen for the shunt resistance—with higher trap values, the increasing recombination of the carriers leads to smaller shunt resistance R_{sh} values. The variations in R_s and R_{sh} are reflected in the fill factor (FF) values; PEDOT:PSS_70 °C devices have the highest FF and PEDOT:PSS_70 °C (vacuum) devices having the lowest FF values.

The J_{SC}, V_{OC} and FF trends contribute to the final PCE values obtained for the devices in different experiment splits. PEDOT:PSS_70 °C devices have the highest PCE values. If we consider the median PCE values, it is nearly 25% higher in PEDOT:PSS_70 °C than in PEDOT:PSS_100 °C devices. The PEDOT:PSS_70 °C (vacuum) devices' median PCE is significantly (40%) lower than that of the PEDOT:PSS_100 °C devices.

Sample size for *each* experimental split consists of 24 devices from 6 substrate fabricated from 2 runs.

38.4 Conclusion

It has been demonstrated that PCDTBT:PCBM PAL OPV cells built with PEDOT:PSS as HTL show superior PCE when the HTL layer is formed with 70 °C drying temperature rather than 100 °C HTL drying temperature. However, drying PEDOT:PSS in vacuum, even at the lower 70 °C ends up degrading the device significantly resulting in poorer solar cells.

References

1. Q. Liu, Y. Jiang, K. Jin, J. Qin, J. Xu, W. Li, J. Xiong, J. Liu, Z. Xiao, K. Sun, S. Yang, X. Zhang, L. Ding, 18% Efficiency organic solar cells. Sci. Bull. **1**, 001 (2020)
2. J. Hou, O. Ingänas, R.H. Friend, F. Gao, Organic solar cells based on non-fullerene acceptors. Nat. Mater. **17**, 119–128 (2018)
3. W. Zhao, D. Qian, S. Zhang, S. Li, O. Inganäs, F. Gao, J. Hou, Fullerene-free polymer solar cells with over 11% efficiency and excellent thermal stability. Adv. Mater. **28**, 4734–4739 (2016)
4. Y. Liu, J. Zhao, Z. Li, C. Mu, W. Ma, H. Hu, K. Jiang, H. Lin, H. Ade, H. Yan, Aggregation and morphology control enables multiple cases of high-efficiency polymer solar cells. Nat. Commun. **5**, 5293 (2014)
5. H. Bin, L. Gao, Z. Zhang, Y. Yang, Y. Zhang, C. Zhang, S. Chen, L. Xue, C. Yang, M. Xiao, Y. Li, 11.4% efficiency non-fullerene polymer solar cells with trialkylsilyl substituted 2D-conjugated polymer as donor. Nat. Commun. **7**, 13651 (2016)
6. Q. Fan, W. Su, Y. Wang, B. Guo, Y. Jiang, X. Guo, F. Liu, T.P. Russell, M. Zhang, Y. Li, Synergistic effect of fluorination on both donor and acceptor materials for high performance non-fullerene polymer solar cells with 13.5% efficiency. Sci. China Chem. **61**, 531–537 (2018)
7. C. Sun, F. Pan, H. Bin, J. Zhang, L. Xue, B. Qiu, Z. Wei, Z. Zhang, Y. Li, A low cost and high performance polymer donor material for polymer solar cells. Nat. Commun. **9**, 743 (2018)
8. S. Li, L. Ye, W. Zhao, H. Yan, B. Yang, D. Liu, W. Li, H. Ade, J. Hou, A wide band gap polymer with a deep highest occupied molecular orbital level enables 14.2% efficiency in polymer solar cells. J. Am. Chem. Soc. **140**, 7159–7167 (2018)
9. Y. Cui, H. Yao, L. Hong, T. Zhang, Y. Xu, K. Xian, B. Gao, J. Qin, J. Zhang, Z. Wei, J. Hou, Achieving over 15% efficiency in organic photovoltaic cells via copolymer design. Adv. Mater. **31**, 1808356 (2019)
10. Y. Lin, J. Wang, Z. Zhang, H. Bai, Y. Li, D. Zhu, X. Zhan, An electron acceptor challenging fullerenes for efficient polymer solar cells. Adv. Mater. **27**, 1170–1174 (2015)
11. V. Shrotriya, G. Li, Y. Yao, C.W. Chu, Y. Yang, Transition metal oxides as the buffer layer for polymer photovoltaic cells. Appl. Phys. Lett. **88**, 073508 (2006)
12. K. Wang, H. Ren, C. Yi, C. Liu, H. Wang, L. Huang, H. Zhang, A. Karim, X. Gong, Solution-processed Fe_3O_4 magnetic nanoparticle thin film aligned by an external magnetostatic field as a hole extraction layer for polymer solar cells. ACS Appl. Mater. Interfaces **5**, 10325–10335 (2013)
13. M.D. Irwin, D.B. Buchholz, A.W. Hains, R.P.H. Chang, T.J. Marks, Semiconducting nickel oxide as an efficiency-enhancing anode interfacial layer in polymer bulk-heterojunction solar cells. Proc. Natl. Acad. Sci. USA **105**, 2783–2787 (2008)
14. S. Han, W.S. Shin, M. Seo, D. Gupta, S.J. Moon, S. Yoo, Improving performance of organic solar cells using amorphous tungsten oxides as an interfacial buffer layer on transparent anodes. Org. Electron. **10**, 791–797 (2009)
15. H. Yao, Y. Cui, R. Yu, B. Gao, H. Zhang, J. Hou, Design, synthesis, and photovoltaic characterization of a small molecular acceptor with an ultra-narrow band gap. Angew. Chem. Int. Ed. **56**, 3045–3049 (2017)

16. Z. Wang, X. Zhu, J. Zhang, K. Lu, J. Fang, Y. Zhang, L. Zhu, W. Ma, Z. Shuai, Z. Wei, From alloy-like to cascade blended structure: designing high-performance all-small-molecule ternary solar cells. J. Am. Chem. Soc. **2018**(140), 1549 (2018)
17. J.W. Lee, H. Ahn, W.H. Jo, Conjugated random copolymers consisting of pyridine- and thiophene-capped diketopyrrolopyrrole as co-electron accepting units to enhance both J_{SC} and V_{OC} of polymer solar cells. Macromolecules **2015**(48), 7836–7842 (2015)
18. Y. Qin, M.A. Uddin, Y. Chen, B. Jang, K. Zhao, Z. Zheng, R. Yu, T.J. Shin, H.Y. Woo, J. Hou, Highly efficient fullerene-free polymer solar cells fabricated with polythiophene derivative. Adv. Mater. **28**, 9416–9422 (2016)
19. S. Zhang, Y. Qin, J. Zhu, J. Hou, Over 14% efficiency in polymer solar cells enabled by a chlorinated polymer donor. Adv. Mater. **30**, 1800868 (2018)
20. T. Ma, K. Jiang, S. Chen, H. Hu, H. Lin, Z. Li, J. Zhao, Y. Liu, Y.-M. Chang, C.-C. Hsiao, H. Yan, Efficient low-bandgap polymer solar cells with high open-circuit voltage and good stability. Adv. Energy Mater. **5**, 1501282 (2015)
21. D. Baran, T. Kirchartz, S. Wheeler, S. Dimitrov, M. Abdelsamie, J. Gorman, R.S. Ashraf, S. Holliday, A. Wadsworth, N. Gasparini, P. Kaienburg, H. Yan, A. Amassian, C.J. Brabec, J.R. Durrant, I. McCulloch, Reduced voltage losses yield 10% efficient fullerene free organic solar cells with >1 V open circuit voltages. Energy Environ. Sci. **9**, 3783–3793 (2016)
22. S. Qu, H. Wang, D. Mo, P. Chao, Z. Yang, L. Li, L. Tian, W. Chen, F. He, Fine tuning of open-circuit voltage by chlorination in thieno[3,4-*b*]thiophene–benzodithiophene terpolymers toward enhanced solar energy conversion. Macromolecules **50**, 4962–4971 (2017)
23. Z. Li, K. Jiang, G. Yang, J.Y. Lai, T. Ma, J. Zhao, W. Ma, H. Yan, Donor polymer design enables efficient non-fullerene organic solar cells. Nat. Commun. **7**, 13094 (2016)
24. J. Liu, S. Chen, D. Qian, B. Gautam, G. Yang, J. Zhao, J. Bergqvist, F. Zhang, W. Ma, H. Ade, O. Inganäs, K.F. Gao, H. Yan, Fast charge separation in a non-fullerene organic solar cell with a small driving force. Nat. Energy **2016**(1), 16089 (2016)
25. F. Zhao, S. Dai, Y. Wu, Q. Zhang, J. Wang, L. Jiang, Q. Ling, Z. Wei, W. Ma, W. You, C. Wang, X. Zhan, Single-junction binary-blend nonfullerene polymer solar cells with 12.1% efficiency. Adv. Mater. **29**, 1700144 (2017)
26. D. Liu, B. Yang, B. Jang, B. Xu, S. Zhang, C. He, H.Y. Woo, J. Hou, Molecular design of a wide-band-gap conjugated polymer for efficient fullerene-free polymer solar cells. Energy Environ. Sci. **10**, 546–551 (2017)
27. X. Che, Y. Li, Y. Qu, S.R. Forrest, High fabrication yield organic tandem photovoltaics combining vacuum- and solution-processed subcells with 15% efficiency. Nat. Energy **3**, 422–427 (2018)
28. Y. An, X. Liao, L. Chen, Q. Xie, M. Zhang, B. Huang, Z. Liao, H. Guo, A. Jazib, J. Han, F. Liu, A.K.-Y. Jen, Y. Chen, A_1-A_2 type wide bandgap polymers for high-performance polymer solar cells: energy loss and morphology. Sol. RRL **3**, 1800291 (2019)
29. I. Shin, H. Ahn, J.H. Yun, J.W. Jo, S. Park, S.-Y. Joe, J. Bang, H.J. Son, High-performance and uniform 1 cm^2 polymer solar cells with D_1-A-D_2-A-type random terpolymers. Adv. Energy Mater. **8**, 1701405 (2018)
30. X. Fan, W. Song, T. Lei, B. Xu, F. Yan, N. Wang, H. Cuia, Z. Ge, High-efficiency robust organic solar cells using transfer-printed PEDOT:PSS electrodes through interface bonding engineering. Mater. Chem. Front **3**, 901–908 (2019)
31. T.S. van der Poll, J.A. Love, T.-Q. Nguyen, G.C. Bazan, Non-basic high-performance molecules for solution-processed organic solar cells. Adv. Mater. **24**, 3646–3649 (2012)
32. Q. Zhang, B. Kan, F. Liu, G. Long, X. Wan, X. Chen, Y. Zuo, W. Ni, H. Zhang, M. Li, Z. Hu, F. Huang, Y. Cao, Z. Liang, M. Zhang, T.P. Russell, Y. Chen, A solution-processed high performance organic solar cells using a small molecule with the thieno[3,2-b]thiophene central unit. Nat. Photonics **9**, 35–41 (2014)
33. L. Meng, Y. Zhang, X. Wan, C. Li, X. Zhang, Y. Wang, X. Ke, Z. Xiao, L. Ding, R. Xia, H.-L. Yip, Y. Cao, Y. Chen, Organic and solution-processed tandem solar cells with 17.3% efficiency. Science **361**, 1094–1098 (2018)

34. Z. Li, K.H. Chiu, R.S. Ashraf, S. Fearn, R. Dattani, H.C. Wong, C.-H. Tan, J. Wu, J.T. Cabral, J.R. Durrant, Toward improved lifetimes of organic solar cells under thermal stress: substrate-dependent morphological stability of PCDTBT:PCBM films and devices. Sci. Rep. **5**, 1–9 (2015)
35. H.C. Wong, Z. Li, C.H. Tan, H. Zhong, Z. Huang, H. Bronstein, I. McCulloch, J.T. Cabral, J.R. Durrant, Morphological stability and performance of polymer-fullerene solar cells under thermal stress: the impact of photoinduced PC60BM oligomerization. ACS Nano **8**(2), 1297–1308 (2014)
36. P.A. Staniec, A.J. Parnell, A.D.F. Dunbar, H. Yi, A.J. Pearson, T. Wang, P.E. Hopkinson, C. Kinane, R.M. Dalgliesh, A.M. Donald, A.J. Ryan, A. Iraqi, R.A.L. Jones, D.G. Lidzey, The nanoscale morphology of a PCDTBT:PCBM photovoltaic blend. Adv. Energy Mater. **1**(4), 499–504 (2011)
37. Editorial, Reporting solar cell efficiencies in solar energy materials and solar cells. Sol. Energy Mater. Sol. Cells **92**, 371–373 (2008)
38. P. Mantri, S.M.H. Rizvi, B. Mazhari, Estimation of built-in voltage from steady-state current–voltage characteristics of organic diodes. Organ. Electron. **14**(8), 2034–2038 (2013)
39. S.M.H. Rizvi, P. Mantri, B. Mazhari, Traps signature in steady state current–voltage characteristics of organic diode. J. Appl. Phys. **115**, 244502 (2014)

Chapter 39
Electrical Performance of Protein-Based Flexible MIM Structure Fabricated at Room Temperature for Proteotronic Applications

Shrabani Guhathakurata, Nabin Baran Manik, and Sandipan Mallik

39.1 Introduction

The cutting-edge hardware systems embedded with flexible electronics widens the avenue of invigorating novel applications like flexible lighting, flexible power storage and show advancements for gadgets, engineering, materials, and wearable sensors observing wellbeing and propensities, lightweight implantable hardware for clinical diagnostics, robotics and automated aircrafts, energy generating gadgets and sensors [1, 2]. Although regular electronic devices are truly equipped for all required capacities, flexibility in hardware tend to increase and enhance the mechanical properties and features to create hybrid novel systems, or independent systems, expected to be profoundly tolerable to be deformed, at minimal expense, thinner, and dispensable [3, 4]. To bend and roll enormous region of photovoltaics for simpler handling, to adjust into unpredictable structures, collapsing, curving, extending, and distorting that is what expected for gadgets on skin, while keeping up with gadget execution, quality and performance [5–7]. Be that as it may, there is extraordinary scope in fabrication of novel structures which allows nano-electronic gadgets to be produced on flexible platform. This has additionally reached out for adaptability as well as stretchability and flexibility with possibility to be accomplished by using elastomers [8, 9].

S. Guhathakurata (✉) · S. Mallik
Department of Electronics and Communication Engineering, NIST (Autonomous), Brahmapur, Odisha 761008, India
e-mail: sguha.phy.2018@gmail.com

S. Guhathakurata · N. B. Manik
Department of Physics, Condensed Matter Physics Research Centre, Jadavpur University, Kolkata 700032, India

R. Singh et al. (eds.), *The Physics of Semiconductor Devices*, Springer Proceedings in Physics 306, https://doi.org/10.1007/978-981-97-1571-8_39

Organic biomaterials would dominate the next generation electronic industry, owing to good electronic, mechanical and optical properties comparable to their inorganic counterpart. Moreover organic materials are environment friendly and often require room temperature processing and thus have wide range of applications in biosensors [10], proteomics [11], pharmaceutical [12], biocompatible coatings [13]. Biocompatible and self-degradable materials such as cellulose [14], chitosan [15], starch [16], proteins [17] etc. are thus extensively studied to meet the growing demand in organic electronic industry. The natural plant based protein Papain commonly extracted from the green papaya fruit latex has great potential as an organic material due to its simple extraction process and easy availability. It has molecular weight of 23,406 Da, pH value of 5–8 [18], thermally stable [19] and has optimum temperature range up to 65 °C which is considerably higher. Many researchers have reported electrical characteristics of Papain as well as its device applications.

György et al. performed matrix assisted synthesis of Papain thin films using pulse laser deposition technique. Frozen Papain solution was used as deposition target instead of conventional powder deposition target [20]. The bulk dielectric constant of Papain was determined by conductivity study of Papain pellet and was reported as 74.44 F/m by Muthulakshmi et al. [21]. Study of the impact of temperature on dielectric constant of Papain has shown that increase in temperature enhances dielectric properties of Papain [22]. It is seen that AC conductivity of Papain is frequency dependent and Papain obeys large polaron tunneling model.

By looking in to the context of the potential of Papain and its possibility in device electronics, in this study, a flexible MIM structure with Papain thin film was fabricated using ITO as working electrodes. Physical parameters as well as chemical structure of Papain was studied. Electrical properties were estimated by studying the capacitance–voltage (CV) characteristics. The charge storage mechanism in the protein film is discussed on the basis of the experimental results.

39.2 Experimental

Flexible indium tin oxide (ITO)-coated polyethylene terephthalate (PET) obtained from Sigma Aldrich (CAS No.: 639303, surface resistivity 60 Ω/sq.) was used as substrate for the fabrication of Papain based MIM capacitor. Prior fabrication procedure the substrate was cleaned using 2-propanol and rinsed with D.I. water. The substrate was dried under UV exposure for 2 h. Aqueous solution (1.5 mg/ml) of Papain (Aldrich, CAS No.: 9001-73-4) was spin-coated at 2500 rpm for 60 s. on the ITO-PET substrate and allowed to dry for 2 h. at 40 °C. Papain film thickness was estimated using FESEM (Zeiss Supra 40). Raman Spectroscopy (Horiba T64000) was employed to study the chemical bond structures of the Papain film. The surface morphology was studied with AFM (Nano surf easyscan2), and UV–Vis Spectrophotometer (Shimadzu UV-1800) was employed to estimate the optical band gap of the film. Photoluminescence (PL) spectra were measured using a He-Cd laser, operating at 325 nm. Finally, ITO of thickness 100 nm was sputtered using (Hind High Vacuum sputtering unit BC-300), with a mask of area 2×10^{-3} cm^2 for fabrication of top electrode. Electrical characteristics of the device were measured using Agilent E4980A precision LCR meter and Keithley 2604B source meter.

39.3 Result and Discussion

Figure 39.1a shows the surface morphology of the spin coated Papain films using AFM with a scan area of $10 \times 10\ \mu m^2$. From surface scan, it can be observed that the surface of the film had large grains and some islands that may be considered as polycrystalline surface [23, 24]. The average surface roughness (RMS) was estimated to be ~ 8.21 nm. Figure 39.1b represents the cross sectional FESEM image where the thickness was found to be ~ 3 μm. A homogenous deposition of spin-coated surface was observed from the image where presence of small valleys correlates with the AFM study. The room temperature PL spectrum of Papain film is shown in Fig. 39.1c. There is one broad peak at 340 nm. The deconvoluted spectra show two peaks at 340 and 372 nm that shows presence of some defects in the dielectric. Absorption spectra of Papain was obtained in the wavelength range of 200–800 nm using UV–Vis spectroscopy and corresponding Tauc plot is illustrated in Fig. 39.1d to determine the optical bandgap of the film. Tauc plot extrapolation shows the bandgap is around 4.06 eV. Transmittance spectra of Papain film on ITO coated PET is shown in Fig. 39.1e. It is seen that percentage transmittance of Papain film is more than 85%. Thus, it is evident that the structure can be used in transparent device for optoelectronic applications.

The Raman spectra of papain is shown in Fig. 39.1f. Raman spectroscopy analyses vibrational modes in Papain structure. Amide I (C=O stretching vibration), amide II and amide III (associated with C–N stretching and N–H bending) are of main interest. The other amides that were not visible correspond to amide IV, V,VI associated with OCN bending, out-of-plane NH bending, and out-of-plane C=O bending respectively [25]. The peak visible at 1650 cm^{-1} corresponds to amide I band and a weak amide II band (substantial C-N stretching) is seen around 1550 cm^{-1}, whose intensity depends on α-helical content. Amide III (α helix, random coil, β sheet) bands are moderately visible around 1300 cm^{-1}.

The plot of normalized capacitance of Papain film against applied voltage is shown in Fig. 39.2a. The normalized capacitance is varied in the frequency range of 100 kHz to 1 MHz. Normalized capacitance of MIM capacitor is given as,

$$\frac{\Delta C}{C_0} = \alpha V^2 + \beta V \tag{39.1}$$

where the zero bias capacitance is given as C_0, ΔC is the difference between applied bias capacitance and C_0, V is the applied bias, α is the quadratic voltage coefficient of capacitance and β is the linear voltage coefficient of capacitance. $\Delta C/C_0$ is described

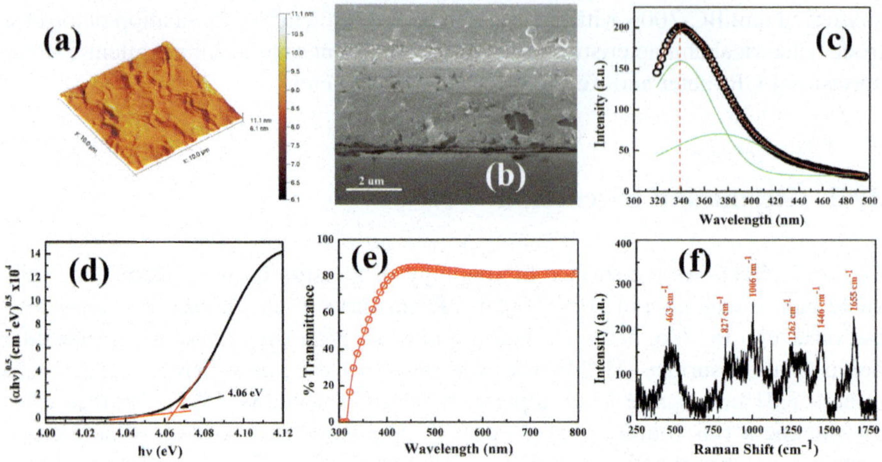

Fig. 39.1 **a** AFM image of Papain showing surface morphology, gives the r.m.s roughness ~ 8.21 nm, **b** the cross-sectional FESEM image of Papain shows thickness of 3 μm, **c** the PL spectra of Papain shows absorption peak (340 nm) as well as a defect peak, **d** Tauc plot of Papain obtained from UV–Vis spectroscopy shows band gap at 4.06 eV, **e** 85% of transmittance in papain film was observed. **f** Raman spectrum of Papain shows peaks describing its secondary structure as obtained in amide I, amide II and amide III regions

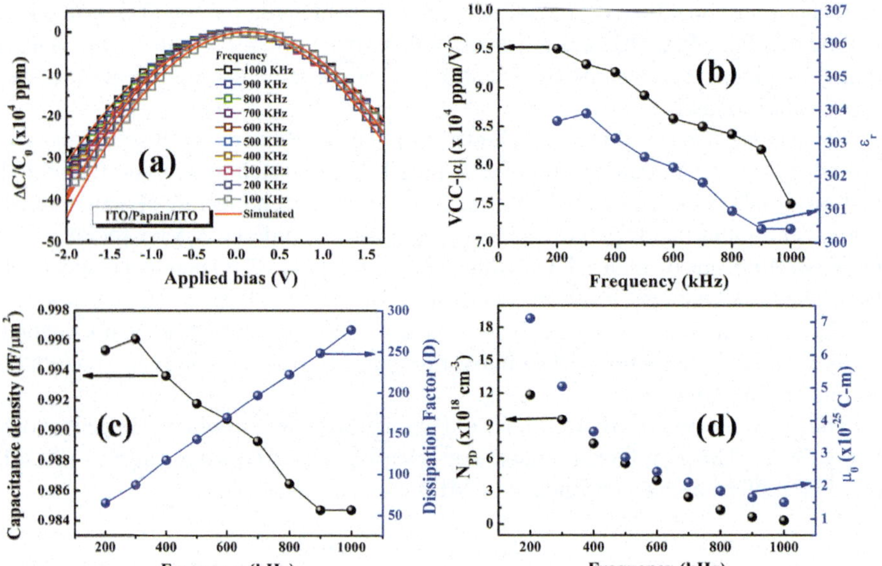

Fig. 39.2 **a** Normalized capacitance with negative slope of ITO/Papain/ITO device plotted against applied voltage, **b** frequency dispersion of VCC-|α| and dielectric constant of Papain, **c** capacitance density variation with frequency and change of dissipation factor with frequency, **d** variation of number of dipoles (N_{PD}) and permanent dipole moment (μ_0) with frequency

as parts per million (ppm). The value of VCC-α is of great significance in the capacitance curve and here a downward curvature is observed in the capacitance curve that attributes to the negative value of VCC-α.

The variation of magnitude of VCC-|α| with change in frequency is shown in Fig. 39.2b. The value of VCC-|α| at 200 kHz frequency is 9.5×10^4 ppm/V^2 that changes to 7.5×10^4 ppm/V^2 at 1 MHz. Polarization of the dielectric film that gives rise to different dipole orientation results in negative value of VCC-α. According to "The International Technology Roadmap for Semiconductors (ITRS)" lower value of VCC-α is desirable (lower than 100) [26, 27]. Thus, Papain having negative VCC-α value can be stacked with other inorganic dielectric having high value of VCC-α, such that the net VCC-α can be lowered within desired limit [28, 29]. The relative permittivity is varied with frequency and also shown in Fig. 39.2b. The value of permittivity is very high and is 300.4 at 1 MHz.

The capacitance density and the dissipation factor are plotted with applied frequency and shown in Fig. 39.2c. The capacitance density non-linearly degrades with frequency that demonstrates the presence of traps in the dielectric, at higher frequency as the traps fail to follow change in the applied ac signal, the capacitance value decreases. At 200 kHz the capacitance density is 0.995 fF/μm^2 that decreased to 0.9847 fF/μm^2 at 1 MHz. The dissipation factor increases with frequency and a similar phenomenon is reported and explained in earlier literature [30–33]. The reasons of dielectric loss are mainly high surface roughness, electrode resistance, improper grain boundaries [34].

The equation of VCC-α is reported by Sanal et al. [27] with the help of an analytical model,

$$\alpha_{VCC} = -\frac{N_{pd}\lambda^3\mu_0^4}{81\varepsilon(kT)^3t^2} \tag{39.2}$$

where, N_{pd} is the number of dipoles, μ_0 is the permanent dipole moment, λ is the field correction factor, ε is the static permittivity, t is the thickness of the dielectric. Using the relations, $\frac{\chi}{\varepsilon_0} = \varepsilon - 1$ and $\chi = \frac{\mu_0^2 N_{pd}}{3kT}$, the dipole moment is given as,

$$\mu_0 = \sqrt{\frac{3kT\varepsilon_0(\varepsilon_r - 1)}{N_{pd}}} \tag{39.3}$$

The number of dipoles can be written as,

$$N_{pd} = \frac{-\lambda^3\varepsilon_0(\varepsilon_r - 1)^2}{9\varepsilon_r kTt^2\alpha_{VCC}} \tag{39.4}$$

Using these two equations the number of dipoles and permanent dipole moment of papain was estimated as 3.328×10^{17} cm^{-3} and 1.5×10^{-25} C-m respectively at 1 MHz frequency. Variation of number of dipoles and permanent dipole moment with frequency is shown in in Fig. 39.2d.

The electrical characteristics of protein can be explained using ionic double layer model. The peptide bond is a covalent bond that join amino group and carboxyl group of two adjacent amino acids by removing water molecules. In an aqueous solution of protein the acidic and basic amino acids react with water and can generate mobile ions [35]. Upon interaction with water, the carboxyl group deprotonates and produce H_3O^+ ions, while the amino group protonates to produce OH^- mobile ion [36]. Both positive and negative side chains of papain fail to respond to applied voltage due to long chain structure and thus the H_3O^+ and OH^- ions move towards negative and positive electrodes resulting in ionic double layer. This may take place using Grotthuss transfer that is explained as transport of free proton (H^+) that hops by exchanging hydrogen bond with covalent bond towards the electrodes that may be considered analogous to semiconductor transport of electrons and holes [37].

39.4 Conclusion

In summary papain based organic, transparent, flexible MIM capacitor has been fabricated with ITO as top and bottom electrode. Negative VCC-α and very high dielectric constant ~ 300.4 was observed. The dissipation factor was found to be considerably high which may be due to poor contact adhesion and dielectric defects. The number of dipoles with their permanent dipole moment is estimated using analytical modeling. The charge storage mechanism resembles ionic double layer model that can be further explained using Grotthuss transfer and shows analogy with inorganic MIM capacitors. The thickness of the protein layer can be further minimized by reducing the spin speed, but this may enhance the defects in the film, hence the process of protein deposition can be modified by layer by layer assembling method. The mechanical stability of the device and its impact on its electrical property needs to be studied by doing flexibility test using repetitive bending method in future.

Acknowledgements Sandipan Mallik and Shrabani Guhathakurata would like to sincerely thank the Department of Science and Technology-Science and Engineering Research Board (DST-SERB), New Delhi, India, for providing research funding (Project No.: ECR/2018/ 002477) under the DST-ECR scheme.

References

1. S.S. Bhore. Formulation and evaluation of resistive inks for applications in printed electronics (2013)
2. W. Gao, H. Ota, D. Kiriya, K. Takei, A. Javey, Flexible electronics toward wearable sensing. Acc. Chem. Res. **52**(3), 523–533 (2019)
3. J.E. Mates, I.S. Bayer, J.M. Palumbo, P.J. Carroll, C.M. Megaridis, Extremely stretchable and conductive water-repellent coatings for low-cost ultra-flexible electronics. Nat. Commun. **6**(1), 1–8 (2015)

4. C. Yang, W. Lin, Z. Li, R. Zhang, H. Wen, B. Gao, G. Chen, P. Gao, M.M.F. Yuen, C.P. Wong, Water-based isotropically conductive adhesives: towards green and low-cost flexible electronics. Adv. Funct. Mater. **21**(23), 4582–4588 (2011)
5. S. Park, S.W. Heo, W. Lee, D. Inoue, Z. Jiang, K. Yu, H. Jinno, D. Hashizume, M. Sekino, T. Yokota, Self-powered ultra-flexible electronics via nano-grating-patterned organic photovoltaics. Nature **561**(7724), 516–521 (2018)
6. P. Kumar, S. Guhathakurata, A. Choudhury, A. Sharma, A.R. Tripathy, S. Sachin Kumar, P.P. Pancham, P. Das, S.S. Mahato, S. Mahata, Flexible BSA MIM capacitor with negative voltage coefficient for RF applications. Appl. Phys. Lett. **116**(17), 171904 (2020)
7. A.R. Tripathy, A. Choudhury, A. Dash, P. Panigrahi, S.S. Kumar, P.P. Pancham, S.K. Sahu, S. Mallik, Polymer matrix composite engineering for PDMS based capac-itive sensors to achieve high-performance and broad-range pressure sensing. Appl. Surf. Sci. Adv. **3**, 100062 (2021)
8. F. Wang, H.-J. Kim, S. Park, C.-D. Kee, S.-J. Kim, I.-K. Oh, Bendable and flexible supercapacitor based on polypyrrole-coated bacterial cellulose core-shell composite network. Compos. Sci. Technol. **128**, 33–40 (2016)
9. J. Kang, D. Son, G.N. Wang, Y. Liu, J. Lopez, Y. Kim, J.Y. Oh, T. Katsumata, J. Mun, Y. Lee, Tough and water-insensitive self-healing elastomer for robust electronic skin. Adv. Mater. **30**(13), 1706846 (2018)
10. S. Dong, B. Wang, Electrochemical biosensing in extreme environment. Electroanal. Int. J. Fundam. Pract. Asp. Electroanal. **14**(1), 7–16 (2002)
11. H. Miyachi, A. Hiratsuka, K. Ikebukuro, K. Yano, H. Muguruma, I. Karube, Application of polymer-embedded proteins to fabrication of DNA array. Biotechnol. Bio-eng. **69**(3), 323–329 (2000)
12. M.A. Cooper, Optical biosensors in drug discovery. Nat. Rev. Drug Discov. **1**(7), 515–528 (2002)
13. V. Aubakirova, R. Farrakhov, A. Sharipov, V. Polyakova, L. Parfenova, E. Parfenov, Investigation of biocompatible PEO coating growth on Cp-Ti with in situ spectro-scopic methods. Materials (Basel) **15**(1), 9 (2021)
14. Y.H. Jung, T.-H. Chang, H. Zhang, C. Yao, Q. Zheng, V.W. Yang, H. Mi, M. Kim, S.J. Cho, D.-W. Park, High-performance green flexible electronics based on biodegradable cellulose nanofibril paper. Nat. Commun. **6**(1), 1–11 (2015)
15. P. Feng, P. Du, C. Wan, Y. Shi, Q. Wan, Proton conducting graphene oxide/chitosan composite electrolytes as gate dielectrics for new-concept devices. Sci. Rep. **6**(1), 1–9 (2016)
16. R.L. Shogren, Starch: properties and materials applications, in *Biopolymers from Renewable Resources*. ed. by D.L. Kaplan (Springer, Berlin, 1998), pp.30–46
17. E. Alfinito, L. Reggiani, J. Pousset, in *Proteotronics: Electronic Devices Based on Proteins*. Lecture Notes in Electrical Engineering (2015)
18. L.A.A. Sluyterman, Product inhibition of Papain action. Biochim. Biophys. Acta **85**(2), 316–321 (1964)
19. K. Sangeetha, T.E. Abraham, Chemical modification of papain for use in alkaline medium. J. Mol. Catal. B Enzym. **38**(3–6), 171–177 (2006)
20. E. György, J. Santiso, A. Figueras, G. Socol, I.N. Mihailescu, Biomolecular papain thin films growth by laser techniques. J. Mater. Sci. Mater. Med. **18**, 1643–1647 (2007)
21. S. Muthulakshmi, P. Chithralekha, M. Balaji, G. Sanjeev, D.P. Padiyan, in *Effect of Exposure to Electron Beam Irradiation in Biopolymer Papain and Their Electrical Behaviour* (2013)
22. S. Muthulakshmi, S. Iyyapushpam, D.P. Padiyan, Effect of temperature on the AC impedance of protein and carbohydrate biopolymers. Bull. Mater. Sci. **37**(7), 1575–1582 (2014)
23. A. Bag, M.K. Hota, S. Mallik, C.K. Maìti, Bipolar resistive switching in different plant and animal proteins, in *Proceedings of the 21th International Symposium on the Physical and Failure Analysis of Integrated Circuits (IPFA)* (IEEE, 2014), pp. 203–206
24. A. Bag, M.K. Hota, S. Mallik, C.K. Maìti, Observation of long term potentiation in papain-based memory devices, in *Proceedings of the 21th International Symposium on the Physical and Failure Analysis of Integrated Circuits (IPFA)* (IEEE, 2014), pp. 199–202

25. A. Rygula, K. Majzner, K.M. Marzec, A. Kaczor, M. Pilarczyk, M. Baranska, Raman spectroscopy of proteins: a review. J. Raman Spectrosc. **44**(8), 1061–1076 (2013)
26. T.H. Phung, P. Steinmann, R. Wise, Y.-C. Yeo, C. Zhu, Modeling the negative quadratic VCC of SiO_2 in MIM capacitor. IEEE Electron Device Lett. **32**(12), 1671–1673 (2011)
27. A. Sanal, P. Sathyanarayanan, V. Velmurugan, D. Kannadassan, Negative VCC in MIM capacitors: modeling and experiments. J. Comput. Electron. **17**(1), 458–462 (2018)
28. S.J. Kim, B.J. Cho, M.-F. Li et al., Improvement of voltage linearity in high-/spl kap-pa/MIM capacitors using HfO_2–SiO_2 stacked dielectric. IEEE Electron Device Lett. **25**(8), 538–540 (2004)
29. C. Jorel, C. Vallée, P. Gonon, E. Gourvest, C. Dubarry, E. Defay, High performance metal–insulator–metal capacitor using a $SrTiO_3/ZrO_2$ bilayer. Appl. Phys. Lett. **94**(25), 253502 (2009)
30. M.K. Hota, C.K. Sarkar, C.K. Maiti, Frequency-dependent dielectric response of $HfTaO_x$-based metal–insulator–metal capacitors. Semicond. Sci. Technol. **27**(8), 85002 (2012)
31. R.K. Goyal, A.B. Kulkarni, Electrical properties of novel three-phase polymer nanocom-posites with a high dielectric constant. J. Phys. D Appl. Phys. **45**(46), 465302 (2012)
32. C.Y. Kim, H.S. Lee, J.K. Woo et al., Frequency-dependent capacitance–voltage and conductance–voltage characteristics of low-dielectric-constant SiOC(-H) thin films deposited by using plasma-enhanced chemical vapor deposition. J Korean Phys. Soc. **57**, 1976–1982 (2010)
33. A. Goswami, A.P. Goswami, Dielectric and optical properties of ZnS films. Thin Solid Films **16**(2), 175–185 (1973)
34. R. Thomas, D.C. Dube, M.N. Kamalasanan, S. Chandra, A.S. Bhalla, Structural, electrical, and low-temperature dielectric properties of sol–gel derived SrTiO 3 thin films. J. Appl. Phys. **82**(9), 4484–4488 (1997)
35. A.L. Lehninger, in *Lehninger Principles of Biochemistry: David L. Nelson, Michael M. Cox* (Recording for the Blind & Dyslexic, New York, 2004)
36. J.E. Nielsen, G. Vriend, Optimizing the hydrogen-bond network in Poisson-Boltzmann equation-based PKa calculations. Proteins Struct. Funct. Bioinform. **43**(4), 403–412 (2001)
37. A. Noy, Bionanoelectronics. Adv. Mater. **23**(7), 807–820 (2011)

Chapter 40
Estimation of Asymmetrically Distributed Anti-phase Domains Ratio in GaAs/Si(100) Epitaxial Layers Using High Resolution X-Ray Diffraction

Ravi Kumar, R. Roychowdhury, V. K. Dixit, and T. K. Sharma

40.1 Introduction

It is of great importance and interest to grow polar layers on non-polar substrates for making various types of devices exploring the electronic and optical properties of both the materials [1–10]. GaAs on Silicon (Si), GaP on Germanium (Ge), InP on Ge, GaAs on Si are some of the examples of material combination which fall under this category. However, issues related to lattice mismatch, difference in thermal expansion coefficients, and generation of antiphase domains (APD) arise in this type of growth [11–17]. Antiphase domains (APD) are two 90° in-plane rotated domains that grow on the substrate since the substrate is unable to provide preferential atom sites to any one of the atoms in the polar layer. The usage of group IV materials particularly Si is of paramount importance when it comes to integrate compound semiconductors with silicon technology. On the other hand, the group III–V zincblende materials particularly GaAs are very important for making optoelectronic devices. Therefore, monolithic integration of Si with GaAs is essential for the realization of several devices including lasers, high-speed electronics, and photovoltaics [1–10].

However, monolithic integration of III–V compound semiconductors on Silicon faces several inherent challenges, for example in the case of GaAs on Si, lattice mismatch is $\approx 4.1\%$, and the difference in thermal expansion coefficient is $\approx 110\%$, which lead to the generation of antiphase domains (APD) in the GaAs layers grown on

---

R. Kumar (✉) · R. Roychowdhury · V. K. Dixit · T. K. Sharma
Semiconductor Materials Laboratory, Materials Science Section, Raja Ramanna Centre for Advanced Technology, Indore, Madhya Pradesh 452013, India
e-mail: ravi@rrcat.gov.in

Homi Bhabha National Institute, Training School Complex, Anushakti Nagar, Mumbai, India

T. K. Sharma
e-mail: tarun@rrcat.gov.in

R. Singh et al. (eds.), *The Physics of Semiconductor Devices*, Springer Proceedings in Physics 306, https://doi.org/10.1007/978-981-97-1571-8_40

Si [11–17]. Large lattice mismatch and difference in thermal expansion coefficients are the inherent obstructions in the epitaxial growth of GaAs on Si. Efforts are made by various researchers to avoid generation of APD in the grown layer by changing the growth parameters, and also by choosing the substrate miscut [18–20]. Huang et al. [18] have shown that by choosing 5° miscut in substrate, the density of anti-phase domains is reduced. Recently, Petrushkov et al. [21] have shown that anti-phase domain density is less in GaSb($\bar{1}$00) layer in comparison to GaSb(100) layer while both are grown on substrates with 6° miscut. It is reported in literature that the presence of APD in the epilayer has a major impact on the in-plane electronic transport properties of such structures [22]. Non miscut substrates are commonly used in complementary metal–oxide–semiconductor processing [23] whereas both the non-miscut and miscut substrates are used in development of multi-junction solar cells, lasers and other optoelectronic devices.

It is therefore important to quantify the presence of APD as quickly as possible with the existing tools readily available in a laboratory to understand the growth mechanism. APD can be studied through various techniques such as Raman scattering [24], anisotropic chemical etching [25–27], reflection high-energy electron diffraction [25], transmission electron microscope [1, 28–30] and X-ray diffraction (XRD) techniques [14, 19, 31, 32]. Among these, XRD can provide the average information of the microstructure of the grown layer. However, no XRD study is available in the literature which addresses the issue of estimation of APD ratio of two sub-lattices that might be present in III–V polar layer grown on non-polar (100) substrate. In case of III–V layers grown on non-polar (111) substrates, some XRD reports are available that addresses the issue in context of 60° rotated twin around surface normal [111], and the same is depicted in Fig. 40.1a and is also discussed in subsequent section of this article [33, 34].

In order to estimate the ratio of population of the two sublattices from XRD for nearly lattice matched (111) epitaxy case, a method is proposed by Christian Koppka et al. [33] which uses an asymmetric φ scan diffraction pattern of the sample that is recorded in the range of 0°–360°. Here, three peaks appear from one sublattice and additional three peaks arise from the convolution of layer and substrate. Note that the six peaks appear at an interval of 60° in an asymmetric φ scan. Thereafter, two ω/2θ scans are recorded at the φ positions of two 60° separated consecutive peaks in φ scan. The deconvolution is performed to separate the layer peak information from the substrate wherever necessary. The areas under the layer peak in both the scans are estimated and their ratio gives the relative population of the two sublattices. This approach cannot be used for estimating the population ratio of two sublattices for (100) epitaxy as depicted in Fig. 40.1b due to the following reason. In case of (100) epitaxy, anti-phase domains give rise to 90° rotated sublattices around surface normal [100] viz. GaAs-IPD and GaAs-APD sublattices where IPD (APD) stand for in-phase (anti-phase) domain in which in-plane directions are parallel (90° ahead) with substrate direction [30, 34]. If we record a φ-scan (pole figure) from 0° to 360° for an asymmetric reflection e.g. (111) for (100) epitaxy case, only four peaks are observed at the interval of 90° as shown in Fig. 40.2a. The peaks arising from the two sub-lattices along with substrate peak convolute with each other and the APD ratio

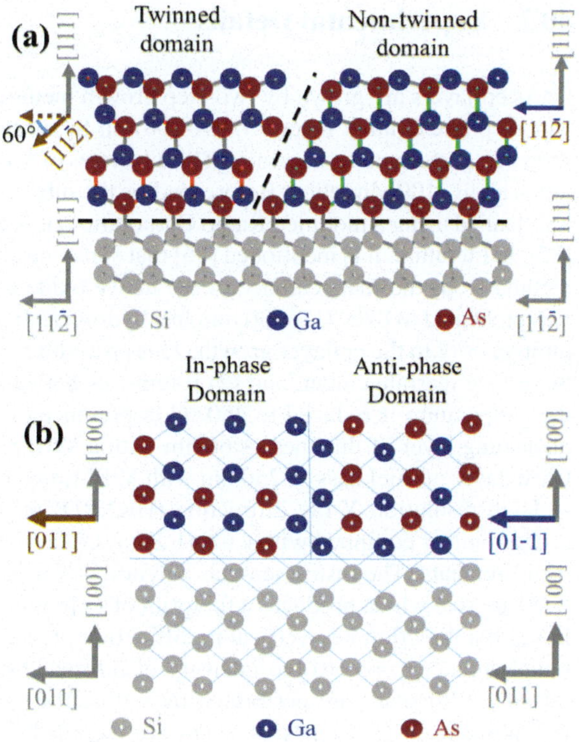

Fig. 40.1 **a** Formation of micro-twins in (111) epitaxy. **b** Formation of APD in (100) epitaxy

cannot be estimated [34]. In this article, a method for estimating the APD ratio in a polar layer viz. GaAs grown on a non-polar non-miscut (100) oriented substrate viz. Si (100) substrate is presented. The proposed method is applicable to any polar layer grown on (100) oriented non-polar substrate independent of the substrate miscut.

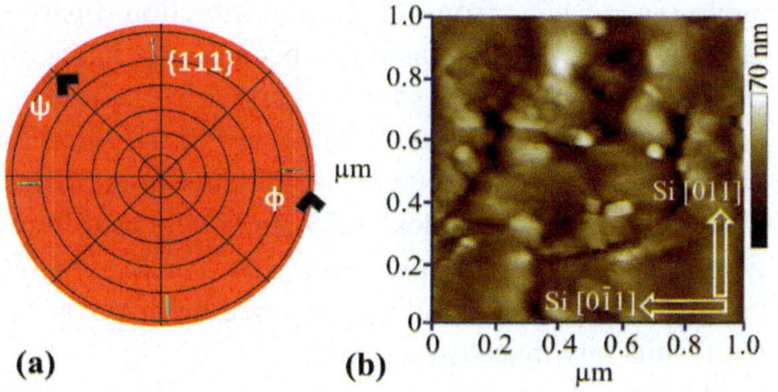

Fig. 40.2 **a** Pole figure of the sample acquired for (111) reflection, **b** atomic force microscopy image of GaAs epitaxial layer grown on Si substrates

40.2 Experimental Details

GaAs epilayers are grown by two-step growth method in a horizontal metal organic vapour phase epitaxy reactor (AIX-200) on Si substrate at 50 mbar pressure. The growth is performed on non-miscut Si substrates of ≈ 600 μm thickness having orientation along [100] direction. Before loading the substrate into the reactor, Si substrates are cleaned using a modified Radio Corporation of America (RCA) cleaning method [35]. In literature, it is mentioned [36] that substrate cleaning using above mentioned method stops the reappearance of the native oxide even after 1–2 h. The substrate was preheated to 870 °C for 30 min in a hydrogen (H_2) flow of ≈ 8 standard litre per minute, prior to the epilayer growth. This procedure has two benefits: (1) it promotes Si surface rearrangement, and (2) it removes native oxide from Si substrate. After the temperature is reduced to 450 °C in presence of a high flow of arsine, a GaAs nucleating layer of thickness ≈60 nm with a V/III ratio ≈340 is grown. After that GaAs layer of thickness ≈ 250 nm with V/III ratio ≈100 is grown at 670 °C.

High resolution X-ray diffraction (HRXRD) instrument, PANalytical X'PERT diffractometer equipped with CuKα1 X-rays (wavelength = 1.54056 Å) is used for measurements. The instrument has a hybrid 4 × monochromator (beam divergence ≈ 20 arc sec) which produces a footprint of ≈ 15 mm × 2 mm on the sample surface in sagittal direction i.e., vertical to diffraction plane, and a xenon gas filled proportional counter to record the intensity of diffracted beam. Omega (ω) and omega/2theta (ω/2θ) scans are performed by setting the opening of detector equal to 1°. For Williamson-Hall analysis, scans are recorded by inserting a triple bounce Ge (220) analyser crystal (acceptance angle ≈ 12 arc sec) in front of the detector. In this article, ω, θ, and 2θ are defined as the angles which incident X-ray beam makes with the sample surface, reflecting plane and diffracted X-ray respectively. On the other hand, φ is the azimuthal angle that lies between the projection of the incident X-ray beam on sample surface and sample in-plane $[01\bar{1}]$ direction whereas ψ is the angle between the growth direction i.e. [100] and surface normal of the corresponding diffraction plane. Pole figure is acquired for (111) planes which are inclined at an angle ($\psi = 54.7°$) from the growth direction. Instrumental broadening effects have been neglected in data analysis because the values of full width at half maximum (FWHM) of diffraction patterns is about five times larger than the instrumental broadening [37]. Wafer bowing [38] occurs in the samples where large difference in the values of thermal expansion coefficients of substrate and layer exist. Therefore, it may dominate for samples under study here as difference in thermal expansion coefficient is ≈ 110% for GaAs layer grown on Si substrate. However, wafer bowing can be neglected in the present case since the layer is of mosaic nature as discussed in the results and discussion section.

Reciprocal space maps (RSM) are recorded in asymmetric reflection geometry at 0° and 90° azimuths by keeping [011] and $[0\bar{1}1]$ directions parallel to sagittal direction respectively. For Williamson-Hall (WH) analysis, HRXRD scans are recorded in symmetric reflections geometry or in skew-symmetric reflection geometry if $\psi \neq 0°$ [13, 39]. Atomic force microscopy is performed on a Bruker (MultiMode 8-h) machine. A conductive tip with radius of curvature ≈ 10 nm made of platinum/ iridium is used to scan over $1.0 \times 1.0\ \mu m^2$ area over the film surface.

40.3 Results and Discussion

Figure 40.1a shows the formation of micro-twins in (111) epitaxy case, where the micro-twins formation arises due to two different type of atom's stacking sequences in the layer along [111] direction, out of them one is similar to the stacking sequence present in the layer. Therefore, in φ-scan for any allowed asymmetric reflection, three peaks at 120° interval appear from one sublattice and additional three peaks appear at 120° interval from another sublattice. Note that the angular separation between the adjacent peaks of two sublattices is 60° [34]. Therefore, the peaks from both the sub-lattices do not superimpose and allow us to determine of their relative population. Figure 40.1b shows the formation of the APD due to mixed domain nucleation within one terrace [1]. This makes Ga-prelayer and As-prelayer domains on the same terrace. One sublattice can be called as in-phase domain (IPD) since it carries the directions of the substrate and another sublattice is called as anti-phase domain (APD) as in-plane direction is 90° ahead of the corresponding direction of in-phase domains. Therefore, two domains are formed in the layer and are rotated by 90° in-plane with respect to the substrate normal [100] [1, 40, 41]. For ease of presentation, one of the domains is marked as GaAs-IPD (GaAs-APD) whose [011] ($[01\bar{1}]$) direction is parallel with Silicon [011] direction.

Figure 40.2a shows the pole figure of the sample acquired for (111) skew-symmetric reflection. At $\psi = 54.7°$, four distinct peaks at the interval of 90° are observed in the φ circle. In an earlier publication from our group [34], a similar information was obtained for GaP/Si(001) case where it was concluded that the anti-phase domain peaks cannot be recorded separately in HRXRD. The reason can be described as for GaAs-IPD case where peaks will appear at 0°, 90°, 180° and 270° for any allowed asymmetric reflection and for GaAs-APD case the peaks will appear at 90°, 180°, 270° and 0° for the same asymmetric reflection. Thus, the peaks will overlap and the possibility of the estimation of relative population vanishes. Figure 40.2b shows the AFM image of the surface of GaAs/Si sample where distinct camouflage patterns of the grains along with trenches are observed. We have already seen such type of patterns in GaAs/Si case earlier [42]. Recently, we have shown that the AFM picture of tensile strained polar layer grown on polar substrate contains cracks running along one directions [13] and in case of compressively strained polar layer grown on polar substrate, no major difference in the topography of layer between the two in-plane orthogonal directions is seen and the trenches are less visible [43]. It is clear that we do not observe any camouflage pattern in those layers. Recently Ostheim et al. [22] have shown that lack of camouflage pattern is indicative of a less number of APD since such patterns arise due to APBs that pierce through the surface of the polar layer in the form of trenches. They also observed that when the polar layer is APD free, no trenches can be seen on the surface, and the surface does not exhibit the pronounced camouflage pattern. With this discussion, it can be fairly assumed that the camouflage pattern is indicative of the presence of APD in the grown layer.

After observing the camouflage pattern in the AFM image, which is having alphabet "L" shape, we recorded several ω scans for (400) symmetric reflections by varying the value of azimuth from 0° to 360° in the step of 30° to map in-plane crystalline quality of the sample. In a recent article, we have also mapped in-plane crystalline quality of the layer using this method but those samples were APD free (polar layer on polar substrate) [13, 39, 43]. Figure 40.3 shows the variation of FWHM of the rocking curves recorded for (400) symmetric reflections as a function of azimuth. A sinusoidal type variation of FWHM with respect to azimuth indicates about the existence of an anisotropy in the microstructures between the two in-plane orthogonal directions (0° and 90°). In our experiment at 0° and 90° azimuths, the directions perpendicular to diffraction plane i.e., sagittal directions are aligned with Silicon substrate directions [011] and $[0\bar{1}1]$ respectively. The symmetric omega scans reveal the information primarily along sagittal direction [13, 39, 43]. Therefore, the observed difference in the FWHMs for 0° (180°) and 90° (270°) azimuths shown in Fig. 40.3 indicates that there exists an anisotropy in the microstructures of GaAs epilayer along Si[011] and Si $[0\bar{1}1]$ directions. It is observed that the value of FWHM is larger along Si[011]/Si $[0\bar{1}\bar{1}]$ direction in comparison to Si $[0\bar{1}1]$/Si $[01\bar{1}]$ direction. One can therefore infer that the density of defects/dislocation is large along Si[011] direction in comparison to other in-plane orthogonal direction.

Keeping in the mind that the density of defects/dislocation is linked with the layer relaxation, reciprocal space map (RSM) of asymmetric {422} reflection at 0° (90°) azimuth is recorded, where X-ray scattering plane is aligned along $[01\bar{1}]$ ([011]) as shown in Fig. 40.4. The relaxation [13] values estimated from the RSM show that the layers are fully relaxed. However, the variation of counts shown in Fig. 40.4 looks slightly different along the two azimuths indicating one of the domains is slightly more relaxed. It is important to note that the critical thickness of GaAs epilayer on Si substrate is below 20 Å [44, 45] while the thickness of GaAs epilayer in our

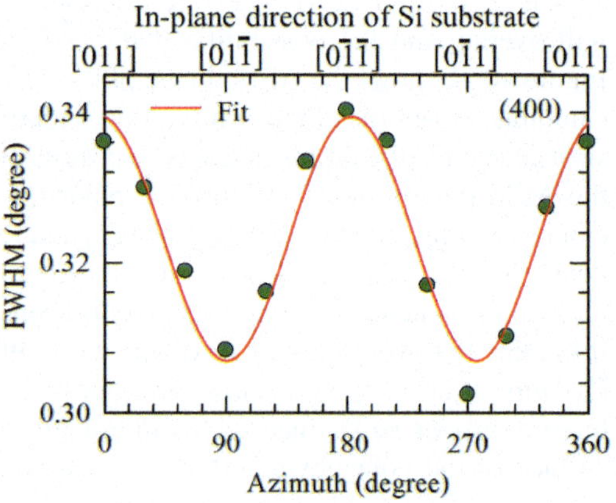

Fig. 40.3 FWHMs of rocking curves of GaAs epitaxial layer grown on Si substrate, plotted as a function of azimuth where solid line is a guide to the reader's eye

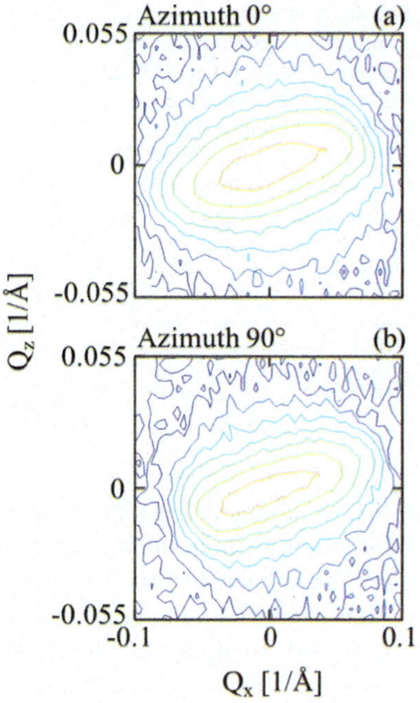

Fig. 40.4 Reciprocal space maps of GaAs/Si layer recorded at 0° and 90° azimuths for {422} reflection

case is 0.3 μm. Therefore, the GaAs layer is expected to be fully relaxed through the generation of misfit dislocations and other factors arising from the polar-on-nonpolar growth.

We have recently shown for tensile strained layers viz. GaP/GaAs and compressively strained layers viz. InP/GaAs that the initial relaxation of the layers occurs due to formation of α-dislocations [13, 43]. For compressively strained layer having only one sublattice type the initial relaxation occurs along $[0\bar{1}\bar{1}]$ where line of dislocation for α dislocations is along $[01\bar{1}]$ [43]. As discussed in earlier part of this article, in GaAs/Si two sublattices GaAs-IPD and GaAs-APD are present, therefore initial relaxation occurs along $[0\bar{1}\bar{1}]$ direction of each GaAs sublattice. Therefore, initial relaxation occurs along Si $[0\bar{1}\bar{1}]$ for GaAs-IPD sublattice and along Si $[01\bar{1}]$ for GaAs-APD sublattice knowing that the $[0\bar{1}\bar{1}]$ direction of GaAs-IPD/GaAs-APD sublattice is along Si $[0\bar{1}\bar{1}]$/Si $[0\bar{1}1]$ respectively. Such an understanding finds its support from two observations previously mentioned; first the RSM study where layer is fully relaxed along both in-plane directions, and second the presence of camouflage pattern, that is typically L shaped, in the AFM image of the sample.

The observation that variation of counts/contours in RSM scan along q_x is different along the two azimuths as shown in Fig. 40.4 indicates that the defects/distribution is different along the two directions. The large broadening along q_x observed at RSM recorded at azimuth 0° in comparison to the one recorded at azimuth 90° is indicative of smaller lateral grain size (LCL) along Si[011]. As the broadening along q_z in RSM scans seems similar for both the azimuths indicates that the vertical coherence length

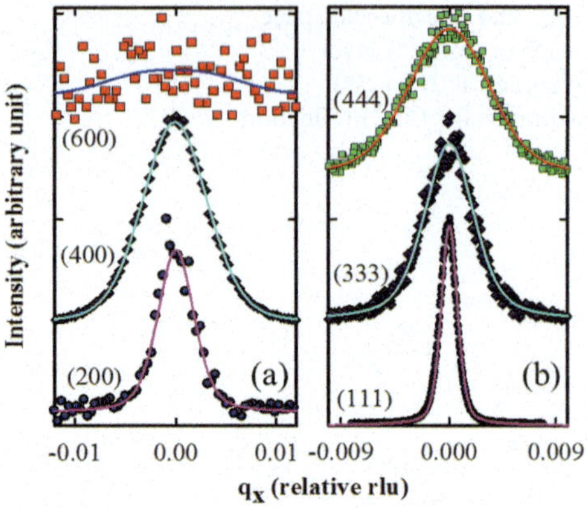

Fig. 40.5 Intensity versus q_x plot **a** for (200), (400) and (600) reflections, **b** for (111), (333) and (444) reflections for GaAs/Si sample. The overlaying solid lines show the pseudo-Voigt fitting of the experimental data

(VCL) is similar and is independent of azimuthal variation. This is in agreement with the literature where the measured values of VCL are found to be equivalent to the layer thickness [46–50].

It is available in literature that APD influence the width of certain reflections [14, 15, 42, 46]. To appreciate the effect of APD on certain reflections, we recorded two sets of ω scans: one set of symmetric (L00) reflections at $\psi = 0°$ viz. (200), (400), (600) and another set of skew-symmetric (LLL) reflections at $\psi = 54.7°$ viz. (111), (333), (444). Figure 40.5a,b shows the diffraction patterns corresponding to these reflections along with overlaying line arising out of the pseudo-Voigt fitting of the curves. Out of which (200) and (600) reflections were weak and (600) reflection was not at all measurable due to poor signal to noise (S/N) ratio. Rest of the reflections seems to show good S/N ratio. Inherently (600) reflection ($2\theta = 109.7°$) is a weak reflection as the structure factor has the term ($f_{Ga} - f_{As}$) and form factors f_{Ga} and f_{As} has minimal contrast at $2\theta = 109.7°$.

To complicate the problem further, the presence of APD in the system broadens the (600) reflection because the structure factor for these reflections depends upon the relative positions of Ga and As atoms [14]. We were unable to measure FWHM of (600) reflection with accuracy. The (200) reflection ($2\theta = 31.6°$) is also affected by APD but it is measurable as form factors f_{Ga} and f_{As} have large contrast at $2\theta = 31.6°$ in comparison to $2\theta = 109.7°$. The (111), (333) and (444) reflections are intense and unaffected by presence of APD as for (111) and (333) reflections, the structure factor (s) is proportion to $\sqrt{f_{Ga}^2 + f_{As}^2}$, and for (444) reflection $s \propto (f_{Ga} + f_{As})$ [51]. All the skew-symmetric reflections i.e. (111), (333) and (444) are intense and FWHM of all the peaks can be precisely measured.

As discussed in the introduction section that for (111) epitaxy case the area under the layer peaks in asymmetric ω/2θ scans have been related with the population of sublattices. However, it seems inappropriate to use asymmetric ω/2θ scans for such type of estimation due to the following reasons. In asymmetric ω/2θ scans, the reciprocal lattice vector does not traverse along in-plane direction [52, 53], therefore the intensity of the HRXRD scan cannot be precisely linked with the lateral grain size or lateral coherence length (LCL). It is important to note that the vertical grain size or vertical coherence length (VCL) remains uniform across the sample which we have also discussed in the previous paragraph [46–50]. Therefore, LCL of the sublattices is the parameter that can be used to estimate the relative population of the sublattices.

To find the LCL of sublattices, Williamson-Hall analysis using symmetric scans is a well-established procedure [13, 39, 54]. As discussed earlier, for a Cu Kα_1 source only two reflections i.e. (200) and (400) are available for WH analysis since (600) reflection is not measurable. In WH analysis, a linear fit of the respective values of the FWHM i.e., Δq_x for ω scans is made using the following formula,

$$(\Delta q_x)^n = (\Delta q_{LCL})^n + (\alpha_{\psi=0^\circ} q)^n \tag{40.1}$$

where Δq_{LCL} is lateral coherence length in reciprocal units, $\alpha_{\psi=0^\circ}$ is the tilt angle between the mosaic blocks present in the layer, and n is defined as $n = 1 + (1 - f)^2$ where f is the fraction of the Lorentzian component in the pseudo-Voigt profile such that value of n lies between 1 and 2. From this exercise, the values of LCL and tilt can be obtained from a set of ω scans. It is obvious that at least three symmetric reflections are required for a good accuracy in WH analysis. Thus, the accuracy of the conventional WH analysis is limited by the strength of the (600) symmetric reflection for GaAs epilayer grown on Si substrates [15, 42, 46].

In view of such a limitation, modified Williamson-Hall analysis is used where a set of parallel planes inclined to surface plane viz. (111), (333) and (444) are used [15, 46]. In modified WH analysis, a linear fit of the respective values of the FWHM i.e. Δq_x for ω scans is made using the following formula to find LCL and misorientation of the grains $\alpha_{\psi = 54.7^\circ}$,

$$(\Delta q_x)^n = (\Delta q_{LCL})^n + (\alpha_{\psi=54.7^\circ} q)^n \tag{40.2}$$

Modified WH analysis based on ω scans of (111), (333) and (444) skew-symmetric reflections are carried out and the results are shown in Fig. 40.6a for both the azimuths. As obvious from Fig. 40.6a, the experimental data can be accurately fitted with a straight line for both the azimuths confirming the usefulness of modified WH analysis for evaluating the micro-structures of Zincblende epilayers. Table 40.1 summaries the values of LCL determined for both the azimuths from the modified WH.

After the determination of $\alpha_{\psi = 54.7^\circ}$, the tilt ($\alpha_{\psi = 0^\circ}$) and twist ($\alpha_{\psi = 90^\circ}$) values are estimated by fitting the data using the following formula [13, 15, 39, 43, 46],

$$\alpha_\psi^n = (\alpha_{\psi=0^\circ} \cos\psi)^n + (\alpha_{\psi=90^\circ} \sin\psi)^n \tag{40.3}$$

The fits to Eq. 40.3 for several reflections at both the azimuths are shown in Fig. 40.6b. Table 40.1 summaries the values of tilt and twist determined for both the azimuths from the fitting. The value of LCL is of sub-micrometer size which shows that layer is mosaic in nature. The values of LCL are also slightly different along Si[011] and Si $[0\bar{1}1]$ directions as shown in Table 40.1. It indicates that the crystalline

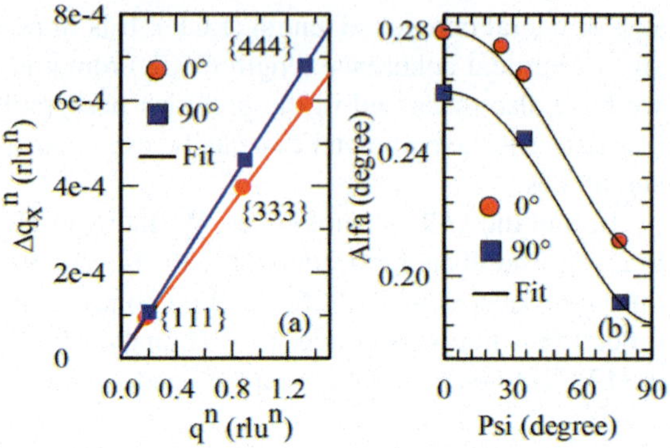

Fig. 40.6 Williamson-Hall analysis: **a** the width (Δq_x) obtained from pseudo-Voigt fitting of the omega scan diffraction peaks, n is related to Lorentzian fraction in the pseudo-Voigt fitting, is plotted as a function of q^n, **b** the values of α_{hkl} are plotted as a function of ψ for 0° and 90° azimuths

Table 40.1 Summary of the microstructural parameters along with the values of dislocation density of GaAs/Si epitaxial layer estimated from the Williamson-Hall analysis

Azimuth (°)	Silicon substrate direction ⊥ to diffraction plane	WH analysis				Dominant dislocations	Density of 60° dislocations (cm^{-2})
		n	LCL (Å)	Tilt and Twist			
				Tilt (Twist) (deg.)	Ratio		
0	[011]	1.4	3016	0.28 (0.2)	1.4	60° Perfect	6.6×10^9
90	[0$\bar{1}$1]	1.4	4317	0.26 (0.18)	1.4	60° Perfect	5.7×10^9

quality of GaAs layer is better along Si [0$\bar{1}$1] direction. Furthermore, the tilt is less along Si [0$\bar{1}$1] direction when compared with Si[011] direction. It also indicates that the crystalline quality is better along Si [0$\bar{1}$1] direction. Note that the values of twist along the two orthogonal directions are slightly different, however, the Tilt/Twist ratio is nearly same (≈1.4) along the two directions as clearly shown in Table 40.1. It is also understood that the dislocations in compressively strained GaAs epilayers grown on Si are mainly 60° perfect dislocations along both the directions since Tilt/Twist ratio remains ≈1.4 [13, 39, 43, 46]. This is in agreement with earlier observations in case of compressively strained InP/GaAs epitaxial layers [43]. Table 40.1 shows the dislocation density along both the azimuths estimated using the following formula [39, 43],

$$D_{60^\circ perfect\ dislocations} = \frac{4\alpha^2_{\psi=0^\circ}}{4.35a^2} \tag{40.4}$$

The value of dislocation density for 60°perfect dislocations is slightly higher for azimuth 0° case, which is consistent with smaller LCL and larger tilt values at azimuth 0° in comparison to azimuth 90° case. As we have seen from the RSM observations that LCL is smaller along Si[011] direction.

From AFM image of the sample, a slight elongation of the grown layer along $[0\bar{1}1]$ direction can be seen. These observations match with the trends seen in LCL values which are shown in Table 40.1. The APD ratio can be estimated by the ratio of LCL. The ratio comes out to be ≈ 1.43 where GaAs-APD sublattice population dominates the GaAs-IPD sublattice population. The GaAs layer was grown on non-miscut substrate; hence it may be expected that the sublattice ratio should be 1:1 if growth parameters are not kept in such a way to promote a particular sublattice population. For the sample under study, high flow of Arsine was used before the growth of GaAs layer. It favours that the first layer of atoms should be of Arsenic atoms which eventually is the case of GaAs-APD sublattice for the mixed nucleation case.

40.4 Conclusion

Camouflage pattern is observed in AFM picture of GaAs polar layer grown on non-polar Si substrate which is observed in APD rich layers. The APD can grow through mixed domain nucleation at the plane terrain and incongruent domain nucleation at surface steps. The anisotropic values of FWHM for (400) reflection as a function of azimuth is observed indicating that the in-plane crystalline quality is not uniform but depends on the azimuth. The reciprocal space maps at orthogonal in-plane directions revealed that the layer is fully relax along both orthogonal directions owing to large thickness of the grown layer in comparison to the critical layer thickness. In compressive or tensile strained polar layers grown on polar substrate, initial relaxation of the layer occurs along a particular direction which depend on the type of the strain in layer. However, it is found that in APD prone layer, initial relaxation occurs simultaneously along both the orthogonal directions as the domains are 90° twisted. The effect of APD is prominent on (600) reflection, hence for the estimation of LCL and other microstructural parameters modified WH analysis is used. From the estimated values of LCL along orthogonal directions, the APD ratio is estimated. The proposed method for the estimation of APD ratio can be used on any (100) epitaxy case.

Acknowledgements The authors thank Dr M. P. Joshi and Dr Rajmohan for the thickness measurements, and Mr. U.K. Ghosh and Mr. A. Khakha for their help during the growth of the samples and Dr. Neha Sharma and Dr. J. Jayabalan for AFM measurements. Authors also acknowledge Dr. Shankar V. Nakhe, Director, RRCAT for his constant support during the course of this work.

References

1. S.Y. Woo, S. Hosseini Vajargah, S. Ghanad-Tavakoli, R.N. Kleiman, G.A. Botton, Direct observation of anti-phase boundaries in heteroepitaxy of GaSb thin films grown on Si(001) by transmission electron microscopy. J. Appl. Phys. **112**, 074306 (2012)

2. Z.J. Yu et al., GaAs/silicon PVMirror tandem photovoltaic mini-module with 29.6% efficiency with respect to the outdoor global irradiance. Prog. Photovolt. Res. Appl. **27**, 469–475 (2019)
3. R.M. Weikle et al., Submillimeter-Wave Schottky diodes based on heterogeneous integration of GaAs onto silicon, in *2019 United States National Committee of URSI National Radio Science Meeting (USNC-URSI NRSM)*, pp. 1–2 (2019). https://doi.org/10.23919/USNC-URSI-NRSM.2019.8713040
4. P. Bhasker, C. Shang, J. Bowers, N. Dagli, Low loss, compact waveguides in GaAs/Oxidized AlGaAs layers directly grown on silicon, in *Conference on Lasers and Electro-Optics* STh3N.7 (Optical Society of America, 2019). https://doi.org/10.1364/CLEO_SI.2019.STh3N.7
5. K.T. VanSant et al., Toward low-cost 4-terminal GaAs/Si tandem solar cells. ACS Appl. Energy Mater. **2**, 2375–2380 (2019)
6. J.A. Carlson, J.M. Dallesasse, Heterogeneous integration of light-emitting transistors on silicon for hybrid electronic-photonic logic circuitry, in *Conference on Lasers and Electro-Optics* JTh2A.60 (Optical Society of America, 2019)
7. J. Yang et al., Thin Ge buffer layer on silicon for integration of III–V on silicon. J. Cryst. Growth **514**, 109–113 (2019)
8. B. Shi, Y. Han, Q. Li, K.M. Lau, 1.55-μm lasers epitaxially grown on silicon. IEEE J. Sel. Top. Quantum Electron. **25**, 1–11 (2019)
9. M. Rio Calvo et al., Molecular-beam epitaxy of GaSb on 6°-offcut (0 0 1) Si using a GaAs nucleation layer. J. Cryst. Growth **529**, 125299 (2020)
10. C.S.C. Barrett et al., Dissolution of antiphase domain boundaries in GaAs on Si(001) via post-growth annealing. J. Mater. Sci. **54**, 7028–7034 (2019)
11. K. Volz et al., GaP-nucleation on exact Si (001) substrates for III/V device integration, in *15th International Conference Metalorganic Vapor Phase Epitaxy (ICMOVPE-XV),* vol. 315, pp. 37–47 (2011)
12. T.J. Grassman et al., Nucleation-related defect-free GaP/Si(100) heteroepitaxy via metal-organic chemical vapor deposition. Appl. Phys. Lett. **102**, 142102 (2013)
13. R. Kumar et al., Observation of anisotropic distribution of microstructure in GaP/GaAs epitaxial layers. J. Appl. Phys. **120**, 135307 (2016)
14. D.A. Neumann, H. Zabel, R. Fischer, H. Morkoç, Structural properties of GaAs on (001) oriented Si and Ge substrates. J. Appl. Phys. **61**, 1023 (1987)
15. R. Kumar, T. Ganguli, V. Chouhan, V.K. Dixit, The study of microstructure of III-V polar on non-polar heterostructures by HRXRD. J. Nano-Electron. Phys. **3**, 17 (2011)
16. A. Georgakilas et al., Generation and annihilation of antiphase domain boundaries in GaAs on Si grown by molecular beam epitaxy. J. Mater. Res. **8**, 1908–1921 (1993)
17. P. Sheldon, K.M. Jones, M.M. Al-Jassim, B.G. Yacobi, Dislocation density reduction through annihilation in lattice-mismatched semiconductors grown by molecular-beam epitaxy. J. Appl. Phys. **63**, 5609 (1988)
18. S.H. Huang, G. Balakrishnan, A. Khoshakhlagh, L.R. Dawson, D.L. Huffaker, Simultaneous interfacial misfit array formation and antiphase domain suppression on miscut silicon substrate. Appl. Phys. Lett. **93**, 071102 (2008)
19. S.F. Fang et al., Gallium arsenide and other compound semiconductors on silicon. J. Appl. Phys. **68**, R31–R58 (1990)
20. R. Kaplan, LEED study of the stepped surface of vicinal Si (100). Surf. Sci. **93**, 145–158 (1980)
21. M.O. Petrushkov et al., Effect of the crystallographic orientation of GaSb films on their structural properties during MBE heteroepitaxy on vicinal Si(001) substrates. Semiconductors **54**, 1548–1554 (2020)
22. L. Ostheim et al., Effect of the interface morphology on the lateral electron transport in (001) GaP/Si heterostructures. J. Appl. Phys. **126**, 215704 (2019)
23. R. Loo et al., Selective area growth of InP on on-axis Si (001) substrates with low antiphase boundary formation. ECS Trans. **41**, 249–263 (2011)
24. R. Fischer et al., Growth and properties of GaAs/AlGaAs on nonpolar substrates using molecular beam epitaxy. J. Appl. Phys. **58**, 374–381 (1985)

25. P.N. Uppal, H. Kroemer, Molecular beam epitaxial growth of GaAs on Si(211). J. Appl. Phys. **58**, 2195–2203 (1985)
26. Y. Li, L.J. Giling, A closer study on the self-annihilation of antiphase boundaries in GaAs epilayers. J. Cryst. Growth **163**, 203–211 (1996)
27. K. Morizane, Antiphase domain structures in GaP and GaAs epitaxial layers grown on Si and Ge. J. Cryst. Growth **38**, 249–254 (1977)
28. J.P. Gowers, TEM image contrast from antiphase domains in GaAs: Ge(001) grown by MBE. Appl. Phys. A **34**, 231–236 (1984)
29. Z. Liliental-Weber, E.R. Weber, L. Parechanian-Allen, J. Washburn, On the use of convergent-beam electron diffraction for identification of antiphase boundaries in GaAs grown on Si. Ultramicroscopy **26**, 59–63 (1988)
30. Y. Li et al., On the formation of antiphase domains in the system of GaAs on Ge. J. Cryst. Growth **163**, 195–202 (1996)
31. A. Létoublon et al, X-ray study of antiphase domains and their stability in MBE grown GaP on Si, in *Proceedings on 16th International Conference on Molecular Beam Epitaxy ICMBE,* vol. 323, pp. 409–412 (2011)
32. W. Guo et al., Thermodynamic evolution of antiphase boundaries in GaP/Si epilayers evidenced by advanced X-ray scattering. Appl. Surf. Sci. **258**, 2808–2815 (2012)
33. C. Koppka et al., Suppression of rotational twin formation in virtual GaP/Si(111) substrates for III–V nanowire growth. Cryst. Growth Des. **16**, 6208–6213 (2016)
34. R. Roychowdhury et al., Role of surface energy on the morphology and optical properties of GaP micro and nano structures grown on polar and non-polar substrates. Appl. Surf. Sci. **419**, 957–967 (2017)
35. V.K. Dixit et al., Effect of two-step growth process on structural, optical and electrical properties of MOVPE-grown GaP/Si. J. Cryst. Growth **310**, 3428 (2008)
36. D.J. Meyer, in *Silicon Epitaxy (Chap. 10- Si-Based Alloys: SiGe and SiGe:C)*, vol. 72 (Academic Press, 2001)
37. J.E. Ayers, The measurement of threading dislocation densities in semiconductor crystals by X-ray diffraction. J. Cryst. Growth **135**, 71–77 (1994)
38. R.M. Lum et al., Effects of misfit dislocations and thermally induced strain on the film properties of heteroepitaxial GaAs on Si. J. Appl. Phys. **64**, 6727–6732 (1988)
39. R. Kumar, V.K. Dixit, T.K. Sharma, Anisotropic distribution of dislocations density in tensile strained GaP/GaAs epilayers. Vacuum **154**, 214–217 (2018)
40. Q.H. Xie et al., Asymmetric distribution of microtwins in a GaAs/Si heterostructure grown by molecular beam epitaxy. Appl. Phys. Lett. **57**, 2803–2805 (1990)
41. G.A. Devenyi et al., The role of vicinal silicon surfaces in the formation of epitaxial twins during the growth of III-V thin films. J. Appl. Phys. **110**, 124316 (2011)
42. R. Kumar et al., Study of the microstructure information of GaAs epilayers grown on silicon substrate using synchrotron radiation. J. Synchrotron Radiat. **23**, 238–243 (2016)
43. R. Kumar, V.K. Dixit, C. Mukherjee, T.K. Sharma, Anisotropic distribution of microstructure in compressively strained InP/GaAs epitaxial layers. Superlattices Microstruct. (2018). https://doi.org/10.1016/j.spmi.2018.06.028
44. R. Hull, A. Fischer-Colbrie, Nucleation of GaAs on Si: experimental evidence for a three-dimensional critical transition. Appl. Phys. Lett. **50**, 851–853 (1987)
45. G. Cohen-Solal, F. Bailly, M. Barbé, Critical thickness in heteroepitaxial growth of zinc-blende semiconductor compounds. J. Cryst. Growth **138**, 68–74 (1994)
46. R. Kumar et al., Evaluation of vertical coherence length, twist and microstrain of GaAs/Si epilayers using modified Williamson-Hall analysis. J. Nano-Electron. Phys. **6**, 02010 (2014)
47. X.L. Wang et al., The effects of LT AlN buffer thickness on the properties of high Al composition AlGaN epilayers. Mater. Lett. **60**, 3693–3696 (2006)
48. E. Arslan, M.K. Ozturk, A. Teke, S. Ozcelik, E. Ozbay, Buffer optimization for crack-free GaN epitaxial layers grown on Si(1 1 1) substrate by MOCVD. J. Phys. Appl. Phys. **41**, 155317 (2008)
49. J. Kozlowski, R. Paszkiewicz, M. Tlaczala, Structure characterization of (Al, Ga)N epitaxial layers by means of X-Ray diffractometry. Phys. Status Solidi B **228**, 415–418 (2001)

50. B.S. Yadav et al., Highly oriented GaN films grown on ZnO buffer layer over quartz substrates by reactive sputtering of GaAs target. Thin Solid Films **517**, 488–493 (2008)
51. J.E. Ayers, *Heteroepitaxy of Semiconductors: Theory, Growth, and Characterization* (CRC Press, Boca Raton, 2007)
52. M. Birkholz, *Thin Film Analysis by X-Ray Scattering* (Wiley-VCH, New York, 2006)
53. P.F. Fewster, *X-Ray Scattering from Semiconductors and Other Materials* (World Scientific, Singapore, 2013)
54. G.K. Williamson, W.H. Hall, X-ray line broadening from filed aluminium and wolfram. Acta Metall. **1**, 22 (1953)

Chapter 41
A Vertical GaN Split Gate Trench MOSFET Device with Reduced Switching Energy Losses

Nilesh Kumar Jaiswal and V. N. Ramakrishnan

41.1 Introduction

Thanks to the outstanding material features of Gallium Nitride (GaN), such as a high critical breakdown field and high electron mobility, GaN transistors have been regarded as a promising solution for next generation power conversion systems [1–3]. GaN power devices are being investigated in lateral and vertical structures. The development of lateral GaN transistors has been challenged by the normally-on behavior [4, 5] and normally-off cascade [6–8] in the device, which result in a large power loss and thus affect the switching performance [9, 10]. Vertical GaN devices are proving to be a favorable applicant for power switches due to their normally-off behavior and ability to withstand large breakdown voltages [11]. One of the major challenges of traditional vertical GaN transistors is the thicker UID-GaN region required in order to keep the 2DEG away from the implanted layer, which has resulted in an increase in specific on-resistance [12, 13].

Several techniques to reduce specific on-resistance have been proposed in traditional vertical GaN transistors, such as vertical channel junction FET (JFET) [14, 15]. The trench based vertical GaN MOSFET (TG-MOSFET) technology makes a more attractive device and provides reduced conduction losses due to the absence of parasitic JFETs. In addition, TG-MOSFETs has high packing densities, which lowers the specific on-resistance (R_{on}) by parallel conduction of a large number of unit cells on one chip [16–20]. As a result of these features, TG-MOSFET is well suited to control switching and converters [21]. There are substantial gate-to-drain capacitance (C_{GD}) and charge (Q_{GD}) issues with TG-MOSFETs because of the

N. K. Jaiswal (✉)
Department of Electronics and Communication Engineering, Shri Vishnu Engineering College for Women, Bhimavaram 534202, India
e-mail: nileshsonu.jaiswal2@gmail.com

V. N. Ramakrishnan
Department of Micro and Nanoelectronics, Vellore Institute of Technology, Vellore 632014, India

R. Singh et al. (eds.), *The Physics of Semiconductor Devices*, Springer Proceedings in Physics 306, https://doi.org/10.1007/978-981-97-1571-8_41

small thickness of oxide that separates from gate to drain, resulting in high switching losses in the circuits [22]. The two important methods to reduce Q_{GD} are: (1) using a thick oxide layer on the bottom of trenches [23] and (2) using split-gate technology [24]. The electric field (E-field) and on-resistance degradation are mainly due to an increase in the bottom oxide thickness. However, the deposition of high quality thick bottom oxide is very challenging. Hence, high power GaN vertical split-gate structures are gaining more attention. The split-gate structure uses an insulated field-plate (FP) between the gate and drain electrode to decrease the C_{GD} and Q_{GD} and, thus, leads to low switching losses.

Despite the fact that the SGT-MOSFET has been demonstrated to have an excellent figure of merit (FOM) [25, 26], the switching loss of the device has not been reported. To analyze the switching loss of the SGT-MOSFET, a simulation study has been carried out using a device-circuit-integrated model in this work. Remarkable improvements in switching energy loss and power loss have been achieved in the SGT-MOSFET compared with the conventional TG-MOSFET at the same voltage rating.

41.2 Simulation Methodology

In order to validate the model, a 600 V TG-MOSFET was designed and simulated using the TCAD Sentaurus simulator [27]. The TG-MOSFET was properly shown with the use of an appropriate physics model, presented in our previous paper [26], to match the experimental results. The simulation methodology has been adopted by computing the Poisson's equation, including the electron and hole continuity equations and 2D drift–diffusion carrier transport equations. The recombination phenomenon was followed into account by using three processes, specifically the Radiative, Auger, and Shockley–Read–Hall (SRH) models for a fixed minority-carrier lifetime. In order to take into consideration the effects of velocity saturation, the Caughey Thomas field dependent mobility model and Masetti model for doping dependency were incorporated into the simulation. In our previous work, we have reported a systematic analysis and comparison in terms of lower on-resistance and higher frequency FOMs of the proposed SGT-MOSFET when compared to conventional TG-MOSFET devices, as shown in Fig. 41.1a, b. Our simulation results show that the proposed SGT-MOSFET has lower FOMs when compared to a conventional TG-MOSFET with a similar breakdown voltage (V_{br}) of 600 V. The device performance parameters of TG-MOSFET and SGT-MOSFET devices are included in Table 41.1.

In this paper, we used the TCAD mixed-mode platform to simulate the switching behavior of both devices using a device-circuit hybrid simulator. The hybrid simulator establishes it easy to connect the device to circuit models, hence providing a platform for simulating emerging devices.

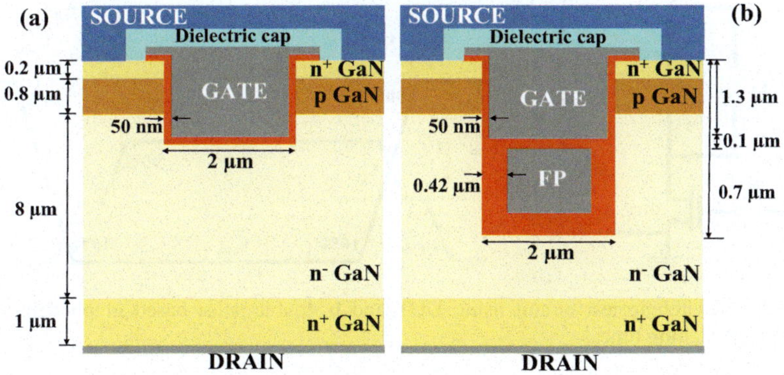

Fig. 41.1 Schematic device structure of **a** TG-MOSFET and **b** SGT-MOSFET

Table 41.1 Comparison of the device performances

Parameter	TG-MOSFET	SGT-MOSFET
R_{on} (mΩ cm^2)	0.68	0.46
V_{br} (V)	600	660
E_{ox} (MV/cm)	3.85	2.3
C_{GD} (nF/cm^2)	0.46	0.09
Q_{GD} (nC/cm^2)	312	154
HFOM ($C_{GD} \times R_{on}$) (mΩ nF)	0.3128	0.0414
HFOM ($Q_{GD} \times R_{on}$) (mΩ nC)	212.16	70.84

41.3 Switching Characteristics Analysis

The switching characteristics of the device under test (DUT) i.e., a TG-MOSFET and a SGT-MOSFET, have been simulated for the double-pulse test circuit, as shown in Fig. 41.2a. We have designed the test circuit with an inductive load ($L_1 = 100\,\mu$H) and freewheeling GaN schottky barrier diode (GaN SBD). A supply voltage (V_{DS} = 480 V) and an external gate resistance ($R_g = 2\ \Omega$) are selected for the numerical study [28]. A pulsed voltage source has been utilised to drive the gate in the simulation, allowing for direct comparison to the switching time calculation as given in Fig. 41.2b. The rise (t_r) and fall time (t_f) along with the on-delay ($t_{d(on)}$) and off-delay time ($t_{d(off)}$) are shown in Fig. 41.2b.

Figure 41.3a,b displays the TG-MOSFET and SGT-MOSFET waveforms during switched-off transients. The SGT-MOSFET has a $t_{d(off)}$ of 0.06 μs, a t_f of 0.07 μs, and total switched-off time (t_{off}) of 0.13 μs, which are about 2 × faster than that of TG-MOSFET. The switched-on transients are shown the TG-MOSFET and SGT-MOSFET waveforms in Fig. 41.4a, b. The SGT-MOSFET has a $t_{d(on)}$ of 0.01 μs, a t_r of 0.05 μs, and a total switched-on time (t_{on}) of 0.06 μs, which are about 2.5 ×

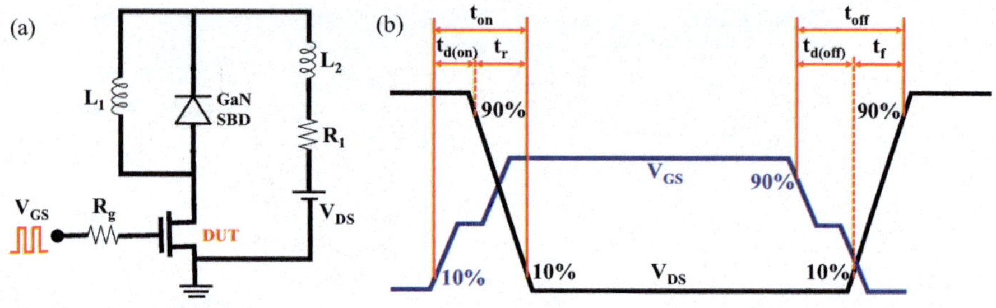

Fig. 41.2 **a** Switching test circuit using DUT and **b** double-pulse based graph definition for calculating switching times

shorter than that of TG-MOSFET. The t_{on} and t_{off} is observed with a similar current rating (24 A). We noted that the SGT-MOSFET has been exhibited faster switched-on and switched-off times due to its lower gate-to-drain capacitance and charge of the device.

During switched-on and switched-off transitions, power is dissipated because the voltage and current across the device are not zero. Switching energy is obtained by integrating this power loss throughout the transition time interval. The reduction of this energy loss increases the efficiency of the system. Switching energy losses at switched-on (E_{on}), switched-off (E_{off}) and total switching energy loss (E_{SW}) are calculated as shown in (41.1) and (41.2). Figure 41.5 compares the energy loss components of the SGT-MOSFET, and TG-MOSFET counterparts. The extracted energy losses of TG-MOSFET during E_{on}, and E_{off} transients are 297.7, and 2007.2 μJ, respectively. The extracted energy losses of SGT-MOSFET during E_{on}, and E_{off} transients are 59.6, and 1146 μJ, respectively. The E_{off} is higher than E_{on} due to longer transition time

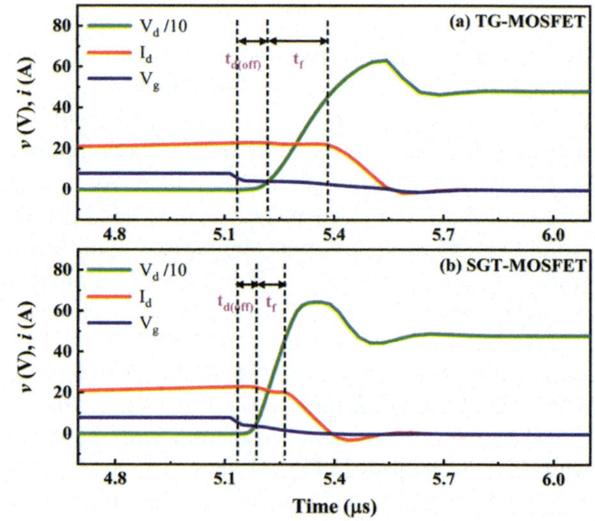

Fig. 41.3 Switched-off transient waveforms of both devices

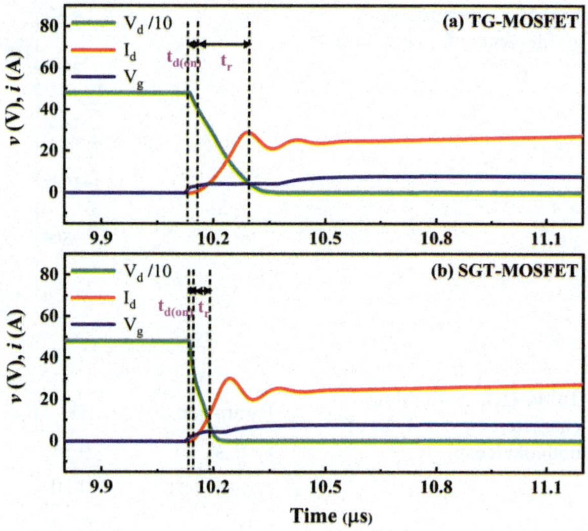

Fig. 41.4 Switched-on transient waveforms of both devices

during the t_{off}. We noted that the total switching energy loss of the SGT-MOSFET is almost cut in half when compared to that of the TG-MOSFET. Finally, the total power loss (P_{SW}) is calculated as a function of total switching energy loss, and respective frequency as mentioned in (41.3) [29]. The P_{SW} for SGT-MOSFET is found to 241 W, which is ~ 47% lower than that of TG-MOSFET. Because of its low power loss, the PSGT-MOSFET can be employed in power conversion systems with high efficiency. The key parameters are given in Table 41.2.

$$E_{off} = \left[\int_0^{t_{off}} V_d(t) \cdot I_d(t) dt \right] \tag{41.1}$$

$$E_{on} = \left[\int_0^{t_{on}} V_d(t) \cdot I_d(t) dt \right] \tag{41.2}$$

$$P_{SW} = (E_{on} + E_{off}) \cdot f \tag{41.3}$$

41.4 Conclusions

In this paper, we have explained the switching characteristics and energy losses of the SGT-MOSFET for the first time and compared them to that of the TG-MOSFET. The device has been simulated and analyzed using a device-circuit hybrid simulator implemented in TCAD simulation. We show that the SGT-MOSFET device has a total switched-on time of 0.06 μs and a switched-off time of 0.13 μs, which is about

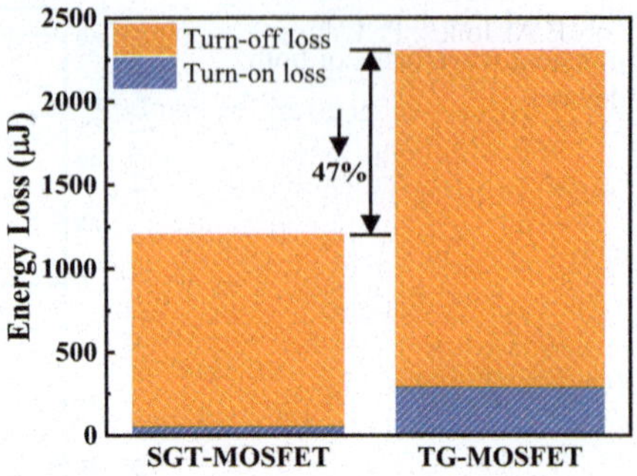

Fig. 41.5 Energy loss components of both devices

Table 41.2 Switching characteristics analysis of both devices

Parameter	TG-MOSFET	SGT-MOSFET
t_f (μs)	0.16	0.07
$t_{d(off)}$ (μs)	0.10	0.06
t_r (μs)	0.14	0.05
$t_{d(on)}$ (μs)	0.02	0.01
E_{on} (μJ)	297.7	59.6
E_{off} (μJ)	2007.2	1146.0
P_{SW} (W)	460	241

2 × faster than a TG-MOSFET with a similar current rating. The total power loss of the SGT-MOSFET has been found to be ~ 47% lower than that of the TG-MOSFET.

Acknowledgements The authors acknowledge Council of Scientific and Industrial Research (CSIR), Government of India for project fellowship [File No: 09/844(0085)/2019-EMR-I].

References

1. Y.-F. Wu, R. Coffie, N. Fichtenbaum, Y. Dora, C.S. Suh, L. Shen, P. Parikh, U.K. Mishra, Total GaN solution to electricity power conversion, in *69th IEEE Device Research Conference, Conference Digest*, pp. 217–218 (2011)
2. W. Zhang, Z. Xu, Z. Zhang, F. Wang, L.M. Tolbert, B.J. Blalock, Evaluation of 600 V cascode GaN HEMT in device characterization and all-GaN-based LLC resonant converter, in *2013 IEEE Energy Conversion Congress and Exposition (ECCE)*, pp. 3571–3578 (2013)
3. Z. Xu, W. Zhang, F. Wang, L.M. Tolbert, B.J. Blalock, Investigation of 600 V GaN HEMTs for high efficiency and high temperature applications, in *IEEE Applied Power Electronics Conference and Exposition (APEC)*, Fort Worth, pp. 131–136 (2014)
4. K.J. Chen, O. Haberlen, A. Lidow, C.L. Tsai, T. Ueda, Y. Uemoto, Y. Wu, GaN-on-Si power technology: devices and applications. IEEE Trans. Electron Devices **64**(3), 779–795 (2017)
5. Y. Cai, Y. Zhou, K.J. Chen, K.M. Lau, High-performance enhancement-mode AlGaN/GaN HEMTs using fluoride-based plasma treatment. IEEE Electron Device Lett. **26**(7), 435–437 (2005)

6. E.A. Jones, F. Wang, B. Ozpineci, Application-based review of GaN HFETs, in *Proceedings of Workshop on Wide Bandgap Power Devices and Applications. Knoxville* (2014)
7. T. Ueda, Recent advances and future prospects on GaN-based power devices, in *Proceedings of International Power Electronics Conference, Hiroshima* (2014)
8. R. Fabrizio, G. Giuseppe, F. Patrick, I. Ferdinando, An overview of normally-off GaN-based high electron mobility transistors. Materials **12**(10), 1599 (2019)
9. E.A. Jones, F.F. Wang, D. Costinett, Review of commercial GaN power devices and GaN-based converter design challenges. IEEE J. Emerg. Select. Topics Power Electron. **4**(3), 707–719 (2016)
10. J. Lu, R. Hou, D. Chen, Loss distribution among paralleled GaN HEMTs, in *IEEE Energy Conversion Congress and Exposition, Portland*, pp. 1914–1919 (2018)
11. T. Oka, Recent development of vertical GaN power devices. Japan. J. Appl. Phys. **58**, SB0805 (2019)
12. I. Ben-Yaacov, Y.-K. Seck, U.K. Mishra, S.P. DenBaars, AlGaN/GaN current aperture vertical electron transistors with regrown channels. J. Appl. Phys. **95**(4), 2073–2078 (2004)
13. S. Chowdhury, B.L. Swenson, M.H. Wong, U.K. Mishra, Current status and scope of gallium nitride-based vertical transistors for high-power electronics application. Semicond. Sci. Technol. **28**(7), 074014 (2013)
14. I.C. Kizilyalli et al.: Method and system for a GaN vertical JFET utilizing a regrown gate. U.S. Patent 9184305 B2 (2015)
15. D. Ji, S. Chowdhury, Design of 1.2 kV power switches with low R_{on} using GaN-based vertical JFET. IEEE Trans. Electron Dev. **62**(8), 2571–2578 (2015)
16. T. Oka, Y. Ueno, T. Ina, K. Hasegawa, Vertical GaN-based trench metal oxide semiconductor field-effect transistors on a free-standing GaN substrate with blocking voltage of 16 kV. Appl. Phys. Exp. **7**(2), 021002 (2014)
17. T. Oka, T. Ina, Y. Ueno, J. Nishii, 1.8 mΩ cm^2 vertical GaN-based trench metal–oxide–semiconductor field-effect transistors on a free-standing GaN substrate for 1.2-kV-class operation. Appl. Phys. Exp. **8**(5), 054101 (2015)
18. R. Li, Y. Cao, M. Chen, R. Chu, 600 V/1.7 Ω normally-off GaN vertical trench metal oxide semiconductor field-effect transistor. IEEE Electron Dev. Lett. **37**(11), 1466–1469 (2016)
19. S. Chowdhury, J. Kim, C. Gupta, S. Keller, S.-H. Chan, U.K. Mishra, in *Trenched Vertical Power Field Effect Transistors With Improved On-Resistance and Breakdown Voltage*. U. S. Patent, 20170125574A1 (2017)
20. N.K. Jaiswal, V.N. Ramakrishnan, Vertical GaN reverse trench-gate power MOSFET and DC-DC converter. Trans. Electr. Electron. Mater. **22**, 363–371 (2021)
21. B.J. Baliga, Power semiconductor device figure of merit for high-frequency applications. IEEE Electron. Devices Lett. **10**(10), 455–457 (1989)
22. D. Ji, W. Li, S. Chowdhury, A study on the impact of channel mobility on switching performance of vertical GaN MOSFETs. IEEE Trans. Electron Dev. **65**(10), 4271–4275 (2018)
23. A. Agarwal, K. Han, B.J. Baliga, Analysis of 1.2 kV 4H-SiC trench-gate MOSFETs with thick trench bottom oxide, in *6th IEEE workshop on Wide Bandgap Power Devices and Applications (WiPDA)*, pp. 125–129 (2018)
24. K. Han, B.J. Baliga, W. Sung, Split-gate 1.2 kV 4H-SIC MOSFET: analysis and experimental validation. IEEE Electron Device Lett. **38**(10), 1437–1440 (2017)
25. N.K. Jaiswal, V.N. Ramakrishnan, S.D. Roy, Optimization of Vertical GaN based SGT-MOSFET for low R_{on}, in *IEEE Workshop on Wide Bandgap Power Devices and Applications in Asia (WiPDA Asia)*, Japan, 1–5 (2020).
26. N.K. Jaiswal, V.N. Ramakrishnan, S.D. Roy, Vertical GaN split gate trench MOSFET with improved high frequency FOM, in *17th IEEE India Council International Conference (INDICON)*, New Delhi, pp. 1–5 (2020)
27. TCAD Sentaurus Device Manual, Synopsys, Inc., Mountain View, CA (2020)
28. Transform: TPH3205WS: 600 V cascade GaN FET in TO-247 (2018)
29. Infineon. How to select the right CoolMOS and its power handing capability: AN-CoolMOS-03. Application Note (2002)

Chapter 42
Optimization of Negative Differential Conductance (NDC) Point for Multi Gate Devices Considering Self Heating Effects Using Surface to Volume Ratio (SVR)

Vivek Kumar, Nitanshu Chauhan, Arnab Datta, and Sudeb Dasgupta

42.1 Introduction

Multi-gate (MG) geometry and new channel material are two primary strategies for improving traditional MOSFETs, minimizing leakage, regulating short channel impact, and increasing drive-current/speed at nodes beyond 20 nm [1, 2]. High-k dielectrics and floating-body design, on the other hand, increase thermal resistance while drive-current increases power dissipation. The high self-heating of transistors impacts their performance, variability, and reliability [3–5]. Performance (i.e., a decrease of ON current) and reliability (i.e., HCI, NBTI, and electromigration) concerns in the front end of the line (FEOL) and back end of the line (BEOL) of modern ICs are caused by higher power density because SHE and joule heating [6–8]. As a result of the severity of the problem, the scientific community is pursuing numerous studies in Self Heating, such as type characterization and measurements of SHE [9, 10]. However, this is a difficult task due to the small dimensions of nanoscale devices and their inability to give sufficient resolution of spatial and temporal parameter fluctuations due to heating. As a result, numerical modelling is the only way to examine temporal and spatial temperature dispersion in nanoscale devices [11–13].

An increase in drain voltage (V_{DS}) causes higher electron energy, leading to more lattice heating. Due to an increase in the lattice temperature at higher V_{DS}, the mobility of charge carriers reduces; as a result, the drain current (I_{DS}) shows a reduction. The drain voltage at which such reversal occurs is the NDC point on the I_{DS}–V_{DS} characteristic, as shown in Fig. 42.2c.

V. Kumar (✉) · N. Chauhan · A. Datta · S. Dasgupta
Indian Institute of Technology Roorkee (IITR), Roorkee 247667, India
e-mail: vivek.kumar@nituk.ac.in

V. Kumar · N. Chauhan
National Institute of Technology Uttarakhand, Srinagar, Garhwal 246174, India

R. Singh et al. (eds.), *The Physics of Semiconductor Devices*, Springer Proceedings in Physics 306, https://doi.org/10.1007/978-981-97-1571-8_42

This NDC point (V_D) is a prospective critical design parameter for Analog and RF circuit applications since the output conductance is negative beyond this point. In this work, for the first time, we have analyzed this NDC point in 14 nm SOI FinFET, and we have provided the design guidelines for Analog and RF circuit designing.

42.2 Simulation Setup and Calibration

Synopsys TCAD [14] commercial FEM software is used to do 3-D electrothermal simulations to perform NDC point variation due to self-heating effect. Simulation setup of 14 nm SOI FinFET designed for analysis is shown in Fig. 42.1a–c. Table 42.1 lists the physical, electrical, and thermal parameters of the device considered for 14 nm SOI FinFET device simulation. For simulation accuracy, the simulation setup is extensively calibrated with 14 nm SOI FinFET experimental data provided in [15]. Details of electrical and thermal models used in the simulation are discussed in the following section.

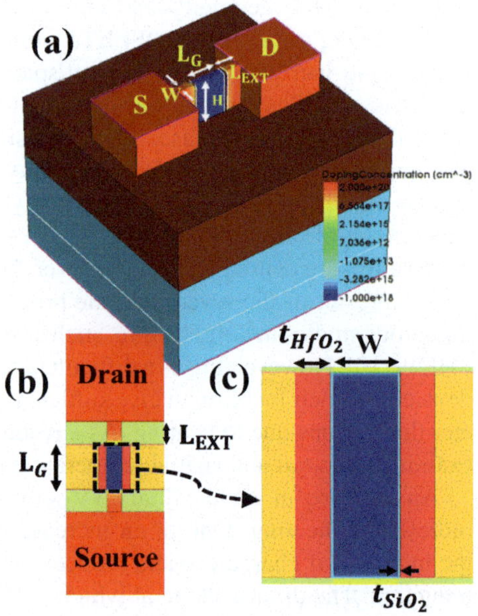

Fig. 42.1 **a** 3D simulation setup; S: Source, D: Drain, **b** 2D cut plane along the channel showing channel and extension region and explaining L_G and L_{EXT}, and **c** magnified channel region showing width, HiK and interfacial oxide layer thickness

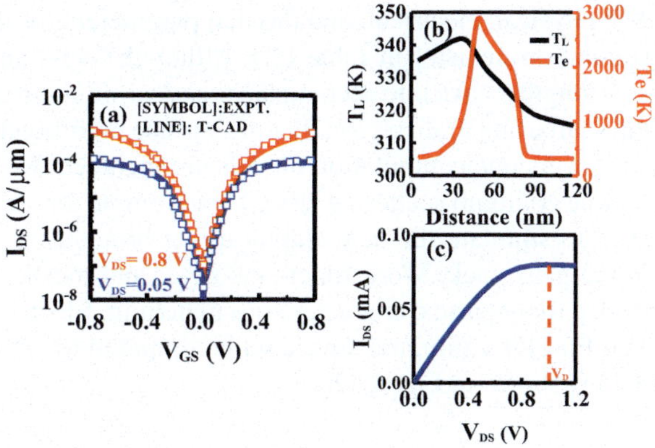

Fig. 42.2 **a,b** I_{DS}–V_{GS} calibration data where red and blue color represents 0.8 and 0.05 V drain to source voltage respectively, **b** T_L and T_e variation, **c** I_{DS}–V_{DS} plot showing NDC point

Table 42.1 Physical, electrical and thermal parameters used for 14 nm SOI FinFET simulation

Parameter	Value
Gate length (L_G)	20 nm
Height of fin (H)	26 nm
Width of fin (W)	6.5 nm
Effective oxide thickness (EOT)	0.85 nm
BOX thickness (t_{BOX})	40 nm
Spacer length (L_{EXT})	10 nm
S/D doping	$2 \times 10^{20}/cm^3$
S/D distribution resistance	$4.5 \times 10^{-8}\ \Omega\ cm^2$
Work function	4.305 eV
Carrier saturation velocity (V_{SAT})	0.8×10^7 cm/s
S/D thermal conductivity	0.62 W/K-cm
Thermal conductivity of channel	0.25 W/K-cm
Thermal conductivity of gate metal	1.62 W/K-cm
Thermal conductivity of nitride	0.185 W/K-cm
Thermal conductivity of HiK	0.014 W/K-cm
Interface thermal resistance (Si/SiO_2)	2×10^{-8} m^2-K/W
Interface thermal resistance (HfO_2/SiO_2)	8.3×10^{-8} m^2-K/W

42.2.1 Electrical Calibration

To account for quantum phenomena, a multi-valley density approximation model was used. To contain the mobility degradation a thin layer inversion and accumulation layer mobility model is employed, as well as a high field saturation model with carrier temperature dependence models to account for short channel effects. In this simulation, the drift–diffusion model was used with the coupled with hydrodynamic and thermodynamics model to include carrier transport in nanoscale device.

The physical, electrical, and thermal parameters for the device that were used in the simulation are listed in Table 42.1. Although Gaussian doping is chosen for the extension length to take into account the practical situation, constant doping is preferred for the source and drain region. For the surface mobility degradation, the high-Lombardi and inversion/accumulation models are included. Remote phonon scattering (RPS), remote coulomb scattering (RCS), and remote dipole scattering (RDS) models were used to simulate the degradation effect brought on by the high-k dielectric found in the gate stack. Work function, stress, and mesh tuning are used to calibrate the device for electrical characteristics matching. Electrical Characteristics of the 14 nm SOI FinFET calibrated device with compared to experimental characteristics given in [15] shown in Fig. 42.2a.

42.2.2 Thermal Calibration

The hydrodynamic model simulates electron temperature since it considers thermal nonequilibrium phenomena between lattice and electron. The thermodynamic model used for lattice temperature calculation. Interface thermal resistance, thickness, and doping dependent thermal conductivity are considered for the various regions of the device, such as fin, source, drain, and gate oxide stack, as can be seen in Table 42.1, for effective thermal modelling of the device. Variation of lattice and electron temperature along the channel of the device are shown in Fig. 42.2b which matches with the data reported in [16].

42.3 Results and Discussion

This work discusses the optimization and variation of VD using the Surface Volume Ratio (SVR) approach [17]. VD is a highly dependent parameter on the temperature of the FinFET, and SVR is an essential parameter in thermal optimization. For Analog and RF circuit design high current drive is preferred, which is achieved by a multi-fin approach. Aspect ratio and SVR are closely related parameters, so the effect of aspect ratio (i.e. Fin Height/Fin Width) of FinFET on maximum lattice temperature in multi-fin FinFET at different fin pitch and extension lengths can be seen in Fig. 42.3a–d. Increasing fin pitch facilitates a better heat transfer rate, so the temperature of multi FinFET is reduced similarly with higher extension length lattice temperature of the device reduces. Figure 42.3e,f show that an increase in gate length (L_G) reduces the NDC point, but it increases with the Increase in extension length.

Moreover, increase in fin height also leads to a reduction in NDC point due to higher drive current, which leads to early self-heating effect occurrence. From these analyses, we can conclude that variation of lattice temperature and NDC point is linked to the device dimension; however, these relationships are not linear in nature.

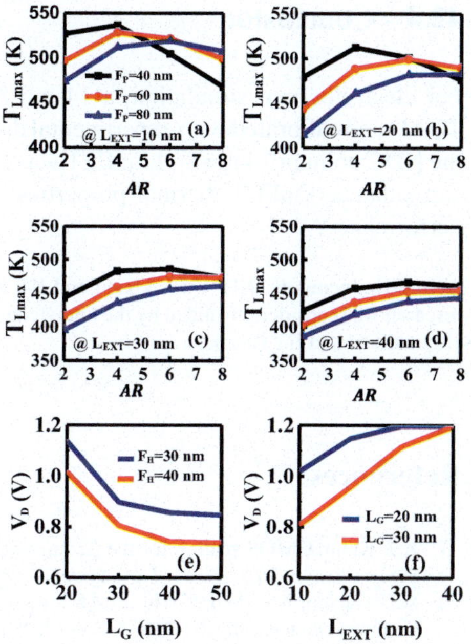

Fig. 42.3 **a–d** Variation of T_{Lmax} at L_{EXT} of 10, 20, 30 and 40 nm, **e,f** the variation of V_D with L_G and L_{EXT}

So, the variation of NDC point with SVR, which is dependent on all device physical parameters, is heuristic in nature, a similar conclusion made from our FEM simulation, which is shown in Fig. 42.4. Point A and B (Fig. 42.4) has minimum and maximum VD, respectively. Increasing SVR always does not lead to higher VD, since an increase in the device's width simultaneously increases SVR, which leads to higher current and lattice temperature. Furthermore, the device dimensions are critical design parameters for Analog and RF applications. Therefore, the optimal selection of SVR is an essential parameter to optimize the device performance considering the thermal effects.

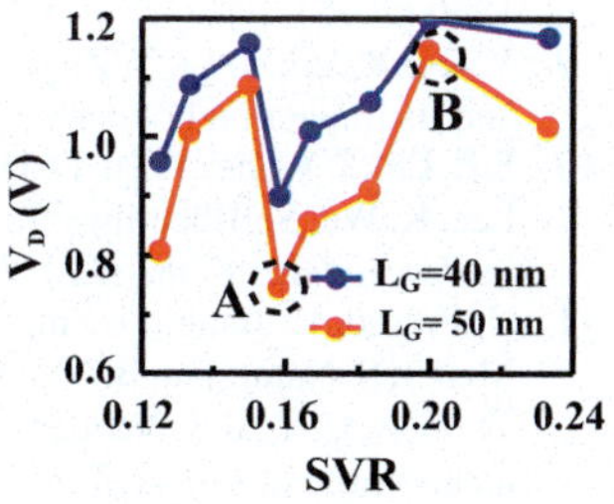

Fig. 42.4 Variation of NDR point (V_D) with surface to volume ratio (SVR) at 40 and 50 nm Gate length of FinFET

42.4 Conclusion

For electrothermal simulation, a 14 nm SOI FinFET was developed in Synopsys TCAD and calibrated with experimental data. The effect of surface to volume ratio on the NDC point in 14 nm SOI FinFET was analyzed and discussed. SVR optimization can enhance FinFET thermal properties and improve the Analog and RF circuits performance.

Acknowledgements NIT Uttarakhand and IIT Roorkee provided funding for this research. The research was considerably aided by the knowledge and expertise supplied by our colleagues at NIT Uttarakhand and IIT Roorkee.

References

1. K.J. Kuhn, CMOS scaling for the 22 nm node and beyond: device physics and technology, in *Symposium on VLSI Technology, Systems and Applications*, pp. 1–2 (2011)
2. A.B. Sachid, M. Chen, C. Hu, FinFET with high-k spacers for improved drive current. IEEE Electron Device Lett. **37**(7), 835–838 (2016)
3. C. Prasad, S. Ramey, L. Jiang, Self-heating in advanced CMOS technologies. IEEE IRPS **6A**, 4.1-4.7 (2017)
4. E. Bury, A. Chasin, B. Kaczer, K.-H. Chuang, J. Franco, M. Simicic, P. Weckx, D. Linten, Self-heating-aware CMOS reliability characterization using degradation maps. IEEE IRPS **2A**, 3-1–3-6 (2018)
5. Y. Zhao, Y. Qu, Impact of Self-Heating Effect on Transistor Characterization and Reliability Issues in Sub-10 nm Technology Nodes. IEEE J. Electron Devices Soc. **7**, 829–836 (2019)
6. M.A. Alam, B.K. Mahajan, Y. Chen, W. Ahn, H. Jiang, S.H. Shin, A device-to-system perspective regarding self-heating enhanced hot carrier degradation in modern field-effect transistors: a topical review. IEEE Trans. Electron Devices **66**(11), 4556–4565 (2019)
7. S. Mishra, H.Y. Wong, R. Tiwari, A. Chaudhary, R. Rao, V. Moroz, S. Mahapatra, TCAD-based predictive NBTI framework for sub-20-nm node device design considerations. IEEE Trans. Electron Devices **63**(12), 4624–4631 (2016)
8. H. Jiang, X. Liu, N. Xu, Y. He, G. Du, X. Zhan, Investigation of self-heating effect on hot carrier degradation in multiple-fin SOI FinFETs. IEEE Electron Device Lett. **36**(12), 1258–1260 (2015)
9. Y. Qu, X. Lin, J. Li, Y. Zhao, Ultra-fast (<1 ns) electrical characterization of self-heating effect and its impact on hot carrier injection in 14 nm FinFETs. IEDM Tech. Dig. (2017)
10. S.E. Liu, J. Wang, Y. Lu, D. Huang, C.F. Huang, W.H. Hsieh, J.H. Lee, Y.S. Tsai, J.R. Shih, Y.-H. Lee, K. Wu, Self-heating effect in FinFETs and its impact on devices reliability characterization, in *Proceedings of the IEEE International Reliability Physics Symposium*, pp. 4 (2014)
11. C. Prasad, L. Jiang, D. Singh, M. Agostinelli, C. Auth, P. Bai, T. Eiles, J. Hicks, C.H. Jan, K. Mistry, S. Natarajan, B. Niu, Packan, P.D. Pantuso, S. Ramey, A. Schmitz, B. Sell, S. Suthram, J. Thomas, C. Tsai, P. Vandervoorn, Self-heat reliability considerations on Intel's 22 nm Tri-Gate technology, in *Proceedings IEEE IRPS*, pp. 5D 1.1–5D.1.5 (2014)
12. M.A. Wahab, S. Shin, M.A. Alam, 3D modeling of spatiotemporal heat-transport in III-V gate-all-around transistors allows accurate estimation and optimization of nanowire temperature. IEEE Trans. Electron Devices **62**(11), 3595–3604 (2015)
13. C.W. Chang, S.E. Liu, B.L. Lin, C.C. Chiu, Y.-H. Lee, K. Wu, Thermal behavior of self-heating effect in FinFET devices acting on back-end interconnects, in *Proceedings of IEEE International Reliability Physics Symposium*, pp. 2F.6.1–2F.6.5 (2015)

14. Sentaurus Device User Guide, Version O-2018.06-SP2, Synopsys, Mountain View, CA, USA (2018)
15. C.-H. Lin et al., High performance 14 nm SOI FinFET CMOS technology with 0.0174 μm^2 embedded DRAM and 15 levels of Cu metallization, in *IEDM Technical Digest*, pp. 3.8.1–3.8.3 (2014)
16. F. Stellari, A.K. Jenkins, J.A. Weger, B. Linder, P. Song, Self-heating characterization of FinFET SOI devices using 2D time resolved emission measurements, in *Proceedings of IEEE International Reliability Physics Symposium*, pp. 2B.1.1–2B.1.6 (2015)
17. S. Makovejev et al., RF extraction of self-heating effects in FinFETs. IEEE Trans. Electron Devices **58**, 3335–3341 (2011)

Chapter 43
Performance of Strained-SiGe in FinFETs and Stacked Nanosheet FETs for Sub-7 nm Technology Node

Eleena Mohapatra, Taraprasanna Dash, Sanghamitra Das, and Devika Jena

43.1 Introduction

Silicon fin field effect transistors (FinFETs) have enabled sustainable technology scaling from planar devices due to superb short channel effects (SCEs) immunity. FinFETs have been scaled down to 10 nm nodes by improving gate-to-channel controllability at reduced gate length and the fin height-to-width aspect ratio [1, 2]. However, in the recent technology node, taller and thinner fins are required to completely control the SCEs which also increases fabrication complexity, layout and cost [3, 4]. Recently, nanosheet FETs (NSFETs) have been considered the successor of FinFETs due to their superior gate electrostatics integrity, higher current density and larger effective channel widths under the same footprint [5]. Furthermore, NSFET allows flexible device design for power performance optimizations due to the tuned width of the nanosheet [6].

On the other hand, strain engineering is one of the key aspects to boost transistor performance. This technique is known to affect the band structure of semiconductor materials, energy bandgap and carrier mobility [7]. This effect can reduce the electron effective mass by changing its subband structure in inversion to obtain improvement in electrical properties with high mobility. Strain is incorporated in Gate-All-Around devices by multi-layered epitaxy of alternating Si/SiGe layers, S/D regrowth epitaxy and stressor deposition [8]. The compressively strained SiGe channels imply the enhancement in the carrier mobility which further improves the device performance.

E. Mohapatra (✉) · T. Dash · D. Jena
Department of ECE, R V College of Engineering, Bengaluru 560059, India
e-mail: eleenamohapatra@gmail.com

S. Das
Department of ECE, Silicon Institute of Technology, Bhubaneswar 751024, India

R. Singh et al. (eds.), *The Physics of Semiconductor Devices*, Springer Proceedings in Physics 306, https://doi.org/10.1007/978-981-97-1571-8_43

In this work, a comparison of the dc performances of FinFETs and NSFETs by strained-SiGe effects is quantitatively investigated through TCAD simulations.

This paper is organized as follows. Section 43.1 introduces the importance of NSFETs and strain engineering technique to boost the device performance. The device design details and physics-based models for the simulation have been discussed in Sect. 43.2. In Sect. 43.3, the comparison of various electrical parameters of FinFETs and NSFETs considering the impact of strained-SiGe has been analyzed. Finally, the conclusion of this work has been presented in Sect. 43.4.

43.2 Simulation Environment

Figure 43.1 shows the schematic diagrams of FinFET and NSFET. The calibration of these simulated devices with the experimental data has been reported in our previous research works [9]. All the geometrical parameters of the two devices are shown in Table 43.1. The gate length, High-k thickness, gate oxide thickness and spacer length of the two devices are 14, 1.5, 1 and 7 nm respectively. The S/D and channel doping concentration of the two purposed devices is 10^{16} and 10^{20} cm^{-3}, respectively.

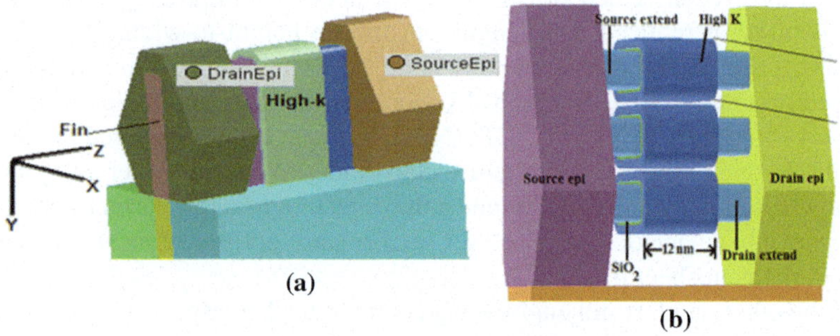

Fig. 43.1. 3D Schematic of **a** FinFET and **b** stacked NSFET

Table 43.1 Device designed parameters

Parameters	FinFET	NSFET
Gate length (L_G)	14 nm	14 nm
Width	5 nm	12 nm
Height	30 nm	6 nm
Thickness of high-k (HfO_2)	1.5 nm	1.5 nm
Thickness of oxide (SiO_2)	1 nm	1 nm
Spacer length (L_{SP})	7 nm	7 nm
S/D doping concentration	10^{20} cm^{-3}	10^{20} cm^{-3}
Channel doping concentration	10^{16} cm^{-3}	10^{16} cm^{-3}

MINIMOS-NT is used for simulations of FinFETs and NSFETs [10]. The drift–diffusion transport equations along with Poisson, carrier continuity equations and density gradient model are calculated self-consistently. Slot boom bandgap narrowing model is considered for all Si and SiGe regions. Lombardi mobility, Shockley–Read–Hall recombination and low field mobility models are also used. The stress-induced changes are also considered. This impact of strain along the channel on Si and SiGe band structure and its models have been reported [11, 12]. The strain integration techniques used in CMOS technology are sorted into two categories (a) Global strain and (b) Local strain. The piezoresistive approach is used to estimate the relationship between strain and no-strain. The demonstration for compliance coefficient (C_{ij}) of Si and Ge is given by: For Si,

$$
\begin{aligned}
c_{11} &= 163.8\ \text{GPa} - T_L.\left(0.0128\frac{\text{GPa}}{\text{K}}\right),\\
c_{12} &= 59.2\ \text{GPa} - T_L.\left(0.0048\frac{\text{GPa}}{\text{K}}\right),\\
c_{44} &= 81.7\ \text{GPa} - T_L.\left(0.0059\frac{\text{GPa}}{\text{K}}\right),
\end{aligned}
\tag{43.1}
$$

where the lattice temperature is represented by T_L. The values of c_{11}, c_{12} and c_{44} for Ge are 126.0GPa, 44.0GPa, and 67.7GPa, respectively [13]. The device's isotropic strain can be calculated using the lattice constants in 3D systems with no stress:

$$
\varepsilon_{xx} = \varepsilon_{yy} = \varepsilon_{zz} = \left(\frac{1+\nu_{SiGe}}{1-\nu_{SiGe}}\right)\left(\frac{a_{SiGe}-a_{Si}}{a_{SiGe}}\right) \tag{43.2}
$$

where Poisson's ratio and strain tensor components are denoted by ν_{SiGe} and ε respectively. Likewise, a_{SiGe} and a_{Si} indicate the lattice constant of SiGe and Si, respectively. Using linear interpolation and the following values of Si and Ge, the lattice constant and Poisson's ratio of SiGe may be linked:

$$
\begin{aligned}
a_{Si} &= 5.43102\,\text{Å} + \left(1.41\times10^{-5}\,\text{Å/K}\right)\cdot(T_L - 300\,\text{K}),\ \text{and}\\
a_{Ge} &= 5.6579\,\text{Å} + \left(3.34\times10^{-5}\,\text{Å/K}\right)\cdot(T_L - 300\,\text{K}),\\
\nu_{Si} &= 0.28\ \text{and}\ \nu_{Ge} = 0.273
\end{aligned}
\tag{43.3}
$$

The increase in mechanical stress in the channel region increases the drain current. The uniaxial stress (σ_{ZZ}) applied in the z-direction is presented. The relation between strain components (ε) and stress tensor (σ) is given by

$$
\sigma = C\cdot\varepsilon \tag{43.4}
$$

where C is the stiffness tensor.

43.3 Result and Discussion

The longitudinal stress distribution across the channel of the proposed devices with 30% Ge content is presented in Fig. 43.2. In this section, the comparative study of the electrical properties of strained-SiGe FinFET and NSFET is presented. The higher stress distribution is observed in the channel of NSFET due to improvement in the hole mobility compared to FinFET. The comparison of the I_D–V_G between FinFET and NSFET is shown in Fig. 43.3. The drain current of SNSFET is higher due to an increase in carrier mobility in the stacked channels than FinFET in the saturation region.

Figure 43.4 shows the variation of V_{TH} and I_{ON} of the two devices at 30% Ge content. The V_{TH} drops significantly for SNSFET compared to FinFET. However, the value of I_{ON} increases for SNSFET which is due to the decrease in energy bandgap. The value of SS of the proposed devices is shown in Fig. 43.5. It is observed that SS also decreases for strained-SiGe SNSFET which is due to the enhancement in carrier mobility.

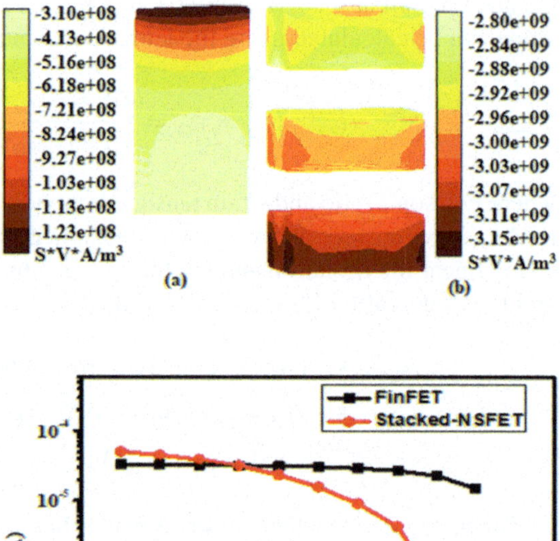

Fig. 43.2 Stress distribution in **a** FinFET and **b** stacked NSFETs

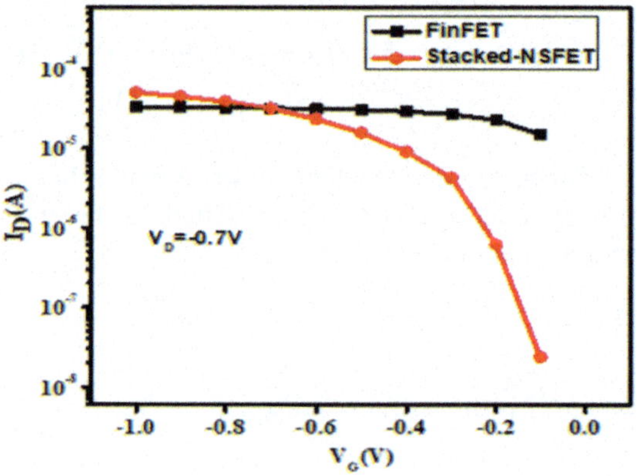

Fig. 43.3 I_D–V_G characteristics in FinFET and stacked NSFET at $V_D = -0.7$ V

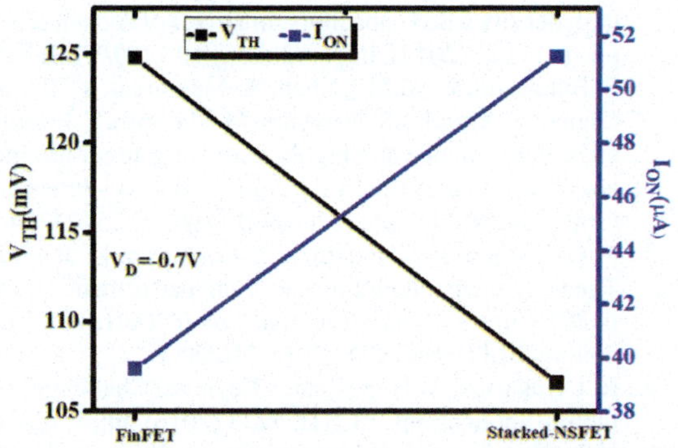

Fig. 43.4 Impact of Strained-SiGe on V_{TH} and I_{ON} variation in FinFET and Stacked-NSFET

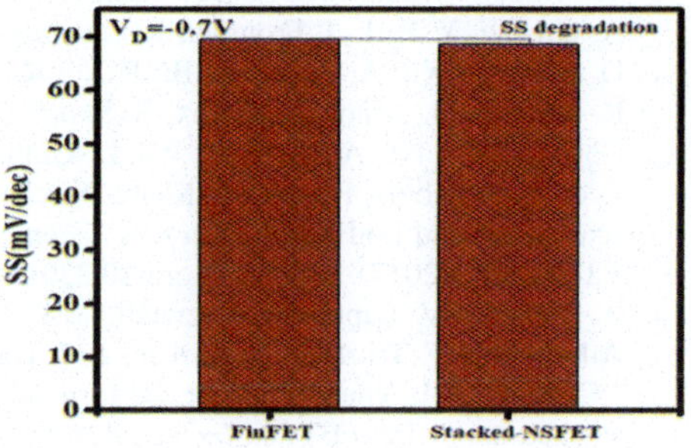

Fig. 43.5 Impact of Strained-SiGe on SS variation in FinFET and Stacked-NSFET

43.4 Conclusion

We have investigated the strain-induced variation in Stacked-NSFET compared to FinFET. Strained-SiGe produces an almost uniform uniaxial strain in the channel which further boosts the carrier mobility and increases the drive current. The enhancement in the I_{ON} revealed that stacked strained-SiGe NSFETs offer better design options for performance and power management. Vertically stacked strained-SiGe NSFET provides a valuable design option for performance improvement in future technology nodes.

References

1. C. Auth, C. Allen, A. Blattner, D. Bergstrom, M. Brazier, M. Bost, M. Buehler, V. Chikarmane, T. Ghani, T. Glassman, R. Grover, W. Han, D. Hanken, M. Hattendorf, P. Hentges, R. Heussner, J. Hicks, D. Ingerly, P. Jain, S. Jaloviar, R. James, D. Jones, J. Jopling, S. Joshi, C. Kenyon, H. Liu, R. McFadden, B. McIntyre, J. Neirynck, C. Parker, L. Pipes, I. Post, S. Pradhan, M. Prince, S. Ramey, T. Reynolds, J. Roesler, J. Sandford, J. Seiple, P. Smith, C. Thomas, D. Towner, T. Troeger, C. Weber, P. Yashar, K. Zawadzki, K. Mistry, A 22nm high performance and low-power CMOS technology featuring fully-depleted tri-gate transistors, self-aligned contacts and

high density MIM capacitors, in *Digest of Technical Papers—Symposium on VLSI Technology*, pp. 131–132 (2012). https://doi.org/10.1109/VLSIT.2012.6242496
2. S. Bangsaruntip, G.M. Cohen, A. Majumdar, Y. Zhang, S.U. Engelmann, N.C.M. Fuller, L.M. Gignac, S. Mittal, J.S. Newbury, M. Guillorn, T. Barwicz, L. Sekaric, M.M. Frank, J.W. Sleight, High performance and highly uniform gate-all-around silicon nanowire MOSFETs with wire size dependent scaling in *Technical Digest—International Electron Devices Meeting, IEDM* (2009). https://doi.org/10.1109/IEDM.2009.5424364
3. M.G. Bardon, P. Schuddinck, P. Raghavan, D. Jang, D. Yakimets, A. Mercha, D. Verkest, A. Thean, Dimensioning for power and performance under 10 nm: the limits of FinFETs scaling, in *2015 International Conference on IC Design and Technology, ICICDT 2015* (2015). https://doi.org/10.1109/ICICDT.2015.7165883
4. R. Divakaruni, V. Narayanan, (Keynote) Challenges of 10 nm and 7 nm CMOS for server and mobile applications. ECSTr. **72**, 3 (2016). https://doi.org/10.1149/07204.0003ECST
5. N. Loubet, T. Hook, P. Montanini, C.W. Yeung, S. Kanakasabapathy, M. Guillom, T. Yamashita, J. Zhang, X. Miao, J. Wang, A. Young, R. Chao, M. Kang, Z. Liu, S. Fan, B. Hamieh, S. Sieg, Y. Mignot, W. Xu, S.C. Seo, J. Yoo, S. Mochizuki, M. Sankarapandian, O. Kwon, A. Carr, A. Greene, Y. Park, J. Frougier, R. Galatage, R. Bao, J. Shearer, R. Conti, H. Song, D. Lee, D. Kong, Y. Xu, A. Arceo, Z. Bi, P. Xu, R. Muthinti, J. Li, R. Wong, D. Brown, P. Oldiges, R. Robison, J. Arnold, N. Felix, S. Skordas, J. Gaudiello, T. Standaert, H. Jagannathan, D. Corliss, M.H. Na, A. Knorr, T. Wu, D. Gupta, S. Lian, R. Divakaruni, T. Gow, C. Labelle, S. Lee, V. Paruchuri, H. Bu, M. Khare, Stacked nanosheet gate-all-around transistor to enable scaling beyond FinFET, in *Digest of Technical Papers—Symposium on VLSI Technology*, pp. T230–T231 (2017). https://doi.org/10.23919/VLSIT.2017.7998183
6. S. Barraud, V. Lapras, B. Previtali, M.P. Samson, J. Lacord, S. Martinie, M.A. Jaud, S. Athanasiou, F. Triozon, O. Rozeau, J.M. Hartmann, C. Vizioz, C. Comboroure, F. Andrieu, J.C. Barbé, M. Vinet, T. Ernst, Performance and design considerations for gate all-around stacked-nanowires FETs, in *Technical Digest—International Electron Devices Meeting, IEDM*, pp. 29.2.1–29.2.4 (2018). https://doi.org/10.1109/IEDM.2017.8268473
7. G. Tsutsui, S. Mochizuki, N. Loubet, S.W. Bedell, D.K. Sadana, Strain engineering in functional materials. AIP Adv. **9**, 030701 (2019). https://doi.org/10.1063/1.5075637
8. S. Barraud, V. Lapras, M.P. Samson, L. Gaben, L. Grenouillet, V. Maffini-Alvaro, Y. Morand, J. Daranlot, N. Rambal, B. Previtalli, S. Reboh, C. Tabone, R. Coquand, E. Augendre, O. Rozeau, J.M. Hartmann, C. Vizioz, C. Arvet, P. Pimenta-Barros, N. Posseme, V. Loup, C. Comboroure, C. Euvrard, V. Balan, I. Tinti, G. Audoit, N. Bernier, D. Cooper, Z. Saghi, F. Allain, A. Toffoli, O. Faynot, M. Vinet, Vertically stacked-NanoWires MOSFETs in a replacement metal gate process with inner spacer and SiGe source/drain, in *Technical Digest—International Electron Devices Meeting, IEDM*. 17.6.1–17.6.4 (2017). https://doi.org/10.1109/IEDM.2016.7838441
9. E. Mohapatra, T.P. Dash, J. Jena, S. Das, C.K. Maiti, Strain induced variability study in gate-all-around vertically-stacked horizontal nanosheet transistors. Phys. Scrip. **95**, 065808 (2020). https://doi.org/10.1088/1402-4896/AB89F5
10. Minimos-NT—User Manual, Global TCAD Solutions, https://www.globaltcad.com/download/minimos-nt-user-manual/. Accessed 30 March 2022
11. J.S. Lim, S.E. Thompson, J.G. Fossum, Comparison of threshold-voltage shifts for uniaxial and biaxial tensile-stressed n-MOSFETs. IEEE Electron Device Lett. **25**, 731–733 (2004). https://doi.org/10.1109/LED.2004.837581
12. W. Zhang, J.G. Fossum, On the threshold voltage of strained-Si-Si1-xGex MOSFETs. IEEE Trans. Electron Devices **52**, 263–268 (2005). https://doi.org/10.1109/TED.2004.842716
13. E. Mohapatra, T.P. Dash, J. Jena, S. Das, J. Nanda, C.K. Maiti, Performance analysis of Si-channel nanosheet FETs with strained SiGe source/drain stressors, in *Lecture Notes in Electrical Engineering*, vol. 665, pp. 329–337 (Springer, Singapore, 2020). https://doi.org/10.1007/978-981-15-5262-5_23

Chapter 44
Thermal Admittance Spectroscopy of AlGaN/GaN HEMT Structure

Shradha Gupta, J. K. Kaushik, Kapil Narang, and Ankur Gupta

44.1 Introduction

The great potential of the Group III nitride devices, such as high breakdown voltage, wide band gap, and high-temperature operability, has led to its extensive production and great interest in high power and RF applications [1]. However, AlGaN/GaN HEMT devices still suffer from current collapse [2, 3], high leakage current [4], threshold voltage instability [5], and kink effect [6, 7] which affect its long term reliability. In addition, the presence of lattice mismatch when GaN is grown on a commonly used substrate such as Si, Saphire, and SiC doesn't allow the fabricated device to be defects free. Therefore, studying deep traps/defects becomes essential to analyse the shortcomings of the wide band gap semiconductor devices. The deep trap study will further assist in improving the device's performance.

Thermal Admittance Spectroscopy (TAS) is a method reported in the literature to characterize deep traps and shallow doping, which is based on the basic study of the space charge region formed in the Schottky contacts or p-n junction [8]. The TAS measurements allow easy seeing of the relatively shallow or fast levels and supplement the methods of Deep Level Transient Spectroscopy (DLTS) and thermally Stimulated Capacitance (TScap) [9]. TAS measurement is mainly useful for semiconductors with high resistivity, samples where capacitance does not change with the voltage, at a test frequency with small values of capacitance which cannot be analysed using the DLTS [10].

S. Gupta · J. K. Kaushik · K. Narang
Solid State Physics Laboratory, Lucknow Road, 110054 Timarpur, India

S. Gupta (✉) · A. Gupta
Indian Institute of Technology, Hauz Khas, New Delhi 110016, India
e-mail: shradhagupta1289@gmail.com

R. Singh et al. (eds.), *The Physics of Semiconductor Devices*, Springer Proceedings in Physics 306, https://doi.org/10.1007/978-981-97-1571-8_44

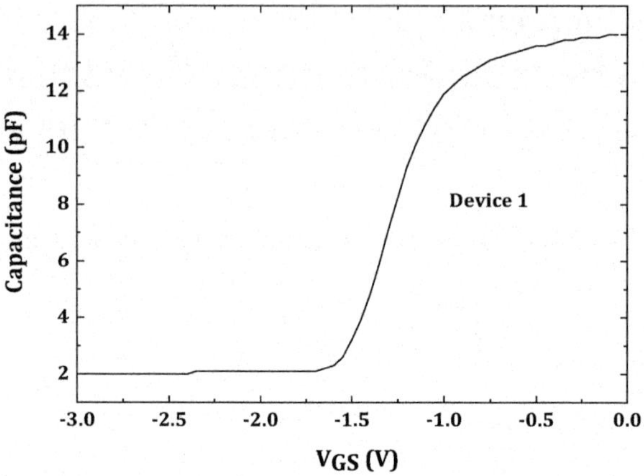

Fig. 44.1 CV curve of device 1 at 300 K at 50 KHz

In this paper, the TAS measurements were performed on different devices and studied trap properties. The experimental details were given in Sect. 44.2, the methodology to obtain the results was explained in Sects. 44.3 and 44.4 discusses the results, and Sect. 44.5 concludes the paper's work.

44.2 Experimental Details

In this work, all TAS measurements were performed on different devices. Device1 (D_1) grown on Si substrate obtained from MACOM (NPTB00004A) "Nitronex Inc. from USA" (Richardson 2008) having 0.5 μm gate-length, 10 × 200 μm width, and 1 μm source field plate (SFP) length. Device2 (D_2) was grown using 4H SiC substrate obtained from CREE [11] having 0.4 μm gate length and 6 × 350 μm width. Moreover, device 3 (D_3) consists of AlGaN/GaN Heterostructure grown on SiC substrate. SEMETROL DLTS (Deep Level Transient spectroscopy) system was used for the TAS measurement with temperatures varying from 80 to 500 K. The TAS measurement was obtained with the frequency varying from 10 to 100 kHz, AC Voltage magnitude of 100 mV, and DC Measured Voltage for D_1 (− 1.5 V), D_2 (− 3.55 V), and D_3(− 0.5 V). The C-V and the respective depth profile of the devices used in this work are shown in Figs. 44.1,44.2,44.3 and 44.4.

44.3 Methodology

The basic study of traps using admittance spectroscopy at different temperature revolves around the complex quantity of admittance and is described as

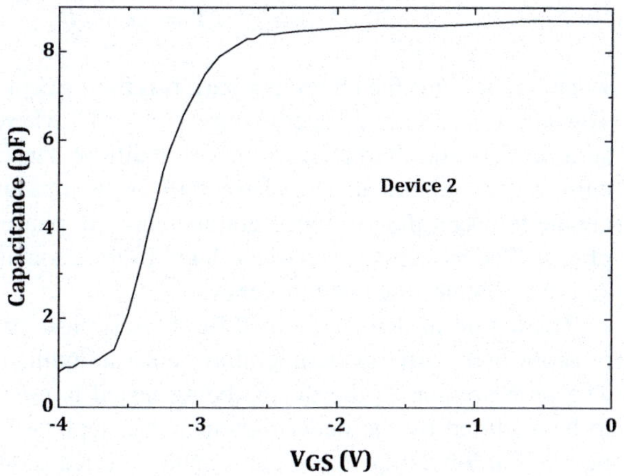

Fig. 44.2 CV curve of device 2 at 300 K at 50 KHz

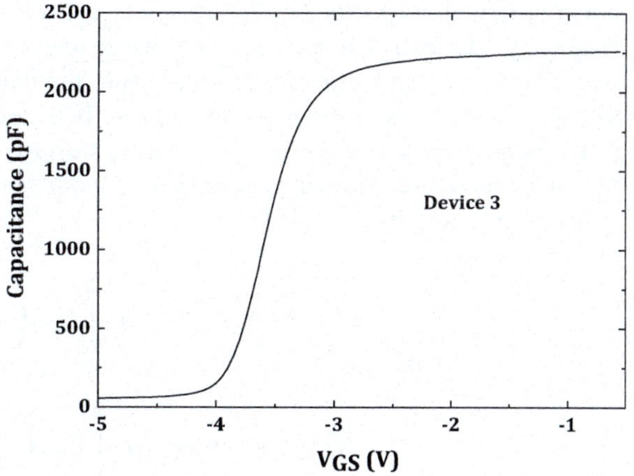

Fig. 44.3 CV curve of device 3 at 300 K at 50 KHz

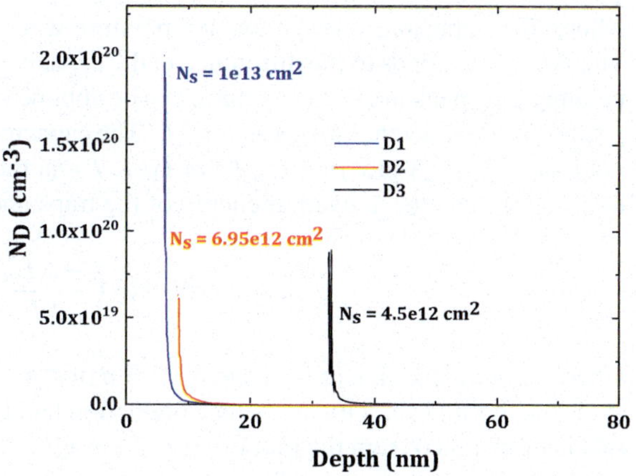

Fig. 44.4 Extracted depth profile from C-V measurement of AlGaN/GaN HEMT device used at 300 K

$$Y = G + iB \tag{44.1}$$

where i represents the imaginary part, B is the susceptance, and G is the conductance. The admittance was also defined by $Y = I/V$, where the measurement device was stimulated by the alternating voltage V at different test frequencies, and subsequently, current I was obtained. The phase shift of the measured current was used to differentiate between the estimated conductance (in phase) and susceptance (quadrature phase). The susceptance B was related to the capacitance as ωC, where $\omega = 2\pi f$, where f denotes the test frequency.

The nature of defects defined by the response arises from the admittance spectroscopy at the different temperatures corresponding to the range of test frequencies. The probe region of the device being tested is governed by the carriers available to be captured by the deep traps near the depletion region edge. The defects will be filled at low temperature and unable to release the trapped charges as they may not gain sufficient thermal energy within the test frequency range. The defect will not contribute to capacitance, as the charge in itself is unable to respond to the test frequency. The energy required by the charge to evacuate from the defects increases with the increase in temperature. As a result, the defects may be able to contribute to capacitance if the emission rate $e_n > f$. It is necessary to determine the e_n at different frequencies, which can be obtained through the conductance peak at a particular temperature. The capacitance and conductance at different frequencies and temperatures are given by [10].

$$C(f,T) = C_{\iota T} + \Delta C\left[1 + \left(\frac{f}{f_t}\right)^2\right]^{-1} \tag{44.2}$$

$$G(f,T) = 2\pi f_t \Delta C\left[1 + \left(\frac{f_t}{f}\right)^2\right]^{-1} \tag{44.3}$$

where, $C_{\iota T}$ capacitance at the low temperature where the emission rate is very less. The ΔC corresponds to the difference in the capacitance at the low and the high temperature. The peak in the conductance can be obtained by differentiating Eq. 44.3. The peak in the conductance arises at $f = f_t$. The capacitance will also have an inflection point at $f = f_t$. A plot of $C(f,T)$ or $G(f,T)$ at the different test frequencies was used to estimate the activation energy of the traps and defined as [10]

$$e_n = c_n N_c \exp\left(\frac{-\Delta E_A}{kT}\right) \tag{44.4}$$

where, capture coefficient is given as $c_n = \sigma_n v_{th}$, σ_n is the capture cross section and v_{th} is the thermal velocity, N_c is the conduction band density of states. The emission rate is determined from the plot of $C(f,T)$ or $G(f,T)$ at the peak temperature with respect to the test frequency. Consequently, the activation energy ΔE_A was obtained from slope between T^2/e_n vs 1/kT according to Eq. 44.4.

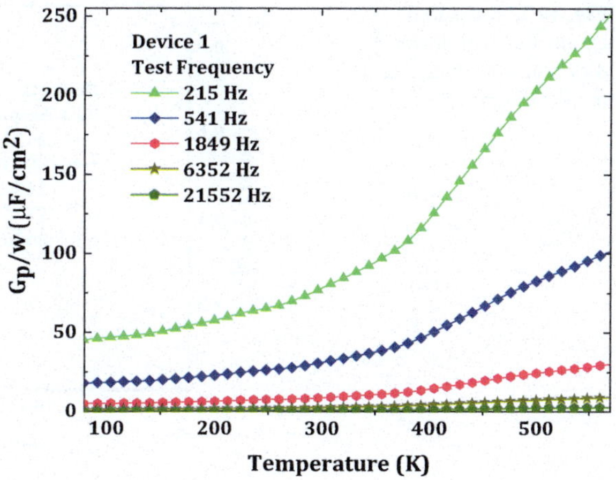

Fig. 44.5 Measured conductance of device 1 at different frequencies and temperatures

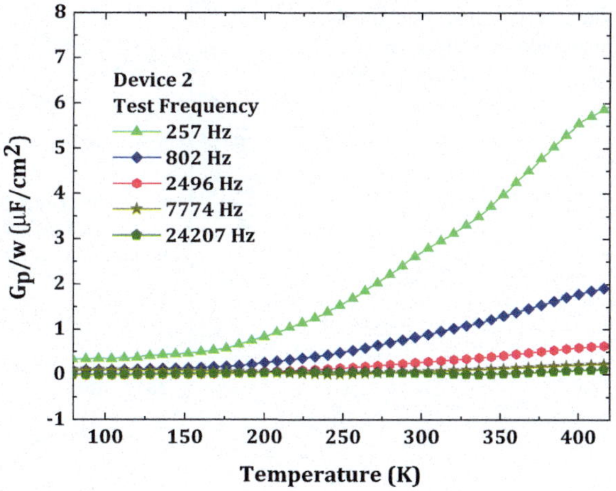

Fig. 44.6 Measured conductance of device 2 at different frequencies and temperatures

44.4 Results and Discussion

The conductance for device 1, 2, and 3 were extracted from the TAS measurement using the methodology explained in section III and is shown in Figs. 44.5,44.6 and 44.7, respectively. It can be worth noting from the figures that the devices have variations in conductance due to the amount of leakage present within the device. Prior to the defect-related study, it is important to analyse the leakage and the change in capacitance with the voltage of the devices. Moreover, the extracted activation energy and the respective capture cross-section are shown in Fig. 44.8.

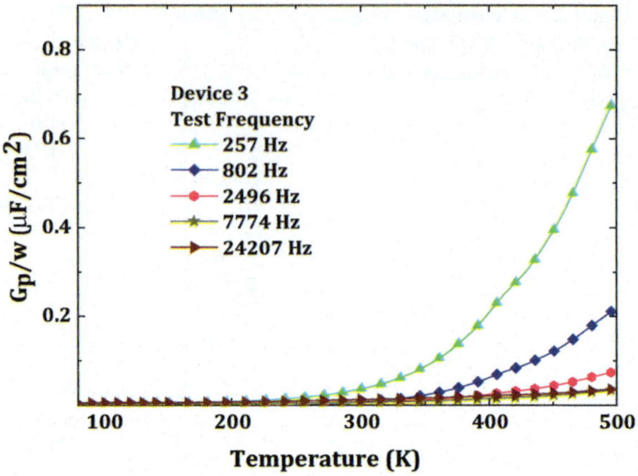

Fig. 44.7 Measured conductance of device 3 at different frequencies and temperatures

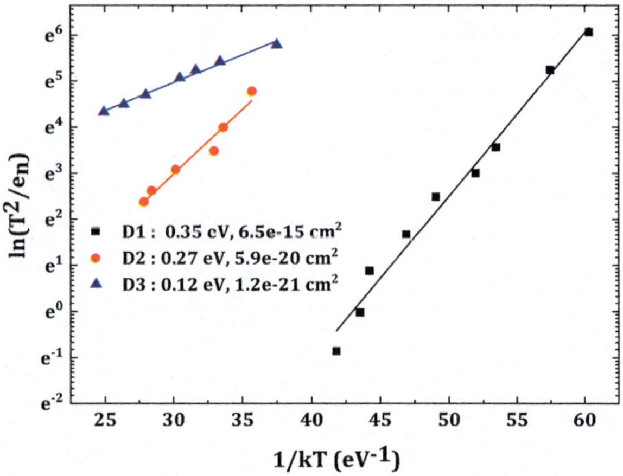

Fig. 44.8 Arrhenius plots of devices 1, 2, and 3 obtained using thermal admittance spectroscopy measurements

Among all devices used in this work, device 1 has the highest conductance, and the defects could respond in the low-temperature range, as observed from Fig. 44.8. The same activation energy was also obtained from the change in series resistance. On the other hand, defects of devices 2 and 3 were able to respond in the high-temperature range due to the comparatively low value of conductance, as can be noted from the Figs. 44.6 and 44.7. The extracted activation energy was similar to the value reported in the literature at the different locations of the devices and is given in Table 44.1.

Table 44.1 The activation energy reported in the literature is based on the location and the substrate used

Substrate	Location/reason	Ea (reported in literature) (eV)
Si	Below the 2DEG channel [12]	0.370
	At surface [13]	0.31
	Due to oxygen impurities substituting for nitrogen Vacancies in GaN buffer [14]	0.38
	Deep barrier/interface traps [15]	0.3
	Identified in the AlGaN barrier [16]	0.33
SiC	Located in the AlGaN barrier [17]	0.31
	At the AlGaN surface [18]	0.27
	Defects located at the AlGaN/GaN interface [19]	0.26

44.5 Conclusion

In this work, we have performed trap studies on different AlGaN/GaN HEMT devices using TAS measurement. The TAS measurement determines the activation energy of 0.35 eV for AlGaN/GaN HEMT on Si, 0.27, and 0.12 eV for AlGaN/GaN HEMT on SiC. The similar activation energy was also reported in the literature at different locations of the AlGaN/GaN HEMT structure.

References

1. U.K. Mishra, P. Parikh, Y.F. Wu, Proceedings of the IEEE **90**, 1022 (2002). https://doi.org/10.1109/JPROC.2002.1021567
2. J.M.M. Tirado, J.L. Sanchez-Rojas, J.I. Izpura, J.L. Sánchez-Rojas, J.I. Izpura, IEEE Trans. Electron Dev. **54**, 410 (2007)
3. S. Gupta, J.K. Kaushik, A. Gupta, V.R. Balakrishnan, *2019 IEEE International Conference on Electrical, Computer and Communication Technologies (ICECCT)* (IEEE, 2019), pp. 1–5. https://doi.org/10.1109/ICECCT.2019.8869264, https://ieeexplore.ieee.org/document/8869264/
4. J.K. Kaushik, V.R. Balakrishnan, D. Mongia, U. Kumar, S. Dayal, B.S. Panwar, R. Muralidharan, Thin Solid Films **612**, 147 (2016)
5. Y.H. Wang, Y.C. Liang, G.S. Samudra, T.F. Chang, C.F. Huang, L. Yuan, G.Q. Lo, Semicond. Sci. Technol. **28** (2013). https://doi.org/10.1088/0268-1242/28/12/125010
6. G. Meneghesso, F. Zanon, M.J. Uren, E. Zanoni, S. Member, F. Zanon, M.J. Uren, E. Zanoni, IEEE Electron Dev. Lett. **30**, 100 (2009). https://doi.org/10.1109/LED.2008.2010067
7. J.K. Kaushik, V.R. Balakrishnan, B.S. Panwar, R. Muralidharan, IEEE Trans. Electron Dev. **60**, 3351 (2013). https://doi.org/10.1109/TED.2013.2279158
8. U. Reislöhner, H. Metzner, C. Ronning, Phys. Rev. Lett. **104**, 226403 (2010)
9. D.L. Losee, J. Appl. Phys. **46**, 2204 (1975)
10. SEMETROL, Thermal Admittance Spectroscopy. Tech. rep., TAS - SEMETROL, LLC (2022). https://semetrol.com/tas/
11. R.S. Pengelly, S.M. Wood, J.W. Milligan, S.T. Sheppard, W.L. Pribble, IEEE Trans. Microw. Theor. Techn. **60**, 1764 (2012)

12. M. Marso, M. Wolter, P. Javorka, P. Kordoš, H. Lüth, Appl. Phys. Lett. **82**, 633 (2003). https://doi.org/10.1063/1.1540239
13. F. Jabli, M.A. Zaidi, N. Ben Hamadi, S. Althoyaib, M. Gassoumi, J. Alloys Compounds **653**, 624 (2015). https://doi.org/10.1016/j.jallcom.2015.09.007
14. H.N. Kim, *Qualitative and Quantative Characterization of Trapping Effects in AlGaN/GaN High Electron Mobility Transistors*. Ph.D. thesis, The Ohio State University (2009). http://rave.ohiolink.edu/etdc/view?acc_num=osu1250612796
15. Y.S. Hwang, L. Liu, F. Ren, A.Y. Polyakov, N.B. Smirnov, A.V. Govorkov, E.A. Kozhukhova, N.G. Kolin, V.M. Boiko, S.S. Vereyovkin, V.S. Ermakov, C.F. Lo, O. Laboutin, Y. Cao, J.W. Johnson, N.I. Kargin, R.V. Ryzhuk, S.J. Pearton, J. Vacuum Sci. Technol. B. Nanotechnol. Microelectron. Mater. Process. Measure. Phenomena **31**, 022206 (2013). https://doi.org/10.1116/1.4795210
16. M.J. Anand, G.I. Ng, S. Arulkumaran, B. Syamal, X. Zhou, Distribution of trap energy level in AlGaN/GaN high-electron-mobility transistors on Si under ON-state stress (2015). https://doi.org/10.7567/APEX.8.104101
17. I. Jabbari, M. Baira, H. Maaref, R. Mghaieth, Physica E Low-Dimensional Syst. Nanostruct. **104**, 216 (2018)
18. M. Gassoumi, J.M. Bluet, C. Gaquière, G. Guillot, H. Maaref, Microelectron. J. **40**, 1161 (2009). https://doi.org/10.1016/j.mejo.2007.02.005
19. J.G. Tartarin, G. Soubercaze-Pun, J.L. Grondin, L. Bary, J. Mimila-Arroyo, J. Chevallier, AIP Conf. Proc. **922**, 163 (2007). https://doi.org/10.1063/1.2759658

Chapter 45
Numerical Modelling of WO_3-Based Memristive System for Neuromorphic Computation

Sanjay Kumar, Rajan Agrawal, Pawan Kumar, and Shaibal Mukherjee

45.1 Introduction

A memristor is a non-volatile electronic memory device that has been theorized as the fourth fundamental passive circuit element [1]. Artificial neural networks, which are inspired by the remarkable efficiency of biological systems, can be realized utilizing memristors [2]. The neuromorphic application requires low power, high density, and complementary-metal oxide semiconductor (CMOS) compatible devices [2]. All these properties are offered by memristive devices, making them highly suitable for neuromorphic computation [3], synaptic computation [4], and resistive random-access memory (RRAM) [5]. The most significant application among these is the synaptic function as synapse. In a neuromorphic computational system, memristors will act as synapses and neuron blocks are built by CMOS processes due to their high versatility for the neuromorphic system.

Most of the experimental data reported in the literature for various material systems such as WO_3 [6], Y_2O_3 [7], ZnO [8] etc. shows asymmetric behavior in the pinched hysteresis loop in the current–voltage (I-V) characteristics.

Here, a new parabolic window function is proposed to study and analyze the performance of a memristive system. The proposed model is validated with the experimental report on WO_3-based memristive system [6] and window function fulfills all the necessary conditions reported by Prodromakis et al. [9]. The proposed window function leads to better non-linear behavior at the device boundaries. Moreover, in

S. Kumar (✉) · R. Agrawal · P. Kumar · S. Mukherjee
Hybrid Nanodevice Research Group, Department of Electrical Engineering, Indian Institute of Technology Indore, Indore, MP 453552, India
e-mail: phd1701202005@iiti.ac.in

S. Mukherjee
Centre for Advanced Electronics (CAE), Indian Institute of Technology Indore, Indore, MP 453552, India

R. Singh et al. (eds.), *The Physics of Semiconductor Devices*, Springer Proceedings in Physics 306, https://doi.org/10.1007/978-981-97-1571-8_45

order to eliminate some inconsistencies in the [10], certain modifications are made so as to achieve better correlation with the experimental data.

45.2 Analytical Model

The existing [10] non-linear model, as reported in the literature, is described by (45.1).

$$I(t) = W^n \beta \sinh(\alpha V_i(t)) + X\{e^{\gamma V_i(t)} - 1\} \tag{45.1}$$

Here, W is the state variable of the memristor, n is the free parameter to control the non-linear dependence of the drift velocity on the applied voltage, αandβ are the fitting constants used to characterize the ON state, X and γ denote the net electronic barrier in the OFF state of the device. The first term leads to flux-controlled memristor with symmetric behavior of the device for both positive and negative polarities of the applied voltage. Thus, it is necessary to make certain improvements in the model to achieve better resemblance with the experimental data, having asymmetric characteristics in positive and negative cycles of applied voltage. The I-V relationship for proposed model is given by Eq. (45.2) [11].

$$I(t) = \begin{Bmatrix} b_1 W^{a_1}\left(e^{\alpha_1 V_i(t)} - 1\right) + X\left(e^{\gamma V_i(t)} - 1\right), V_i(t) \geq 0 \\ b_2 W^{a_2}\left(e^{\alpha_2 V_i(t)} - 1\right) + X\left(e^{\gamma V_i(t)} - 1\right), V_i(t) < 0 \end{Bmatrix} \tag{45.2}$$

The first term describes the asymmetric behavior of the device for opposite voltage polarities which is neglected by the previous model and thus introduces parameters a_1 and a_2. These parameters a_1 and a_2 are the degrees of influence of the state variable on current for positive and negative voltage polarities, respectively. The second term of (45.2) represents an ideal diode equation which plays a significant role when the state variable approaches zero i.e., the device is near the OFF state. In this equation, some fitting parameters such as b_1, b_2, α_1, and α_2 are used to fit the output characteristics of the memristor device. b_1 and b_2 are the conductivity control parameters of the device for both positive and negative applied voltages, respectively. The conductivity of the device is determined by the migration of oxygen vacancies towards the opposite side. Thus, the current starts flowing as a result of the conducting path formed inside the switching material. Here, V_i is the applied triangular waveform as an input voltage.

Equation (45.3) represents the rate of change of state variable with time. The new parabolic window function $f(W)$, as depicted in Fig. 45.1a, ensures that the state variable is bounded between 0 and 1 and it is given by (45.4).

$$\frac{dW}{dt} = A * {V_i}^m * f(W) \tag{45.3}$$

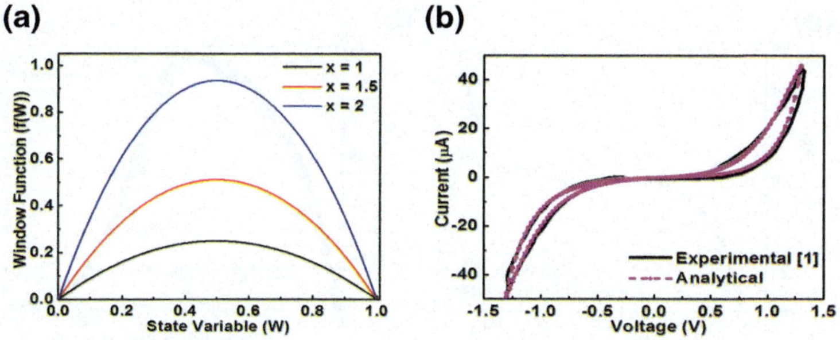

Fig. 45.1 **a** Parabolic window function implemented in the analytical model, **b** analytical resemblance of I-V response of a memristive system

Table 45.1 Physical significance and range of parameters for analytical modeling

Parameters	Range	Physical Significance
b_1, b_2	10^{-7}–10^{-5}	Experimental fitting parameters
γ	4–6	Diode parameters like thermal voltage and ideality factor
α_1, α_2	± 2 to ± 6	Experimental fitting parameters
a_1, a_2	1–5	Hysteresis loop broadness control parameters
X	10^{-10}	Magnitude of ideal diode behavior
A	Arbitrary	Control the effect of the window function
m	Odd integer	Control the effect of input on the state variable
x	$1 \leq x \leq 2$	Parameter for bounding window function between 0 and 1

$$f(W) = \left[2^x - (W^2 - W + 2)^x\right] \tag{45.4}$$

where, A and m are the parameters to determine the dependence of the state variable on voltage input and m is always an odd integer to ensure that the opposite polarity of the applied voltage leads to opposite change in the rate of change of state variable (Table 45.1).

45.3 Results and Discussion

To study the various characteristics of WO_3-based memristive system, similar electrical conditions are used which have been experimentally reported elsewhere [6]. To study the I-V characteristics of the WO_3-based memristive system, a triangular voltage waveform is applied on the device with peak-to-peak voltage of − 1.3 to 1.3 V with a compliance current of 10 μA. Figure 45.1b presents the analytical fitting of the *I*-*V* characteristics, which shows a better resemblance with the reported experimental data of the device.

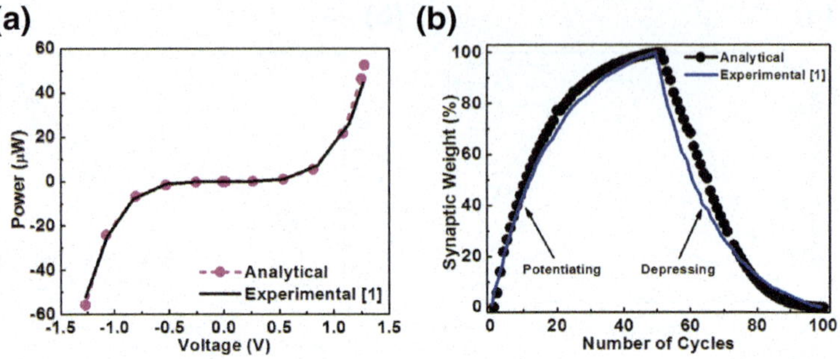

Fig. 45.2 **a** Actual power consumption by the device at different voltages, **b** synaptic behavior of the memristive device

Thus, this observation concludes that the analytical model successfully imitates the memristive behavior. Figure 45.2a presents the actual power consumption by the device. Both parameters such as applied voltage and generated current determine the actual power consumption by the device.

The variation in the current and voltage can lead to change in the power. The negative direction of power shows the negative polarity of the voltage applied to the device under operation. The device utilizes maximum power when the state variable is close to unity i.e., the device is in the ON state.

Figure 45.2b illustrates the potentiation (P) and depression (D) process of the device while studying the synaptic behavior. During the potentiation process, a train of 50 consecutive positive rectangular pulses followed by 50 rectangular negative pulses with amplitude of 1.2 V and pulse width of 5 μs is applied on the device. Under positive pulses, the synaptic weight i.e., the device conductance increases continuously with the number of stimulating positive pulses and decreases subsequently for the negative pulses.

This behavior is similar to the learning behavior of the human brain where the synaptic strength of the neuromorphic system changes according to the number of stimuli received during the learning process.

Thus, the continuous strengthening and weakening in synaptic weight defined by the potentiation and the depression processes where the former relates to the learning function of the brain and the latter is analogous to the forgetting process of the human memory. Figure 45.3 shows the direct-current (DC) characteristic of the memristive device. As results show that the analytical current has good agreement with the experimental current.

The amplitude of modeled and experimentally-obtained current increases for the positive sweep, as shows in Fig. 45.3a, and decreases for the negative sweep, as in Fig. 45.3b. This is because the device remembers its state when the input is turned OFF.

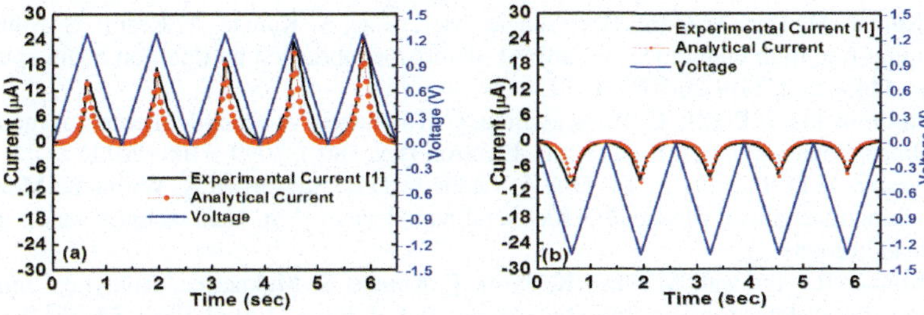

Fig. 45.3 DC characteristic of memristive device for **a** positive sweep and **b** negative sweep

45.4 Conclusion

Here, WO_3-based memristive switching characteristics have been examined using an analytical model with a new parabolic window function and a modified I-V relationship for neuromorphic applications. The previous model had several discrepancies with the experimental data, such as absence of a parameter to account for asymmetric nature of loop in the positive and negative voltage cycles. Our model successfully eradicates this shortcoming by defining suitable parameters. The model is found to imply considerable similitude with the learning and forgetting processes of the human brain through the potentiating and the depressing process respectively. It is expected that these properties of the memristor will prove to be useful in future evolution of the device for realizing the hardware implementation of neural systems.

Acknowledgements Sanjay Kumar would like to thank DST for providing DST-INSPIRE (Inspire code: IF170791) PhD fellowship. This work is also partially supported by CSIR (No. 22(0841)/20/EMR-II).

References

1. L. Chua, Memristor—the missing circuit element. IEEE Trans. Circuit Theor. **18**(5), 507–519 (1971)
2. T. Chang, *Tungsten Oxide Memristive Devices for Neuromorphic Applications*. Ph. D thesis (University of Michigan, 2012)
3. S.H. Jo, T. Chang, I. Ebong, B. Bhavitavya, P. Mazumder, W. Lu, Nanoscale memristor device as synapse in neuromorphic systems. Nano Letter **10**, 1297 (2010)
4. G.S. Snider, *Cortical Computing with Memristive Nanodevices* (SciDAC Review, Winter, 2008), pp.58–65
5. S.H. Jo, K.H. Kim, W. Lu, High density cross-bar array based on a silicon memristive systems. Nano Lett. **9**, 870 (2009)
6. T. Chang, S.H. Jo, K.H. Kim, P. Sheridan, S. Gaba, W. Lu, Synaptic behaviors and modeling of a metal oxide memristive device. Appl. Phys. A **102**, 857–863 (2011)
7. M. Das, A. Kumar, R. Singh, M.T. Htay, S. Mukherjee, Realization of synaptic learning and memory functions in Y_2O_3 based memristive device fabricated by dual ion beam sputtering. Nanotechnology **29**(055203), 1–9 (2018)

8. A. Kumar, M. Das, V. Garg, B.S. Sengar, M.T. Htay, S. Kumar, A. Kranti, S. Mukherjee, Forming-free high-endurance Al/ZnO/Al memristor fabricated by dual ion beam sputtering. Appl. Phys. Lett. **110**(253509), 1–5 (2017)
9. T. Prodromakis, B.P. Peh, C. Papavassiliou, C. Toumazou, A versatile memristor model with nonlinear dopant kinetics. IEEE Trans. Electron Dev. **58**(9), 3099–3105 (2011)
10. J.J. Yang, M.D. Pickett, X. Li, D.A.A. Ohlberg, D.R. Stewart, R.S. Williams, Memristive switching mechanism for metal/oxide/metal nanodevices. Nat. Nanotechnol.Nanotechnol. **3**, 429–433 (2008)
11. S. Kumar, R. Agrawal, M. Das, K. Jyoti, P. Kumar, S. Mukherjee, Analytical model for memristive systems for neuromorphic computation. J. Phys. D Appl. Phys. **54**, 355101 (2021)

Chapter 46
Flexible Organic Field-Effect Transistor Using Natural Protein as a Gate Dielectric for Green Electronics

Gargi Konwar, Pulkit Saxena, Vivek Raghuwanshi, Sachin Rahi, and Shree Prakash Tiwari

46.1 Introduction

The rapid advancement of electronics towards low-cost, large-volume, and disposable applications has enormously increased the electronic-waste issue over the years and caused serious environmental and human health risks. Green electronics has been considered as a possible solution to mitigate the adverse effect of technology on the environment [1–3]. Numerous biodegradable and nature-inspired materials have recently emerged as crucial candidates in the production of high-performance sustainable devices [2]. Organic field-effect transistor (OFET) is an essential active device for flexible electronics due to its potential use in several applications including wearable, implantable, and memory applications [4, 5]. The efficiency of OFETs is influenced by several factors including materials and interfaces and a suitable selection of these is highly recommended. Natural proteins like gelatin have been demonstrated as prominent gate dielectric components to exhibit high-performance OFET, leaving a minimal footprint on the environment after the lifespan of devices [6–8]. Though high performance is achieved from these devices, bilayer dielectrics are known to provide improved electrical characteristics. Egg albumen is another potential gate dielectric that can be extracted readily and cheaply without needing any further processes. Pristine egg albumen was studied earlier mostly with oxide-based or thermally evaporated semiconductors operating at higher voltages [9–11]. Here, a thin gelatin layer is coated over egg albumen film, which acts as a smoothing layer to facilitate better organic semiconductor growth and suppresses the leakage issue of egg albumen dielectric. On applying an external electric field at the gate terminal, the mobile ions migrate towards the dielectric/semiconductor interface and

G. Konwar · P. Saxena · V. Raghuwanshi · S. Rahi · S. P. Tiwari (✉)
Department of Electrical Engineering, Indian Institute of Technology Jodhpur, Jodhpur, Rajasthan 342037, India
e-mail: sptiwari@iitj.ac.in

R. Singh et al. (eds.), *The Physics of Semiconductor Devices*, Springer Proceedings in Physics 306, https://doi.org/10.1007/978-981-97-1571-8_46

resulting in an electric double layer (EDL), which enhances the gate capacitance density and enables the device to work at a lower voltage.

Here, in this paper, bilayer combinations of egg albumen/gelatin as gate dielectric was utilized to develop high-performance flexible OFET devices with an active layer of the blend of TIPS-pentacene and Polystyrene (PS). The combined effect of gate dielectric and blend active layer help to attribute average and maximum field-effect mobility (μ) of 0.9 and 1.8 $cm^2 V^{-1} s^{-1}$ respectively, and current on–off (I_{on}/I_{off}) ratio of 10^4 at a voltage of − 5 V.

46.2 Experimentation

Figure 46.1a, b show the schematic and real picture of an array of OFET devices fabricated on flexible indium tin oxide (ITO) coated PET substrate.

Initially, a conventional method was used to clean the substrates by putting in a hot solution of 2-propanol, acetone, and methanol respectively; and finally dried with N_2 blow. To prepare the first layer of dielectric, the extracted egg albumen was blended with deionized water (DI) at a 1:1 ratio and did ultrasonication for 30 min to thoroughly mix them. The prepared solution was then spin-coated at 2000 rpm for 60 s and further annealed at 105 °C for 10 min. To prepare the second layer, 10 wt% gelatin was dissolved in DI for continuously stirring at 60 °C for 2.5 h. Further, the gelatin solution was spin-coated at 2500 rpm over the first layer of albumen for

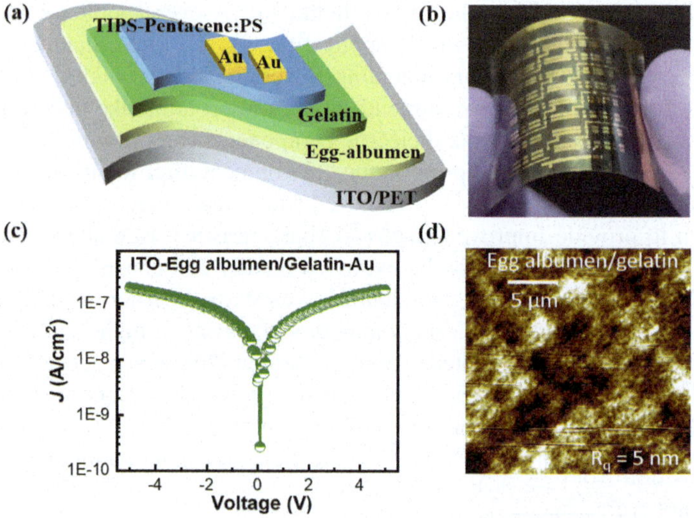

Fig. 46.1 **a** OFET schematic and **b** real view of an array of OFET structures on a PET substrate with bilayer of egg albumen and gelatin gate dielectric. **c** Current density (*J-V*) for MIM structure, and **d** AFM image of the surface morphology of bilayer dielectric

60 s and annealed at 50 °C for 36 h. The approximate thickness of egg albumen and gelatin were found to be 260 nm and 300 nm respectively.

To form the blend layer of polymer and organic semiconductor, both TIPS-Pentacene and polystyrene with a similar concentration of 0.5 wt% were dissolved in toluene at 65 °C for 2.5 h. Further, mixed them at 1:1 ratio to prepare the blend solution at the same temperature for another 1 h. It was then drop cast on the dielectric film and created a closed environment for proper semiconductor growth. Finally, a vacuum thermal evaporation process was used to deposit gold (Au) top source/drain contacts.

Keithley 4200 SCS analyzer was used to characterize the electrical characteristics of fabricated devices in lab environment. Device parameters were derived from the transfer plot in the saturation regime. From the metal–insulator-semiconductor-metal (MISM) structure of ITO/egg albumen/gelatin/TIPS-PEN:PS/Au, the gate capacitance density (C_i) was estimated at 1 kHz and found to be 7 nF/cm^2. This value was further used to extract the μ.

46.3 Results and Discussion

For any high-performance OFET device, dielectric layer has a vital role as the charge accumulation happens near dielectric semiconductor interfaces. Properties of dielectric material should be suitable for transistor operation. In that case, smooth surface morphology with lower gate leakage is preferable for better semiconductor growth as well as lower charge trapping sites at interfaces. Initially, the property of bilayer gate dielectric combination was studied by observing the current density (J-V) across metal–insulator-metal (MIM) structure of ITO/egg albumen/gelatin/Au as shown in Fig. 46.1c. Voltage sweeps between −5 and 5 V were applied to study the electrical behavior of dielectric material and observe current density in the range of 10^{-7} A, which indicates a good insulating property of proposed bilayer dielectric. In addition, atomic force microscopy (AFM) images of a top surface of dielectric films were obtained to examine the surface roughness. Figure 46.1d shows the AFM image of egg albumen/gelatin bilayer. Coating of gelatin suppresses the roughness from 0.9 to 0.5 nm and makes it suitable for proper growth of organic semiconductors. Hence, bilayer dielectric layer showed much lower roughness as compared to pristine egg albumen, eventually helping towards improvement in device performance.

The operation of p-channel OFET device was studied through the electrical characteristics. Input gate voltage (V_{GS}) sweep between 0 and −5 V and $V_{DS} = -5$ V were applied to observe the transfer and output plots as shown in Fig. 46.2a, b. A sharp transition from off to on in I_{DS} vs. V_{GS} plot and excellent saturation in I_{DS} vs. V_{DS} plot were observed from these devices. μ and V_{TH} were calculated from the maximum slope of $(I_{DS})^{1/2}$ vs. V_{GS} plot using saturation regime I_{DS} formula. All other crucial electrical parameters including sub-threshold slope (SS), I_{on}/I_{off}, and interface trap density (D_{it}) were extracted. Maximum and average μ values of of 1.8 $cm^2\ V^{-1}\ s^{-1}$ and 0.9 (±0.5) $cm^2\ V^{-1}\ s^{-1}$ along with the I_{on}/I_{off} of ~ 10^4, SS

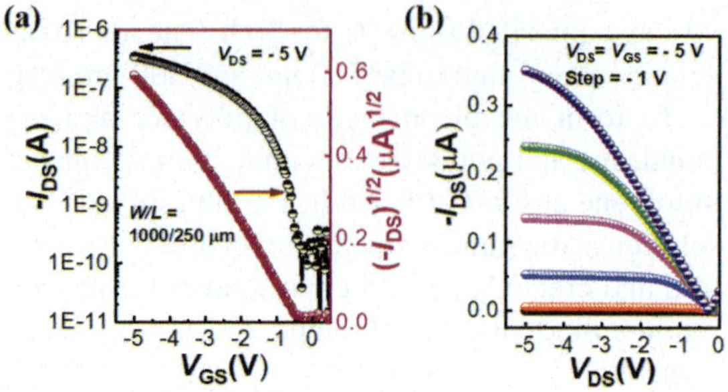

Fig. 46.2 p-channel OFET characteristics **a** transfer and **b** output curve of a fabricated device

of 0.26 V/dec, and D_{it} of 1.5 ($\pm$0.6) $\times$ 10^{11} cm^{-2} eV^{-1} were exhibited. This small value of *SS* and D_{it} signifies the good quality of dielectric: semiconductor interface. This excellent electrical performance was a result of the combined effects of bilayer gate dielectric and the polymer semiconductor composite layer.

Further device operational stability under different circumstances was also studied to examine their suitability in real-world applications. In this, repeatability and bias-stress behaviors were evaluated and shown in Fig. 46.3a, b. In repeatability, transfer curves were scanned continuously for 100 no. of cycles. It was observed that plots were overlapped with one other showing very low variation in the characteristics. For bias-stress study, constant biasing was applied at the gate terminal for 1 h, and observed the I_{DS} during stress period. A typical decay is observed for normalized value of I_{DS}, which is very common in organic devices due to the availability of charge-trapping sites at the interfaces. V_{TH} was shifted and increased due to the charge trapping as shown in Fig. 46.3b.

In addition, cyclic stability was performed to investigate the reliability of these devices in real applications. Continuous input pulses between the range of −5 V and 5 V were applied over a period of time to gate terminal and observed the corresponding output response I_{DS}. Figure 46.4a represents the switching behavior of device characteristics and a relatively stable cyclic behavior was observed for these devices.

Moreover, circuit applicability was studied by connecting an external load resistance (R_L) with the OFET devices. Resistive load inverter circuit for various R_L values from 10 to 40 MΩ was implemented and characterized for their voltage transfer

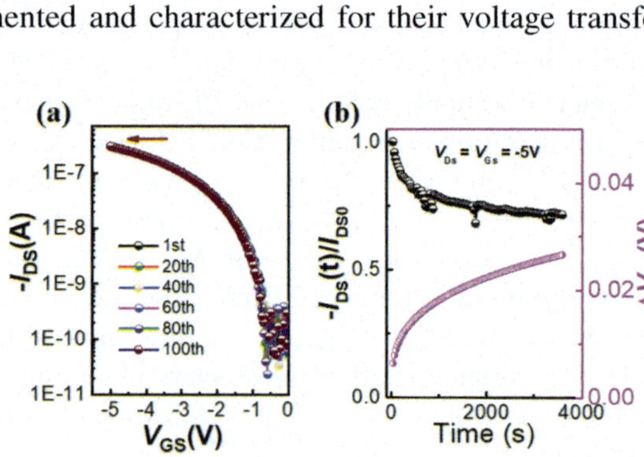

Fig. 46.3 **a** Repeatability of transfer curve for 100 no. of cycles, and **b** bias-stress stability for 1 h

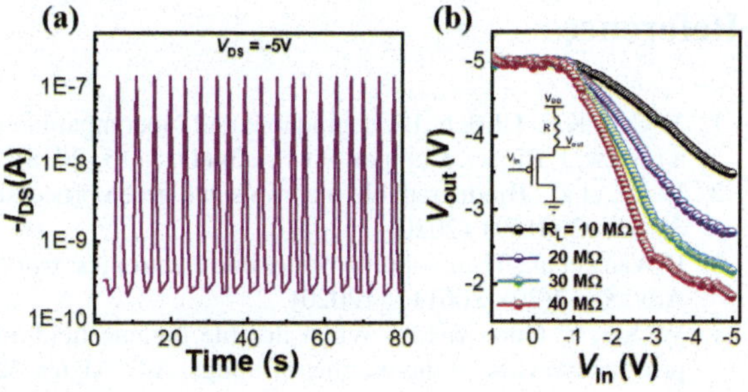

Fig. 46.4 **a** Switching response of I_{DS} for a pulse input, and **b** resistive load-based voltage transfer characteristics with different load resistance

curves. Corresponding output response (V_{OUT}) for each R_L value was obtained and shown in the typical voltage transfer characteristics of Fig. 46.4b. At low input (V_{IN}), V_{OUT} was almost equal to the V_{DD} of -5 V. For a higher V_{IN} value, V_{OUT} becomes low showing a good inverter characteristic. The switching characteristics of resistive load inverter is more pronounced for high R_L values as the voltage drop across the resistance becomes higher. This novel bilayer gate dielectric combinations could be utilized further in future sustainable and green electronics.

46.4 Conclusion

Flexible solution-processed eco-friendly OFET devices were demonstrated with bilayer gate dielectric combinations of egg albumen/gelatin. An excellent p-channel performance for low operating voltage of $-$ 5 V was achieved from these devices with maximum and average mobility values of 1.8 and 0.9 ($\pm$ 0.5) $cm^2\ V^{-1}\ s^{-1}$. In addition, very high electrical and operational stability was confirmed for these devices through repeatability, bias-stress, and cyclic stability tests. Utilization of these devices in circuit application was also demonstrated using external load resistances providing promising results for switching. These findings indicate that natural proteins such as egg albumen and gelatin can be employed as promising gate dielectric components in various future-generation smart and biodegradable applications.

Acknowledgements Authors would like to acknowledge Sir Visvesvaraya Young Faculty Research Fellowship (YFRF), Ministry of Electronics and Information Technology (MeitY), Govt. of India.

References

1. Y. Cao, K.E. Uhrich, Biodegradable and biocompatible polymers for electronic applications: a review. J. Bioact. Compat. Polym. **34**(1), 3–15 (2018)
2. W. Li, et al., Biodegradable materials and green processing for green electronics. Adv. Mater **32**(33) e2001591 (2020)
3. P. Wang et al., The evolution of flexible electronics: from nature, beyond nature, and to nature. Adv. Sci. **7**(20), 2001116 (2020)
4. W. Shi, Y. Guo, Y. Liu, When flexible organic field-effect transistors meet biomimetics: a prospective view of the internet of things. Adv. Mater. **32**(15), 1901493 (2020)
5. H. Sirringhaus, 25th anniversary article: organic field-effect transistors: the path beyond amorphous silicon. Adv. Mater. **26**(9), 1319–1335 (2014)
6. Y. He et al., Solution-processed natural gelatin was used as a gate dielectric for the fabrication of oxide field-effect transistors. Org. Electron. **38**, 357–361 (2016)
7. S. Mandal et al., Low operating voltage organic field-effect transistors with gelatin as a moisture-induced ionic dielectric layer: the issues of high carrier mobility. ACS Appl. Mater. Interfaces **12**(17), 19727–19736 (2020)
8. V. Raghuwanshi, P. Saxena, S. Rahi, A.K. Mahato, I. Varun, S.P. Tiwari, Solution-processed flexible organic field-effect transistors with biodegradable gelatin as the dielectric layer: an approach toward biodegradable systems. ACS Appl. Electron. Mater. **2**(10), 3373–3379 (2020)
9. J.W. Chang, C.G. Wang, C.Y. Huang, T.D. Tsai, T.F. Guo, T.C. Wen, Chicken albumen dielectrics in organic field-effect transistors. Adv. Mater **23**(35), 4077–4081 (2011)
10. D.B. Jeon, J.Y. Bak, S.M. Yoon, Oxide thin-film transistors fabricated using biodegradable gate dielectric layer of chicken albumen. Jpn. J. Appl. Phys. **52**(12R), 128002 (2013)
11. S.J. Kim et al., Nonvolatile memory thin-film transistors using biodegradable chicken albumen gate insulator and oxide semiconductor channel on eco-friendly paper substrate. ACS Appl. Mater Interfaces **7**(8), 4869–4874 (2015)

Chapter 47
Design and Analysis of $Cs_3Sb_2Br_9$/Si Tandem Solar Cell Using SCAPS-1D

Sachchidanand, Akash Patnaik, Anil Kumar, and Pankaj Sharma

47.1 Introduction

The performance of single-junction (SJ) photovoltaic (PV) devices or solar cells (SCs) is limited to achieving the theoretical efficiency limit of Shockley-Quiesser (SQ limit) due to absorption losses with energies lower than bandgap, E_g, and relaxation losses with energies higher than E_g. The perovskite (ABX_3: where '*A*' and '*B*' are organic/inorganic and metal cations, respectively, and '*X*' is an anion) based SJ-SC expresses an excellent performance in the field of emerging PV technology with a power conversion efficiency (PCE) of 26.1% [1] because of various inbuilt features such as large absorption coefficient, tunable E_g, the longer diffusion length of carriers, excellent charge mobility, lower non-radiative recombination rate, and simple synthesis processing [2]. On the other hand, perovskite-based SJ-SCs still struggle with thermal instability and toxicity which degrades the SC performance [3]. The issue of toxicity of perovskite is suppressed by choosing a suitable non-toxic metal '*B*' cation such as tin (Sn), germanium (Ge), titanium (Ti), antimony (Sb), and bismuth (Bi) [4–6]. The thermal instability of perovskite is improved by choosing an appropriate inorganic '*A*' cation [7].

A multi-junction (MJ) structure can be an effective tactic for surpassing the SQ limit. A two-terminal (2-T) tandem solar cell (TSC) effectively reduces the spectral losses by incorporating multiple E_g structures, from lower E_g i.e. 1.1 eV to higher E_g i.e. 2.1 eV, in the series arrangement [8]. By this type of arrangement, TSC utilizes a wider range of spectrum to produce a higher photocurrent as compared

Sachchidanand (✉) · A. Patnaik · P. Sharma
Advance Semiconductor Materials and Nanoelectronics Research Group, PDPM Indian Institute of Information Technology, Design and Manufacturing, Jabalpur, MP 482005, India
e-mail: 1922604@iiitdmj.ac.in

A. Kumar
Department of Electronics and Communication Engineering, PDPM Indian Institute of Information Technology, Design and Manufacturing, Jabalpur, MP 482005, India

R. Singh et al. (eds.), *The Physics of Semiconductor Devices*, Springer Proceedings in Physics 306, https://doi.org/10.1007/978-981-97-1571-8_47

to SJ-SC. In the context of MJ-TSC, III-V semiconductors incorporated with Si express the highest PCE of ~ 39% [9]. Despite the impressive performance of III-V semiconductors and/or Si-based TSCs, these struggles with the high processing cost and complex fabrication methods of III-V semiconductors.

A primary focus of recent TSC development is to enhance the PCE along with the reduction the cost-per-watt of PV technologies. In this effort, perovskite became the prime choice in TSCs as the top subcell with either Si or perovskite as the bottom subcell. In between both TSCs combination, the perovskite/Si TSCs become an appropriate combination because (1) the coupling of perovskite with Si limits the drawback of perovskite, and (2) pairing of perovskite with Si enables to take advantage of the > 95% market share currently held by Si [8]. Thus, combining low-cost, quickly developed perovskite with Si has the potential to make TSC a more cost-effective technology. Recently, lead-free perovskite/Si TSC has been successfully explored to improve PCE while lowering the technology's levelized cost of electricity (LCOE). Recently, Islam et al. and Amri et al. have studied the 2-T perovskite/Si TSC with the help of a one-dimension solar cell capacitance simulation tool (SCAPS-1D) [10, 11]. In the search for an appropriate wide bandgap (WBG) absorber, a novel lead-free $Cs_3Sb_2Br_9$ perovskite has been shown as a suitable alternative due to its higher E_g, better structural stability, and nontoxic. However, lead-free $Cs_3Sb_2Br_9$ perovskite has been reported for the optoelectronics applications like LEDs, photodetectors, and various nanostructures like nanoflakes, nanocrystals, etc. [12–15]. Recently, our group numerically simulated the PSC incorporation the $Cs_3Sb_2Br_9$ as an absorber [16–18]. To the best of the author's knowledge, there have been no reports of perovskite/Si TSC incorporating $Cs_3Sb_2Br_9$ as the top subcell, till date. This work aims to design a stable, and cost-effective combination of $Cs_3Sb_2Br_9$/Si TSC for effective long-term commercial viability.

47.2 Device Structure and Methodology

In general, two different configurations are utilized to design a tandem device; a stack configuration or four-terminal (4-T), and a series configuration or 2-T. In the conventional method for designing the 2-T tandem device, the WBG material is arranged in series with the narrow bandgap (NBG) material in such a way that both can absorb the maximum incident solar spectrum in a broad range. In the presented work, perovskite/Si TSC is designed by utilizing $Cs_3Sb_2Br_9$ perovskite (WBG material, having $E_g = 2.0$ eV) as the top absorber subcell and Si (NBG material, having $E_g = 1.12$ eV) as the bottom absorber subcell. Figure 47.1a, b shows the schematic of the standalone $Cs_3Sb_2Br_9$ PSC, and Si SC. For the $Cs_3Sb_2Br_9$ PSC, Spiro-OMeTAD as holes transport layer (HTL) and TiO_2 as electron transport layer (ETL) are considered with a thickness of 0.15 μm and 0.10 μm, respectively. For the Si SC, highly doped n- and p-type Si are used as the connecting layer with a thickness of 0.02 μm and 0.1 μm, respectively. Recently, the standalone $Cs_3Sb_2Br_9$ PSC is simulated in our

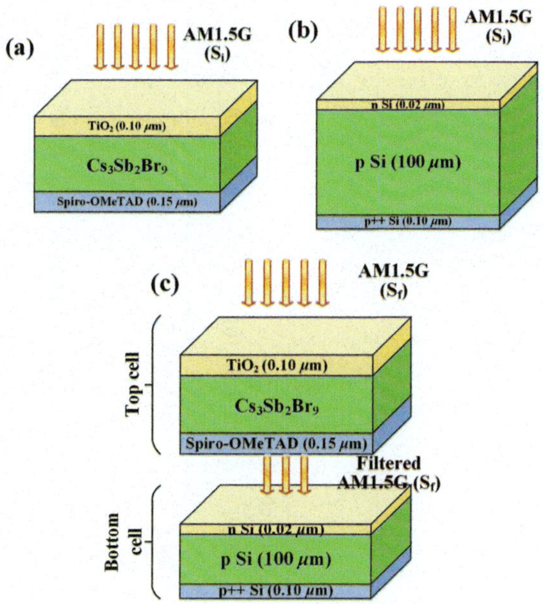

Fig. 47.1 Schematic of single absorber-based **a** perovskite and **b** Si solar cell, and **c** 2-T perovskite/Si TSC

group and the standalone Si SC is simulated with one of the default models in the SCAPS-1D simulation package [16, 19].

The numerical simulation of the proposed 2-T $Cs_3Sb_2Br_9$/Si TSC, schematically shown in Fig. 47.1c, is performed by using the SCAPS-1D. The working method of the proposed 2-T $Cs_3Sb_2Br_9$/Si TSC involves the implementation of the top subcell under illumination with AM1.5G spectrum (S_i) under 1 sun with 100 $mWcm^{-2}$ followed by the bottom subcell under illumination with the filtered AM1.5G spectrum (S_f), drained after the top subcell. An ideal interconnecting layer (ICL) or tunnel junction (TJ), having neither optical loss nor electrical loss is assumed in between the top and bottom subcells. Due to the insufficiency of the SCAPS-1D to simulate the MJ structure, the top and bottom subcells are simulated separately. The current-matching condition was obtained between the top and bottom subcells by varying the top absorber's thickness to get the short-circuit current density (J_{SC}) value at each variation. For the current-matching condition for complete TSC, the filtered AM1.5G spectrum (S_f) is applied to the bottom subcell by using Eq. (1), which is given as:

$$S_f(\lambda) = S_i(\lambda).\exp(-\alpha_{ETL}(\lambda).d_{ETL}) \\ .\exp(-\alpha_{Absorber}(\lambda).d_{Absorber}).\exp(-\alpha_{HTL}(\lambda).d_{HTL}) \quad (47.1)$$

where $\alpha(\lambda)$ is the absorption coefficient of materials and d is the thickness of materials. The $\alpha(\lambda)$ of each layer is adapted from the literature [13, 18, 20, 21]. The simulation input parameters of each layer are listed in Table 47.1 extracted from the experimented reports [10, 16].

Table 47.1 Parametric input of each layer for SCAPS-1D simulation

Parameters	Top subcell			Bottom subcell		
	TiO_2 [16]	$Cs_3Sb_2Br_9$ [16]	Spiro-OMeTAD [16]	n^+ Si [10]	p Si [10]	p^+ Si [10]
E_g/χ (eV)	3.2/4.2	2.0/3.98	3.1/2.1	1.12/4.5	1.12/4.5	1.12/4.5
ε_r	10	10	3	11.9	11.9	11.9
N_A/N_D (cm^{-3})	$0/1 \times 10^{17}$	$1 \times 10^9/1 \times 10^9$	$1 \times 10^{19}/0$	$0/1 \times 10^{20}$	$1 \times 10^{16}/0$	$1 \times 10^{19}/0$
N_C/N_V (cm^{-3})	$1 \times 10^{19}/1 \times 10^{19}$	$1 \times 10^{14}/1 \times 10^{14}$	$2.5 \times 10^{18}/1.8 \times 10^{19}$	$2.8 \times 10^{19}/1.04 \times 10^{19}$	$2.8 \times 10^{19}/1.04 \times 10^{19}$	$2.8 \times 10^{19}/1.04 \times 10^{19}$
μ_n/μ_p (cm^2V^{-1} s^{-1})	20/10	22.1/1.1	$2 \times 10^{-4}/2 \times 10^{-4}$	$1.5 \times 10^3/4.5 \times 10^2$	$1.5 \times 10^3/4.5 \times 10^2$	$1.5 \times 10^3/4.5 \times 10^2$
Defect type	Neutral	Neutral	Neutral	Neutral	Neutral	Neutral
N_t (cm^{-3})	1×10^{15}	1×10^{14}	1×10^{15}	1×10^{12}	1×10^{12}	1×10^{12}
α (cm^{-1})	[18]	[13]	[20]	[21]	[21]	[21]

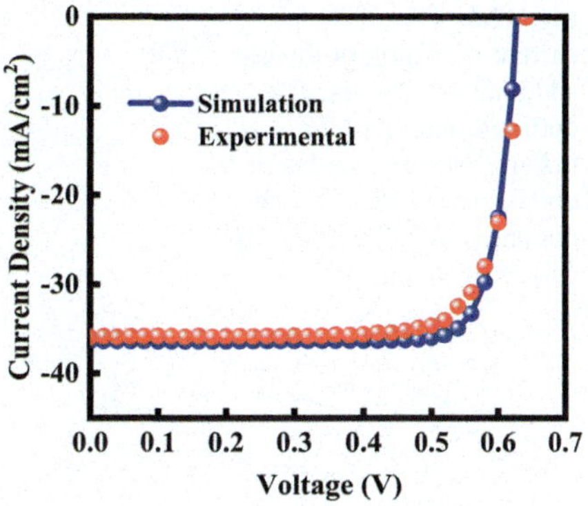

Fig. 47.2 The simulation and experimental J-V characteristics of the standalone Si SC

Table 47.2 Comparison table of the simulation and experimental results of the standalone Si SC

	PCE (%)	V_{OC} (V)	J_{SC} ($mAcm^{-2}$)	FF (%)
Simulation	18.83	0.63	36.29	82.74
Experimental [22]	18.30	0.635	35.71	79.20

47.3 Results and Discussion

47.3.1 Validation of Standalone SCs with the Experimental Result

To the best of the author's knowledge, there is no experimental report has been found in the literature for lead-free inorganic $Cs_3Sb_2Br_9$ PSC as an absorber. Furthermore, the SCAPS-1D simulation of the standalone Si SC is validated by comparing the simulation result with the experimental result in [22]. The structure of the standalone Si SC is taken as reported in the literature. Figure 47.2 shows the current density vs voltage (J-V) characteristics of the simulation and experimental results of the standalone Si SC structure. It is evident from Fig. 47.2 that the simulation result matches well with the experimental result. Table 47.2 lists the comparison data between the simulation and experimental results of the standalone Si SC.

47.3.2 Current-Matching Strategy Between the Top and Bottom Subcells

The proposed 2-T $Cs_3Sb_2Br_9$/Si TSC is designed by using the series configuration of the $Cs_3Sb_2Br_9$ PSC as the top subcell and the Si SC as the bottom subcell. The current-matching between both subcells is obtained, because of the series configuration of both subcells, by varying the top absorber's thickness till the current density in both

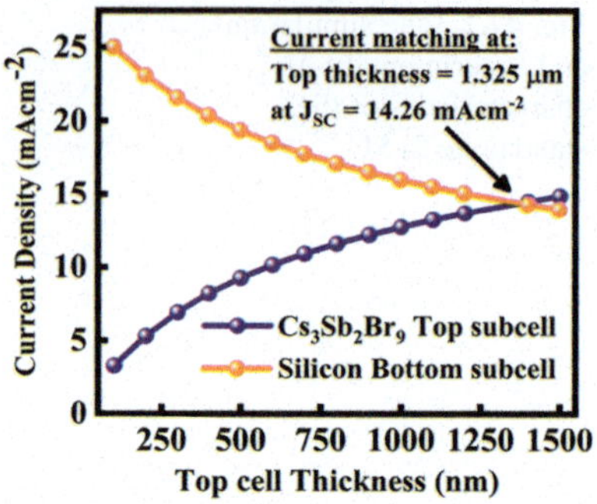

Fig. 47.3 The current-matching of the top and bottom subcells. The optimum J_{SC} of 14.26 mAcm^{-2} for both subcells is obtained at the top absorber thickness of 1.325 μm after current-matching

subcells is identical. The optimum value for the current-matching of the 2-T tandem device is decided by the minimum valued J_{SC} from either subcell. The maximum open-circuit voltage (V_{OC}) for the 2-T tandem device is calculated by the summed value of V_{OC} of both subcells. Figure 47.3 shows the current-matching result of the proposed 2-T $Cs_3Sb_2Br_9$/Si TSC where J_{SC} is shown as the function of the top absorber's thickness. Figure 47.3 depicts that there is a continuous decay of J_{SC} with the top subcells' thickness variation due to a lower power density of the filtered AM1.5G spectrum (S_f), drained after the top subcell. In conclusion, the optimum value of J_{SC} after the current-matching strategy is 14.26 mAcm^{-2} at the optimum top absorber's thickness of 1.325 μm.

47.3.3 2-T $Cs_3Sb_2Br_9$/Si TSC Simulation

The straightforward strategy is used to simulate the 2-T $Cs_3Sb_2Br_9$/Si TSC. First, the incident AM1.5G spectrum through the top subcell is calculated using Eq. (47.1) with the absorption coefficient profile of each layer as listed in Table 47.1. Then, the filtered AM1.5G spectrum is applied to the bottom subcell to perform the simulation. For simplicity, the optical losses at all layers and interfaces are not taken into consideration. Here, only the absorption coefficient profile of all layers is taken care-off. However, other factors of optical losses, such as the transmission of light, scattering, and reflections suggest ignoring due to the sophisticated optical design of the tandem device.

In the present work, with the current-matching strategy, discussed in Sect. 47.3.2, the 2-T $Cs_3Sb_2Br_9$/Si TSC has been investigated. Figure 47.4a shows the J-V of the top subcell, filtered bottom subcell, and the proposed 2-T $Cs_3Sb_2Br_9$/Si TSC. Figure 47.4b expresses the EQE-λ characteristics of the top and bottom subcells. From Fig. 47.4a for the 2-T $Cs_3Sb_2Br_9$/Si TSC, the J_{SC} is dominated by the top subcell while the V_{OC} expresses the value as the addition of both subcells. Table 47.3 lists the photovoltaic parameters of the proposed 2-T $Cs_3Sb_2Br_9$/Si TSC and also shows the comparison with other experimental perovskite/Si TSCs. Thus, the result and analysis of this work highlight the feasibility of lead-free $Cs_3Sb_2Br_9$ perovskite

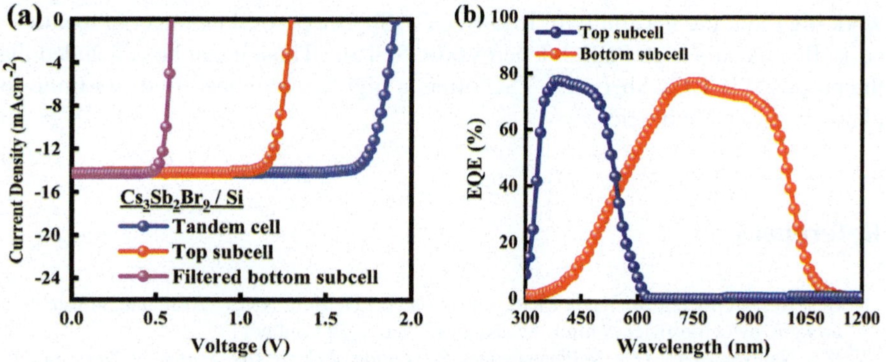

Fig. 47.4 **a** J-V characteristics of the proposed 2-T $Cs_3Sb_2Br_9$/Si TSC along with the top and the filtered bottom subcells and **b** external quantum efficiency (EQE) vs wavelength (λ) characteristics of the top and bottom subcells

Table 47.3 Photovoltaic parameters of perovskite/Si-based TSC and comparison with experimental perovskite/Si TSCs

Tandem cell	PCE (%)	V_{Oc} (V)	J_{SC} ($mAcm^{-2}$)	FF (%)	References
$FA_{0.5}MA_{0.38}Cs_{0.12}PbI_{2.04}Br_{0.96}$/Si	13.70	1.58	11.50	75.00	[23]
$MAPbI_3$/Si	17.10	1.68	16.10	78.00	[24]
$MAPbI_3$/Si	18.10	1.76	14.00	77.00	[25]
$(FAPbI_3)_{0.83}(MAPbBr_3)_{0.17}$/Si	21.10	1.71	15.80	78.30	[26]
$Cs_3Sb_2Br_9$/Si	22.94	1.90	14.26	84.66	This work

material to attain a cost-effective and high PCE 2-T perovskite/Si TSC. Thus, it can be concluded that the proposed $Cs_3Sb_2Br_9$/Si TSC offers a highly viable lead-free substitute for photovoltaics technologies.

47.4 Conclusion

The lead-free $Cs_3Sb_2Br_9$/Si TSC has been successfully explored to improve PCE while lowering the technology's LCOE. The top and bottom subcells are simulated individually by using the SCAPS-1D due to the restriction of SCAPS-1D for the single junction. The current-matching strategy is obtained for the same J_{SC} in both subcells. The optimum thickness of the top absorber for current-matching is found as 1.325 μm with a J_{SC} of 14.26 $mAcm^{-2}$. Finally, the proposed 2-T $Cs_3Sb_2Br_9$/Si TSC exhibits the simulation results as PCE of 22.94% followed by V_{OC} of 1.90 V, J_{SC} of 14.26 $mAcm^{-2}$, and fill factor (FF) of 84.66%. The result and analysis of this work highlight the feasibility of lead-free $Cs_3Sb_2Br_9$ perovskite material to attain a cost-effective and high PCE 2-T perovskite/Si TSC. Thus, it can be concluded that the proposed 2-T $Cs_3Sb_2Br_9$/Si TSC offers a highly viable lead-free substitute for photovoltaics technologies.

References

1. Best Research-Cell Efficiency Chart. Photovoltaic Research, NREL. Available: https://www.nrel.gov/pv/cell-efficiency.html. Accessed: 12 Mar 2024 [Online]
2. S.A. Narayanasamy, D.N. Vellalapalayam, *Revolution of Perovskite: Synthesis, Properties and Application* (Springer, 2020)
3. P. Le, The quest for stability of perovskite solar cells: understanding degradation, improving lifetimes and towards experimental standards. THÈSE N O 8106 École Polytechnique Fédérale De Lausanne **8106**, 218 (2018)
4. A.K. Baranwal et al., Lead-free perovskite solar cells using Sb and Bi-based A3B2X9 and A3BX6 crystals with normal and inverse cell structures. Nano Converg. **4**(1) (2017)
5. X. Li et al., Perovskite solar cells employing an eco-friendly and low-cost inorganic hole transport layer for enhanced photovoltaic performance and operational stability. J. Mater. Chem. A **7**(12), 7065–7073 (2019)
6. M. Chen et al., Highly stable and efficient all-inorganic lead-free perovskite solar cells with native-oxide passivation. Nat. Commun.Commun. **10**(1), 1–8 (2019)
7. Z. Qi et al., Highly stable lead-free Cs3Bi2I9 perovskite nanoplates for photodetection applications. Nano Res. **12**(8), 1894–1899 (2019)
8. S.E. Sofia, H. Wang, A. Bruno, J.L. Cruz-Campa, T. Buonassisi, I.M. Peters, Roadmap for cost-effective, commercially-viable perovskite silicon tandems for the current and future PV market. Sustain. Energy Fuels **4**(2), 852–862 (2020)
9. M. Green, E. Dunlop, J. Hohl-Ebinger, M. Yoshita, N. Kopidakis, X. Hao, Solar cell efficiency tables (version 57). Prog. Photovoltaics Res. Appl. **29**(1), 3–15 (2021)
10. T. Islam et al., Investigation of CsSn0.5Ge0.5I3-on-Si Tandem solar device utilizing SCAPS simulation. IEEE Trans. Electron Dev. **68**(2), 618–625 (2021)
11. K. Amri, R. Belghouthi, M. Aillerie, R. Gharbi, Device optimization of a lead-free perovskite/silicon tandem solar cell with 24.4% power conversion efficiency. Energies **14**(12), 1–20 (2021)
12. Z. Ma et al., Electrically-driven violet light-emitting devices based on highly stable lead-free perovskite Cs3Sb2Br9 quantum dots. ACS Energy Lett., 385–394 (2020)
13. P. Liu et al., Lead-free Cs3Sb2Br 9 single crystals for high performance narrowband photodetector. Adv. Opt. Mater. **8**(21), 1–9 (2020)
14. Z. Zheng et al., Submillimeter and lead-free Cs3Sb2Br 9 perovskite nanoflakes: Inverse temperature crystallization growth and application for ultrasensitive photodetectors. Nanoscale Horizons **4**(6), 1372–1379 (2019)
15. C. Lu et al., Synthesis of lead-free Cs3Sb2Br9 perovskite alternative nanocrystals with enhanced photocatalytic CO_2 reduction activity. Nanoscale **12**(5), 2987–2991 (2020)
16. V. Sachchidanand, A.K. Garg, P. Sharma, Numerical simulation of novel lead-free Cs3Sb2Br9 absorber-based highly efficient perovskite solar cell. Opt. Mater. **122**(PA), 111715 (2021)
17. A.K. Sachchidanand, P. Sharma, Performance investigation of organic/inorganic bottom cell on lead-free Cs3Sb2Br9 based all-perovskite tandem solar cell. IEEE Trans. Electron Dev. **69**(6), 3462–3469 (2022)
18. B.G. Obeid, A.S. Hameed, H.H. Alaaraji, Structural and optical properties of TiO_2:MgOo thin films preparing at 373K. Dig. J. Nanomater. Biostruct.Nanomater. Biostruct. **12**(4), 1239–1246 (2017)
19. Available http://scaps.elis.ugent.be/. Accessed 24 Jan 2021 [online]

20. M. Filipič et al., CH3NH3PbI3 perovskite/silicon tandem solar cells: characterization based optical simulations. Opt. Express **23**(7), A263 (2015)
21. S. De Wolf et al., Organometallic halide perovskites: sharp optical absorption edge and its relation to photovoltaic performance. J. Phys. Chem. Lett. **5**(6), 1035–1039 (2014)
22. K. Kim et al., Simulations of chalcopyrite/c-Si tandem cells using SCAPS-1D. Sol. Energy **145**, 52–58 (2017)
23. T. Leijtens, K.A. Bush, R. Prasanna, M.D. McGehee, Opportunities and challenges for tandem solar cells using metal halide perovskite semiconductors. Nat. Energyergy **3**(10), 828–838 (2018)
24. J. Zheng et al., Large area efficient interface layer free monolithic perovskite/homo-junction-silicon tandem solar cell with over 20% efficiency. Energy Environ. Sci. **11**(9), 2432–2443 (2018)
25. J.P. Mailoa et al., A 2-terminal perovskite/silicon multijunction solar cell enabled by a silicon tunnel junction. Appl. Phys. Lett. **106**(12) (2015)
26. W. Li et al., High-performance solar flow battery powered by a perovskite/silicon tandem solar cell. Nat. Mater.Mater. **19**, 1326–1331 (2020)

Chapter 48
One-Pot Preparation of Fe/Cu Catalytic Solution for the Growth of Carbon Nanotubes for Use in Gas Sensor and Field Emission Devices

Mohd Sarvar, Mohammad Moeen Hasan Raza, Shah Masheerul Aalam, Mohd Sadiq, Mohd. Shahid Khan, and Javid Ali

48.1 Introduction

Carbon nanotubes (CNTs) have a wide range of possible applications due to their unique characteristics. Carbon nanotubes are divided into two types: SWCNTs and MWCNTs. SWCNTs are graphite roll layers that are one atom thick and have diameters ranging from 1 to 3 nm. MWCNTs, on the other hand, are made up of many layers of graphite roll with diameters ranging from 4 to 100 nm. Both CNTs are many microns thick [1–5].

Some elements of thin film properties, such as electrics and optics, as well as surface morphology, have recently been explored [6–12]. Both copper oxide CuO and Cu_2O, can act as interesting catalysts in electrical and other applications [13–15]. Iron catalysts, on the other hand, can be used as magnetic recording. Fe and Cu Nano-composite are now intriguing materials in the field of catalyst research due to their novel uses [16, 17].

Catalysis as a referent object of nano science and technology [18] by metal Nano composite is growing rapidly, as indicated by the high number of publications in the previous years. It can be shown that transition metal Nano-composites are especially inspiring to utilise as catalysts due to their high surface to volume ratio and high surface energy, which makes the atoms on their surface very active [19–22]. As a result, metal Nano-composite are attractive catalysts for a wide range of processes, including oxidation, hydrogenation, and C–C coupling reactions [23]. Metal oxide

M. Sarvar · M. M. H. Raza · S. M. Aalam · M. Sadiq · Mohd. S. Khan · J. Ali (✉)
Department of Physics, Jamia Millia Islamia, New Delhi 110025, India
e-mail: javi-reslab@rediffmail.com

M. Sadiq
Department of Physics, ARSD College, University of Delhi, New Delhi 110019, India

R. Singh et al. (eds.), *The Physics of Semiconductor Devices*, Springer Proceedings in Physics 306, https://doi.org/10.1007/978-981-97-1571-8_48

composite are intriguing materials because they may be made with exceptionally high surface area and highly ionic, as well as unique crystallography that demonstrates multiple surface reaction sites [24].

Because the quality and structure of carbon nanotubes (CNTs) are strongly reliant on the qualities and type of catalyst materials utilised, it is critical to investigate the catalysts. Controlling the diameter and number of layers in a CNT is critical since the catalyst particles serve as the foundation for the growing CNTs. The synthesis of CNTs intends to synthesize some in a high temperature environment directly on the substrate, which is placed in the high temperature field. CNTs can be produced in a variety of techniques, including laser ablation [25], arc discharge [26], and several types of chemical vapour deposition (CVD) [27]. The use of metal composite act as catalysts for LPCVD is appealing due to the huge volume of low-cost accessible particles and the low particle concentrations required [28].

Because of their Electrical Conductivity, CNTs can exist in both metallic and semi-conducting phases, depending on whether they are single or multi walled. Other qualities, like as elasticity strength and strength, make them suitable for use as probe tips in very high-resolution scanning probe microscopy. Furthermore, both CNTs as SWCNTs and MWCNTs are used in electronic devices [29, 30].

The generated nanocomposites coated with Si substrate by spin coating unit for generating MWCNTs, the mechanism of CNT growth, and the effect of catalyst on the CNT growth will be discussed in detail in this work, which used Cu/Fe contents on the silicon substrate. All applied characteristic analysis, including SEM, and Raman spectroscopy, will also be shown and discussed.

48.2 Experimental

The catalyst is prepared using ferric nitrate Fe $(NO_3)_3{\cdot}9H_2O$, aluminium nitrate Cu $(NO_3)_3{\cdot}3H_2O$, PVA, and ethanol.

48.2.1 Preparation of Catalyst

48.2.1.1 Iron Catalyst

For 60 min, 20 ml of a 5% (W/V) weight per volume ratio PVA solution in distilled water was combined and agitated. The solution was then treated with 10 ml of 100% ethanol. After agitating the solution for 30 min, we produced the 2:1 (V/V) volume to volume ratio solution of PVA and ethanol. The ferric nitrate (Fe (NO3)3.9H2O) 0.1 mol/L solution was then added to the PVA and ethanol solution to generate a concentration colloidal solution. After stirring for 30 min, the mixture was sonicated for another 60 min.

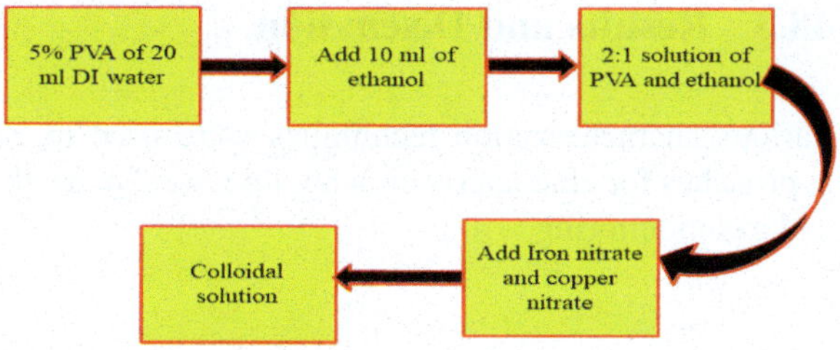

Fig. 48.1 Diagram of catalyst preparation

48.2.1.2 Iron/Copper Catalyst

20 ml of a 5% (W/V) weight per volume ratio PVA solution in distilled water was blended and agitated for 60 min. After that, 10 mL of pure ethanol was added to the solution. We obtained the 2:1 (V/V) volume to volume ratio solution of PVA and ethanol after stirring the solution for 30 min. The ferric nitrate (Fe (NO3)3.9H_2O) solution was then added in 0.1 mol/L increments to the PVA and ethanol solutions to generate a colloidal solution of concentration. After stirring for 30 min, 0.1 mol/L copper nitrate was added to the solution. After 30 min of stirring, the mixture was sonicated for another 60 min.

As a result, the ideal solution for spin coating was created. The colloidal solution was applied to the silicon wafers by spin-coating. The colloidal solution-coated wafers were simply placed into the dryer below the melting points of iron hydroxide and copper oxide, and then into the furnace for quick heating at roughly 100 °C to generate the catalyst in oxide form, evaporate the ethanol, and oxidise the PVA. Figure 48.1 demonstrates the systematic steps of catalyst preparation.

48.2.2 Growth of CNTs

Carbon nanotubes were created on flat silicon surfaces by decomposing acetylene gas. It was revealed that spin-coated Fe $(NO_3)_3.9H_2O$ with concentrations of 0.1 mol/L dissolved in a colloidal solution with a 2:1 (v/v) absolute ethanol to PVA ratio produced iron nanoparticles. To synthesize the catalyst composite for the formation of MWCNTs, a 0.1 mol/L concentration of copper nitrate was added to this colloidal solution. PVA's role was to interact with and bind iron and silver ions. CNTs were synthesized on a silicon substrate using a LPCVD at a temperature of 700 °C for 30 min with acetylene as the main gas stream. Substrates with a thin film layer of Fe/Cu catalyst composite were put into the quartz tube, and as reactant gases, a mixture of $C_2H_2/H_2/NH_3$ gases with flow rates of 30/100/50 sccm was conveyed into the quartz tube. The source gas was acetylene (C_2H_2), the carrier gas was hydrogen (H_2), and the amorphous surface was etched using ammonia (NH_3). These gases are absorbed, and a chemical reaction takes place, leading in the growth of CNTs on the substrate surface; therefore, CNTs were formed using LPCVD.

48.3 Results and Discussion

Various characterization techniques were used to characterise the materials. The approaches for characterisation are reviewed, as well as the results of field emission and gas monitoring.

48.3.1 Scanning Electron Microscopy (SEM)

We used SIGMA VP model FESEM to examine the surface morphology of CNTs. According to Fig. 48.2, MWCNTs with diameters ranging from 10 to 50 nm were effectively produced using the LPCVD process.

In Fig. 48.2a, b, MWCNTs were grown on iron-coated Si substrate whose diameter is less than the MWCNTs which was grown on iron/copper coated Si substrate which is clearly seen in Fig. 48.2c, d. The as-grown MWCNTs which were grown on Fe coated catalyst is less defective as compared to the Fe/Cu-coated catalyst which is clearly seen in Fig. 48.2.

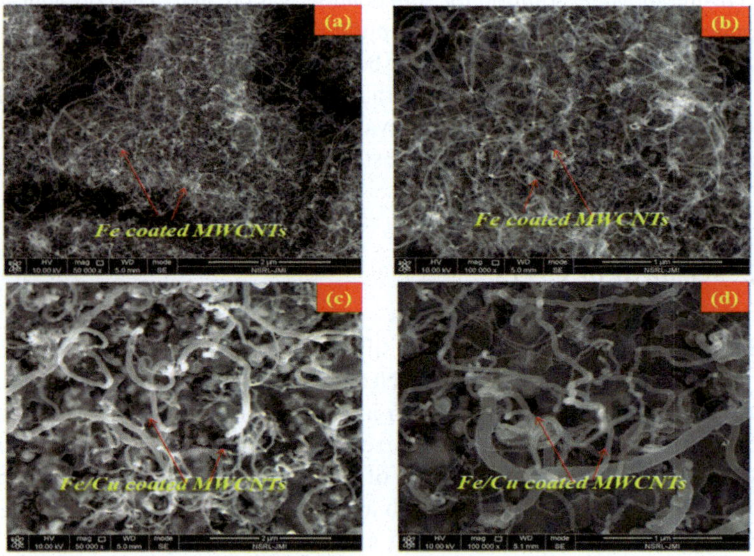

Fig. 48.2 FESEM study of as-grown MWCNTs **a**, **b** (Fe coated), **c**, **d** (Fe/Cu coated)

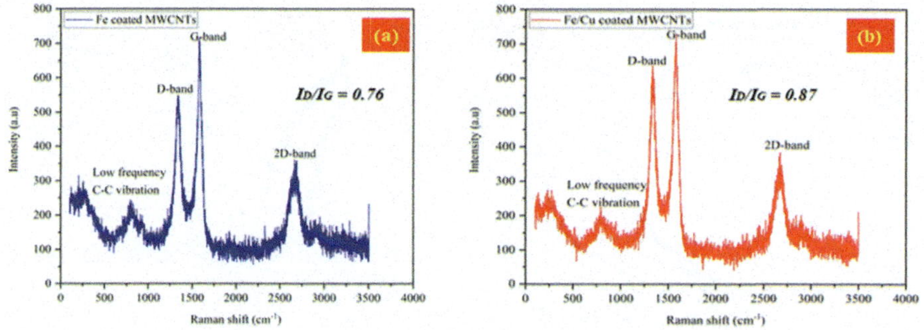

Fig. 48.3 Raman study of as-grown MWCNTs **a** (Fe coated), **b** (Fe/Cu coated)

Table 48.1 Calculated various parameters

Sample	D-band cm^{-1}	G-band cm^{-1}	G′-band cm^{-1}	I_D/I_G	I_{2D}/I_G
Fe coated MWCNTs	1343	1577	2690	0.76	0.50
Fe/Cu coated MWCNTs	1334	1585	2676	0.87	0.52
MWCNTs (Standard)	~ 1300	~ 1600	~ 2700		

48.3.2 Raman Spectroscopy

Raman spectroscopy is used to evaluate the quality of carbon nanotubes. To analyse the as-grown CNTs with different catalyst concentrations, a laser beam with an excitation wavelength of 532 nm was used. Figure 48.3 shows a standard Raman spectrum of CNTs grown on Si using the LPCVD technique. Raman features have two dominant modes, such as radial breathing mode at low frequencies and tangential G-band at higher frequencies, but there are no RBM (100–300 cm^{-1}) modes present in the spectrum, which leads to the formation of MWCNTs. The G-band, also known as the high energy band, corresponds to the graphite plane's stretching mode, which ranges from 1500 to 1600 cm^{-1}. The D-band, which has a wavelength of 1200–1400 cm^{-1}, is created by defect-induced double resonant Raman scattering.

The intensity ratio of the D-band to the G-band can be analysed to measure the defective quality of MWCNTs. Table 48.1 lists the I_D/I_G values associated with various samples (Fig. 48.4).

48.4 Field Emission, F-N Plot and Stability Measurements

Using a diode configuration, the properties of field emission were examined. An anode plate that is generally linked to a positive voltage and CNTs thin films that behave as a cold cathode that is generally grounded form the diode assembly. Screws were applied to maintain a minimum distance of 500 μm between the anode and

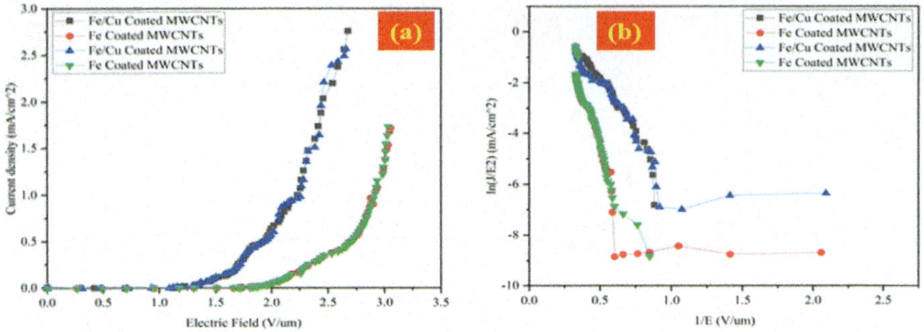

Fig. 48.4 Field emission characteristics and F-N plot measurement of MWCNTs **a** (Fe coated), **b** (Fe/Cu coated)

cathode. While the electrical contacts were being formed, the diode assembly was kept in the field emission chamber. After achieving a vacuum of 10^{-6} torr, current versus voltage measurements were done with a Keithley multimeter. Field emission studies revealed that the sample Fe/Cu catalysed MWCNTs was an outstanding field emitter with a lower turn-on voltage (1.49 V/m) and higher current density (2.75 mA/cm^2) than Fe catalysed MWCNTs. Depicts this Fig. 48.5a. The F-N plot measurement reveals that the Fe/Cu catalysed MWCNTs has a higher field enhancement factor than Fe catalysed MWCNTs, as shown in Fig. 48.5b. We repeat thc analysis by two times and it shows good repeatability, which shown in Fig. 48.5, and found a good result all measurement data are shown in Table 48.2.

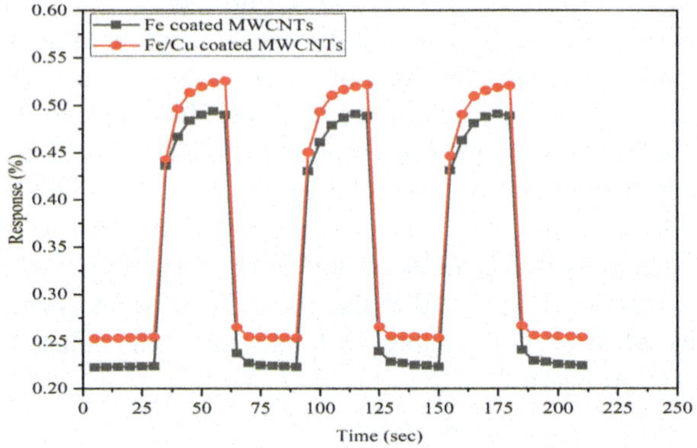

Fig. 48.5 Gas sensing response of as-grown MWCNTs **a** (Fe coated), **b** (Fe/Cu coated)

Table 48.2 Calculated various parameters of field emission measurement

Sample	Turn-on voltage (V/μm)	Threshold voltage (V/μm)	Current density (mA/cm^2)	Field enhancement factor (ϕ)
Run-1				
Fe coated MWCNTs	1.83	2.00	1.71	6.138×10^3
Fe/Cu coated MWCNTs	1.49	1.73	2.75	7.538×10^3
Run-2				
Fe coated MWCNTs	1.84	2.06	1.73	5.799×10^3
Fe/Cu coated MWCNTs	1.49	1.73	2.57	7.236×10^3

48.5 Magnitude of Resistance Variation in Presence of NH_3 Gas

Two sets of chemiresistive sensors were designed to detect various harmful gases. On exposure to the target gas NH_3, the as-fabricated chemiresistive sensors were tested. The amplitude of the gas sensor response revealed that sensors with Fe/Cu coated catalysed MWCNTs are more sensitive than sensors with Fe coated catalysed MWCNTs. The measurement was taken at 20 ppm. The Fe/Cu coated MWCNTs give more adsorption sites for the target gas molecules, resulting in a significant rise in sensor resistance, confirming the larger magnitude of sensor response. The increased defects produced by the Fe/Cu coated catalyst particles are also to blame for the good sensor response. Figure 48.5 depicts the response % of MWCNTs sensors, which confirms the quick response with a quick recovery and good repeatability.

48.6 Conclusion

This is crucial in terms of materials engineering because the procedure is far easier and less expensive than existing preparation methods. As a result, we hope that it would be easily adaptable to industry for low-cost and large-area coating catalysts for such growth of MWCNTs. Our findings show that Fe/Cu coated MWCNTs are good field emitters with low turn-on voltage and high current density. Our results indicate a remarkably straightforward process for the production composite catalyst materials for CNTs growth and application in field emitters and gas sensors.

Acknowledgements Mohd Sarvar would like to express his gratitude to the DST, New Delhi, for the DST-INSPIRE Fellowship (Ref No. DST/INSPIRE FELLOWSHIP/2019/IF-190286).

References

1. M.Y. Lone, A. Kumar, N. Ansari, S. Husain, M. Zulfequar, R.C. Singh, M. Husain, Mater. Res. Bull. **119**, 110532 (2019)
2. Majzlíková, J. Petra, J. Sedláček, J. Prášek, V. Pekárek, G. Al Svatoš, G. Bannov, OndřejJašek et al., Sensors **15**(2), 2644–2661 (2015)
3. Ajayan, M. Pulickel, Z. Otto, Carbon nanotubes 391–425 (2001)
4. H. He, L.A. Pham-Huy, P. Dramou, D. Xiao, P. Zuo, C. Pham-Huy, BioMed Res. Int. **2013** (2013)
5. R.H. Baughman, C. Changxing, A. Anvar Zakhidov, Z. Iqbal, N. Joseph Barisci, M. Geoff Spinks, G. Gordon, Wallace et al., Carbon nanotube actuators. Science **284**(5418), 1340–1344 (1999)
6. S. Ţălu. *Micro and Nanoscale Characterization of Three Dimensional Surs. Basics and Applications*. (Napoca Star Publishing House, Cluj-Napoca, Romania, 2015)
7. S. Ţălu, M. Bramowicz, S. Kulesza, A. Ghaderi, V. Dalouji, S. Solaymani, M. Fathi Kenari, M. Ghoranneviss. J. Microsc. **264**, 143–52 (2016).
8. D. Dastan, P.U. Londhe, N.B. Chaure, J. Mater. Sci. Mater. Electron. **25**, 3473–3479 (2014)
9. V. Dalouji, S.M. Elahi, S. Solaymani, A. Ghaderi, Eur. Phys. J. Plus **131**, 84 (2016)
10. D. Dastan, N. Chaure, M. Kartha, J. Mater. Sci. Mater. Electron. **28**, 7784–7796 (2017)
11. S. Solaymani, A. Ghaderi, N. Beryani Nezafat, J. Fusion, Energy **31**, 591–591 (2012)
12. S. Ţa˘lu, M. Bramowicz, S. Kulesza, V. Dalouji, S. Solaymani, S. Valedbagi. Microsc. Res. Tech. **79**(12), 1208–13 (2016).
13. S. Stach, Z. Garczyk, S. Ţălu, S. Solaymani, A. Ghaderi, R. Moradian, N. Beryani Nezafat, S.M. Elahi, H.J. Gholamali, Phys. Chem. C. **119**(31), 17887–98 (2015).
14. T. Ghodselahi, S. Solaymani, M. Akbarzadeh Pasha, M.A. Vesaghi, Eur. Phys. J. D **66**, 299 (2012)
15. s. Ţălu, Solaymani, M. Bramowicz, S. Kulesza, A. Ghaderi, S. Shahpouri, S.M. Elahi, J. Mater Sci. Mater Electron **27**(9), 9272–7 (2016).
16. S. Ţălu, M. Bramowicz, S. Kulesza, A. Ghaderi, V. Dalouji, S. Solaymani, Z. Khalaj, Electron Mater Lett. **12**(5), 580–8 (2016).
17. H. Lashgari, A. Boochani, A. Shekaari, S. Solaymani, E. Sartipi, R.T. Mendi, Appl. Surf. Sci. **369**, 76–81 (2016)
18. J. Grunes, J. Zhu, G.A. Somorjai, Chem. Commun. 2257–60 (2003).
19. N. Toshima, T. Yonezawa, New J. Chem. **22**, 1179–1201 (1998)
20. M. Moreno-Mañas, R. Pleixats, Acc. Chem. Res. **36**, 638–643 (2003)
21. A. Roucoux, J. Schulz, H. Patin, Chem. Rev. **102**, 3757–3778 (2002)
22. M.P. Pileni, I. Lisiecki, Colloids Surf. A **80**(1), 63–68 (1993)
23. M. Valden, X. Lai, D.W. Goodman, Science **281**, 1647–1650 (1998)
24. P.K. Stoimenov, R.L. Klinger, G.L. Marchin, K.J. Klabunde, Langmuir **18**, 6679–6686 (2002)
25. Y. Saito, T. Nakahira, S. Uemura, J. Phys. Chem. B **107**, 931–934 (2002)
26. A. Thess, R. Lee, P. Nikolaev, H. Dai, P. Petit, J. Robert, C. Xu, Y.H. Lee, S.G. Kim, A.G. Rinzler, D.T. Colbert, G.E. Scuseria, D. Tomanek, J. Fischer, R.E. Smalley, Science **273**, 483–487 (1996)
27. C.J. Lee, J.H. Park, J. Park, Chem. Phys. Lett. **323**, 560–565 (2000)
28. H. Ago, K. Nakamura, S. Imamura, M. Tsuji, Chem. Phys. Lett. **391**, 308–313 (2004)
29. B.Q. Wei, R. Vajtai, P.M. Ajayan, Appl. Phys. Lett. **79**, 1172 (2001)
30. B. Gao, Chem. Phys. Lett. **327**, 69 (2000)

The manufacturer's authorised representative in the EU is Springer Nature Customer Service Centre GmbH, Europaplatz 3, 69115 Heidelberg, Germany. If you have any concerns regarding our products, please contact ProductSafety@springernature.com

Printed and bound by CPI Group (UK) Ltd, Croydon, CR0 4YY

01/10/2026

02225905-0002